SURFACTANTS IN CHEMICAL/ PROCESS ENGINEERING

SURFACTANT SCIENCE SERIES

CONSULTING EDITORS

MARTIN J. SCHICK
Consultant
New York, New York

FREDERICK M. FOWKES
Department of Chemistry
Lehigh University
Bethlehem, Pennsylvania

Volume 1: NONIONIC SURFACTANTS, edited by Martin J. Schick

Volume 2: SOLVENT PROPERTIES OF SURFACTANT SOLUTIONS, edited by Kozo Shinoda (*out of print*)

Volume 3: SURFACTANT BIODEGRADATION, by R. D. Swisher *(See Volume 18)*

Volume 4: CATIONIC SURFACTANTS, edited by Eric Jungermann

Volume 5: DETERGENCY: THEORY AND TEST METHODS (*in three parts*), edited by W. G. Cutler and R. C. Davis

Volume 6: EMULSIONS AND EMULSION TECHNOLOGY (*in three parts*), edited by Kenneth J. Lissant

Volume 7: ANIONIC SURFACTANTS (*in two parts*), edited by Warner M. Linfield

Volume 8: ANIONIC SURFACTANTS–CHEMICAL ANALYSIS, edited by John Cross

Volume 9: STABILIZATION OF COLLOIDAL DISPERSIONS BY POLYMER ADSORPTION, by Tatsuo Sato and Richard Ruch

Volume 10: ANIONIC SURFACTANTS–BIOCHEMISTRY, TOXICOLOGY, DERMATOLOGY, edited by Christian Gloxhuber

Volume 11: ANIONIC SURFACTANTS–PHYSICAL CHEMISTRY OF SURFACTANT ACTION, edited by E. H. Lucassen-Reynders

Volume 12: AMPHOTERIC SURFACTANTS, edited by B. R. Bluestein and Clifford L. Hilton

Volume 13: DEMULSIFICATION: INDUSTRIAL APPLICATIONS, by Kenneth J. Lissant

Volume 14: SURFACTANTS IN TEXTILE PROCESSING, by Arved Datyner

Volume 15: ELECTRICAL PHENOMENA AT INTERFACES: FUNDAMENTALS, MEASUREMENTS, AND APPLICATIONS, edited by Ayao Kitahara and Akira Watanabe

Volume 16: SURFACTANTS IN COSMETICS, edited by Martin M. Rieger

Volume 17: INTERFACIAL PHENOMENA: EQUILIBRIUM AND DYNAMIC EFFECTS, by Clarence A. Miller and P. Neogi

Volume 18: SURFACTANT BIODEGRADATION, Second Edition, Revised and Expanded, by R. D. Swisher

Volume 19: NONIONIC SURFACTANTS: CHEMICAL ANALYSIS, edited by John Cross

Volume 20: DETERGENCY: THEORY AND TECHNOLOGY, edited by W. Gale Cutler and Erik Kissa

Volume 21: INTERFACIAL PHENOMENA IN APOLAR MEDIA, edited by Hans-Friedrich Eicke and Geoffrey D. Parfitt

Volume 22: SURFACTANT SOLUTIONS: NEW METHODS OF INVESTIGATION, edited by Raoul Zana

Volume 23: NONIONIC SURFACTANTS: PHYSICAL CHEMISTRY, edited by Martin J. Schick

Volume 24: MICROEMULSION SYSTEMS, edited by Henri L. Rosano and Marc Clausse

Volume 25: BIOSURFACTANTS AND BIOTECHNOLOGY, edited by Naim Kosaric, W. L. Cairns, and Neil C. C. Gray

Volume 26: SURFACTANTS IN EMERGING TECHNOLOGIES, edited by Milton J. Rosen

Volume 27: REAGENTS IN MINERAL TECHNOLOGY, edited by P. Somasundaran and Brij M. Moudgil

Volume 28: SURFACTANTS IN CHEMICAL/PROCESS ENGINEERING, edited by Darsh T. Wasan, Martin E. Ginn, and Dinesh O. Shah

Volume 29: THIN LIQUID FILMS, edited by I. B. Ivanov

Volume 30: MICROEMULSIONS AND RELATED SYSTEMS: FORMULATION, SOLVENCY, AND PHYSICAL PROPERTIES, edited by Robert S. Schecter and Maurice Bourrel

Volume 31: CRYSTALLIZATION AND POLYMORPHISM OF FATS AND FATTY ACIDS, edited by Nissim Garti and K. Sato

Volume 32: INTERFACIAL PHENOMENA IN COAL TECHNOLOGY, edited by Gregory D. Botsaris and Yuli M. Glazman

SURFACTANTS IN CHEMICAL/ PROCESS ENGINEERING

edited by

Darsh T. Wasan
Department of Chemical Engineering
Illinois Institute of Technology
Chicago, Illinois

Martin E. Ginn
Stuart School of Business Administration
Illinois Institute of Technology
Chicago, Illinois

Dinesh O. Shah
Department of Chemical Engineering and
Center for Surface Science and Engineering
University of Florida
Gainesville, Florida

Marcel Dekker, Inc.   New York and Basel

Library of Congress Cataloging in Publication Data

Surfactants in chemical.

   (Surfactant science series ; v. 28)
   Includes index.
   1. Surface active agents. I. Wasan, Darsh T.
II. Ginn, Martin E.   III. Shah, Dinesh O.
(Dinesh Ochhavlal)   IV. Series.
TP994.S874 1988   668'.1   88-3869
ISBN 0-8247-7830-8

Copyright © 1988 by MARCEL DEKKER, INC. All Rights Reserved

Neither this book nor any part may be reproduced or transmitted in any form or by any means, electronic or mechanical, including photocopying, microfilming, and recording, or by any information storage and retrieval system, without permission in writing from the publisher.

MARCEL DEKKER, INC.
270 Madison Avenue, New York, New York 10016

Current printing (last digit):
10 9 8 7 6 5 4 3 2 1

PRINTED IN THE UNITED STATES OF AMERICA

Preface

Surfactants are an interesting class of materials, and it is their dual characteristics (hydrophilic-lipophilic nature) that make them so useful. Surfactants play an important role in many processes ranging from the very mundane (washing cloth) to the very sophisticated (microelectronics). In the last several years, there has been a great deal of research activity in understanding the role of surfactants in multibillion-dollar chemical and processing industries. Rates of organic reactions have been shown to be accelerated by many orders of magnitude by carrying out the reactions in appropriate micellar or microemulsion media. New examples include the application of surfactants in unit operations for industrial separations, environmental protection, pharmaceuticals, herbicides and pesticides, engineered materials, and enhanced oil recovery. (The reader is referred to Volume 26 of this series, *Surfactants in Emerging Technologies*, edited by Milton J. Rosen.) All these new developments in the field of surfactants obviously cannot be covered in one volume. However, we have made an attempt to focus on a few of the recent developments in industrial applications of surfactants and to bridge the gap from engineering science to application technology.

In recent years, there has been more involvement by chemical engineers and applied chemists in the process application of surfactants and hence greater willingness on their part to address this important subject. Therefore, it should not be very surprising that most of the contributors to this volume are engineering scientists.

The various chapters may be outlined as follows. The field of surface rheology has largely developed over the past two decades. Edwards and Wasan, in the first chapter, introduce the reader to surface rheological properties such as surface shear and dilatational viscosities and elasticities of fluid-fluid interfaces containing surfactants by using a simplified "pillbox" approach. They then proceed to show the importance of these properties in practical dynamic surface problems with particular emphasis on the role that the dilatational surface properties play in a number of applications involving the use of surfactants such as foam rheology, foam flow in enhanced oil recovery, and foam and emulsion stability.

Distillation is a complex vapor-liquid contacting operation, but it lies at the heart of many process industries. Berg describes the observed effects of surfactants in distillation processes. Surface-tension effects, and their modification by surface-active agents, play a significant role in the performance of distillation equipment. Traces of surfactants have been shown to be capable of improving the performance of packed distillation towers by as much as 80%. Enhancement is achieved through the stabilization against rupture of the thin films of reflux liquid flowing over the irrigated packing, thereby increasing available interfacial area. Similar area increases can be achieved for tray columns through the stabilization of froths. These effects are described, explained, and quantified by both hydrodynamic analyses and laboratory experiments using bench-scale equipment.

Scamehorn and Harwell provide an in-depth examination of how surfactant systems can be used in the treatment of aqueous process streams to achieve important industrial separations. Five important separation techniques are examined, namely, micellar-enhanced ultrafiltration, foam separations, surfactant-enhanced carbon regeneration, extraction into reverse micelles, and admicellar chromatography. These methods are of special interest because they typically require low energy for operation. They are also quite useful in selectively removing organics or multivalent ions from water, in addition to other separations. Accordingly, surfactant-based separations are useful for separating contaminants or for enrichment of value components. Technologies in which these techniques may be widely used include biotechnology applications, pollution control or waste water treatment, and separation of metals.

Li invented the liquid surfactant membrane concept in 1968. Over the last 20 years, considerable efforts by chemists and chemical engineers have been devoted to developing the liquid surfactant membranes for many potential applications, including separation of hydrocarbons, hydrometallurgy, waste water treatment, biomedicine, and biochemical engineering. To date, this concept has been successfully commercialized for the extraction of metals. Gu, Wasan, and Li specifically review the various extraction systems for metal ion recovery and the dominant mechanisms for their extraction.

Sonntag examines the use of surfactants in the aqueous separation of two important fatty acids, oleic and stearic acids, from mixtures of tallow fatty acids. He includes a discussion of some fundamental properties such as adsorption, wetting, and hydrophilization in treating his topic. Aqueous emulsion separations are shown to offer advantages over solidification and solvent crystallization methods based on melting-point differences. Adsorption and wetting are shown to be especially important in separating oleic from stearic acid. Successful separations are achieved by treating the fatty acid mixtures with solutions of 0.5% (by weight) sodium lauryl sulfate and 2% (by weight) magnesium sulfate, followed by centrifuging or rotary vacuum filtration. Surfactant concentrations are regarded as critical since highly emulsified mixtures are found to separate only with great difficulty. The results are discussed in relation to key surface and wetting phenomena.

Somasundaran and Ramachandran review the role of surfactants in flotation, an important unit operation. The principles that govern surfactant adsorption and particle-bubble attachment are discussed with relevant examples from the literature. Selective adsorption of surfactants on particles is dependent on the surface and solubility properties of the solid and the solution chemistry of the surfactants, as well as on the dissolved inorganics and polymers present in the system. The importance of the interactions of the different species of the surfactants with inorganics, as well as with polymers in bulk and at various interfaces, is emphasized. Effects of the different variables such as pH, ionic strength, and temperature are also discussed.

A second chapter by Sonntag deals with the role of surfactants in herbicide emulsions. Surfactants are shown to play a vital role in weed control technology by increasing the effectiveness of herbicides. The action of surfactants relates to their ability to accomplish a more uniform distribution over the surface of leaf foliage (and also by increasing penetration). Factors affecting surfactant usefulness in herbicide dispersion include surface activity (surface-tension lowering), wetting properties, contact angle, micelle formation, and hydrophobic-lipophilic balance (HLB), particularly for nonionic surfactants. The increased use of vegetable oils in herbicide dispersions as replacements for hydrocarbons (and water) is discussed in developing an understanding of these complex systems. Factors favoring use of vegetable oils such as soybean oil include ready availability, improved functionality, and better environmental acceptance. Such factors tend to override costs in many situations.

Kine and Redlich provide a thorough treatment of the role of surfactants in emulsion polymerization. The product of such polymerization is termed "emulsion polymer," a latex or a polymeric dispersion, and consists of minute plastic spheres (10^{-4} mm in diameter) dispersed in a continuous phase, usually water. Each cubic centimeter of latex contains about 100 trillion of these spheres, and each sphere has a layer of a surface-active moiety on its surface. The use of surfactants is shown to be vital in the preparation of the emulsion polymer and, at times, of even greater importance in its formulation and application (e.g., in protective coatings). Factors affecting the preparation of the dispersion, the resulting physical properties, and ultimate application are discussed in developing a greater understanding of interrelationships. Whenever possible, the authors strive to develop both theoretical and practical implications pertinent to various industrial applications. Such applications of emulsion polymers include use in soil stabilizers, protective coatings, sealants, adhesives, tile and paper finishing agents, and specialty concretes, and for the treatment of automotive and aircraft tires.

The late Dr. J. Schulman, mentor of Shah, coined the word "microemulsion." A microemulsion can be defined as a thermodynamically stable, isotopically clear dispersion of two immiscible liquids, consisting of microdomains of one or both liquids stabilized by an interfacial film of surface-active molecules. Since the introduction of microemulsions in 1943, significant advances have been made to explain the formation, structure, properties, and phase behavior of microemulsions. Leung, Hou, and Shah review the various theoretical and experimental aspects, as well as novel applications of microemulsions.

The last chapter, by Woods and Diamadopoulos, reviews the fundamentals of surface phenomena that affect stability of dispersions. Included are equilibrium theories of electrochemical double layer and adsorbed macromolecule and rate theories of coalescence, adsorption, and coagulation. Eleven strategies that affect stability are outlined. The role of surfactants, inorganics and polymers, and solubility are discussed. Methods are given to predict the drop size and to characterize dispersions. Overall strategies and rules of thumb for separation dispersions are presented. Details are given for chemical destabilization, coagulation and flocculation, and deep-bed filtration and flotation. Numerous practical sample calculations are given.

We wish to thank all the authors for their contributions and for providing us with the summaries of their work that are included in this preface. We have attempted to cover some recently selected developments to highlight the importance of surfactants and surface phenomena, and the use of surfactants and the role they play in various applications. We hope this volume, although far from being comprehensive, will be judged as definitive and will stimulate other works to further enhance our knowledge in this technologically important field.

Darsh T. Wasan

Martin E. Ginn

Dinesh O. Shah

Contributors

John C. Berg Department of Chemical Engineering, University of Washington, Seattle, Washington

Evan Diamadopoulos Department of Chemical Engineering, McMaster University, Hamilton, Ontario, Canada

D. A. Edwards Department of Chemical Engineering, Illinois Institute of Technology, Chicago, Illinois

*Z. M. Gu** Department of Chemical Engineering, Illinois Institute of Technology, Chicago, Illinois

Jeffrey H. Harwell Institute for Applied Surfactant Research, University of Oklahoma, Norman, Oklahoma

Mean Jeng Hou Department of Chemical Engineering and Center for Surface Science and Engineering, University of Florida, Gainesville, Florida

Benjamin B. Kine Research Labs, Rohm and Haas Company, Spring House, Pennsylvania

Roger Leung † Department of Chemical Engineering and Center for Surface Science and Engineering, University of Florida, Gainesville, Florida

---

Present affiliation:
*Institute of Atomic Energy, Academia Sinica, Beijing, China.
†Aluminum Company of America, Alcoa Center, Pennsylvania.

Norman N. Li Allied-Signal, Inc., Des Plaines, Illinois

R. Ramachandran Henry Krumb School of Mines, Columbia University, New York, New York

George H. Redlich Research Labs, Rohm and Haas Company, Spring House, Pennsylvania

John F. Scamehorn Institute for Applied Surfactant Research, University of Oklahoma, Norman, Oklahoma

Dinesh O. Shah Department of Chemical Engineering and Center for Surface Science and Engineering, University of Florida, Gainesville, Florida

P. Somasundaran Henry Krumb School of Mines, Columbia University, New York, New York

Norman O. V. Sonntag Consultant, Red Oak, Texas

Darsh T. Wasan Department of Chemical Engineering, Illinois Institute of Technology, Chicago, Illinois

Donald R. Woods Department of Chemical Engineering, McMaster University, Hamilton, Ontario, Canada

Contents

Preface   iii

Contributors   vii

1. DILATATIONAL PROPERTIES OF ADSORBED SURFACTANT INTERFACES AND THEIR APPLICATIONS   1

   D. A. Edwards and Darsh T. Wasan

2. THE EFFECT OF SURFACE-ACTIVE AGENTS IN DISTILLATION PROCESSES   29

   John C. Berg

3. SURFACTANT-BASED TREATMENT OF AQUEOUS PROCESS STREAMS   77

   John F. Scamehorn and Jeffrey H. Harwell

4. LIQUID SURFACTANT MEMBRANES FOR METAL EXTRACTIONS   127

   Z. M. Gu, Darsh T. Wasan, and Norman N. Li

5. SURFACTANTS IN AQUEOUS EMULSIFICATION SEPARATION OF OLEIC AND STEARIC ACIDS   169

   Norman O. V. Sonntag

6. SURFACTANTS IN FLOTATION   195
   P. Somasundaran and R. Ramachandran

7. SURFACTANTS IN HERBICIDE DISPERSIONS   237
   Norman O. V. Sonntag

8. THE ROLE OF SURFACTANTS IN EMULSION POLYMERIZATION   263
   Benjamin B. Kine and George H. Redlich

9. MICROEMULSIONS: FORMATION, STRUCTURE, PROPERTIES, AND NOVEL APPLICATIONS   315
   Roger Leung, Mean Jeng Hou, and Dinesh O. Shah

10. IMPORTANCE OF SURFACTANTS AND SURFACE PHENOMENA ON SEPARATING DILUTE OIL-WATER EMULSIONS AND DISPERSIONS   369
    Donald R. Woods and Evan Diamadopoulos

Index   541

SURFACTANTS IN CHEMICAL/ PROCESS ENGINEERING

1
Dilatational Properties of Adsorbed Surfactant Interfaces and Their Applications

D. A. EDWARDS and DARSH T. WASAN *Illinois Institute of Technology, Chicago, Illinois*

I.	Introduction	1
II.	Dynamic Surface Properties (Surface Rheology)	2
III.	Surface Dilatational Properties	11
IV.	Applications	14
	A. Foam Rheology	14
	B. Enhanced Oil Recovery	17
	C. Foam Stability	20
	D. Emulsion Stability	23
V.	Summary	25
	References	26

I. INTRODUCTION

The adsorption of surfactant at a fluid-fluid interface has long been known to alter equilibrium interfacial properties. However, in practice, equilibrium is an inevitable abstraction: the limiting case of actual dynamic processes which occur near, at, and across an interface. Dynamic interfacial properties, which are required to quantify practical dynamic surface problems, are less well known.

Dynamic interfacial properties may be classified in terms of shear properties and dilatational properties. The former, which typically include the surface shear viscosity and the shear elasticity, have been shown relevant to such dynamic phenomena as foam and emulsion stability, coalescence, and interfacial mass transfer [1-3].

We have recently undertaken a fundamental research program to investigate the less understood but more significant dilatational properties, such as the dynamic surface/interfacial tension, the dilatational modulus, and the surface dilatational viscosity. Our objective in this research has been first, to develop techniques to measure the dilatational properties, and second, to investigate (both experimentally and theoretically) the correlation between the dynamic surface/interfacial properties and foam rheology, foam flow in EOR processes, foam stability, and emulsion stability.

In this chapter we summarize the current results from our research efforts, beginning with a simple derivation of the relation between surface and bulk-phase hydrodynamics (in which context dynamic surface properties are defined), followed by a detailed consideration of surface dilatational properties, and concluding with the various theoretical and experimental results that we have obtained for dilatational surface properties and their application to dynamic interfacial phenomena.

II. DYNAMIC SURFACE PROPERTIES (SURFACE RHEOLOGY)

Surface rheology is the study of interfacial response to deformation. Often we consider the deformational response of the surface as defined by a single stress coefficient: the surface tension. However, when surfactant adsorbs to the fluid-fluid interface, an intrinsic rigidity may arise, introducing beyond the tensile response to deformation, a damping or viscous response which cannot be defined completely by the surface tension, requiring coefficients of surface viscosity.

Although the origin of surface tension and surface viscosities is quite intuitive, the surface stress equations that relate these stress coefficients to the deformational response of the surface soon lose intuitive appeal, particularly when the surface becomes curved. Therefore, to advance an intuitive understanding of surface rheology we will throughout this section consider the rheology of the planar fluid surface. The reader interested in the rheology of curved fluid surfaces is referred to Refs. 4 and 5.

Consider, then, a rectangular fluid element containing a portion of two continuous fluid phases, as in Fig. 1. The thickness of the element (2ℓ) is sufficiently small that the inhomogeneous layer of fluid between the two fluid phases loses any discontinuous appearance. This inhomogeneous layer, the interfacial region, is completely contained between the surfaces $z = \pm\ell$.

Properties of Surfactant Interfaces

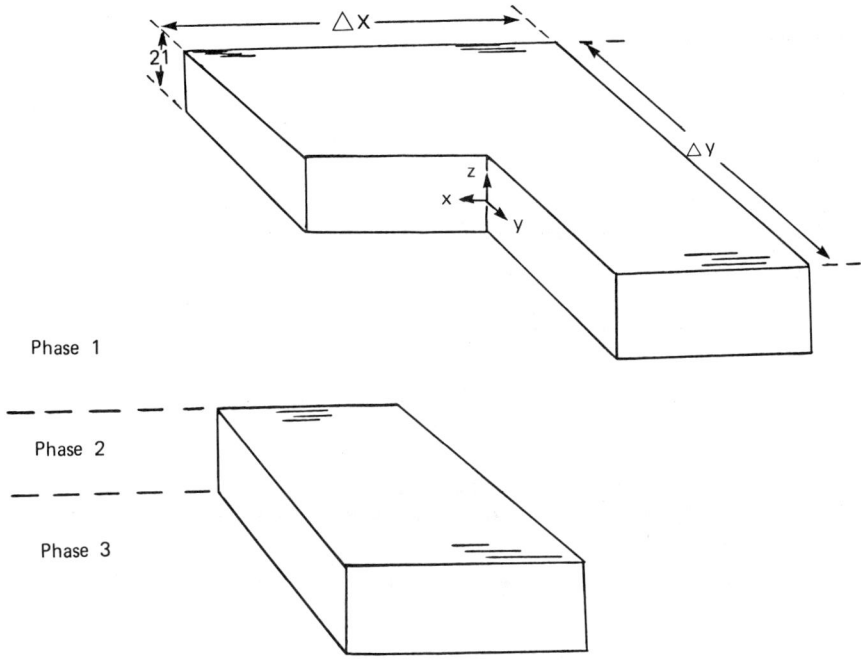

FIG. 1 Microscopic volume enclosing the interfacial region.

The interfacial region contains a large concentration of surfactant molecules such that the viscosity in the region is also large. Inertial momentum transfer, which is inevitably small within the very thin transition region, is then considered negligible compared to viscous momentum transfer.

A balance of the linear momentum on this fluid element may be written for the y component (see Fig. 2) as

$$\int_{-\ell}^{\ell} \left(\tau_{xy} \big|_{\Delta x} - \tau_{xy} \big|_0 \right) \Delta y \, dz + \int_{-\ell}^{\ell} F_y \, \Delta x \, \Delta y \, dz$$

$$+ \int_{-\ell}^{\ell} \left(\tau_{yy} \big|_{\Delta y} - \tau_{yy} \big|_0 \right) \Delta x \, dz$$

$$+ \left(\tau_{zy} \big|_{\ell} - \tau_{zy} \big|_{-\ell} \right) \Delta x \, \Delta y = 0 \quad (1)$$

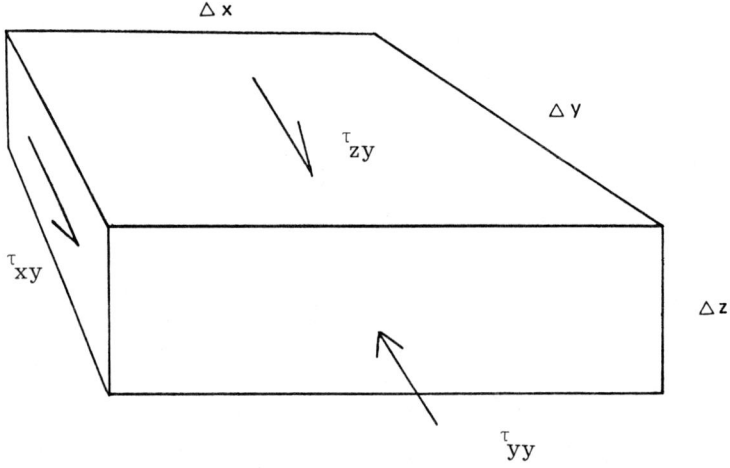

FIG. 2 Linear momentum transfer in the y direction for the microscopic interfacial volume.

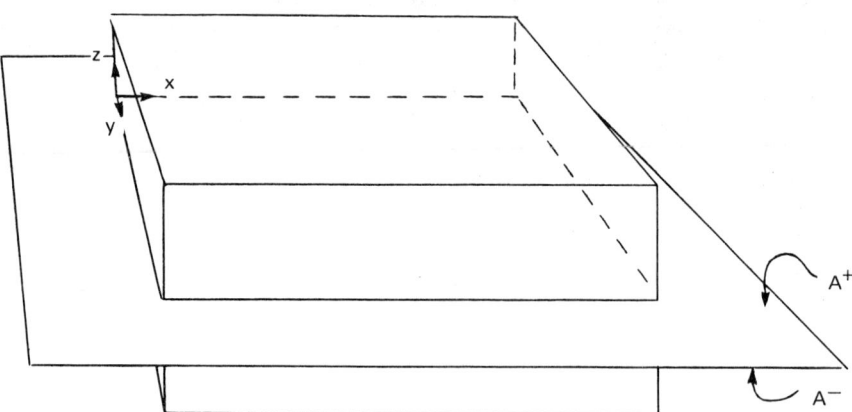

FIG. 3 Macroscopic "dividing surface" approach to the interfacial region.

Here τ_{xy}, τ_{yy}, and τ_{zy} are the contact stresses and F_y is the body force density in the y component containing both local and nonlocal force interactions.

However, most of our observations of the fluid-fluid interface occur at a much larger length scale then 2ℓ, for which the interfacial region appears to have virtually zero thickness. For this and other reasons it is often useful to model the statics and dynamics of the region as those of a singular "dividing surface."

Using this approach, the fluid element may be envisioned quite artificially, as in Fig. 3. We shall call this approach the "macroscopic approach." From the macroscopic approach, the properties of the bulk phases are assumed to possess the same values from the outer boundaries of the interfacial region up to the dividing surface, at which they generally possess singularities. The difference between the macroscopic approach and the true approach is depicted in Fig. 4.

The balance of the y component of linear momentum using the macroscopic approach to the surface becomes

$$\int_{-\ell}^{\ell} \left(\tau_{xy}^m \big|_{\Delta x} - \tau_{xy}^m \big|_0 \right) \Delta y \, dz + \int_{-\ell}^{\ell} F_y^m \Delta x \, \Delta y \, dz$$

$$+ \int_{-\ell}^{\ell} \left(\tau_{yy}^m \big|_{\Delta y} - \tau_{yy}^m \big|_0 \right) \Delta x \, dz$$

$$+ \left(\tau_{zy}^{m1} \big|_{\ell} - \tau_{zy}^{m1} \big|_{A+} \right) \Delta x \, \Delta y$$

$$+ \left(\tau_{zy}^{m2} \big|_{A-} - \tau_{zy}^{m2} \big|_{-\ell} \right) \Delta x \, \Delta y = 0 \tag{2}$$

where

$$\int_{-\ell}^{\ell} \left(\tau_{xy}^m \big|_{\Delta x} - \tau_{xy}^m \big|_0 \right) \Delta y \, dz = \int_{-0}^{\ell} \left(\tau_{xy}^{m1} \big|_{\Delta x} - \tau_{xy}^{m1} \big|_0 \right) \Delta y \, dz$$

$$+ \int_{-\ell}^{0} \left(\tau_{xy}^{m2} \big|_{\Delta x} - \tau_{xy}^{m2} \big|_0 \right) \Delta y \, dz \tag{3}$$

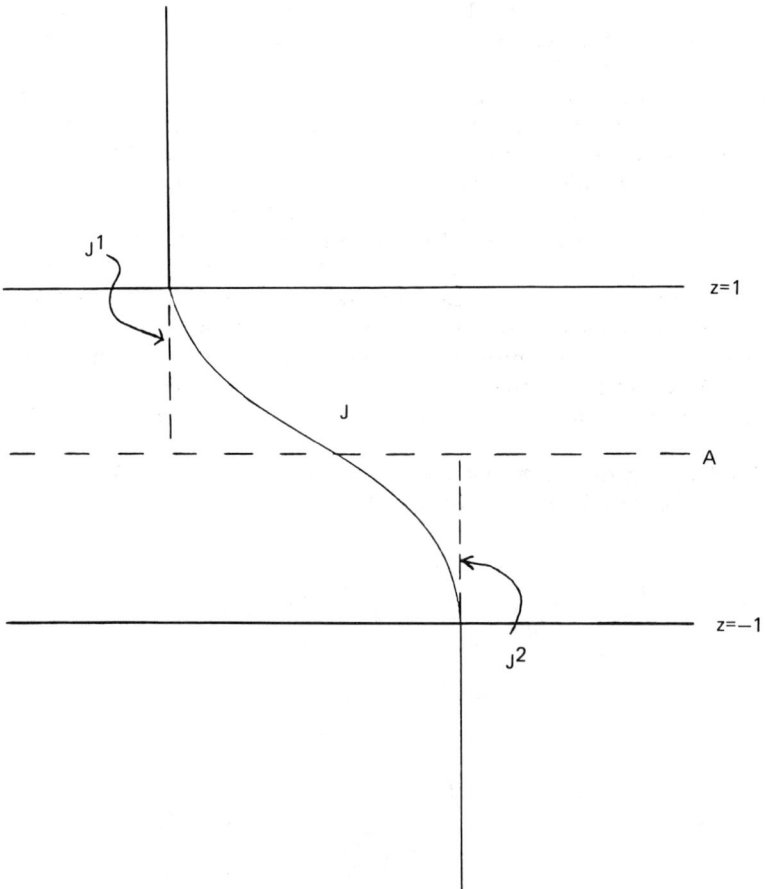

FIG. 4 Difference between the macroscopic and true approaches to the interface for a generic field quantity J.

and where m1 refers to the macroscopically perceived variable relative to phase 1 "at the surface," and similarly, m2 is with respect to phase 2.

What we wish to assign to our dividing surface is the difference between the artificial assessment of the linear momentum content of the element $\Delta x \, \Delta y \, \Delta z$ and the true linear momentum content. The resulting equation is the surface excess linear momentum equation. Subtracting Eq. (2) from Eq. (1) and dividing by the area element $\Delta x \, \Delta y$ yields

$$\int_{-\ell}^{\ell} \left(\frac{\tau_{xy}|_{\Delta x} - \tau_{xy}|_0}{\Delta x} - \frac{\tau_{xy}^m|_{\Delta x} - \tau_{xy}^m|_0}{\Delta x} \right) dz + \int_{-\ell}^{\ell} \left(F_y - F_y^m \right) dz$$

$$+ \int_{-\ell}^{\ell} \left(\frac{\tau_{yy}|_{\Delta x} - \tau_{yy}|_0}{\Delta y} - \frac{\tau_{yy}^m|_{\Delta y} - \tau_{yy}^m|_0}{\Delta y} \right) dz$$

$$+ \tau_{zy}^{m1}|_{A+} - \tau_{zy}^{m2}|_{A-} = 0 \quad (4)$$

where we have noted

$$\tau_{zy}|_{\ell} = \tau_{zy}^{m1}|_{\ell} \quad \text{and} \quad \tau_{zy}|_{\ell} = \tau_{zy}^{m2}|_{-\ell} \quad (5)$$

Finally, taking the limit as

$$\Delta x \, \Delta y \to 0$$

we obtain

$$\frac{\partial}{\partial x} \int_{-\ell}^{\ell} \left(\tau_{xy} - \tau_{xy}^m \right) dz + \frac{\partial}{\partial y} \int_{-\ell}^{\ell} \left(\tau_{yy} - \tau_{yy}^m \right) dz$$

$$+ \int_{-\ell}^{\ell} \left(F_y - F_y^m \right) dz = [\tau_{zy}] \quad (6)$$

Where the "jump" in τ_{zy} is defined as

$$[\tau_{zy}] = \tau_{zy}^{m2}|_{A-} - \tau_{zy}^{m1}|_{A+} \quad (7)$$

We may now define the surface excess pressure components as

$$\tau_{xy}^s \equiv \int_{-\ell}^{\ell} \left(\tau_{xy} - \tau_{xy}^m \right) dz \quad (8a)$$

and

$$\tau^s_{yy} \equiv \int_{-\ell}^{\ell} \left(\tau_{yy} - \tau^m_{yy} \right) dz \tag{8b}$$

and the y component of the surface excess force density as

$$F^s_y \equiv \int_{-\ell}^{\ell} \left(F_y - F^m_y \right) dz \tag{8c}$$

so that the surface excess linear momentum equation for the y component of linear momentum becomes

$$F^s_y + \frac{\partial}{\partial x} \tau^s_{xy} + \frac{\partial}{\partial y} \tau^s_{yy} = [\tau_{zy}] \tag{9}$$

Similarly it may be shown that

$$F^s_x + \frac{\partial}{\partial x} \tau^s_{xx} + \frac{\partial}{\partial y} \tau^s_{yx} = [\tau_{zx}] \tag{10}$$

and

$$[\tau_{zz}] = 0 \tag{11}$$

Equations (9–11) are general surface excess stress equations for the planar fluid surface.

The Newtonian constitutive model for the pressure tensor τ_{ij}, where

$$\tau_{ij} = \begin{bmatrix} \tau_{xx} & \tau_{xy} & \tau_{xz} \\ \tau_{yx} & \tau_{yy} & \tau_{yz} \\ \tau_{zx} & \tau_{zy} & \tau_{zz} \end{bmatrix} \tag{12}$$

is given by

$$\tau_{ij} = \mu \Delta_{ij} - \left(\frac{2}{3}\mu - \kappa \right) v_{k,k} \delta_{ij} - p_t t^\alpha_i t^\beta_j \delta_{\alpha\beta} - p n_i n_j \tag{13}$$

Where p_t is the tangential component of the pressure, p the normal component, μ the shear viscosity, and κ the dilatational viscosity.

Properties of Surfactant Interfaces

Here the rate of deformation tensor Δ_{ij} is given in cartesian coordinates as

$$\Delta_{ij} = \left(\frac{\partial v_i}{\partial x^j} + \frac{\partial v_j}{\partial x^i}\right) \tag{14}$$

where the summation convention is used over the indices i, j, as i = 1,2,3; $v_{k,k}$ is the divergence of the velocity v_k; δ_{ij} is the Kronecker delta; t_i^α is the hybrid tensor of the surface, where α and β represent surface coordinates, as α = 1,2; and n_i is the unit normal of the surface. From Eq. (13), then,

$$\tau_{xy} = \mu\left(\frac{\partial v_x}{\partial y} + \frac{\partial v_y}{\partial x}\right) \tag{15a}$$

and

$$\tau_{yy} = p_t + 2\mu\frac{\partial v_y}{\partial y} - \left(\frac{2}{3}\mu - \kappa\right)\left(\frac{\partial v_x}{\partial x} + \frac{\partial v_y}{\partial y} + \frac{\partial v_z}{\partial z}\right) \tag{15b}$$

Using Eq. (15) in Eq. (8), with the continuity of velocity conditions (see Ref. 6)

$$v_\alpha^{m1}(0) = v_\alpha^{m2}(0) = v_\alpha^s$$

and

$$v_z^{m1}(0) = v_z^{m2}(0) = 0 \tag{16}$$

the last of which is valid for small rates of mass transfer, provides

$$\tau_{xy}^s = \mu^s\left(\frac{\partial v_x^s}{\partial y} + \frac{\partial v_y^s}{\partial x}\right) \tag{17}$$

and

$$\tau_{yy}^s = \sigma + 2\mu^s\frac{\partial v_y^s}{\partial y} + (\kappa^s - \mu^s)\left(\frac{\partial v_x^s}{\partial x} + \frac{\partial v_y^s}{\partial y}\right) \tag{18}$$

where we have defined

$$\mu^s \equiv \int_{-\ell}^{\ell} (\mu - \mu^m) \, dz \tag{19}$$

as the surface excess shear viscosity,

$$\kappa^s \equiv \int_{-\ell}^{\ell} (\kappa - \kappa^m) \, dz + \frac{1}{3}\mu^s \tag{20}$$

as the surface excess dilatational viscosity, and

$$\sigma \equiv \int_{-\ell}^{\ell} (p^m - p_t) \, dz \tag{21}$$

as the surface tension.

Substituting Eqs. (17) and (18) into Eq. (9), we have an explicit relation between interfacial stress and deformation:

$$[\tau_{zy}] = F_y^s + \frac{\partial \sigma}{\partial y} + (\kappa^s + \mu^s) \frac{\partial}{\partial y} \frac{\partial v_x^s}{\partial x} + \frac{\partial v_y^s}{\partial y}$$

$$- \mu^s \frac{\partial}{\partial x} \left(\frac{\partial v_x^s}{\partial y} - \frac{\partial v_y^s}{\partial x} \right) \tag{22}$$

Similarly, it may be shown from Eqs. (10) and (11) that

$$[\tau_{zz}] = F_x^s + \frac{\partial \sigma}{\partial x} + (\kappa^s + \mu^s) \frac{\partial}{\partial x} \left(\frac{\partial v_x^s}{\partial x} + \frac{\partial v_y^s}{\partial y} \right)$$

$$+ \mu^s \frac{\partial}{\partial y} \left(\frac{\partial v_x^s}{\partial y} - \frac{\partial v_y^s}{\partial x} \right) \tag{23}$$

and

$$[\tau_{xx}] = 0 \tag{24}$$

Interfacial stress behavior for the Newtonian surface is thus defined in terms of three stress coefficients μ^s, κ^s, and σ.

III. SURFACE DILATATIONAL PROPERTIES

Surface dilatational properties are those properties that are necessary to define fully a pure expansion or compression of a fluid surface. According to the Newtonian surface stress model (22–24), these properties may be identified as σ and κ^s.

This may clearly be seen by presenting Eqs. (22–24) in the more general vector form as follows. We let $\underline{i}_x$, $\underline{i}_y$, and $\underline{i}_z$ denote base vectors of a cartesian space, possessing the following tensor and differential properties:

$$\underline{\underline{I}} = \underline{i}_x \underline{i}_x + \underline{i}_y \underline{i}_y + \underline{i}_z \underline{i}_z \tag{25}$$

is the unit tensor of the cartesian space,

$$\underline{\underline{I}}_s = \underline{i}_x \underline{i}_x + \underline{i}_y \underline{i}_y \tag{26}$$

is the unit surface tensor of the surface A (with normal $\underline{i}_z$),

$$\nabla = \underline{i}_x \frac{\partial}{\partial x} + \underline{i}_y \frac{\partial}{\partial y} + \underline{i}_z \frac{\partial}{\partial z} \tag{27}$$

is the spatial gradient operator, and

$$\nabla_s = \underline{i}_x \frac{\partial}{\partial x} + \underline{i}_y \frac{\partial}{\partial y} \tag{28}$$

is the surface gradient operator.

Then Eqs. (22) to (24) may be expressed in the vector form

$$\underline{n} \cdot [\underline{\underline{P}}] = \nabla_s \cdot \underline{\underline{\Sigma}} \tag{29}$$

where $\underline{\underline{P}}$ is the bulk phase pressure tensor and $\underline{\underline{\Sigma}}$ is the surface tension tensor:

$$\underline{\underline{\Sigma}} = \sigma \underline{\underline{I}}_s + (\kappa^s - \mu^s) \nabla_s \cdot \underline{v}^s \underline{\underline{I}}_s + \mu^s \left(\nabla_s \underline{v}^s + \nabla_s \underline{v}^{s\dagger} \right) \tag{30}$$

For a purely expanding or compressing surface Eq. (30) becomes

$$\underline{\underline{\Sigma}} = \left(\sigma + \kappa^s \nabla_s \cdot \underline{v}^s \right) \underline{\underline{I}}_s \tag{31}$$

revealing the dilatational properties of the surface to be the surface tension σ and the surface dilatational viscosity κ^s.

Many of our results will be expressed in terms of a "dynamic" surface tension that is measured over a spherical fluid surface. Mathematically, the dynamic surface tension may be expressed as

$$\bar{\sigma} = \frac{1}{2} \underline{\underline{I}}_s : \underline{\underline{\Sigma}} \qquad (32)$$

where (:) denotes the double scalar product operation. Using Eq. (31) in Eq. (32), we have that the dynamic surface tension is composed of both thermodynamic and viscous parts:

$$\bar{\sigma} = \sigma + \kappa^s \nabla_s \cdot \underline{v}^s \qquad (33)$$

Identifying the reversible and irreversible contributions to the dynamic surface tension is a difficult and open question. Numerous attempts have been made, including postulating the functionality of $\bar{\sigma}$ upon surfactant mass transfer [7–9], assuming negligible surface tension gradient (zeroth-order solution) [10], or defining the apparent dissipation in $\bar{\sigma}$ as the viscous, irreversible contribution [11–15]. The particular difficulty with the latter approach, as we will discuss, is that much of the observed "dissipation" in the dynamic surface tension is reversible energy transfer between surface and bulk phases. The former approaches, although apparently more rational, do not guarantee that the measured "viscous" contribution (κ^s) is not a "parameter" of the mass transfer model, rather than a true viscosity.

Still another class of techniques has been proposed based on the propagation of thermally induced microscopic capillary waves in the interface [16–19]. However, such techniques raise the fundamental question of whether microscopic capillary waves may be modeled by means of a macroscopic capillary wave theory.

In the present work we do not speculate on the contributions to the dynamic surface tension but rather, investigate the directly measured dynamic tension $\bar{\sigma}$ and the dilatational modulus

$$E^* \equiv -\frac{d\bar{\sigma}}{d \ln A} \qquad (34)$$

which measures the change in tension with surface area expansion. The dilatational modulus is a very important measure of dilatational surface stress as it expresses in a certain sense the gradient of dynamic surface tension. It is, of course, the gradient of dynamic surface tension that plays a direct role in the surface stress equations [see Eqs. (22) and (23)].

The several techniques that have been proposed to measure the surface dilatational viscous effect implicit in Eq. (33) on the basis of dissipation or phase lag measurements [11–15] have done so by transforming the dilatational modulus [34] into the complex plane, where the imaginary contribution to the modulus is defined as arising from a surface viscous mechanism.

Lucassen and van den Tempel [11,12] and Djabbarah and Wasan [13] measured wave properties of a mechanically driven longitudinal surface wave. In the analysis of the wave motion, the amplitude of the wave ξ was expressed naturally in the complex plane, such that

$$\frac{\partial \sigma}{\partial x} = E^* \frac{\partial^2 \xi}{\partial x^2}$$

defines the complex dilatational modulus, where

$$E^* = E' + i\omega\eta$$

The real portion E' is called the surface elasticity and the term η is called the surface dilatational viscosity, where ω is the wave frequency.

A similar treatment has been used by Rambhadran et al. [14], who measured the pressure drop across an oscillating droplet to determine the surface elasticity and dilatational viscosity.

A technique proposed by Clint et al. [15] is based on the same interpretation of the dilatational modulus. A droplet is formed on the tip of a capillary. In a brief instant the droplet is expanded to a new radius and the pressure drop over the droplet is then measured as a function of time at a fixed radius. The dilatational modulus, which may be inferred directly from the pressure versus time trace, is then transformed into the imaginary plane using a Fourier analysis, and the definition of real and imaginary parts, as with the damped wave and oscillating droplet experiments, allows the determination of the surface dilatational viscosity.

It is, however, not at all clear that the surface dilatational viscosity determined in this way is in fact the parameter defined in Eq. (33) as the surface dilatational viscosity. In fact, the damping quantified by the complex contribution to the dilatational modulus is controlled in part by an exchange of surfactant between the surface and bulk phases, which is not a dissipative mechanism.

IV. APPLICATIONS

A. Foam Rheology

As an application of surface dilatational properties, we may consider the rheology of a "wet" foam, or a foam with a small gas volume fraction. Because of the presence of gas phase, foam flow may often be accompanied by a significant dissipation of energy due to compression or expansion of the gas phase. This dissipation of energy is quantified in the Newtonian constitutive model with the lesser known "dilatational" viscosity κ.

Much effort has been given to the measurement of this second coefficient of viscosity for pure fluids [20–24] and has revealed that the dilatational viscosity is generally larger than the shear viscosity. Still, since most pure fluid viscous flow displays a small degree of compressibility, the dilatational viscosity of pure fluids is primarily of academic interest.

However, the degree of compressibility possible in foam flow makes the effective dilatational viscosity of a foam of general practical interest.

The first theoretical calculation of the dilatational viscosity of a wet foam was made by Taylor [25], who derived the following relation for a dilute wet foam containing spherical bubbles:

$$\kappa^* = \frac{4\mu}{3\phi} \tag{35}$$

where the asterick denotes effective foam properties. Here μ is the liquid phase shear viscosity and ϕ is the gas volume fraction.

Prud'homme and Bird [26] extended the result of Taylor to first order in ϕ with the relation

$$\kappa^* = \frac{4\mu}{3\phi}(1 - \phi) \tag{36}$$

Recently, we extended this result to include surface rheological effects [27] as

$$\kappa^* = \frac{4\mu}{3\phi}\left(1 - \phi + \frac{\kappa^s}{\mu a}\right) \tag{37}$$

where κ^s is the surface dilatational viscosity and a is the bubble radius.

It is certainly not surprising that the dilatational viscosity of a foam would be directly related to the dilatational viscosity of individual surfaces within that foam. In fact, as one would expect, as

Properties of Surfactant Interfaces

liquid is removed from the foam (this limit must be taken in a qualitative sense) the single remaining source of the foam dilatational viscosity is the surface dilatational viscosity.

A relevant application of the surface dilatational viscosity through the effective foam dilatational viscosity Eq. (37) may be found in the process of foam flow in porous media in the presense of oil (i.e., EOR).

Dilatation (compression or expansion) in EOR may occur by several mechanisms, among which are:

1. Absolute pressure changes as foam flows from high-pressure to low-pressure zones in the porous medium result in an expansion of the foam.
2. Contact of foam with oil phase may result in a rapid expansion of the foam due to surface instability.
3. Geometrical nonuniformities in the capillary geometry (e.g., expansion/compression in the capillary diameter) may result in expansion/compression of the foam.
4. Gas transfer from smaller to larger bubbles results in a dilatation of the foam.

A simple model may be proposed by which these dilatational effects in EOR may be gauged in significance relative to shearing effects. In Fig. 5 we show a cross section of a fully formed foam structure of uniform bubble size flowing through a smooth capillary with an assumed Poiseuilleian profile as a model of foam flow through microscopic (generally nonuniform) capillaries within a porous medium. The bubbles are allowed to oscillate with a definite frequency about a mean radius to simulate the effect of dilatation caused by mechansims such as those discussed above.

The dissipation of energy due to the shear viscosity of the foam and the dissipation of energy due to the dilatational viscosity of the foam may be compared for specific values of the capillary radius, bubble size, frequency, and so on, to estimate the significance of the dilatational effect.

The dissipation of energy due to the shear viscosity of the foam may be estimated by the Poiseuille relation

$$\varepsilon_\mu^* = \frac{(\Delta p/L)^2 R^2}{8\mu^*} \tag{38}$$

where $\Delta p/L$ is the pressure drop over the capillary and R is the capillary radius.

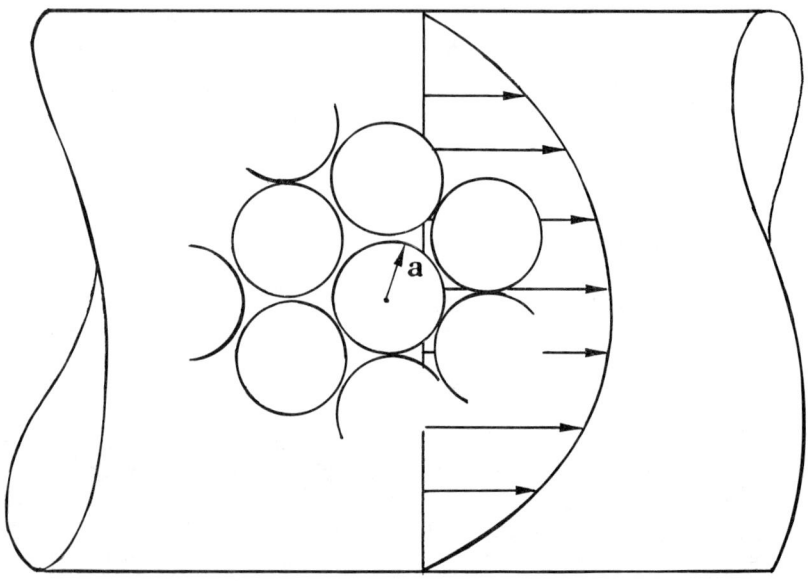

FIG. 5 Cross section of a foam with oscillating bubble surfaces flowing within a smooth capillary.

The dissipation of energy due to the dilatational viscosity of the foam may be approximated as

$$\varepsilon_\kappa \sim \kappa^* \omega^2$$

As physical constants we choose for the capillary radius R a value of 10 μm, for the pressure drop 10 psi/ft, for the mean bubble radius 1 μm, for the foam shear viscosity 100 cP, for the dilatational viscosity of the foam as $\kappa^* \simeq (4/3)(\kappa^s/a)$ and conservatively for the frequency of surface expansion 2 Hz. For this particular geometry, then,

$$\frac{\varepsilon_\kappa^*}{\varepsilon_\mu^*} \simeq \kappa^s \times 10^3 \qquad (39)$$

where κ^s has the units of sP. As mentioned previously, much work remains to establish the absolute magnitude of κ^s, but the work to

Properties of Surfactant Interfaces 17

date indicates that the surface dilatational viscosity is larger in general than the surface shear viscosity and may often approach a value of 1 sP [7–9]. Given the qualitative nature of this model calculation, at most one can speculate that dilatational effects are comparable in significance to shearing effects in the EOR process.

B. Enhanced Oil Recovery

To determine experimentally the significance of surface dilatational properties in EOR, we have performed oil displacement experiments in Berea sandstone cores using aqueous foams stabilized by straight-chain α-olefin sulfonates. Saturation profiles were determined using both microwave and gamma-ray absorption techniques. (A complete discussion of the microwave and gamma-ray experiments together with the results summarized here may be found in Ref. 28.)

The Berea cores (4 in. x 0.75 in. x 12 in.) used as porous media were vacuumed to displace interstitial air and then saturated with 1% NaCl aqueous solution. About six pore volumes of the same solution was injected to stabilize the clay as well as to determine the absolute permeability and porosity (average permeability = 380 millidarcy, average porosity = 0.19). The porous medium was then flooded with Salem crude oil (viscosity at room temperature = 6cP) to irreducible water content. Waterflood (1% NaCl) was injected into the porous medium until the rock adsorption was completed.

Two types of displacement experiments were conducted to analyze foam behavior in porous media. In one set of experiments nitrogen gas was injected following the surfactant solution to form the foam in situ with an imposed pressure differential of 45 psi. In the second set of experiments the nitrogen gas was injected at a constant flow rate of 10 ft/day. During the constant flow rate experiments the dynamic fluid saturations along the core length were measured by the combined gamma ray/microwave technique.

Table 1 shows the results of the foam-enhanced oil recovery tests for the constant-pressure and the constant-flow-rate experiments, respectively. The foam-enhanced oil recovery results from both sets of experiments are similar in that the recovery efficiency and gas breakthrough time increased in the same order: C_{12} AOS, C_{14} AOS, and C_{16} AOS.

On the basis of these experiments and the discussion of Section IV. A, one would expect the dilatational viscosity of the C_{16} AOS-stabilized surface to be the largest, followed by C_{14} AOS and then C_{12} AOS (since large foam viscosity results in a lower foam mobility and generally a higher oil displacement efficiency).

As discussed previously [see Eqs. (33) and (34)], the dilatational modulus, which is a directly measurable quantity, contains not only

TABLE 1 Saturation Data for Constant-Pressure and Constant-Flow-Rate Experiments

Experiment:	Under constant pressure			Under constant flow rate		
	1 C_{12}	2 C_{14}	3 C_{16}	4 C_{12}	5 C_{14}	6 C_{16}
Initial oil saturation	73.0	71.5	74.0	74.0	73.8	75.2
Waterflood oil saturation	42.1	39.8	41.5	42.1	43.2	43.5
Surfactant flood oil saturation	39.0	37.0	39.5	39.8	40.6	41.7
Percent recovery[a]	4.3	3.9	2.7	3.1	3.5	2.4
Foam-flood oil saturation	28.5	22.5	23.5	33.9	33.1	30.6
Percent recovery[a]	14.4	20.2	21.6	8.0	10.2	14.8
Breakthrough time (min)	23.0	35.0	39.0	36.0	48.0	62.0

[a]Percent oil recovery is based on the initial oil saturation.

Properties of Surfactant Interfaces

the surface dilatational viscosity effect but also the surface tension gradient effect, which might also be considered very relevant to foam mobility in EOR.

We have recently developed a method for measuring dynamic surface tension and the dilatational modulus based on the maximum bubble pressure technique for surface tension measurement. The experiment is discussed in detail elsewhere [29]. Also, see the work of Mysels [43].

Using this technique, we have measured the dilatational modulus for the three straight-chain α-olefin sulfonate systems used in the foam-flooding experiments (see Fig. 6).

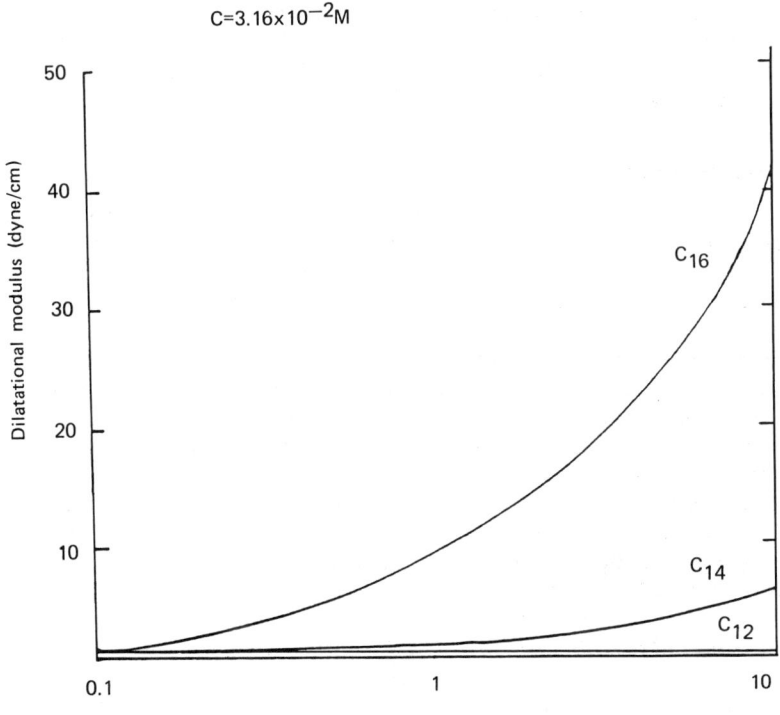

FIG. 6 Dilatational modulus versus frequency of surface expansion for $C_{12}AOS$, $C_{14}AOS$, and $C_{16}AOS$ surfaces.

The results show that the largest dilatational modulus is that of the $C_{16}AOS$ system, followed by $C_{14}AOS$ and $C_{12}AOS$, the exact order suggested by the core-saturation experiments, indirectly confirming the significance of dilatational surface properties in EOR.

The significance of the dynamic surface tension gradient in EOR has been confirmed by Hirasaki and Lawson [30], who concluded that the apparent viscosity of a foam flowing through smooth capillaries in a porous medium increases with increasing surface tension gradient. The increasing foam viscosity results in a low foam mobility and therefore a greater oil displacement efficiency.

However, most of the research to date concerning the role of surface tension and viscosities in EOR has focused on the oil-water interface rather than the gas-water surface. Slattery [31] has concluded, based on a qualitative model of dynamic interfacial effects in EOR, that for low-tension systems, decreasing interfacial viscosity may significantly increase oil displacement, although Flumerfeldt et al. [32] have since concluded that the interfacial viscous effect is a smaller-order contribution than the effect of interfacial tension. In any case, the desirability of a low interfacial tension in EOR appears to be unanimous [31-35].

C. Foam Stability

The efficiency of oil displacement by aqueous foams stabilized by the straight-chain α-olefin sulfonates suggests a further correlation between foam stability and the dilatational modulus. Such a correlation is justified by theoretical studies of thin foam film drainage and stability.

Since the original work of Reynolds [36], who determined the drainage of a liquid film between two flat immobile surfaces, both theoretical and experimental research has shown that drainage between two foam surfaces is generally much more rapid due in part to a fluidic mobility within the bounding surfaces of the film. In fact, much of the thin-film drainage research in the past three decades has focused on quantifying the relevant parameters within a thin film which determine whether the film will drain rapidly (promote instability of the foam) or slowly (promote stability), largely on the basis of the mobility of the bounding surfaces.

Johannes and Whitaker [37] investigated the gravitational thinning of a foam film stabilized by surface-active agents. Comparing both theoretical and experimental results, they concluded that an important foam stabilizing mechanism is a large dilatational modulus, or in their terms, either a large surface dilatational viscosity or a large surface tension gradient. [Note again the difficulty in distinguishing between the source terms in Eq. (33).]

More recently, Zapryanov et al. [38] have concluded that the interfacial tension gradient is a governing factor in determining whether a film drains as an immobile or a mobile bounded film. They predict that at high interfacial tension gradient the film drains according to the Reynolds model, while at low interfacial tension gradient the film becomes significantly effected by parameters such as interfacial viscosity (sum of μ^S and κ^S). This result is illustrated in Fig. 7 where the draining rate of an emulsion film is predicted as a function of the interfacial tension gradient ($K_\sigma = -\partial\sigma/\partial C$, where C is the bulk-phase surfactant concentration) and interfacial viscosity. The dilatational modulus has also been shown significant to foam film stability, emulsion stability [39], and the stability of a liquid film on a solid surface [40]. For a more complete review of thin-film studies, see Ref. 3.

To confirm the significance of the dilatational modulus on foam stability and to show the direct relation to the thin-film drainage, both drainage and foam stability tests were performed for the three surfactant-stabilized aqueous foams $C_{12}AOS$, $C_{14}AOS$, and $C_{16}AOS$ [41]. The drainage time of the foam films was studied using the

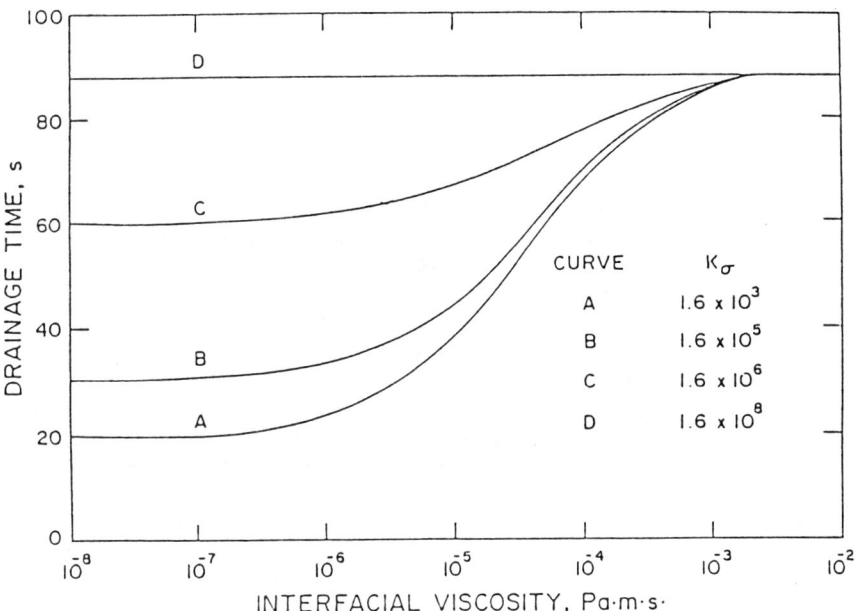

FIG. 7 Effect of interfacial viscosity and interfacial tension gradient on thin-film drainage time. (From Ref. 38.)

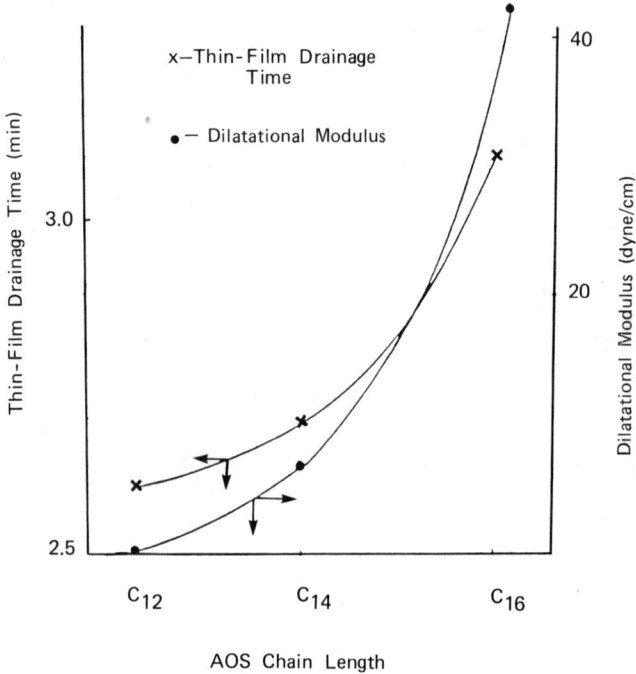

FIG. 8 Thin-film drainage time versus dilatational modulus.

interference microscopic technique described in Ref. 42. Thin films of any desired radius are formed in a specially designed glass cell containing a double concave meniscus. Monochromatic light (wavelength of 546.1 nm) is exposed to the draining film and the reflected light is measured by a fiber-optic probe. As the film thins, the variable thickness produces a series of interference patterns which are recorded through the fiber-optic probe upon a strip-chart recorder. The values of the photocurrent versus time enable an estimation of both the thinning rate and the thickness of the film. To prevent evaporation the films were formed in a completely closed cell and sufficient time was allowed to saturate the cell atmosphere.

Figure 8 reveals the results of the drainage tests, illustrating the proportionality between the dilatational modulus and drainage time of the foam film. The drainage time is seen to increase with the identical order of surfactants as the order of increasing oil displacement in EOR, $C_{12}AOS$, $C_{14}AOS$, and $C_{16}AOS$. The foam stability tests were performed using both static and dynamic methods. The

Properties of Surfactant Interfaces

static method entails foam generation in a graduated cylinder by shaking the cylinder and then monitoring the foam height with time. The foam half-life is determined by the first-order rate equation

$$-\frac{dV_f}{dt} = K_f V_f$$

where V_f is the foam volume and K_f the rate constant. The half-life $t_{1/2}$ is given by

$$t_{1/2} = \frac{0.693}{K_f}$$

The dynamic method is to produce uniform size bubbles at the end of a capillary immersed in a surfactant solution. The bubbles rise to the top of the solution and form a foam that with time reaches an equilibrium level.

In Fig. 9 the results of the static and dynamic foam stability test are plotted versus a dilatational modulus revealing a correlation with the film drainage measurements as foam stability is seen to increase with increasing dilatational modulus.

D. Emulsion Stability

Due to the analogous nature of emulsion stability phenomena to foam stability phenomena, it may be expected that the dilatational modulus also possesses a strong relevance to emulsion stability. To confirm this relevance, we have conducted emulsion stability experiments on oil-in-water emulsions with dodecane as the oil phase and 1 wt% sodium alkylnapthalene dissolved within the aqueous phase. The systems were varied with the addition of 2.5mM calcium chloride and/or 2 wt% gelatin within the aqueous phase.

The emulsion stability tests were conducted using a photomicrographic technique. The emulsion is placed within a Howard cell and photographed through a Nikon microscope with a magnification of 400×. The microscope has a maximum magnification of 1000× and is equipped with interference phase contrast, polarizing light, and slow and fast-speed cinematographic equipment.

The first photographs were taken 20 s after the sample had been emulsified. Additional photographs were taken over a 24-h period. The droplet size distribution was determined using a Zeiss MOP image analyzer. The size distribution was used to determine a first-

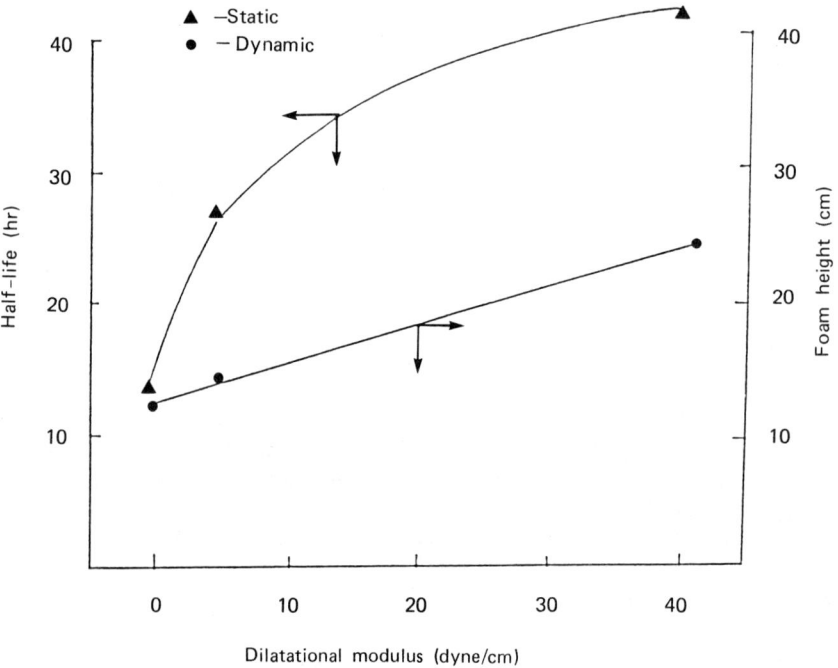

FIG. 9 Foam half-life and equilibrium foam height versus dilatational modulus.

order rate constant K where

$$N = N_0 \exp(-Kt)$$

Here N is the number density of droplets at time t and N_0 is the initial droplet number density.

In Table 2 the rate constants are given for the four emulsion systems. The constants reveal the gelatin to be a stabilizing component and the calcium salt to be a destabilizing component. In Fig. 10 the dilatational moduli of the four systems are plotted, showing the identical correlation observed with the foam systems. The larger the dilatational modulus of the oil/water interface, the greater the stability of the emulsion.

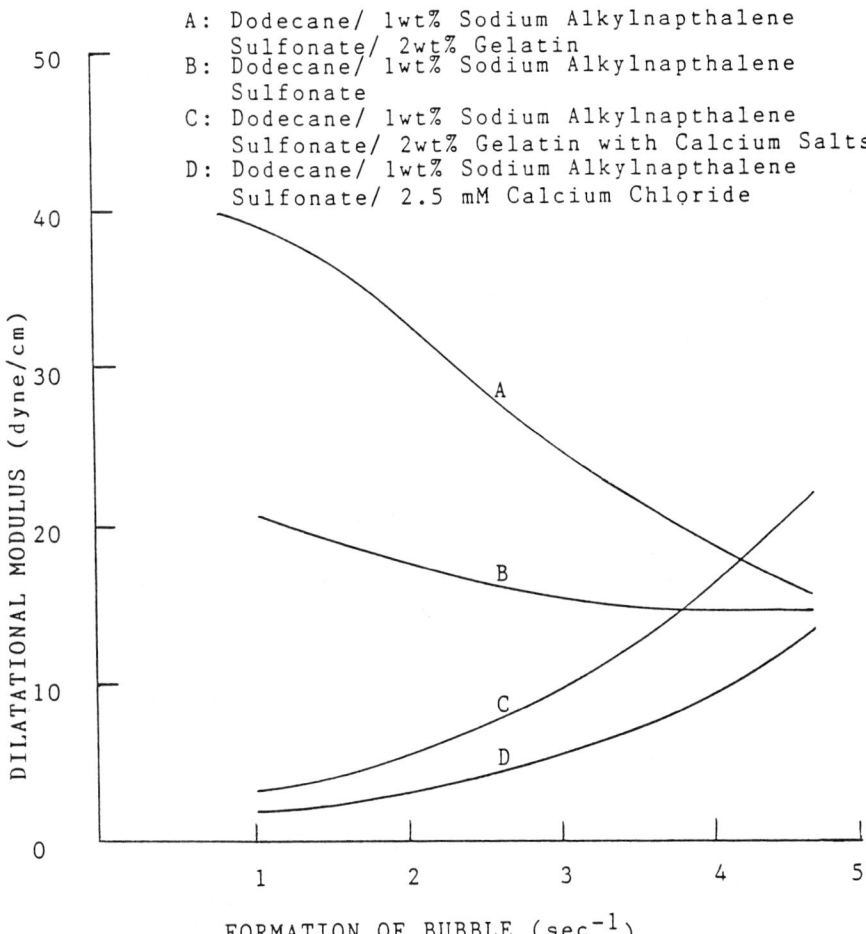

FIG. 10 Dilatational modulus versus frequency of bubble creation for four oil-water interfaces.

IV. SUMMARY

Dilatational surface properties arising from the presence of surface-active agents at the fluid/fluid interface may have both a thermodynamic (σ) and a viscous (κ^s) origin. The dilatational modulus [Eq. (34)] (or dynamic surface tension gradient) contains both the thermodynamic and viscous contributions, is directly measurable, and

TABLE 2   Coalescence Rate Data for the Water-in-Oil Emulsions

System	Rate constant (h^{-1})
A: Dodecane/1 wt% sodium alkylnapthalene sulfonate/2 wt% gelatin	~ 0.0
B: Dodecane/1 wt% sodium alkylnapthalene sulfonate	3.3×10^{-2}
C: Dodecane/1 wt% sodium alkylnapthalene sulfonate/2 wt% gelatin with calcium salts	8.7×10^{-2}
D: Dodecane/1 wt% sodium alkylnapthalene sulfonate/2.5 mM calcium chloride	1.1×10^{-1}

is significant to foam rheology, foam stability, emulsion stability, and oil displacement by foam flooding in porous media. Results presented indicate that a large dilatational modulus enhances foam stability, emulsion stability, and oil displacement in the foam-flooding EOR process. We are currently extending this research to conduct a systematic investigation of the dilatational modulus, including the effects of temperature, electrolyte, surfactant structure, and bulk phase, as well as an investigation of the significance of the dilatational modulus to other dynamic interfacial phenomena.

ACKNOWLEDGMENTS

The financial support provided by the National Science Foundation and the Department of Energy is gratefully acknowledged.

REFERENCES

1. M. Joly, in *Surface and Colloid Science*, Vol. V (E. Matijevic, ed.), Wiley Interscience, New York, 1962.
2. F. C. Goodrich in *Solution Chemistry of Surfactants*, Vol. 2 (K. L. Mittal, ed.), Academic Press, New York, 1979, p. 733.
3. A. K. Maholtra and D. T. Wasan, in *Thin Liquid Films*, Marcel Dekker, New York, in press.

4. D. A. Edwards and D. T. Wasan, submitted to J. Rheol. (1986).

5. D. A. Edwards and D. T. Wasan, submitted to J. Rheol. (1987).

6. D. A. Edwards and D. T. Wasan, submitted to J. Rheol. (1987).

7. K. Miyano, B. M. Abraham, L. Ting, and D. T. Wasan, J. Colloid Interface Sci., 92:297 (1983).

8. L. Ting, D. T. Wasan, K. Miyano, and S. Q. Xu, J. Colloid Interface Sci., 102:248 (1984).

9. L. Ting, D. T. Wasan, and K. Miyano, J. Colloid Interface Sci., 107:345 (1985).

10. L. Wei, W. Schmidt, and J. C. Slattery, J. Colloid Interface Sci., 48:1 (1974).

11. J. Lucassen and M. van den Tempel, J. Colloid Interface Sci., 41:491 (1972).

12. J. Lucassen and M. van den Tempel, Chem. Eng. Sci., 27:1283 (1972).

13. N. F. Djabbarah and D. T. Wasan, Chem. Eng. Sci., 37:175 (1982).

14. T. E. Rambhadran, C. H. Byers, and J. C. Friendly, AIChE J., 22:872 (1976).

15. J. H. Clint, E. L. Neustadter, and T. J. Jones, paper presented at the 2nd European EOR Symposium, 1985.

16. S. Hard and H. Lofgren, J. Colloid Interface Sci., 60:529 (1977).

17. D. Byrne and J. C. Earnshaw, J. Phys. D: Appl. Phys., 12:1145 (1979).

18. S. Hard and R. Neuman, J. Colloid Interface Sci., 82:315 (1981).

19. S. Hard and R. Neuman, J. Colloid Interface Sci., 82:315 (1981).

20. E. F. Fox and G. D. Rock, Phys. Rev. 70:68 (1946).

21. L. N. Liebermann, Phys. Rev. 75:1415 (1949).

22. L. Tisza, Phys. Rev., 61:531 (1942).

23. J. Quinn, J. Acoust. Soc. Am., 18:185 (1946).
24. S. B. Gurevich, C. R. U.S.S.R., 55:17 (1947).
25. G. I. Taylor, Proc. R. Soc. London Ser. A, 226:34 (1954).
26. R. K. Prud'homme and R. B. Bird, J. Non-Newt. Fluid Mech., 3:261 (1977).
27. D. A. Edwards, H. Brenner, and D. T. Wasan, manuscript in preparation.
28. D. W. Huang, Ph.D. thesis, Illinois Institute of Technology, 1985.
29. R. Kao, A. Nikolov, D. A. Edwards, and D. T. Wasan, manuscript in preparation.
30. G. J. Hirasaki and J. B. Lawson, paper presented at the 58th Annual SPE Technical Conference, San Fransisco, 1983.
31. J. C. Slattery, AIChE J., 20:1145 (1974).
32. R. W. Flumerfeldt, J. P. Oppenheim, and J. R. Son, AIChE Symp. Ser., 78:113 (1982).
33. J. C. Melrose, Can. J. Chem. Eng., 48:638 (1970).
34. O. R. Wagner and R. O. Leach, Soc. Pet. Eng. J., 6:335 (1966).
35. D. T. Wasan, S. M. Shah, N. Aderangi, M. S. Chan, and J. J. McNamara, Soc. Pet. Eng. J., 18:409 (1978).
36. O. Reynolds, Philos. Trans. R. Soc. London Ser. A, 77:157 (1886).
37. W. Johannes and S. Whitaker, J. Phys. Chem., 69:1471 (1965).
38. Z. Zapryanov, A. K. Maholtra, N. Aderangi, and D. T. Wasan, Int. J. Multiphase Flow, 9:105 (1983).
39. J. Lucassen, M. van den Tempel, A. Vrij, and F. T. Hesselink, Proc. K. Ned. Akad. Wet., $B73$:108 (1970).
40. E. Ruckenstein and R. Jain, J. Chem. Soc. Faraday Trans. 2, 70:132 (1974).
41. D. W. Huang, A. Nikolov, and D. T. Wasan, Langmuir, 2:672 (1986).
42. A. A. Rao, D. T. Wasan, and E. D. Manev, Chem. Eng. Commun., 15:63 (1982).
43. K. J. Mysels, Langmuir, 2:428 (1986).

2
The Effect of Surface-Active Agents in Distillation Processes

JOHN C. BERG *University of Washington, Seattle, Washington*

I.	Introduction	29
II.	The Distillation Process	30
III.	The Effect of Surface Tension on Flow Structure	33
IV.	Surface Tension Gradients: The Results of Zuiderweg and Harmens	38
	A. Unsupported-Area Equipment	41
	B. Supported-Area Equipment	47
	C. The Stability Analysis of Wang et al.	53
V.	Surfactant Effects	59
	A. Unsupported-Area Equipment	60
	B. Supported-Area Equipment	65
	References	73

I. INTRODUCTION

Distillation is a complex vapor-liquid contacting operation involving large interfacial area-to-volume ratios. It is therefore not surprising that surface-active agents, whose characteristic feature is their ability to concentrate themselves at interfaces, should have the potential for profoundly influencing distillation. The literature contains ample evidence of such influence, but the subject is rarely examined in a systematic way. There are three major reasons for this. First, the consequences of the presence of surfactants in distillation can be extremely varied and complex, manifesting themselves in such completely different ways in different circumstances that it is difficult to identify them as arising from a common origin. On the one hand, their presence may lead to the foam flooding of a column, rendering it totally inoperable. On the other hand, it has been shown how additions of surfactant may nearly double the efficiency of a packed tower [1] and significantly enhance the efficiency of a tray tower [2]. Second, the unraveling of the explanations of the effects of surfactants relies on the relatively new subject of interfacial hydrodynamics. Until relatively recently, the underlying principles of this discipline were not widely understood or appreciated. Indeed, the subject is in a state of evolution as ongoing basic research is bringing new insights. The topic is missing entirely from most books on hydrodynamics or fluid mechanics. Finally, the effects of surfactants are greatest in smaller-scale equipment, so the economic incentive for understanding them is not as great as it might otherwise be.

The motivation for an examination of the effects of surfactants on distillation at this time is twofold. First, there are situations for which sufficient understanding exists that practical advantage may be taken of efficiency-enhancing surfactant addition, or so that, on the other hand, surfactant-induced problems may be properly diagnosed. Second, the present state of knowledge concerning surfactant effects is highly incomplete. A review at this time, pointing out some of the present gaps, may stimulate further research.

In order to describe the observed effects of surfactants in distillation it is necessary to begin with a brief description of the distillation process itself. Next is given a discussion of interfacial and capillary phenomena in distillation in the absence of surfactants. This forms the basis for understanding the consequences of the presence of surfactants. Finally, the role of surfactants is discussed, and recommendations for further study are given.

II. THE DISTILLATION PROCESS

The objective of distillation is the separation or partial separation of the components of a liquid mixture by means of partial vaporization. It is necessary, of course, that the equilibrium composition of the vapor mixture differ from that of the liquid. In a single-stage continuous distillation in which a given liquid mixture is fed to a chamber where a given fraction is vaporized and separated from the liquid, it is the thermodynamic vapor-liquid equilibrium that dictates the extent of separation achievable. The degree of separation may be increased by connecting such single-stage units in series with countercurrent flow of the vapor and liquid. These "stages" are connected vertically in order to exploit gravity for phase disengagement. The result is a distillation "column," as shown schematically in Fig. 1. Heat is generally supplied only at the bottom, where the liquid is partially vaporized and sent up through the column. At the top, some of the vapor is condensed into liquid (called reflux) and sent back down the column, while the rest is removed as "overhead product." Feed may enter at any stage along the column. Separation of the more volatile from the less volatile components is achieved during the contact of the rising vapor with the descending reflux. On any given stage in the column, part of the vapor stream is condensed and part of the reflux stream is vaporized. The result of this exchange of (mostly latent) heat and mass between the streams is that the vapor stream emerging is enriched in the more volatile components, while the reflux liquid descending has been partially depleted of these components. Under conditions of ideal contact, the vapor and liquid streams emerging from a given stage are in equilibrium, but under no conditions can the streams being contacted on a given stage be

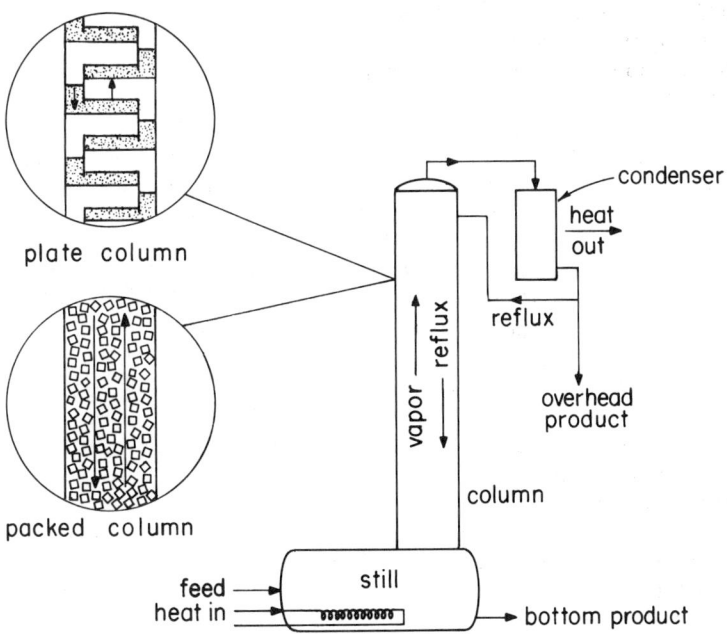

FIG. 1 Schematic of distillation equipment. Insets show a plate column (unsupported-area) and packed column (supported-area) configurations.

in equilibrium. There is thus a mass transfer driving force for the desired exchange of components between the streams. More details on the principles of multistage distillation may be found in a variety of textbooks [3–5].

A wide variety of equipment exists for carrying out distillation. Columns vary in size from heights as small as 100mm and diameters as small as 10mm, for bench-scale separations, up to giant columns with heights greater than 100m and diameters greater than 10m, found in large petroleum refineries. The design of column internals varies widely and is the object of continuing research and development, but its objective is always the same: to provide intimate and effective contact between the rising vapor and the descending liquid. Intimate and effective contact requires a large interfacial area between the vapor and the liquid. Both phases must also have good access to the interface. Thus even if a superficial froth provides a large interfacial area, the contact may be ineffective if most of the bulk liquid is far from the interface. Another requirement is that there be adequate time of contact between the phases to permit mass transfer. This means that there must be sufficient liquid holdup. It is also important that both phases be in a state of sufficent agitation to assure efficient mass transfer within the phases. Finally, it is important to prevent vapor or liquid passing through a stage without adequate interphase contact on that stage (i.e., entrainment).

Meeting all the criteria above requires the establishment of a complex flow structure within a distillation column. While an almost infinite variety of specific designs are used, most column configurations can be divided into two general types: discrete-stage devices and continuous-contact devices. Both are illustrated schematically in the insets of Fig. 1. Discrete-stage devices usually consist of a series of trays with openings to permit the rise of the vapor and the descent of the liquid. A liquid pool of a certain depth resides on each tray, while vapor from the stage below bubbles up through it. In some smaller distillation columns, the liquid may descend by draining through the same openings, but in larger columns (as well as many smaller ones) the reflux descends through larger openings or tubes (called downcomers) located either at the sides or the center of the column cross section. Downcomer locations are offset from one stage to the next so that reflux liquid must flow across the cross-sectional area of the tray before descending to the next tray. Under proper operating conditions, the trays provide for adequate liquid holdup and contact time, while the bubbling action of the rising vapor provides large interfacial area and good intraphase agitation. The tray spacing must be adequate to preclude significant liquid spraying or frothing on one tray from rising to the next.

Surface-Active Agents in Distillation

Continuous-contact devices consist of an open column (undivided by trays) filled with packing elements (rings, saddles, spheres, helices, etc.) or gridworks of various kinds to provide a large interfacial area. The presumption is that descending reflux will wet out the packing or gridwork with a uniformly distributed thin liquid film. The slowed rate of liquid descent provides adequate holdup and contact time with the rising vapor. Although discrete stages are not physically provided in such columns, countercurrent staging nonetheless exists.

Perhaps the most important point of distinction between the discrete-stage and continuous-contact configurations is with respect to the mode of interfacial area formation. In the former case, area is formed as bubbles of vapor are forced through pools of reflux liquid. The amount of area formed bears no direct relationship to the area of the tray or of the holes, but depends more on the bubble size and rate of bubbling. We thus refer to a tray column as an unsupported-area device. It is to be contrasted with the continuous-contact, packed column in which the vapor-liquid interfacial area does depend directly on the amount of solid surface area made available. A packed column is thus referred to as a supported-area device.

For either type of column construction, the internal flow scheme is highly complex, and vapor and liquid flow rates must be maintained within careful ranges for satisfactory operation. If the flow rate is so great that the countercurrent flow of the opposite phase is impeded, the column is said to be flooded. On the other hand, if the flow rates are too low, phase distribution and contact will usually be poor, and in any event, the column is operated at below its design capacity. Flow rates are generally adjusted to a certain percentage (50 to 80%) of the corresponding flooding values for optimal operation.

III. THE EFFECT OF SURFACE TENSION ON FLOW STRUCTURE

Because of the complexity of the flow structure existing in distillation equipment, it is useful to subdivide it into simplified flow elements which might be amenable to more quantitative discussion or to analysis. Such a breakdown makes it possible to show how surface tension or capillary effects influence the flow structure. We first consider how liquid surface tension level effects the flow structure, ignoring for the moment the existence of surface tension gradients which necessarily exist in the multicomponent interfaces present in distillation. Thus in this section, attention is focused on the behavior of pure liquids or liquid mixtures which do not produce significant surface tension gradients in order to separate out the influence of capillary effects in the absence of tension gradients. These will be discussed in detail in succeeding sections.

In tray columns (unsupported-area devices) the primary processes are the formation, detachment, and rise of vapor bubble through liquid pools, as pictured in Fig. 2. When the bubbles reach the surface, they may either burst immediately into a spray of droplets, or they may exist briefly in a froth layer atop the liquid pool. When the bubbles burst, small jets are formed that pinch off into droplets under the influence of surface tension by the mechanism described originally by Rayleigh [6]. Rayleigh theory predicts that the jet length varies inversely with the square root of surface tension, but the drop size is independent of it. The bubble size and frequency in a given case are found to depend in complicated ways on the hole size, liquid depth, vapor velocity and liquid density, viscosity, and surface tension [7]. The effect of surface tension alone is difficult to separate out, but in general a decrease in surface tension level leads to larger bubbles, other factors remaining the same. Fane and Sawistowski [8] claim that the magnitude of the surface tension plays an important role in determining the efficiency of tray columns operating in the spraying regime. They produced data showing tray efficiency inversely proportional to surface tension, explaining some of the low efficiencies obtained in vacuum distillation (lower temperature, hence higher surface tension). Andrew [9], on the other hand, investigated the froth height developed on a small glass sieve plate by a variety of liquids and found it to be independent of surface tension. The existence of a froth or foam on

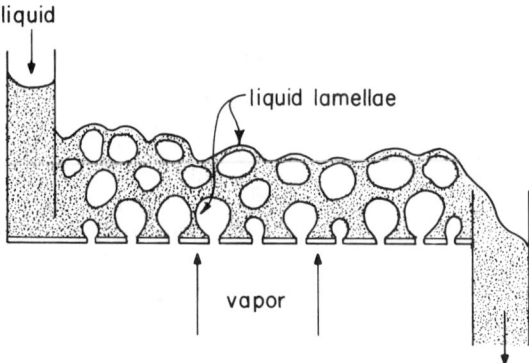

FIG. 2 Schematic of vapor and liquid flows on a tray. Liquid downcomers are shown at left and right. Liquid lamellae between rising bubbles and between bubbles and vapor above the tray are indicated.

top of the liquid on a tray cannot be supported by a pure liquid, but is generally associated with multicomponent systems in which surface tension gradients exist [10], as explained later, and in particular with the presence of surface-active contamination. The existence of any frothing at all with "pure" liquids, as reported for example by Andrew, is probably attributable to such contamination. In summary, attempts to correlate the flow structure on or the distillation efficiency of tray columns with the surface tension of the distillation mixture have not produced convincing generalizations.

In packed columns (supported-area devices) the flow element is that of a thin liquid film flowing downward over a solid surface. The film may be coherent and of essentially uniform thickness, or it may be punctuated with dry patches of varying size. In the extreme, the dry patches may occupy most of the solid surface area, with the liquid flowing downward in rivulets or drops. If the liquid film is coherent, the surface tension level plays no role in its flow properties (thickness, velocity profile), but it does play an important role in determining whether dry patches will develop and persist (i.e., whether or not they will be spontaneously rehealed). This may be understood in the following way. A liquid film is always subject to disturbances which will lead to the existence of local thin spots. Also, the irregularities introduced by the sharp edges of the packing elements always produce local thin spots in the film. A hydrodynamic stability analysis by Jain and Ruckenstein [11] examines the fate of such disturbances. The results show that whenever the thickness of the film is of the order of a few hundred angstroms or less, it will spontaneously thin itself and rupture, leaving a dry patch. This process can occur very rapidly. Depending on the nature of the liquid-vapor-solid interline, the dry patch may reheal itself so quickly that its occurrence is scarcely noticeable, or it may persist and even grow in size. The factor that dictates the existence of dry patches is thus the ability of the film to reheal itself.

For pure-liquid isothermal films, dry patch persistence is determined by a balance between surface tension forces and those of fluid pressure. Performing such a force balance, Hartley and Murgatroyd [12] determined that the minimum wetting rate required to reheal a dry patch on a vertical surface is given by

$$\frac{\dot{m}}{W} = K \left(\frac{\mu \rho}{g}\right)^{1/5} [\sigma (1 - \cos \theta)]^{3/5} \qquad (1)$$

where $\dot{m}/W$ is the mass flow rate of the liquid per unit width; g the gravitational constant; μ, ρ, and σ the viscosity, density, and surface tension of the liquid, respectively; and θ the contact angle.

K is a constant that depends on the system of units chosen. Comparison of Eq. (1) with experimental data bears out the trends of its predictions.

The contact angle, θ, is a quantitative measure of the physical interaction between a liquid and a solid. As indicated in Fig. 3, it is the angle made by the liquid against the solid, measured in the liquid. If θ = 0°, the liquid "wets out" the solid; if 0 < θ < 90°, the solid is "wet" by the liquid; and if θ > 90°, the solid is said not to be wet by the liquid. Under equilibrium conditions, the contact angle may be related to the surface tension of the liquid and the surface energies of the solid-vapor and solid-liquid interfaces by Young's equation [13]:

$$\cos \theta = \frac{\sigma_{SV} - \sigma_{SL}}{\sigma_{LV}} \qquad (2)$$

where $\sigma_{SV} = (\partial F/\partial A_{SV})_{T,V}$ and $\sigma_{SL} = (\partial F/\partial A_{SL})_{T,V}$, in which F is the Helmholz free energy of the system, A_{ij} is the area of the appropriate interface, and the derivatives are taken under conditions of constant temperature and volume. It is also presumed in defining these "interfacial tensions," σ_{SV} and σ_{SL}, that adsorption equilibrium exists. $\sigma_{LV} \equiv \sigma$ = the surface tension of the liquid.

Equation (1) implies that dry patch persistence is extraordinarily sensitive to the contact angle. The critical wetting rate at θ = 20°, for example, is more than twice the rate at θ = 10° and more than five times the rate at θ = 5°, at the same surface tension. It should be noted that the contact angle generally exhibits a significant difference in value depending on whether the liquid is advancing or receding over the solid surface, the advancing angle being larger than the receding angle. Such hysteresis is well understood in terms of solid surface roughness and chemical (energetic) heterogeneity

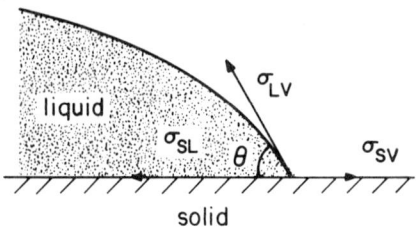

FIG. 3   The contact angle θ, of a liquid against a solid.

[14], but raises some ambiguities in the application of Young's equation. It is clear, however, that it is the advancing contact angle that is relevant to the question of rewetting a dry patch. It is interesting to note in retrospect that while the value of this contact angle plays no role in determining the initial instability of the film to dry patch formation, it is the dominant factor in their removal, as shown in the experiments of Silvi and Dussan [15].

The role of liquid surface tension level is seen by substitution Eq. (2) into Eq. (1) to obtain

$$\frac{\dot{m}}{W} = -K \left(\frac{\mu \rho}{g}\right)^{1/5} [\sigma_{SV} - (\sigma_{SL} + \sigma_{LV})]^{3/5} \qquad (3)$$

where the quantity in brackets is the "spreading coefficient" of the liquid on the solid. It is negative for liquids that do not wet out the solid. For liquids that do spontaneously wet out a solid, no flow need be imposed to assure the destruction of the dry patch. Other factors remaining unchanged, a reduction in the surface tension will lead to more even wetting of the packing (i.e., to fewer and smaller dry patches). On the other hand, solid surfaces of high energy (high σ_{SV}), such as clean metals and most ceramic materials, should be more easily wet than lower-energy materials such as the various plastics.

The consequences of liquid film differences (i.e., whether the liquid film is coherent or broken up into dry patches and rivulets) in a packed distillation column are multifold. Of greatest importance is the reduction in available interfacial area accompanying the breakup of a film into dry patches and rivulets. This can produce a marked drop in the mass transfer efficiency of the unit. Coughlin [16] investigated the mass transfer efficiency and other characteristics of a 15mm x 76mm column packed with 10-mm ceramic Raschig rings (which were wet out by the liquid) compared to those of the same column packed with rings of Saran polymer and pothyethylene, both of which were poorly wet. The operation studied was the absorption of oxygen from air into an aqueous solution of sodium sulfite. The measured contact angles of the liquid test solution against the Saran and polyethylene were 66° and 78°, respectively, as compared to 0° for the ceramic material. The wetting differences between the polymeric materials and the ceramic packing produced little difference in the pressure drop or the operating holdup, but produced a marked decrease in the drainage time for the operating holdup and increased the tendency of the column to exhibit loading. Most important, the poorer wetting of the polymeric packing produced a 20% reduction in overall mass transfer efficiency. One would expect interfacial area reductions caused by film breakup to have the greatest effect on efficiency on those systems for which the mass transfer is vapor-phase controlled (i.e., for which the major resistance to mass transfer lies in the vapor phase). This situation is characteristic of most distillation operations. When the mass transfer resistance of the liquid film dominates, the process of film breakup may produce sufficient agitation of the liquid to increase the mass transfer coefficient in that phase, partially offsetting the effect of reduced area. Coughlin's results are consistent with those of Sherwood and Holloway [17], who compared the efficiencies of gas stripping in columns packed with paraffin-coated packing elements with those using clean packing.

King and Walmsley [18] compared the efficiency of a stainless steel Stedman packing for the distillation of a water-deuterium mixture against that of an organic test mixture and found good wetting and high efficiency for the latter but not the former. The surface tension of the organic liquid was sufficiently low to produce good wetting. The performance of the packing was significantly improved by cleaning it with potasium permanganate solution. It is well known that unclean metal surfaces generally bear a thin layer of organic contaminant which is poorly wet by most liquids, in contrast to clean metal surfaces, which are wet out by most liquids.

One apparent resolution to the problem of poor wetting in a packed column is to increase the liquid rate, as suggested by the results of Hartley and Murgatroyd [12]. Generally, however, it cannot be raised sufficiently without exceeding the flooding limit. It is thus important to choose a packing material that is wet out by the liquid to be distilled. Reducing the surface tension level and thereby improving wetting through the use of surfactants (wetting agents) may be useful, but the consequences of their presence are more complex than a simple surface tension reduction (as explained later) and may entail undesirable side effects, such as premature flooding of the column.

IV. SURFACE TENSION GRADIENTS: THE RESULTS OF ZUIDERWEG AND HARMENS

In a landmark paper published in 1958, Zuiderweg and Harmens [19] demonstrated that in distillation, it is more the development of surface tension gradients than the surface tension level that plays a dominant role in determining the flow configuration and efficiency in both unsupported-area and supported-area equipment. They examined a wide variety of binary distillations in bench-scale equipment of both types. It is instructive to look at a sample of their results: a comparison of a distillation of a mixture of n-heptane and toluene with a distillation of a mixture of benzene and n-heptane. When the n-heptane/

Surface-Active Agents in Distillation

toluene mixture, at an average n-heptane concentration of 40 mol %, was distilled in a Oldershaw sieve tray column at vapor velocities between 0.13 and 0.29 m/s, the froth height was observed to be 60mm, and the tray efficiency was 97 to 100%. In contrast, when the benzene/n-heptane mixture, at an average benzene concentration of 40 mol % was distilled in the same column at vapor velocities between 0.14 and 0.62 m/s, the froth height varied between 10 and 20mm, and the tray efficiency was 52 to 55%. What froth was observed in the latter system seemed to consist of larger bubbles which tended to burst into a spray of droplets. This type of spraying was not evident in the more frothy, efficient systems. The observation that the efficiency of a tray column operating with a "foaming" system was often double that of a column operating with a spraying system is consistent with results reported by several earlier investigators [20-22].

A similar comparison was made when the two mixtures, both with an average concentration of the more volatile component equal to 65 mol %, were distilled in a column packed with Fenske helices at a vapor velocity of 0.06 m/s. The height of packing equivalent to a theoretical plate (HETP), a measure of the mass transfer efficiency in the column (low HETP = high efficiency), was 15mm for the n-heptane/toluene system but 30mm for the benzene/n-heptane system. It was difficult to see a large difference in the flow behavior between the two cases in the packed column, but when a similar comparison was made in a wetted-wall column, it was noted that in the n-heptane/toluene system (system A), the condensate film was uniform and stable, whereas in the benzene/n-heptane system (system B), the film broke up into distinct rivulets. The latter type of behavior in the packed column would lead to reduced interfacial area and reduced mass transfer efficiency, as was in fact observed.

Differences similar to those detailed above were observed for several other systems, including one that has an azeotrope near the middle of the composition range (viz., ethanol and 2,2,4-trimethylpentane). At ethanol concentrations below the azeotrope (53 mol %), the system behaved like that of system B above, while at ethanol concentrations above 53 mol %, it behaved like system A.

Zuiderweg and Harmens explained these differences by examining the properties of the two different systems, as shown in Table 1. In system A, the more volatile component has the lower surface tension, while in system B this is reversed. The high-efficiency characteristics of systems of type A compared to the low efficiency of those of type B were observed for all the systems studied, including the azeotrope. These two types of systems were designated as surface tension positive (σ^+) and surface tension negative (σ^-), respectively, as shown in Fig. 4. For σ^+ systems, surface tension in-

TABLE 1 Comparison of Component Properties

	System A	
Component	n-Heptane	Toluene
Boiling point	98.4°C	110.6°C
Surface tension	20.2 mN/m	28.5 mN/m
	System B	
Component	Benzene	n-Heptane
Boiling point	80.1°C	98.4°C
Surface tension	28.8 mN/m	20.3 mN/m

creases as one moves downward in the column, while for σ^- systems, it decreases moving downward. Systems for which the component surface tensions are essentially equal (within 1 or 2 mN/m of each other) were designated surface tension neutral. Neutral systems behaved essentially as σ^- in tray columns and σ^+ in packed towers. The explanation of Zuiderweg and Harmens for the difference in behavior between the two types of systems relies on the development of surface tension gradients as described below for both unsupported-area and supported-area equipment. Other investigators have since added to the data base of Zuiderweg and Harmens. For example, Medina et al. [23] measure foaming and efficiency of several addition-

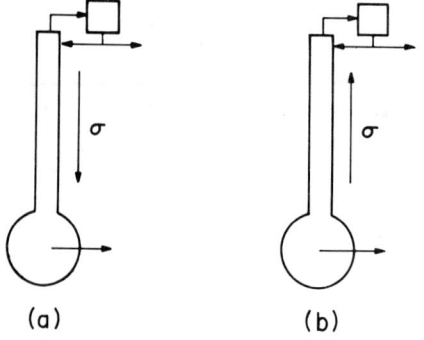

FIG. 4 Schematic of distillation column showing the variation of surface tension in (a) a σ^+ system and (b) a σ^- system.

Surface-Active Agents in Distillation

al binary systems and one ternary system in a sieve plate column and found the same trends as those reported by Zuiderweg and Harmens. With respect to the ternary system, two distinct zones of composition were observed, in one of which the system was σ^+ and the other in which it was σ^-, as determined by whether surface tension was increasing or decreasing, respectively, moving down the column. Moens [24,25], on the other hand, looked at additional binary systems in packed tower distillation columns and also found results consistent with Zuiderweg and Harmens.

A. Unsupported-Area Equipment

Consider the thin lamella of liquid between adjacent vapor bubbles on a tray or the lamella that encloses a vapor bubble as it emerges from the liquid pool on a tray, as shown schematically in Fig. 2. The enlarged view of a section of such a liquid film is shown in Fig. 5, which emphasizes that such films will always develop nonuniformities in thickness. Zuiderweg and Harmens argued that the thinner sections of such films would be closer to being in equilibrium with the vapor phase than would the thicker portions, and therefore relatively leaner in the more volatile component. In σ^+ systems, this would result in the thinner areas of the lamella having higher surface tension than the adjacent thicker areas, so that the surface tension gradient developed would tend to restore uniformity in film thickness, as shown. In σ^- systems, the surface tension gradients would develop in just the opposite sense, so that rapid film drainage, bubble coalescence, and bursting would be favored. Liquid films in the absence of any surface tension gradients will very rapidly thin down and rupture. Whenever a film is formed, as shown in Fig. 6, a capillary pressure driving force is set up to cause thinning in the following way. The pressure difference, ΔP, which exists across a fluid interface, is given by the Young-Laplace equation [13],

$$\Delta P = \sigma \kappa \tag{4}$$

where κ is the curvature of the interface. The pressure is always greater on the concave side of the interface. Thus at point A, where the interface is flat ($\kappa = 0$), $\Delta P = P_{vap} - P_{liqA} = 0$, but at point B, $P_{liqB} < P_{vap} = P_{liqA}$. Thus since $(P_{liqA} - P_{liqB}) > 0$, there will be a spontaneous flow toward the edge of the film, resisted only by negligible fluid inertia. When the lamella thickness reaches a value of a few hundred angstroms, it becomes unstable with respect to disturbances that lead to its rupture, as predicted by the stability analysis of Vrij et al. [26] and confirmed by the experiments of Sheludko [27]. The process of thinning and rupture is virtually instantaneous under the circumstances described. If countering surface

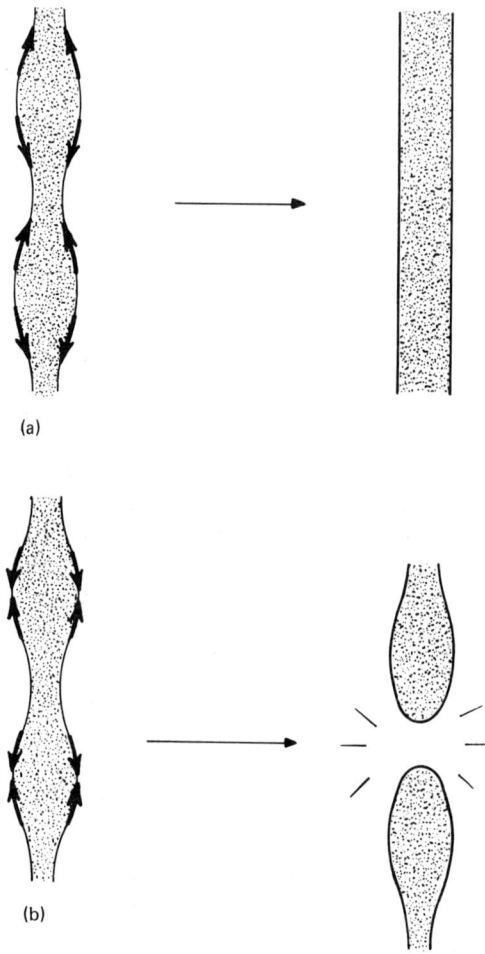

FIG. 5 Surface tension forces set up in a liquid lamella of varying thickness leading to (a) film stabilization in a σ^+ system or (b) film rupture in a σ^- system.

Surface-Active Agents in Distillation

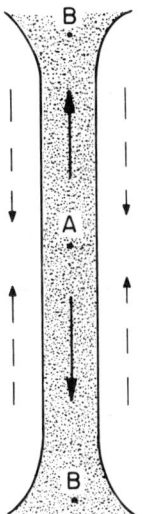

FIG. 6 The drainage of a liquid lamella due to the capillary pressure drop between regions A and B. Dashed arrows show the direction of surface tension forces developed in a σ^+ system or in the presence of a surfactant.

tension forces are set up directed from the edges (thick portions) toward the center (thin portions), however, as suggested by the dashed arrows in Fig. 6, the film drainage will be resisted to some extent and the lamella may have a finite lifetime. This is the situation existing in σ^+ systems. It is thus clear that surface tension neutral systems, incapable of developing stabilizing surface tension gradients, should act like σ^- systems. In fact, Zuiderweg and Harmens indicate that for frothing behavior to be observed, potential surface tension differences of the order of 1 to 2 mN/m must be developed on a tray. This requires not only that the pure component surface tensions be sufficiently far apart, but also that the mass transfer driving force must be adequate. Thus all systems become surface tension neutral at terminal compositions or near pinch points, where the mass transfer driving force goes to zero.

For σ^+ systems, several attempts have been made to correlate quantitatively the foam height or tray efficiency with the magnitude of the surface tension gradient developed. In the course of such an effort, Andrew [9] proposed a frothing mechanism different from that of Zuiderweg and Harmens but essentially equivalent to the traditional explanation for foam stabilization by surface-active agents (see Rosen

[28]). In any multicomponent system at equilibrium, the interfacial layer will be enriched by adsorption of the components that lower the surface tension. When subjected to a rapid stretching process, the dilated area will be temporarily denuded of these components, resulting in a local increase in the surface tension. This sets up surface tension forces directed toward the expanded area, tending to counter the expansion. Andrew cites evidence from high-speed photographs to show that bubble growth and bursting consists of just such a stretching process. As shown in Fig. 7, a local area of weakness on the surface of the bubble blows out into a secondary bubble. The wall of the secondary bubble rapidly expands until rupture occurs. The mechanism resisting stretching is essentially an elastic one, described first by Gibbs and Marangoni (see Ref. 28) over 100 years ago and referred to as Gibbs-Marangoni elasticity. Andrew developed an expression for the magnitude of such elasticity, E_{GM}, for binary solutions,

$$E_{GM} = \theta(x)\left(\frac{d\sigma}{dx}\right)^2 \qquad (5)$$

where x is the mole fraction of the adsorbing component (component of lower surface tension) and $\theta(x)$ is a function that approaches zero as x approaches either zero or unity. Measured froth heights for a variety of binary solutions correlated well with the Gibbs-Marangoni elasticity, as shown in Fig. 8. It should be emphasized that these experiments were carried out in the absence of mass transfer, and the correlation was good for both σ^+ and σ^- systems. The direct relevance of such results to distillation must therefore be questioned. It does appear, however, that Andrew's index for frothing might provide a valid necessary but not sufficient criterion for frothing in distillation systems. Furthermore, it gives insight into one of the mechanisms by which surfactants may influence frothing on distillation trays, as discussed later.

Lowry and van Winkle [29] examined the froth heights obtained over a range of distillation operating conditions in a small perforated plate column using five different σ^+ systems. They found, however, that under mass transfer conditions, the complexity of interaction between the surface tension gradient, vapor velocity, fluid densities, and other properties prevented them from obtaining any correlation that could describe foam height or isolate the effect of the magnitude of the surface tension gradient. Zuiderweg et al. [30] compared the performance of several different tray designs in a 0.45-m-diameter column and found in most cases distinct but somewhat smaller differences between σ^+ and σ^- systems which paralleled the results obtained in the bench-scale columns of the earlier study. As one moves

Surface-Active Agents in Distillation

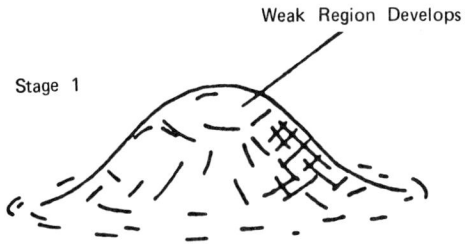

Stage 1 — Weak Region Develops

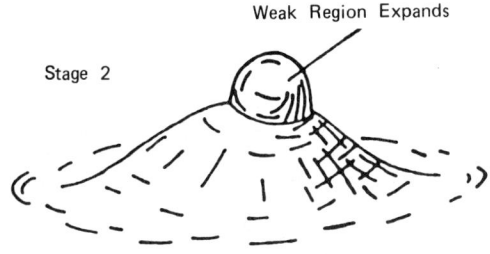

Stage 2 — Weak Region Expands

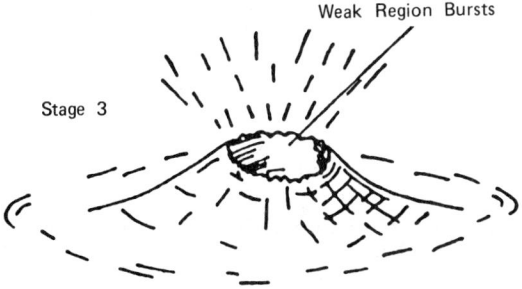

Stage 3 — Weak Region Bursts

FIG. 7 Stages in the bursting of a bubble. (From Ref. 9, by permission.)

into still larger columns with high vapor velocities, the froth formed is broken up into spray, so that the distinction between the two types of systems disappears.

In contrast to the formation of froth in σ^+ systems, σ^- systems, as mentioned earlier, form sprays of droplets. The propensity of σ^- systems to form sprays was later explained by Bainbridge and Sawistowski [31] with reasoning similar to that of Zuiderweg and Harmens. An emerging jet, as shown in Fig. 9, would be assisted

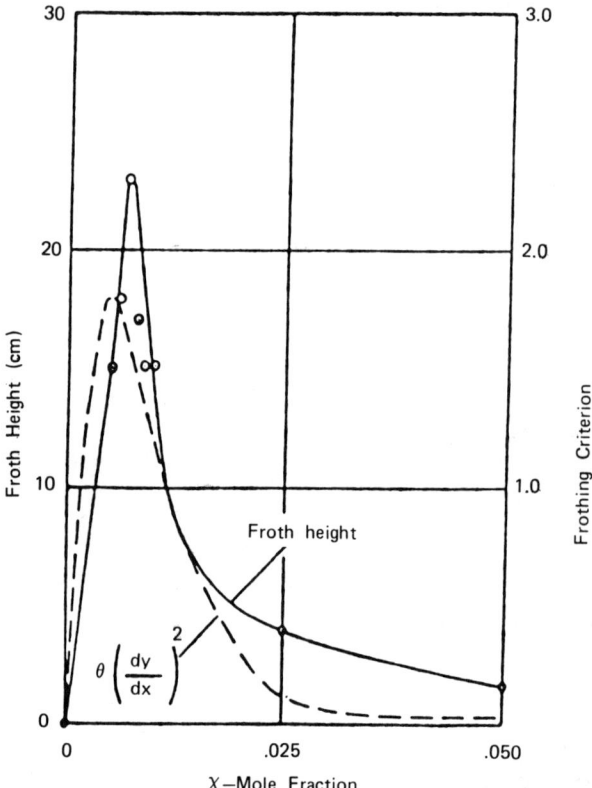

FIG. 8 Comparative values of the froth height and Andrew's frothing criterion for the propionic acid/water system. (From Ref. 9, by permission.)

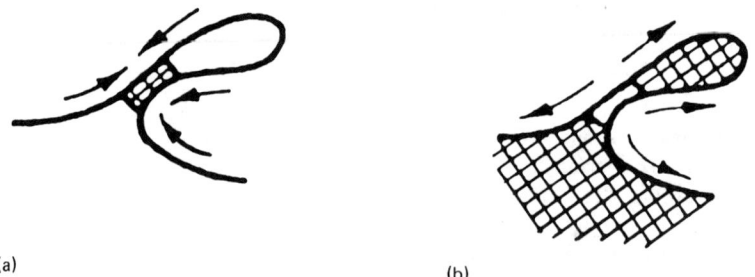

(a)   (b)

FIG. 9 An emerging jet of liquid, showing surface tension forces developed in (a) a σ^+ system and (b) a σ^- system. (From Ref. 31, by permission.)

in breaking up into droplets since the thin portions of the jet would be enriched in the less volatile component and hence have lower surface tension. Surface tension forces favoring breakup thus develop. Later, a quantitative hydrodynamic stability analysis of jet breakup under mass transfer conditions, together with corroborating experiments, was undertaken by Burkholder and Berg [32,33] and Tarr and Berg [34]. These studies revealed that mass transfer in σ^- systems yielded substantially shorter jets and smaller drops than the corresponding σ^+ systems. There is an important difference between the qualitative argument of Bainbridge and Sawistowski and the analysis of Refs. 32–34. The mechanism of Bainbridge and Sawistowski requires that the thinned portion of the emerging jet achieve quasi-equilibration with the vapor phase in order that its surface composition differ from that of the thicker portions. During the short contact times existing, however, it is unlikely that the mass transfer penetration depth would be greater than a few microns, much less than the radius of the emerging jet. Thus it would be difficult to generate surface tension differences. The analysis of Refs. 32–34 however, does not rely on any quasi-equilibration. It reveals that the necking-down process assured by the existence of a finite surface tension (Rayleigh mechanism) establishes secondary flows that generate axial surface tension variations during mass transfer even when the penetration depth is very slight.

The spraying action characteristic of σ^- systems in trays may contribute to entrainment problems, but aside from that, seems not to influence the tray efficiency significantly. Ellis and Legg [35] compared the performance of two widely different σ^- systems in an Oldershaw tray column. An ethylene dichloride/toluene system, with a pure component surface tension difference under operating conditions of less than 2/mN/m performed indistinguishably from an ethylene dichloride/2,2,4-trimethylpentane system with a surface tension difference of more than 100 mN/m.

B. Supported-Area Equipment

The contrast between the stable, coherent liquid films observed in wetted-wall columns with σ^+ systems and the broken pattern of rivulets observed with σ^- systems was explained by Zuiderweg and Harmens by a mechanism which paralleled that given for free films in unsupported-area equipment. As pictured in Fig. 10, the liquid films inevitably develop nonuniformities in thickness, and the local thin spots were thought to approach equilibrium with the vapor more quickly than were the neighboring thicker regions. The thin spots would then be relatively richer in the less volatile component. If this component has the higher surface tension, as in σ^+ systems, surface tension will pull liquid toward the thin regions and cause

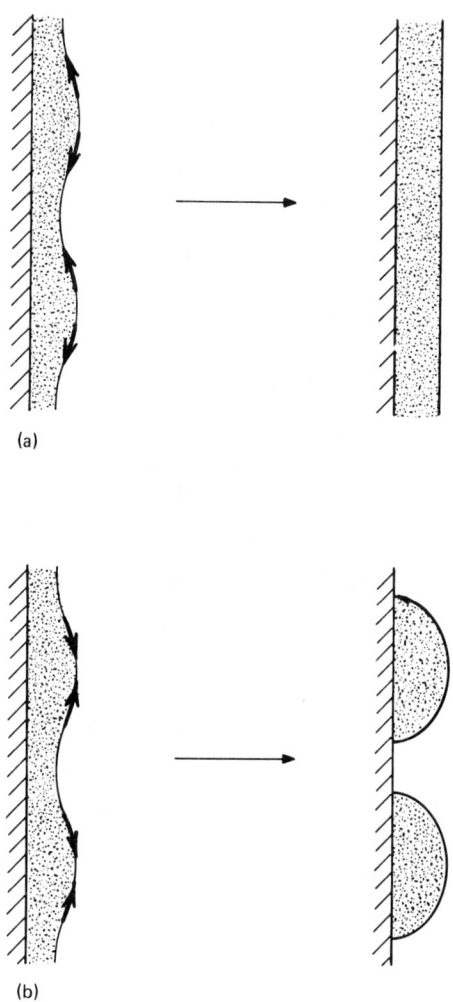

FIG. 10 Surface tension forces set up in a supported liquid film of varying thickness leading to (a) film stabilization in a σ^+ system or (b) film breakup into rivulets or drops in a σ^- system.

them to rethicken, so that the film remains coherent. On the other hand σ^- systems will respond by drawing more liquid away from the already thin spots, leading to film rupture and the formation of dry patches, rivulets, or drops.

Once a dry patch is formed, its growth or shrinkage is governed by the wettability of the solid support by the descending liquid, as discussed earlier in terms of contact angle. The contact angle under distillation (i.e., mass transfer conditions), however, is substantially different from that existing under equilibrium conditions. Figure 3 pictures a solid-vapor-liquid junction. If mass transfer between the vapor and liquid is occurring, the thin wedge of liquid adjacent to the interline will become depleted of the more volatile component. Thus in σ^+ systems, a surface tension gradient is established which promotes the spreading of the liquid. The apparent contact angles existing under mass transfer conditions are substantially different from the equilibrium contact angles that would correspond to systems of identical composition with no mass transfer.

In a series of papers [36-42], Ponter and co-workers report contact angles measured under a variety of heat and mass transfer conditions, including those of distillation. In Ref. 38, an equation very nearly identical to that of Hartley and Murgatroyd was derived, showing the dependence of the liquid film flow rate needed to heal a dry patch on the effective surface tension and the effective contact angle, corresponding to the conditions of mass transfer. Reasonable agreement of these predictions with experimental wetting rates was obtained in all cases. Effective contact angles under total reflux distillation conditions were measured using the sessile drop technique for a variety of systems. Figure 11 shows results obtained for the n-propanol/water system, which has an azeotrope near the middle of the composition range. On the water-rich side of the azeotrope, the system is σ^+, and on the n-propanol-rich side, it is σ^-. The corresponding equilibrium contact angle variation is shown for comparison. It is seen that on the σ^+ side, the surface tension gradient favors spreading and thus produces a reduction in the effective contact angle, whereas on the σ^- side, the surface tension gradients developed cause retraction of the interline and increased effective contact angles. Shown also is the measured minimum wetting rate, and its variation is seen to track closely the variation of the effective contact angle. These wetting rate data are in close agreement with those obtained earlier for the same system by Norman and Binns [43] using a wetted-wall distillation column.

It should be emphasized that although the effective contact angles measured by Ponter and co-workers correlate well with wetting rate data and provide useful insights into the role of surface tension gradients in determining the liquid film characteristics in supported-area equipment, they have no thermodynamic status nor is it evident

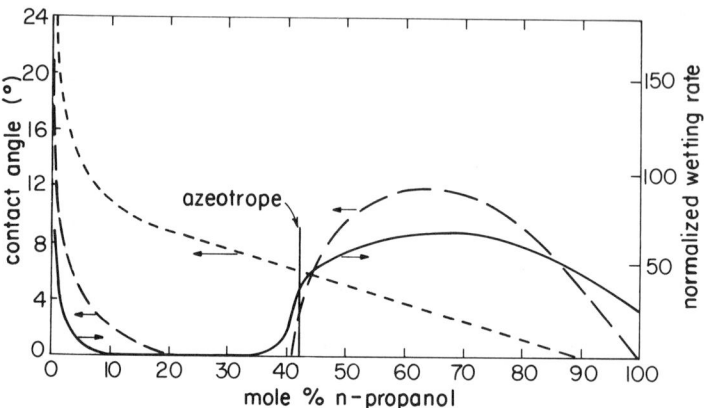

FIG. 11 Comparison of effective contact angle (— —) and equilibrium contact angle (- - -) in distillation of the n-propanol/water system. Also shown is the normalized (dimensionless) minimum wetting rate, defined as the mass flow rate per unit width divided by the liquid viscosity. Data points are omitted for clarity. (Redrawn from Ref. 42.)

how they might be calculated a priori. Indeed, the trend of their variation with compositions appears to violate Young's equation [Eq. (2)]. In a σ^+ system, spreading is promoted because the local surface tension of the liquid adjacent to the interline is increased relative to the bulk. In view of Eq. (2), this should lead to a lower value of cos θ, hence a higher contact angle under mass transfer conditions, opposite to what is observed. Detailed interferometric observations of both advancing interline (σ^+) and receding interline (σ^-) systems reported by Bascom et al. [44] indeed reveal interface shapes qualitatively like those shown in Fig. 12. The true contact angles are consistent with Young's equation, whereas the macroscopically observed apparent angles differ in the opposite sense from the equilibrium contact angle. These interface shapes were predicted in a recent analysis by Neogi [45]. Neogi's study also revealed that the equilibrium wettability plays an important role and that lack of wetting ability can give rise to receding contact lines even when the surface tension gradient opposes such a movement. An extension of Neogi's analysis to include the effect of an oncoming flow might generalize the results of Hartley and Murgatroyd for dry patch healing to the case where surface tension gradients exist.

The next important step is to link the effective contact angle and critical wetting rate results to mass transfer efficiency in supported-area distillation equipment. The general result to be anticipated is

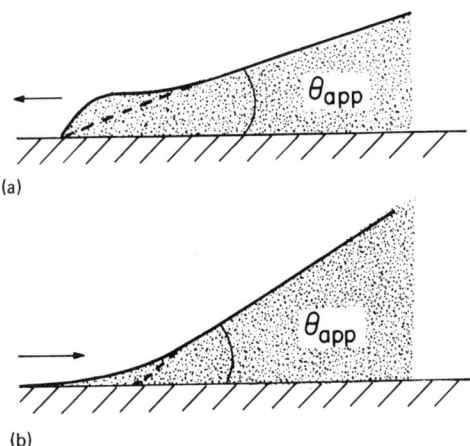

FIG. 12 Apparent contact angles, θ_{app}, observed when the interline is (a) advancing, as in a σ^+ system, or (b) receding, as in a σ^- system.

that the better the wetting is, the higher will be the available interfacial area and the more efficient will be the distillation. Other factors besides interfacial area enter into the determination of overall efficiency, that is, the degree of agitation in the bulk vapor and bulk liquid phases. Most distillations, however, are vapor-phase controlled over nearly all of the composition range, and vapor-phase agitation should be little effected by whether the liquid film breaks up into rivulets or remains coherent. Therefore, one expects that interfacial area changes should play a dominant role. This is borne out by the experiments of Norman et al. [46], who measured the efficiency of a wetted-disk distillation column for a number of binary test systems, including that of n-propanol/water. The column internals consisted of a single vertical row of disks threaded on a rod along the axis of the column with adjacent disks at right angles to each other. This arrangement was desirable because the flow characteristics were much closer to those of a packed column than to those of a wetted-wall column, and the wetting characteristics existing were clearly visible. The overall mass transfer efficiency was expressed in terms of H_{OG}, the overall height of a transfer unit based on the vapor phase. The results obtained for the n-propanol/water system are shown in Fig. 13, and it is seen that they parallel qualitatively the wetting-rate and effective contact angle data of Fig. 11, particularly with respect to the abrupt difference in distillation efficiency on opposite sides of the azeotrope.

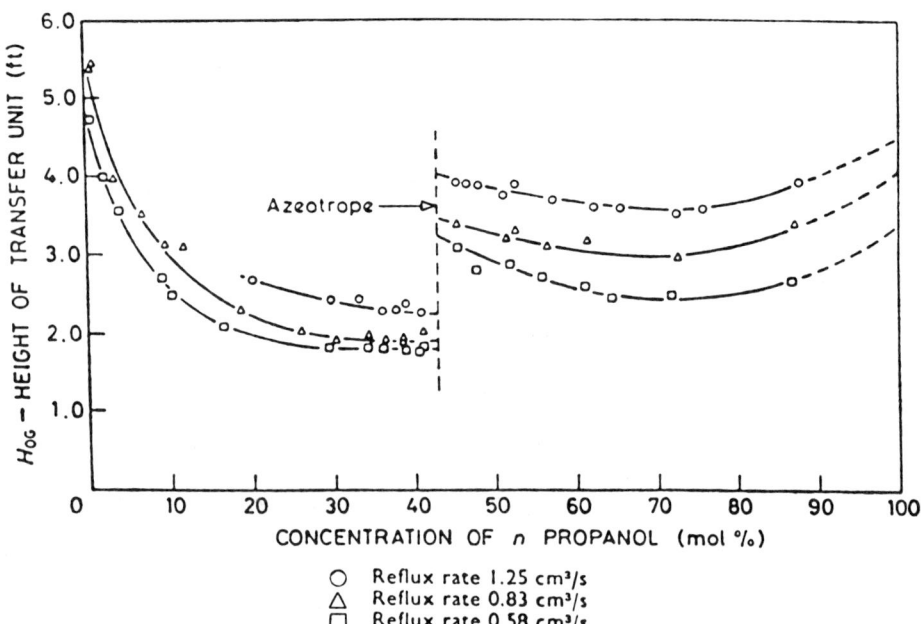

FIG. 13 Experimental values obtained for H_{OG} with the n-propanol/water system being distilled in a disk column. (From Ref. 46, by permission.)

The relatively low efficiencies at the end of the σ^+ region of the system (the lowest n-propanol concentrations) are explained in part by the relatively poor wetting that exists there, as characterized by either the equilibrium or the effective contact angle. It is also partially explained by the fact that at these terminal conditions, liquid-phase resistance becomes important and even dominant. The height of an overall transfer unit may be expressed as [4]

$$H_{OG} = H_G + \frac{mG_M}{L_M} H_L \qquad (6)$$

where H_G is the height of a gas-film transfer unit; G_M and L_L are the gas-phase and liquid-phase superficial velocities, respectively; and m is the distribution equilibrium constant (slope of the y-x vapor-liquid equilibrium line) for the more volatile component. Excellent correlations for H_G and H_L exist in terms of the fluid properties and flow rates. The liquid-phase contribution to the overall mass transfer resistance is represented by the last term in Eq. (6), and it will be large when m is large. The coefficient m becomes large in the n-propanol/water system when the n-propanol concentration is low. In fact, the predicted values of H_{OG} under these conditions are substantially larger even than the experimental values shown in Fig. 13. It is suggested that the film breakup under these terminal conditions actually creates agitation in the liquid film, which partially offsets the effects of the high distribution coefficient.

A second possible explanation for the better-than-expected (albeit poor) mass transfer performance observed in σ^+ systems under conditions where liquid-side resistance is important is the existence of "interfacial turbulence." It is well known [47] that during the desorption of a surface-tension-lowering component from a liquid solution (a σ^+ system), the system may become unstable with respect to small disturbances. A region of local surface dilation (caused by some small local disturbance) brings liquid more rich in the volatile component to the surface. In σ^+ systems this will be liquid of lower surface tension. Surface forces are then established which reinforce the original disturbance, bringing still more volatile-component-rich solution to the surface, and so on, until the disturbance has amplified itself into convection of a macroscopic scale without destroying the coherence of the film. This "micro-Marangoni" effect is dealt with in the pioneering hydrodynamic stability analysis of Pearson [48] and Sternling and Scriven [49] published about the same time as the work of Zuiderweg and Harmens. σ^+ systems, subject to such instability, might exhibit higher liquid-phase mass transfer coefficients, by a factor of 1.5 to 4, than predicted by correlations not taking the effect into account.

C. The Stability Analysis of Wang et al.

Although qualitatively reasonable, the explanation of Zuiderweg and Harmens and others cited so far for the behavior of σ^- systems in supported-area equipment in terms of reduction in available interfacial area is still incomplete. It does not adequately explain the mechanism by which the liquid film is initially broken up into dry patches and rivulets by virtue of differences between the surface composition of the thin and thick regions. Over the relatively short lifetimes of film irregularities, diffusional penetration is not generally great enough to cause such composition differences. It might be argued that there are permanent local thin and thick regions formed as the liquid flows over irregular packing elements (which is indeed the case), but this would not explain the equal propensity of liquid films in σ^- systems to break up in wetted-wall columns without any such irregularities.

In 1971, Wang et al. [50] published the first essentially correct explanation of the film instability and breakup into rivulets occurring in σ^- systems. They formulated a linear hydrodynamic stability analysis of a vertical falling liquid film undergoing mass exchange with the adjoining vapor phase, as pictured in Fig. 14. The analysis, extending an earlier study of vertical laminar films subject to upward-directed surface tension gradients induced by temperature gradients [51], examines in particular the fate of small disturbances in the form of longitudinal waves. If the film were indeed to be unstable with respect to such waves, it would pull apart into rivulets. The key is then to examine the difference between σ^+ and σ^- systems with respect to such instability. Regardless of the type of system, the thicker portions of the film will flow more rapidly downward because there is a larger gravitational force inducing such flow there. In fact, the downward velocity is proportional to the square of the film thickness. Now the key difference is that in σ^- systems, the surface tension is decreasing downward. Thus the faster-flowing, thicker stripes will have higher surface tension than will their laterally adjacent neighbors. The thick stripes will thus draw liquid toward themselves, and the film will break up into rivulets. The σ^+ systems have surface tension increasing downward and will thus be stabilized with respect

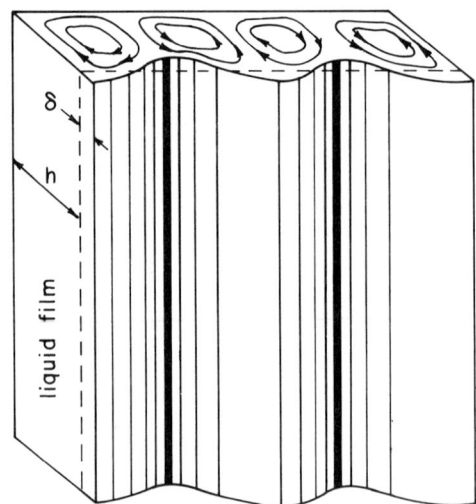

FIG. 14 Schematic of a falling film subject to a transverse disturbance leading to a variation in film thickness. (Redrawn from Ref. 51.)

Surface-Active Agents in Distillation

to such disturbances. This mechanism proposed by Wang et al. is essentially different from that proposed by Zuiderweg and Harmens and does not depend on any quasi-equilibration between the liquid and vapor phases. Performance of the stability analysis revealed the predicted details of the breakup process, such as the degree of "surface tension negativeness" required to induce breakup, and the expected wavelength of the preferentially amplified disturbances.

The undisturbed film is of uniform thickness and has a velocity profile as shown in Fig. 15. The nature of the profile near the surface depends on whether the system is σ^+ or σ^-. The case shown is negative. The figure also shows the coordinate system used. In the model, the surface concentration and the surface tension are assumed to vary linearly with z. The system is then made to suffer perturbations in all of the dependent variables (velocity components, v_x, etc., and concentration of the more volatile component, C_A of the foam

$$\begin{bmatrix} v_x \\ \vdots \\ C_A \end{bmatrix} = \begin{bmatrix} \bar{v}_x(\eta) \\ \vdots \\ \bar{C}_A(\eta) \end{bmatrix} e^{i\alpha\xi} e^{\beta\tau} \tag{7}$$

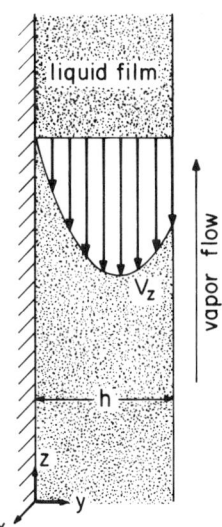

FIG. 15 Velocity profile in falling film, characteristic of a σ^- system. (Redrawn from Ref. 50.)

where η and ξ are nondimensionalized position coordinates; x/h and y/h, respectively; h is the undisturbed film thickness; α is the dimensionless wave number of the disturbance, $2\pi h/A$; A is the wavelength of the disturbance; τ is dimensionless time, t/t_0 (t_0 is an unspecified scale factor which drops out of the final analysis); and β is the dimensionless growth constant of the disturbance amplitude. $\bar{v}_x(\eta)$, and so on, refer to the z-dependent factors of the disturbance variables. These variables are made to satisfy the linearized Navier-Stokes equations and the convective diffusion equation, as summarized below:

$$\nabla \cdot \underline{v} = 0 \tag{8}$$

$$\left(\frac{\partial}{\partial t} - \nabla^2\right) \nabla^2 \underline{v} = 0 \tag{9}$$

$$\left(\frac{\partial}{\partial t} - \nabla^2\right) C_A = \left(\frac{\partial C_{AO}}{\partial Y} v_y + \frac{\partial C_{AO}}{\partial Z} v_z\right) \tag{10}$$

where $\underline{v}$ is the vector velocity disturbance and C_{AO} is the concentration of the more volatile component in the undisturbed state. Equations (8) to (10) must be satisfied subject to the appropriate boundary conditions. These require no slip and no transfer at the solid boundary (y = 0), while the surface tension effects are embodied in the conditions drawn at the vapor-liquid interface:

$$v_y = \frac{\partial \delta}{\partial t} \tag{11}$$

$$p - 2\mu \frac{\partial v_y}{\partial Y} + \sigma \nabla_{II}^2 \delta = 0 \tag{12}$$

$$\underline{\tau}_y + \nabla_{II} \sigma = 0 \begin{cases} \mu \left(\frac{\partial^2 v_y}{\partial x^2} - \frac{\partial^2 v_y}{\partial y^2}\right) = \frac{\partial^2 \sigma}{\partial x^2} \\ \mu \left(\frac{\partial v_2}{\partial Y} + \frac{\delta \partial^2 v_z}{\partial Y^2}\right) = 0 \end{cases} \tag{13}$$

$$D_A \left(\frac{\partial C_A}{\partial Y} + \delta \frac{\partial^2 C_{AO}}{\partial Y^2}\right) + K_A m \left(C_A + \delta \frac{\partial C_{AO}}{\partial Y}\right) = 0 \tag{14}$$

Surface-Active Agents in Distillation

where δ is the surface deflection, p the pressure, ∇_{II} the surface del operator, $\underline{\tau}_y$ the viscous shear stress, D_A the diffusivity of A in the liquid phase, and k_A the mass transfer coefficient of A in the vapor phase. Equation (11) expresses the continuity of velocity at the interface; Eq. (12) is the balance of normal stresses; Eq. (13) is the balance of shear stresses, reducing to the two scalar equations shown (and expresses the surface tension forces which lead to film breakup in the case of σ^- systems), and Eq. (14) is the mass transfer balance for species A.

In the stability analysis, one seeks to determine the conditions under which the growth constant, β, of the disturbance is positive (i.e., has a positive real part, the imaginary part having been neglected) and hence will lead to film breakup. Nondimensionalization of the linearized equations and boundary conditions isolates the dimensionless groups that characterize the system properties and parameters. Perturbations in the form of Eq. (7) are then substituted in and the growth constant, β, is set equal to zero (the condition of neutral stability). The resulting differential equations are integrated in terms of a number of coefficients of integration. These are to be evaluated using the boundary conditions, but since the system of equations and boundary conditions is homogeneous, one sets the determinant of these equations equal to zero to determine the required conditions for the existence of nontrivial solutions. This gives a relationship between the disturbance wavelength and the various dimensionless groups known as the "dispersion" or "characteristic" equation. It is of the form

$$\Phi(\alpha, N_i, \ldots) = 0 \tag{15}$$

The important dimensionless groups, N_i, in the characteristic equation are summarized in Table 2. Of central importance is the Marangoni number, Ma, which expresses the degree of "positiveness" or "negativeness" of the system. As defined by Wang et al., a σ^- system yeilds a positive Ma, since z is directed upward. The values of all the dimensionless groups are set to appropriate values for the particular system under study, and the wave number, α, is determined as a function of Ma for neutral stability. This is not strictly appropriate because it is not possible to vary the Marangoni number independently of all the other dimensionless groups, although their variation may be quite small as Ma is varied. One result of such computations is shown in Fig. 16, in which properties corresponding to the water/formic acid system (with a liquid film thickness of 1.0mm) are used. The corresponding values for the various dimensionless groups are shown in parentheses in Table 2. The top part of the stability diagram corresponds to positive Marangoni numbers and hence to σ^- systems. It shows a region of instability over a range of physically realistic Marangoni numbers for dimensionless wave numbers less than approximately 0.1. A dimensionless wave number of about unity corresponds to a disturbance wavelength of about six times the liquid film depth. This is physically reasonable for the type of rivuleting that is observed. When Ma gets very large, the situation corresponds to a climbing film, not generally observed in distillation.

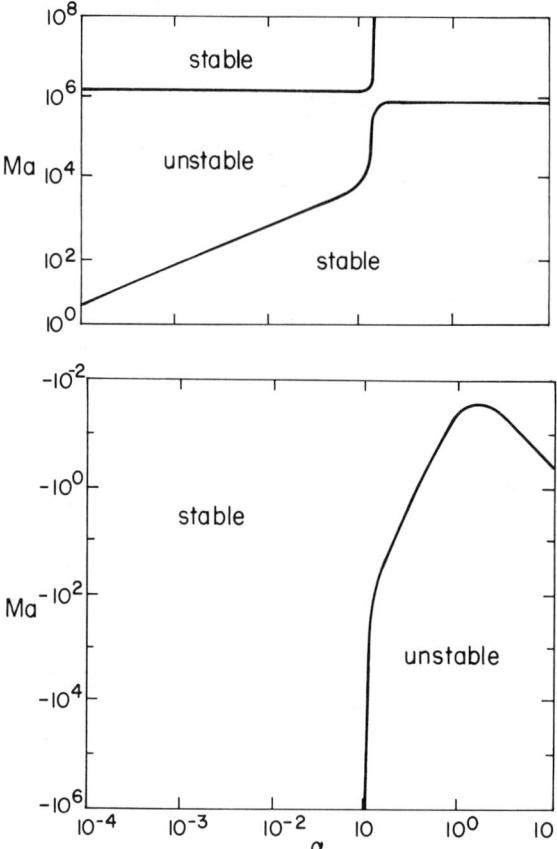

FIG. 16 Stability results of Wang et al. [50] for a falling film expressed as critical Marangoni number versus dimensionless wave number, α.

Table 2 Dimensionless Groups Arising in Stability Analysis of Falling Film

$Ma = \dfrac{(\partial\sigma/\partial z)h^2}{\mu D_A}$	Marangoni no.	
$Re = \dfrac{4m}{W\mu}$	Reynolds no.	(13)
$Sh = \dfrac{k_A m_A h}{D_A}$	Sherwood no.	(0.2)
$Cr = \dfrac{\mu D_A}{\sigma h}$	Crispation no.	(5×10^{-7})
$N_d = \dfrac{\rho g h^3}{\mu D_A}$		(9800)

σ^+ systems (negative Ma) are indicated to be stable with respect to the rivulet-forming disturbances, but do show a region of instability to short wavelength (large α) disturbances. These appear to be identifiable with the "interfacial turbulence" described earlier.

The analysis of Wang et al. adequately explains the breakup of films in σ^- systems, while the wetting studies in correlation with distillation performance data provide a good basis for understanding capillary effects in supported-area equipment.

V. SURFACTANT EFFECTS

The foregoing discussion of surface effects in both unsupported-area and supported-area vapor-liquid contacting devices with and without mass transfer, and in particular the discussion of distillation, provides a basis for anticipating and explaining the effects of surfactant addition in such equipment. The most widely known consequence of the presence of surfactants in distillation equipment is the generation of foam, which leads to excessive entrainment and often to flooding. The literature contains many case studies of this type, (e.g. Ref. 52). In fact, concern was voiced recently [53] that surfactants used in enhanced oil recovery may result in foaming problems in subsequent refinery operations. In a study at the Signal UOP Research Center, however, it was reported that in both vacuum and atmospheric distillations, the surfactants appeared to cause no significant foaming and to concentrate preferentially, in most cases, in the vacuum resid fractions. Severe foaming may also occur in systems in which no materials identifiable as surfactants are present. In a pioneering set of studies [54—56], Ross and Nishioka have shown that when a liquid mixture is near its critical solution point (i.e., on the verge of splitting into a second liquid phase), the component that will be enriched in the second liquid phase will be strongly adsorbed at the liquid-vapor interface and will act like a surfactant. The adsorbed component will resist dilational distortion by generating opposing surface tension gradients; that is, it will impart a Gibbs-Marangoni elasticity to the surface.

The antidote to excessive foaming in distillation is often the purposeful addition of another surfactant, often a silicone or a fluorocarbon liquid. These are believed to spread out rapidly over the foam lamellae, displacing the original surfactant and rupturing the films. The mechanisms of both foam stabilization and antifoam action by surfactants is described in detail elsewhere [10,57] and will not be the focus of this report on surfactant effects in distillation. Instead, we seek to examine the possibility of enhancing distillation efficiency with the use of such materials. The possibility, in particular, of rendering distillation performance with σ^- systems more like that obtained with σ^+ systems appears promising.

A. Unsupported-Area Equipment

Brumbaugh and Berg [2] examined the effect of n-decanol addition in the distillation of water/formic acid system in an Oldershaw sieve tray column. Water and formic acid form an atmospheric azeotrope at 43.5 mol % water, as shown in Fig. 17. On the formic-acid-rich side of the azeotrope, the system is σ^+, and on the water-rich side it is σ^-. In the absence of surfactant, the overall column efficiency was 65% on the σ^+ side and 50% on the σ^- side of the azeotrope. Frothing was observed in the σ^+ regime but none in the σ^- regime. In terms of efficiency, the results of surfactant addition are shown in Fig. 18. The surfactant eliminated 30% of the difference existing between the two regimes. Foaming was observed on the σ^- side with surfactant present, but only to about half the height existing on the σ^+ side in the absence of surfactant. Surfactant addition in the σ^+ system caused no change in efficiency at the lower vapor velocities, but at higher velocities induced foam flooding. Brumbaugh's results

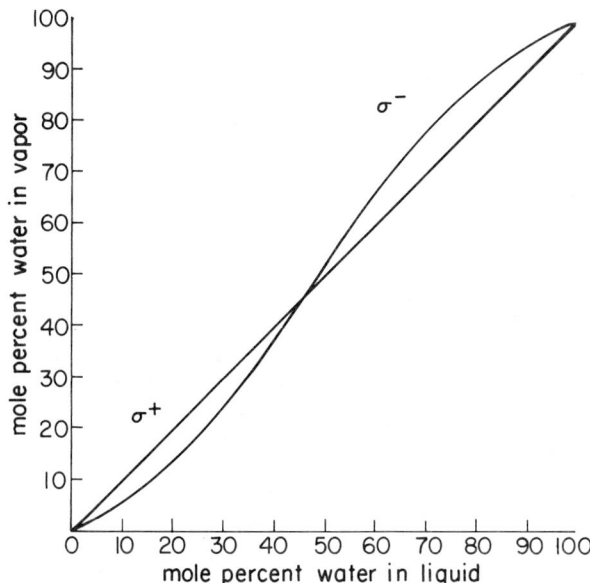

FIG. 17 Vapor-liquid equilibrium data for the water/formic acid system.

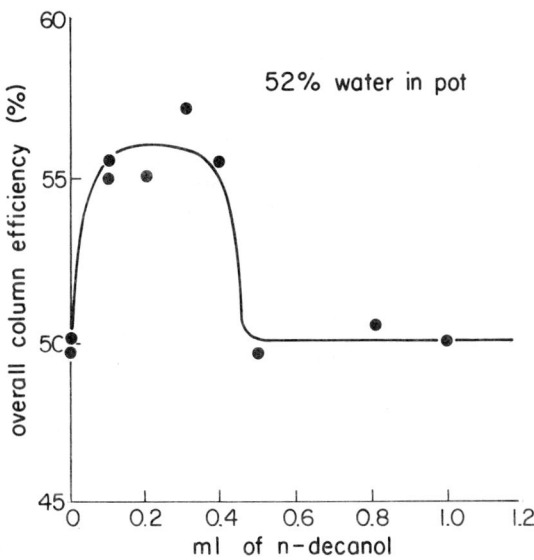

FIG. 18 Effect of n-decanol addition on the tray efficiency in a sieve plate distillation column with the σ^- water/formic acid system. (Redrawn from Ref. 2.)

were later repeated by Rea [58], who extended the study to columns with Teflon trays as well as glass, and examined the effect of sodium tetradecyl sulfate (STS) surfactant as well as n-decanol. STS had to be injected continuously at the top of the column because, unlike the decanol, its volatility is so low that it gradually ends up entirely in the still pot. Frothing and efficiency changes closely paralleled those of Brumbaugh for both the glass and Teflon trays for both surfactants. Additionally, it was noticed that the surfactant led to enhanced weeping of the trays in both cases. An important difference was also observed in the nature of the froth lamellae produced in the σ^+ systems in the absence of surfactant and those produced in the presence of surfactant, in particular with the STS. The lamellae of surfactant-produced foam were so thin that they produced visible interference colors, whereas the froths in the absence of surfactants were much thicker. In order to produce colors, films must be in the thickness range of approximately 0.7 to 10 μm. The thinness of the foam lamellae in the surfactant-containing systems may render the enhanced interfacial area produced proportionately less effective in increasing efficiency. It is likely that the liquid in such lamellae is quickly equilibrated with the vapor, making such an area subsequently ineffective in promoting mass transfer. Rea also extended his study to two σ^- organic systems using fluorocarbon surfactants, but in these found no measurable enhancement in efficiency.

The data presently available on the enhancement of distillation efficiency in tray columns through the use of surfactants seem to indicate little promise. When enhancements are achieved, they are relatively small, and the side effects of increased weeping and possible foam flooding may result in overall poorer operation. It should be acknowledged, however, that the number of systems studied so far is very small and may not warrant such generalization.

Despite the smallness of the efficiency enhancements reported, it is of interest to inquire into the mechanism by which they are achieved. The earlier examination of the role of surface tension gradients in bubble formation, interaction, and stabilization in the absence of surfactants points to two different possibilities for their role: (1) the surfactant may be strongly adsorbed at the vapor-liquid interface, producing a Gibbs-Marangoni elasticity effect, or (2) the surfactant may reverse the surface tension versus composition relationship between the major components of the system, thermodynamically converting a σ^- system into a σ^+ system. Both effects may act in concert. The production of a persistent foam characteristic of the presence of powerful surfactants in the system is the result of the first effect. The existence of foams means that the drainage of the liquid lamellae to critical thinness for rupture has been considerably retarded and that local dilational disturbances that

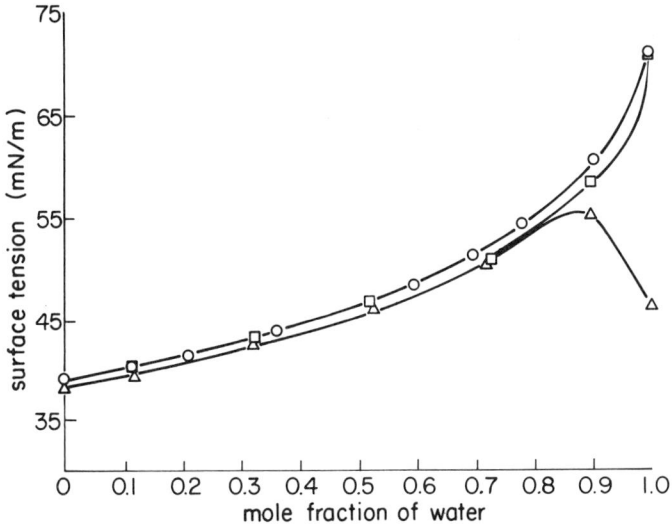

FIG. 19 Variation of surface tension of water/formic acid solutions with composition. n-Decanol concentration equal to (○) 0, (□) 10^{-5} M, and (▽) 10^{-4} M. T = 22°C.

could lead to premature rupture are self-healing. Whether it is Gibbs-Marangoni elasticity or surface tension reversals that are responsible for increased frothing by surfactants is best determined by examining the surface tension dependence on composition for the ternary system of the separating components plus surfactant. Extensive surface tension data for the water/formic acid/n-decanol system as a function of temperature up to 98°C are given by Johnson [59]. These data were obtained by the Wilhelmy technique without detachment. Representative results at room temperature are shown in Fig. 19. They suggest a strong surface tension reversal in the σ^- regime, and very steep surface tension derivatives with decanol concentration at very low formic acid contents. The latter would lead to a strong Gibbs-Marangoni elasticity in that composition range (but not elsewhere). The surface tension reversal would lead to a conversion of the σ^- system to a σ^+ system over the range for which it occurred. Since the data are taken at room temperature, these explanations would be valid for this system only under vacuum distillation conditions. Figure 20 shows the corresponding surface tension data taken at 98°C, as near as possible to atmospheric operating conditions. These data reveal a complete elimination of the surface tension reversal in the σ^- regime and the high Gibbs-Marangoni

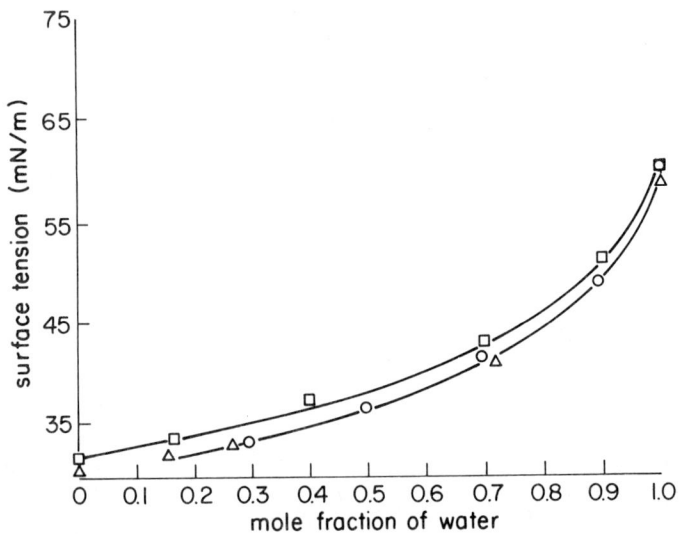

FIG. 20 Variation of surface tension of water/formic acid solutions with composition. n-Decanol concentration equal to (○) 0, (□) 10^{-5} M, and (△) 10^{-4} M. T = 98°C.

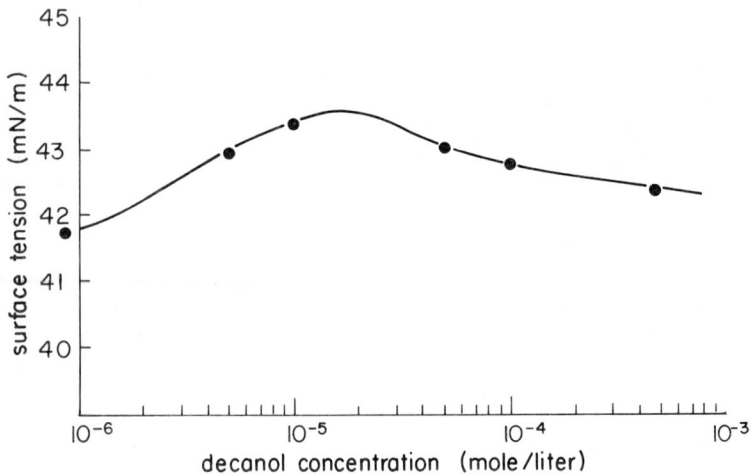

FIG. 21 Surface tension of a 30 mol % aqueous formic acid solution as a function of n-decanol concentration at T = 98°C.

Surface-Active Agents in Distillation

elasticity for the nearly-pure-water systems observed at room temperature. Under the conditions of atmospheric distillation, n-decanol does not act like a surfactant at all. In fact, the data reveal a slight surface tension increase with "surfactant" addition. This is seen more clearly in Fig. 21. It is this slight surface tension increase with "surfactant" concentration coupled with the much greater equilibrium concentration of decanol in the liquid with increased formic acid fraction on the water-rich side of the azeotrope that must be responsible for enhancements in foaming observed with "surfactant" addition in this particular system. This is not to say that the other mechanisms may not be important in other systems under other operating conditions. More distillation performance data, together with surface tension-composition data at the temperature of the distillation are needed to resolve these questions.

B. Supported-Area Equipment

Francis and Berg [1] in 1967 examined the effect that traces of surfactants have on the distillation of a variety of σ^+ and σ^- systems in a bench-scale packed column. The packed section was 40mm in diameter by 0.23m high, and the packing consisted of 4-mm ceramic Berl saddles. The most dramatic results were achieved for the water/formic acid system described above. Figure 22 shows results obtained for distillation efficiency (expressed in terms of HETP) in the presence and absence of n-decanol for both the σ^+ and σ^- sides of the azeotrope. First it is to be noticed that in the absence of decanol the efficiency of the σ^+ system was nearly double that of the σ^- system, as expected. When decanol was added, however, the efficiency of the σ^- system was brought up to nearly that of the σ^+ system. The σ^+ system itself experienced negligible change. The latter result apparently confirmed that interfacial turbulance was not playing a significant role in the mass transfer process in the absence of decanol since it was believed that the presence of a surfactant would suppress such flows [60]. (The surface tension data of Figs. 20 and 21, however, indicate that such "surfactant" action would not occur. Instead, the dominance of vapor-phase resistance to mass transfer in this system renders the existence or absence of interfacial turbulence irrelevant to the overall distillation efficiency in this system).

The dramatic improvement of the σ^- system was thought due to the stabilization of the film against rupture into rivulets, although changes in the film flow were difficult to discern by direct observation, as was the case between σ^+ and σ^- systems in the absence of decanol. The stabilization of the film was readily evident, however, in the neck joining the condenser to the packed section. More recently, Johnson [59] conducted experiments in which the reflux liquid

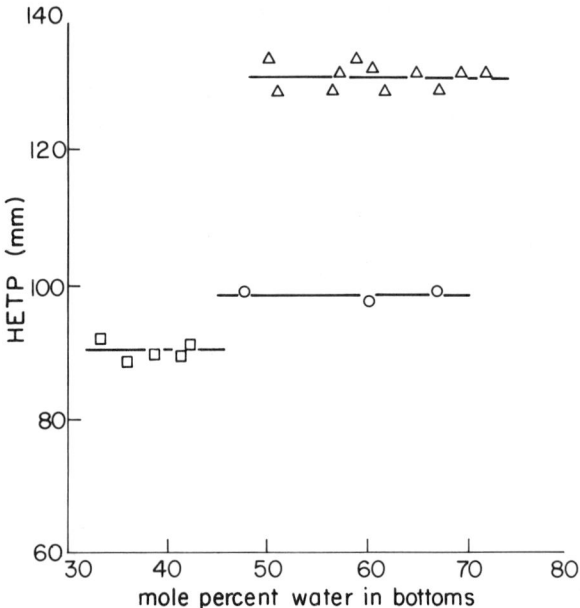

FIG. 22 Separating efficiency of a packed column for water/formic acid test mixtures with and without n-decanol surfactant. ($\triangle$) σ^- system, no surfactant; ($\square$) σ^+ system, no surfactant; ($\bigcirc$) σ^+ system with n-decanol added. (Redrawn from Ref. 1.)

was directed over the surface of a smooth ceramic sphere and observed and photographed directly film stability on the σ^+ side of the azeotrope and instability on the σ^- side in the absence of decanol. Stability was observed on the σ^- side in the presence of decanol. Ponter et al. [61] later measured the effective contact angle as a function of composition under distillation conditions for this system. They found, as expected, significant reductions in contact angle brought about by the presence of n-decanol on the σ^- side of the azeotrope. Only negligible reductions were observed on the σ^+ side.

Francis and Berg found less dramatic improvements in the distillation efficiency by surfactant addition to a number of other (nonaqueous) binary σ^- systems. More recently, Schmidt [62] tested a number of different surfactants in the water/formic acid system, including several long-chain aliphatic alcohols, acids, and ketones. The alcohols generally produced the best results, although the ketones and acids also produced some enhancement of efficiency. The "alcohols," it should be noted, are present as formate esters in the column, esterification in each case being fairly rapid and complete, as verified by capillary chromatography [62].

Moreno and co-workers [63-65] also examined the efficiency in a packed column brought about in the water/formic acid system with n-decanol and found results essentially identical to those of Francis and Berg. Up to a 10% increase in the efficiency on the σ^+ side of the azeotrope was found, however, with the addition of an ethoxylated lauryl alcohol surfactant. This is a puzzling result, and an explanation for it was not given. Another result reported was that a silicone oil (DC 200 fluid, Dow Corning Corp.) decreased the distillation efficiency over the entire composition range, presumably as a result of its adsorption to a solid packing surface, rendering it less than fully wettable under all conditions. This observation is consistent with one made by Schmidt [62] with n-decanol and other surfactants in the same system when distillations were run for a period of several days. The efficiency enhancement in the σ^- system gradually deteriorated with time until it was no longer significant. It was suspected that slow adsorption of the surfactant to the ceramic packing had reduced its wettability, and this was confirmed when the contact angle of the water/formic acid solution against the packing under conditions of no mass transfer was measured at 9°. With cleaned packing, the contact angle was again 0° for all compositions of the water/formic acid system.

Ponter et al. [66] measured increases in efficiency of a column packed with either glass or Teflon Raschig rings upon addition of n-decanol to the σ^+ system of butylamine-water under terminal conditions (nearly pure water). Separate measurements of the effective contact angles under these conditions in the absence of surfactant indicated poor wetting (high contact angles). The measured contact angles were reduced upon addition of surfactant to the system with glass packing but not that with Teflon packing. Surface tension data were not reported. The composition range studied was one in which liquid-phase resistance dominates (very large values of m). It is conceivable that the mass transfer of the surfactant itself led to an enhancement of interfacial turbulence in the liquid.

The most significant of the surfactant effects reported concerns the enhancement of the efficiency in σ^- systems up to nearly that pertaining to a σ^+ system. This can mean nearly doubling the efficiency of the process. It is thus of interest to explain this enhancement in as quantitative terms as possible. It is evident from direct observation that surfactants are capable of stabilizing the reflux film against rupture into dry patches and rivulets. As in the case of unsupported-area devices, there are two possible mechanisms by which this may occur: (1) a Gibbs-Marangoni elasticity effect, or (2) a surface tension reversal.

Although the surface tension data for the water/formic acid/n-decanol system suggest that the Gibbs-Marangoni elasticity mechanism is not the explanation in this particular system, it may be significant under other circumstances and is thus worth examining in some detail. The mechanism is pictured schematically in Fig. 23. It can be shown by computation that the flow beneath the ribbed film consists of an upwelling of liquid at thinning portions of the film, tending to dilate the surface there. This would sweep away surfactant from that region, setting up strong surface tension gradients directed toward the thin regions and tending to restore the film thickness to uniformity. The apparent elasticity results from the speed of surface convection relative to the rate at which compositional uniformity can be restored by transfer from the bulk liquid. Surface diffusion is probably not a factor here. The process of stabilization can be described by expanding the stability analysis of Wang et al. to accommodate the presence of surfactant. The distribution of surfactant in the bulk film is described by the convective diffusion equation for that component, together with the appropriate boundary conditions. Assuming a uniform distribution of surfactant across the depth of the film in the undisturbed state, the convective diffusion equation becomes

$$\left(\frac{\partial}{\partial t} - \nabla^2\right) C_s = 0 \tag{16}$$

where C_s is the surfactant concentration perturbation. The boundary condition for this equation at the vapor-liquid interface contains terms representing adsorption and surface convection and interchange between the surface and the bulk phase. The distribution of the surfactant in the surface then modifies the transverse variation in the surface tension shown in the tangential force balance [Eq. (13)] in accord with the appropriate surface equation of state (assumed here to be of the "two-dimensional ideal gas" type). The adsorption is therefore (consistent with the Gibbs adsorption equation) taken to vary linearly with the subphase concentration of surfactant immediately beneath the surface. The foregoing assumptions are good if the system is sufficiently dilute in surfactant. Under these conditions, the boundary condition to Eq. (16), expressing mass conservation of surfactant at the vapor-liquid interface, is

$$\varepsilon \frac{\partial C_s}{\partial t} + \varepsilon C_{SD} \frac{\partial V_x}{\partial X} + D_S \frac{\partial C_s}{\partial Y} + K_S M_S C_S = 0 \tag{17}$$

Surface-Active Agents in Distillation

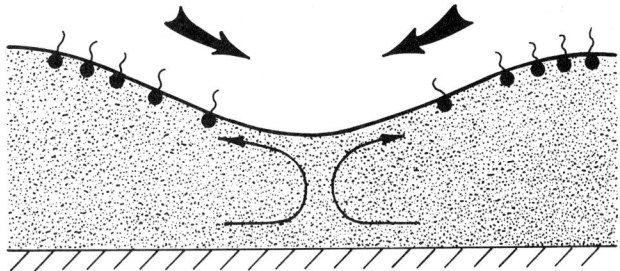

FIG. 23 The origin of Gibbs-Marangoni elasticity as a monolayer of surfactant is dilated, setting up surface tension forces (as indicated by arrows) opposing the dilation.

where ε is the adsorption equilibrium constant, C_{SD} the undisturbed surfactant concentration, D_S the bulk diffusivity of the surfactant in the liquid phase, K_S the surfactant mass transfer coefficient in the gas phase, and m_S the distribution equilibrium coefficient for the surfactant. Nondimensionalization of Eqs. (16) and (17) leads to two new dimensionless groups, characterizing the surfactant, shown in Table 3. The elasticity number, Es, expresses the strength of the Gibbs-Marangoni elasticity effect. The surfactant Sherwood number has its usual significance. In the stability analysis, it has been assumed that the bulk concentration of the surfactant is initially uniform and that the Sherwood number is constant. Both these assumptions are only rough approximations of the true situation, but the

TABLE 3 Additional Dimensionless Groups Arising in Stability Analysis of Falling Film with Surfactant

---

$$Es = \frac{\partial\sigma/\partial C_s \, \varepsilon C_{so}}{\mu D_s} \qquad \text{Elasticity no.}$$

$$Sh_s = \frac{k_s m_s h}{\mu D_s} \qquad \text{Sherwood no. (1)}$$

---

attempt here is to isolate the importance of the Gibbs-Marangoni elasticity effect. The adsorption and the resulting surface tension are permitted to vary with z, however, in accord with the underlying water/formic acid solution concentration and in accord with the assumed linear adsorption equilibrium for the surfactant. The results of the stability analysis are to add the following term to the left-hand side of the characteristic equation of Wang et al. [Eq. (21) in Ref. 50]:

$$\frac{Es(\sinh^2\alpha - \alpha^2)\cosh\alpha}{Ma\,\alpha\sinh\alpha\,(Sh_s\cosh\alpha + \alpha\sinh\alpha)} \tag{18}$$

The inclusion of the Gibbs-Marangoni elasticity effect results in altering the stability diagram as shown in Fig. 24. The computations show that stabilization of σ^- films does indeed occur, but only if Es is very large ($>10^6$). This is larger than might be expected in most systems, and in view of the surface tension data shown earlier for the water/formic acid/n-decanol system, Gibbs-Marangoni elasticity is clearly not the explanation for the stabilization observed by Francis and Berg and others in that system. It may be important in some aqueous systems undergoing vacuum distillation at room temperature or lower.

In retrospect, it is not surprising that disturbances of the long wavelengths involved in rivulet formation are not very strongly affected by the Gibbs-Marangoni elasticity mechanism since these disturbances are not associated with sharp dilations and contractions. It should be noted, however, that the short wavelength instabilities (apparently interfacial turbulence) are significantly suppressed at physically achievable Es values. The general failure of Gibbs-Marangoni elasticity to explain film stabilization in σ^- systems means that we must look to surface tension reversals.

The surface tension reversals shown at room temperature in Fig. 19 cannot explain the film stabilization at the temperature of the distillation (near 100°C). Instead, one must rely on the small surface tension increases wrought by the "surfactant," shown in Fig. 21. These make surface tension reversal at the distillation temperature possible. To show this, it is necessary to recognize that the decanol concentration increases fairly steeply moving down the column. Such a variation is assured by the greatly enhanced solubility of the decanol in the solution as formic acid concentration is increased.

The increase in surface tension with addition of surfactant to a water/formic acid solution is not unique to n-decanol. Some data for other surfactants [59] also show such increases, as indicated in Table 4. We have also performed some experiments with a single

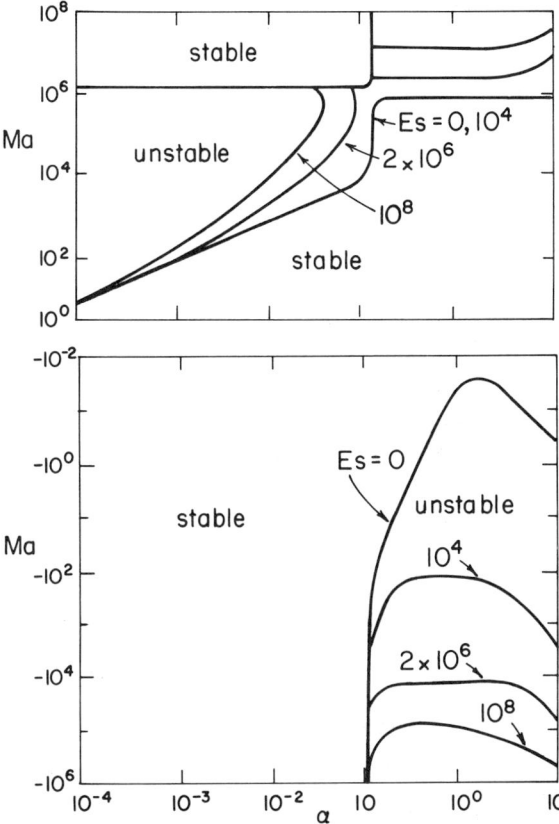

FIG. 24 Stability results for a falling film in the presence of surfactant producing Gibbs-Marangoni elasticity, as expressed by the elasticity number, Es.

ceramic sphere which suggest that such a reversal occurs. A summary of some of the single sphere experimental results is shown in Table 5. The first four lines confirm the stability conditions implied by the packed column results, but the last lines correspond to an interesting result in which an artificial σ^+ system is rendered σ^- through the addition of surfactant. The condensate film of a composition to a σ^- system (from which water would be preferentially desorbing under ordinary conditions) is made σ^+ by contacting it instead with water-saturated, formic-acid-free vapor, so that it is formic acid that must desorb. Such a system is σ^+, and in the

TABLE 4 Surface Tension Dependence of 30 Mol % Aqueous Formic Acis Solution on the Presence of Various Surfactants[a]

Surfactant	Concentration	Surface tension
2- Undercone	10^{-5} (M)	42.5 (mN/,)
Octanoic acid	10^{-5}	42.3
Triton-X 100[b]	Trace	42.3
Neodol 91[c]	Trace	42.3

[a] $T = 98°C$. In the absence of surfactant at 98°C, the surface tension of the solution is 42.1 mN/m.
[b] Octylphenoxy polyethoxy ethanol, Rohm & Haas Co.
[c] C_9–C_{11} primary alcohol, Shell Chemical Co.
Source: Ref. 59.

TABLE 5 Summary of Flow Experiments over Ceramic Sphere

No surfactant
σ^+ conditions ($x_f = 0.78$): rupture flow rate < 0.1 ml/s
σ^- conditions ($x_f = 0.30$): rupture flow rate ~ 1.0 ml/s

With surfactant
σ^+ conditions ($x_f = 0.78$): rupture flow rate < 0.1 ml/s
σ^- conditions ($x_f = 0.30$): rupture flow rate < 0.1 ml/s

Artificial σ^+ system
$x_f = 0.30$, no surfactant, vapor contains no formic acid: rupture flow rate < 0.1 ml/s

$x_f = 0.30$, surfactant, vapor contains no formic acid: rupture flow rate ~ 0.5 ml/s

Source: Ref. 59.

absence of surfactant is quite stable, as shown. When surfactant is added, however, the liquid film is seen to rupture at a rather high liquid rate. The system has been rendered σ^- since formic acid concentration, hence surfactant concentration, and finally, surface tension, all increase, moving upward.

We may conclude from the foregoing studies in supported-area distillation equipment that significant stabilization of liquid films in σ^- systems can be achieved through the addition of small amounts of surfactants, but Gibbs-Marangoni elasticity plays no significant role in the film stabilization. The compounds, which may act as surfactants at room temperature, particularly in water-rich systems, may not act as traditional surfactants at all at the temperature of atmospheric distillation. Film stability may be achieved instead through a small surface tension increase caused by the addition of the "surfactant." The greatest need for the further understanding of "surfactant" effects in distillation is more surface tension data for the systems of interest under operating temperature conditions.

ACKNOWLEDGMENT

This work was supported in part by the National Science Foundation (Grant ENG-7811691).

REFERENCES

1. R. C. Francis and J. C. Berg, Chem. Eng. Sci., 22:685 (1967).
2. K. H. Brumbaugh and J. C. Berg, AIChE J., 19:1078 (1973).
3. W. L. McCabe and J. C. Smith, Unit Operations of Chemical Engineering, 2nd ed., McGraw-Hill, New York, 1967.
4. R. E. Treybal, Mass Transfer Operations, 3rd ed., McGraw-Hill, New York, 1980.
5. C. J. King, Separation Processes, McGraw-Hill, New York, 1971.
6. Lord Rayleigh, Proc. R. Soc., 29:17 (1879).
7. R. Clift, J. C. Grace, and M. E. Weber, Bubbles, Drops, and Particles, Academic Press, New York, 1978.
8. A. Fane and H. Sawistowski, Chem. Eng. Sci., 23:943 (1968).
9. S. P. S. Andrew, Proc. Int. Symp. Distillation, Brighton, England, 1960.

10. S. Ross, in *Kirk-Othmer Encyclopedia of Chemical Technology*, 3rd ed., Vol. 11 (H.F. Mark et al., eds.), Wiley-Interscience, New York, 1980.

11. R. K. Jain and E. Ruckenstein, J. Colloid Interface Sci., *54*: 108 (1976).

12. D. E. Hartley and W. Murgatroyd, Int. J. Heat Mass Transfer, *7*:1003 (1964).

13. R. Defay, I. Prigogine, A. Bellemans, and D. H. Everett, *Surface Tension and Adsorption*, Longman, London, 1966.

14. R. E. Johnson, Jr., and R. H. Dettre, in *Surface and Colloid Science*, Vol. 2 (E. Matejevic and F. R. Eirich, eds.), Academic Press, New York, 1969.

15. N. Silvi and E. B. Dussan V, Phys. Fluids, *28*:5 (1985).

16. R. W. Coughlin, AIChE J., *15*:654 (1969).

17. T. K. Sherwood and F. A. L. Holloway, Trans. AIChE, *36*:21 (1940).

18. P. J. King and P. N. Walmsley, J. Appl. Chem., *15*:98 (1965).

19. F. J. Zuiderweg and A. Harmens, Chem. Eng. Sci., *9*:89 (1958).

20. C. F. Oldershaw, Ind. Eng. Chem. Anal. Ed., *13*:265 (1941).

21. W. M. Langdon and D. J. Tobin, Ind. Eng. Chem. Anal. Ed., *17*:801 (1945).

22. F. C. Collins and V. Lantz, Ind. Eng. Chem. Anal. Ed., *18*:637 (1946).

23. A. G. Medina, C. McDermott, and N. Ashton, Chem. Eng. Sci., *33*:1489 (1978).

24. F. P. Moens, Chem. Eng. Sci., *27*:275 (1972).

25. F. P. Moens, Chem. Eng. Sci., *27*:285 (1972).

26. A. Vrij, F. Th. Hesselink, J. Lucassen, and M. van den Tempel, Proc. K. Ned. Akad. Wet., *1373*:124 (1970).

27. A. Sheludko, Adv. Colloid Interface Sci., *1*:391 (1967).

28. M. J. Rosen, *Surfactants and Interfacial Phenomenas*, Wiley-Interscience, New York, 1978.

29. R. P. Lowry and M. van Winkle, AIChE J., *15*:665 (1969).

30. F. J. Zuiderweg, H. Verburg, and F. A. H. Gilissen, Proc. Int. Symp. Distillation, Brighton, England, 1960.
31. G. S. Bainbridge and H. Sawistowski, Chem. Eng. Sci., *19*:992 (1964).
32. H. C. Burkholder and J. C. Berg, AIChE J., *20*:863 (1974).
33. H. C. Burkholder and J. C. Berg, AIChE J., *20*:872 (1974).
34. L. E. Tarr and J. C. Berg, Chem. Eng. Sci., *35*:1467 (1980).
35. S. R. M. Ellis and R. J. Legg, Can. J. Chem. Eng., *40*:6 (1962).
36. A. B. Ponter, G. A. Davies, W. Beaton, and T. K. Ross, Trans. Inst. Chem. Eng., *45*:T345 (1967).
37. A. B. Ponter, G. A. Davies, W. Beaton, and T. K. Ross, Int. J. Heat Mass Transfer, *10*:1633 (1967).
38. A. B, Ponter, G. A. Davies, T. K. Ross, and P. G. Thornley, Int. J. Heat Mass Transfer, *10*:349 (1967).
39. A. P. Boyes and A. B. Ponter, J. Chem. Eng. Data, *15*:235 (1970).
40. A. P. Boyes and A. B. Ponter, Ind. Eng. Chem. Process Des. Dev., *10*:140 (1971).
41. A. P. Boyes and A. B. Ponter, AIChE J., *18*:935 (1972).
42. A. B. Ponter, W. Peier, and S. Fabre, Chem. Ing. Tech., *50*:444 (1978).
43. W. S. Norman and D. T. Binns, Trans. Inst. Chem. Eng., *38*:294 (1960).
44. W. D. Bascom, R. L. Cottington, and C. R. Singleterry, in *Contact Angle, Wettability, and Adhesion* (R. F. Gould, ed.), Advances in Chemistry Series 43, American Chemical Society, Washington, D.C., 1964.
45. P. Neogi, J. Colloid Interface Sci., *105*:94 (1985).
46. W. S. Norman, T. Cakaloz, A. Z. Fresco, and D. H. Sutcliffe, Trans. Inst. Chem. Eng., *41*:61 (1963).
47. J. C. Berg, in *Recent Developments in Separation Science*, Vol. II (N. N. Li, ed.), CRC Press, Cleveland, Ohio, 1972.
48. J. R. A. Pearson, J. Fluid Mech., *4*:489 (1958).
49. C. V. Sternling and L. E. Scriven, AIChE J., *5*:514 (1959).

50. K. H. Wang, V. Ludviksson, and E. N. Lightfoot, AIChE J., 17:1402 (1971).
51. V. Ludviksson and E. N. Lightfoot, AIChE J., 14:620 (1968).
52. W. L. Bolles, Chem. Eng. Prog., 63:48 (1967).
53. Chem. Eng. News, 62(36):42 (1984).
54. S. Ross and G. Nishioka, J. Phys. Chem., 79:1561 (1975).
55. S. Ross and G. Nishioka, Coll. Polym. Sci., 255:560 (1977).
56. S. Ross and G. Nishioka, in *Foams* (R. J. Akers, ed.), Academic Press, New York, 1976.
57. S. Ross, Chem. Eng. Prog., 63:41 (1967).
58. M. J. Rea, M.S. thesis, University of Washington, 1981.
59. D. K. Johnson, Ph.D. thesis, University of Washington, 1983.
60. J. C. Berg and A. Acrivos, Chem. Eng. Sci., 20:737 (1965).
61. A. B. Ponter, M. Polikar, P. Trauffler, and S. Vijayan, Tenside Deterg., 13:201 (1976).
62. R. Schmidt, M.S. thesis, University of Washington, 1980.
63. A. Rosello and J. M. M. Moreno, Afinidad, 35:459 (1978).
64. J. M. M. Moreno, A. R. Segado, and J. M. R. Patino, An. Quim., 74:1425 (1978).
65. A. R. Segado, E, M. de la Ossa, and J. M. M. Moreno, Grasas Aceites Seville, 30:169 (1979).
66. A. B. Ponter, P. Trauffler, and S. Vijayan, Ind. Eng. Chem. Process Des. Dev., 15:196 (1976).

3
Surfactant-Based Treatment of Aqueous Process Streams

JOHN F. SCAMEHORN and JEFFREY H. HARWELL *Institute for Applied Surfactant Research, University of Oklahoma, Norman, Oklahoma*

I.	Introduction	78
II.	Micellar-Enhanced Ultrafiltration	78
	A. Organic Compound Removal	78
	B. Heavy Metal Removal	94
III.	Foam Separations	98
	A. Particulate Flotation	99
	B. Colligend Flotation	104
	C. Precipitate Flotation	105
	D. Continuous-Flow Foam Separation Process	105
IV.	Surfactant-Enhanced Carbon Regeneration	106
	A. Experimental Results	109
	B. Feasibility	109
V.	Extraction into Reverse Micelles	111
	A. Mechanism and Selectivity	112
	B. Continuous-Flow Application to a Bioseparation	112
VI.	Admicellar Chromatography	115
	A. Adsolubilization	115
	B. Fixed-Bed Operation	117
VII.	Summary	121
	References	122

I. INTRODUCTION

Surfactant-based separation process are emerging as a major class of unit operation for industrial separations. Along with traditional applications, such as mineral flotation, these separations are uniquely well suited to areas of application of emerging or increasing importance.

The first of these areas in biotechnology, where valuable products need to be recovered from dilute aqueous solutions and are easily degraded. Surfactant-based separations can effect a concentration of the biochemical under mild conditions.

A second area where this class of separation methods will be increasingly important is in pollution control. Dissolved toxic organics or heavy metals can be removed from aqueous wastewaters without introducing substantial toxicity from residual surfactant.

Surfactant-based separations generally require little energy and provide an energy-efficient alternative to traditional purification methods such as distillation or distillation followed by extraction.

In this chapter, five surfactant-based separations to treat aqueous process streams are surveyed. These streams may contain a valuable product or be wastewaters. One of these is a traditional technique: foam separation. The other four (micellar-enhanced ultrafiltration, surfactant-enhanced carbon regeneration, extraction into reverse micelles, and admicellar chromatography) are recently developed methods which, in the opinion of the authors, have substantial potential for commercial use. The astute industrial researcher may find one of these methods of value in a specific application.

II. MICELLAR-ENHANCED ULTRAFILTRATION

A. Organic Compound Removal

Wastewater streams containing dissolved organics are a common prob- in the chemical, petroleum, synfuels, and other industries. These organics are often toxic and must be removed before the water can be discharged to the environment or reused in the process. Ordinary ultrafiltration is ineffective in the removal of organic compounds with molecular weights below 300 [1].

Micellar-enhanced ultrafiltration (MEUF) is a recently proposed technique [2,3] which can be used to remove soluble, low-molecular-weight organics from water, as illustrated in Fig. 1. In this process, surfactant is added to the polluted aqueous stream. When the surfactant is present at concentrations greater than its critical micelle concentration (CMC), it forms aggregates called micelles into which the organic pollutant will dissolve or solubilize. The stream is then passed through an ultrafiltration membrane with pore sizes smaller

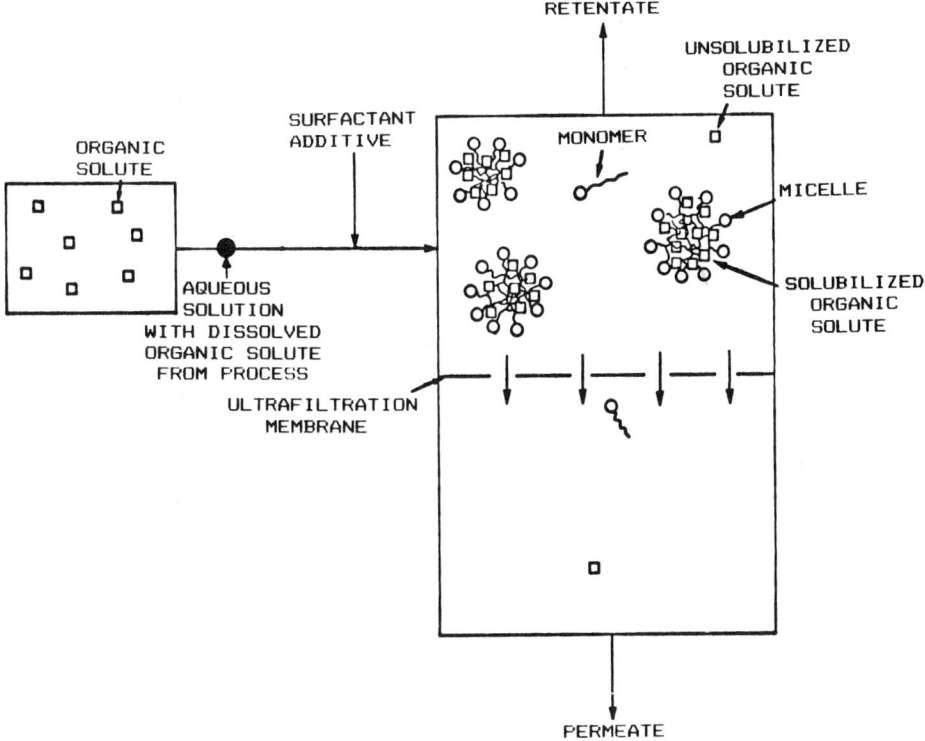

FIG. 1 Schematic of micellar-enhanced ultrafiltration process to remove dissolved organics from water.

than the size of micelles. The micelles containing the organic pollutants are rejected by the membrane. Since the micelles are fairly large (containing about 100 surfactant molecules for the systems used here), a much larger pore size membrane can be used to reject micelles than would be required to reject individual, unaggregated surfactant molecules (monomer).

If the tendency for the pollutant to solubilize in the micelles is large, the concentration of unsolubilized organic in solution will be very small. This concentration corresponds approximately to the concentration of pollutant in the purified or permeate stream if micelles are completely rejected, resulting in very pure permeate. The concentration of surfactant in the permeate can also be very low. A surfactant can be chosen for this application which is biodegradable, so that a low level of surfactant in the permeate does not present environmental problems.

If the solution in which the organics remain after filtration (retentate) can be concentrated to high levels of surfactant and pollutant, a large fraction of the feed will pass through the membrane and a high recycle ratio (permeate/feed) can be attained. Therefore, a small waste stream containing the pollutants at high concentrations will result from the process. Subsequent further purification or disposal of this retentate is less expensive and requires less energy than does treating the original wastewater stream.

The ultrafiltration step requires little energy and relatively low pressure (e.g., 60 psig) and is generally much less expensive than other separation techniques per unit volume of solution processed.

1. Micelle Formation and Solubilization

Surfactants are molecules with a hydrophobic portion (usually a long hydrocarbon chain) and a hydrophilic portion (e.g., sulfate group). Micelles are aggregates formed by these surfactants in which the hydrophobic groups are intertwined together to form a hydrocarbon core from which water is largely excluded and with the hydrophilic groups on the surface of this core exposed to the aqueous solution (see Fig. 1). This configuration allows the hydrocarbon chains to be removed from the aqueous environment while the hydrophilic groups are still exposed to the polar solvent.

When surfactants are present in aqueous solutions at concentrations above the CMC, surfactant is present in two possible states: as unaggregated monomer or as constituents of micelles. As the total concentration of surfactant is increased beyond the CMC, almost all of the added surfactant goes to increase the micelle concentration [4]. In MEUF, the surfactant concentration is designed to be far above the CMC, so that monomer constitutes only a very small fraction of the total surfactant in solution. The vast majority of the surfactant is in micellar form, micelles being the effective separating agent in MEUF.

Nonsurfactant organic compounds may dissolve or solubilize within micelles at four different locations [5-7]: (1) in the hydrocarbon core, (2) in the palisade layer or region between the hydrocarbon core, (3) adsorbed on the micelle surface, and (4) in the polyoxyethylene shell of micelles composed of nonionic surfactants. Organic solute species tend to locate preferentially in regions within the micelle which are similar chemically and in polarity to these molecules.

The solubilization into micelles can be viewed conceptually as being similar to extraction of a dissolved organic from an aqueous phase into an immiscible extractant phase. In MEUF, the extractant phase (micelles) is dispersed in the aqueous phase. As the surfactant concentration is increased (micelle concentration is increased) in a stream of defined organic pollutant content, the fraction of the organic solubilized in micelles increases and the fraction present as unsolubilized species in the water decreases. As an approximation, the ratio of the concentration of pollutant in solubilized form to that in unsolubilized form is proportional to the surfactant concentration constituting micelles. If the surfactant concentration is well above the CMC and the vast majority of the organic solute is solubilized in micelles, the concentration of unsolubilized organic is approximately inversely proportional to the surfactant concentration. For example, one would expect the pollutant concentration in the permeate to be reduced by a factor of 2 if the surfactant concentration were doubled. From a consideration of the solubilization phenomena, the amount of surfactant added per unit volume of solution treated will be expected to be a balance between permeate purity and raw materials cost.

2. Selection of Surfactant

In the selection of a surfactant for use in MEUF, some important desirable characteristics are: (1) high solubilization capacity for the organic pollutant; (2) forms large micelles, so that large pore sizes (with resulting large fluxes) can be used; (3) low monomer concentration, so little surfactant is wasted; and (4) minimal phase-separation problems (macroemulsion formation, precipitation, gelling, etc.). Anionic, cationic, and nonionic surfactants have been considered for application in MEUF [3].

One desirable characteristic of a surfactant is a long hydrocarbon chain, since this results in large micelles, high solubilizations, and low monomer concentrations. Anionic surfactants are restricted to hydrocarbon chain lengths of about 12 carbons or less if the process is to be applied at room temperature. The Krafft temperature is above room temperature for longer hydrophobic groups, so the surfactant precipitates and MEUF is ineffective.

Nonionic surfactants form large micelles, can have high solubilization capacities (per mole of surfactant), and have low monomer concentrations in micellar solutions. These surfactants would appear to be good candidates for use in MEUF. However, it has been found that stable macroemulsions are formed when nonionic surfactants are stirred with some of the organic pollutants of interest. Since the ultrafiltration process occurs under turbulent conditions (on purpose), this is unacceptable. Also, the solubilization capacities of the nonionics are not very high compared to anionics or cationics per pound of surfactant because of their high molecular weights. Since surfactants are sold on a weight basis rather than on a molar basis, nonionics are not economically attractive for use in MEUF.

Cationic surfactants generally have much lower Krafft temperatures than those of anionic surfactants of corresponding hydrophobic group size. We have observed no macroemulsion formation problems with cationics. Phase boundary maps have shown that high concentrations of both surfactant and solute are attainable before phase separation occurs. Therefore, cationic surfactants with large hydrophobic groups can be used in MEUF, resulting in large micelles, high solubilization capacities, and low monomer concentrations. Cationic surfactants do not pose significant environmental risks [8]. *In general, cationic surfactants are the surfactants of choice in MEUF.*

For the experimental studies discussed here, cetylpyridinium chloride monohydrate (CPC) was used because it has a large hydrophobic group and the Krafft temperature is 10.8°C. Therefore, it can be used effectively at room temperature. Also, it is readily available in pure form at low cost. A more detailed discussion of consideration in surfactant selection is given in Ref. 3.

3. Rejection of Pollutant and Surfactant

For practical MEUF use, surfactant would be introduced into the feedstream containing the dissolved organic, and this stream would then be treated in a batch or continuous membrane device. As permeate is removed, the retentate solution becomes more concentrated in the rejected species as the retentate moves through the device for a continuous unit or as a function of time in a batch unit. Therefore, the retentate composition follows a "path." The data reported here are differential data for a given retentate composition. These data can be integrated over the path to predict the average flux (which dictates membrane area requirements) and average permeate concentrations for a continuous unit under specified conditions. For systems for which rejections are high for both pollutant and surfactant, the pollutant/surfactant ratio in the retentate remains essentially unchanged as the retentate becomes more concentrated. Therefore, the data reported will be for a defined pollutant/surfactant ratio.

The surfactant or CPC concentration in the permeate is shown in Fig. 2 for membrane pore sizes varying from 1000 to 50,000 molecular weight cutoff (MWCO) for a path corresponding to a 4-*tert*-butylphenol (TBP) to CPC ratio in the retentate of 1:10 [9]. Also shown in Fig. 2 is the mean ionic molality of the CPC in the retentate. If the micelles were completely rejected and the monomer were not hindered upon passage through the membrane, the CPC concentration in the permeate should correspond approximately to this mean ionic molality.

Let us consider first the retentate concentrations below 220 mM CPC. For membranes with pore sizes of 20,000 or smaller, the micelles are essentially completely blocked. For example, at a retentate CPC concentration of 100 mM, the permeate CPC concentration is 1 mM. If any significant fraction of micelles were leaking through the membrane, this permeate would be at a much higher concentration. In fact, for the 10,000 MWCO membranes and smaller, the permeate surfactant concentration is below the mean ionic molality of the surfactant, indicating that monomer hinderance is significant.

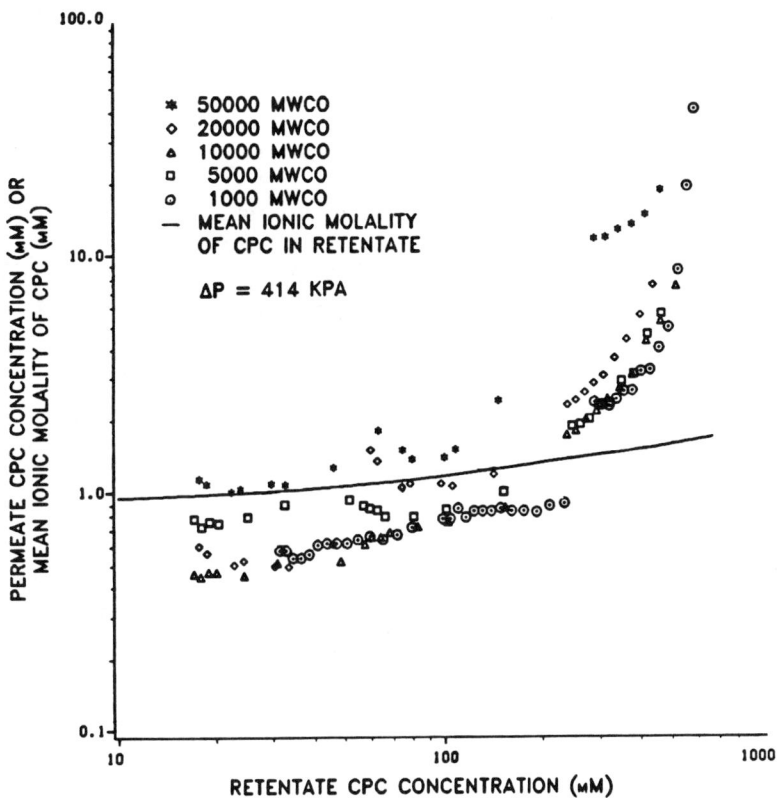

FIG. 2 Effect of membrane pore size on permeate CPC concentration for a retentate TBP/CPC molar ratio of 1:10. (From Ref. 9.)

For the 50,000 MWCO membrane, a very small amount of micelle leakage is occurring. However, the permeate still contains a low level of surfactant concentration. *From these results, we may conclude that the basic concept behind MEUF as a separation technique is valid: that is, micelles are rejected by large pore size membranes.*

It is reasonable to ask why micelles do not form on the permeate side of the membrane, since the monomer can pass through the pores. Given enough time, the surfactant chemical potential difference across the membrane would cause this to happen. If flow rates are slow enough, this will occur [10]. However, under the normal transmembrane pressure drops used in industrial ultrafiltration operations, flow rates are fast enough and contact time between permeate and retentate so short, compared to the time required for this surfactant diffusion process across the membrane to occur, that an insignificant number of micelles form in the permeate under proper conditions.

Above a retentate CPC concentration of about 220 mM, the permeate CPC concentration increases rapidly. It has been hypothesized that this is due to formation of premicellar aggregates by the surfactants (i.e., dimers, trimers, etc.) which are not completely rejected by the membrane [3].

The pollutant (TBP) concentration in the permeate is shown in Fig. 3 for 1:10 and 1:5 TBP/CPC ratio's for a 1000 MWCO membrane. Also shown for comparison are the concentrations of unsolubilized TBP in the retentate as determined by an independent method.

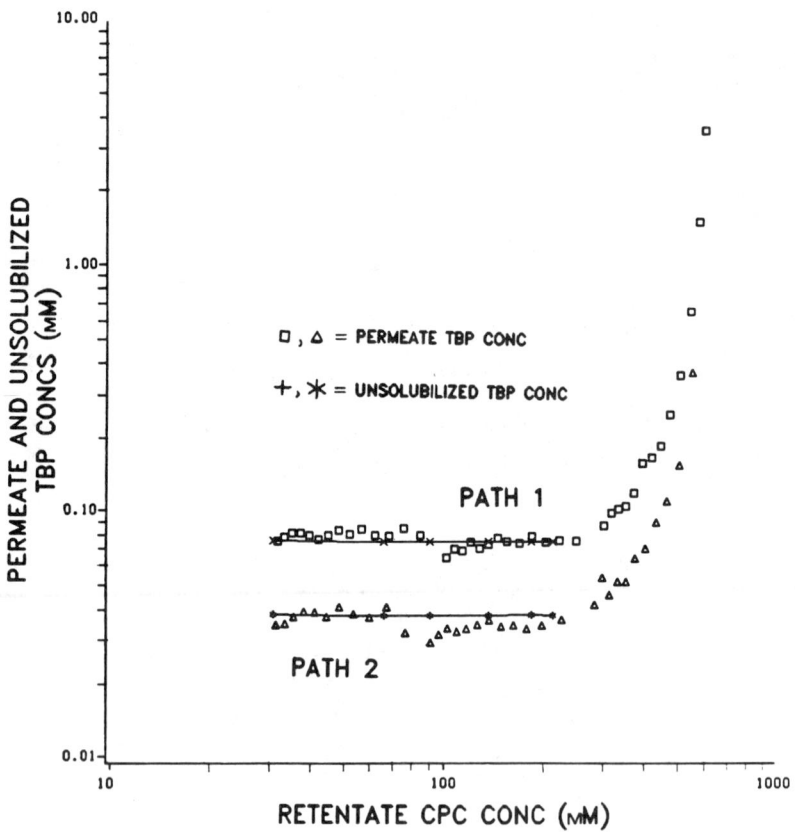

FIG. 3 Concentration of TBP in the permeate for a retentate TBP/CPC molar ratio of 1:5 (path 1) and 1:10 (path 2). (From Ref. 3.)

Surfactant-Based Treatment of Process Streams

First, consider the retentate CPC concentrations below about 220 mM. The TBP concentrations in the permeate are all well below 0.1 mM, while the retentate TBP concentrations are as high as 22 mM, an extremely large concentration decrease on one pass. This translates to TBP rejections from 97.6 to 99.7% along the paths shown in Fig. 3. The permeate TBP concentrations also correspond closely to the unsolubilized concentration in the retentate, further confirming that micelles are completely rejected for this membrane and that the pollutant is not hindered upon passage through the pores, unlike the surfactant. The TBP is smaller than the CPC, so this is not surprising. The permeate TBP concentration is smaller by a factor of almost exactly 2 when the surfactant/pollutant ratio is doubled in Fig. 3, as predicted by the approximation mentioned earlier. *In summary, very good rejection of the organic pollutant can be attained in MEUF, proving the technical feasibility of the technique.*

At CPC retentate concentrations above about 220 mM, the TBP concentration in the permeate increases rapidly, as does that of the surfactant. Some organic solute would be expected to be solubilized or associated with premicellar aggregates. If leakage of these aggregates through the membrane is responsible for the high surfactant concentrations in the permeate under these conditions, it is probably also responsible for the leakage of the TBP. In any case, *operation of MEUF at very high surfactant concentrations in the retentate leads to lower rejections and should be avoided.* However, we will show that these concentrations are so high for CPC as not to limit the process significantly.

4. Flux Rate in MEUF

The flux is shown in Fig. 4 for MEUF of a 1:10 TBP/CPC retentate solution for membrane pore sizes from 1000 to 50,000 MWCO. As the pore size of the membrane increases, the flux increases, as expected. The flux decreases linearly as the logarithm of the retentate concentration increases at high retentate concentration, indicating that this is a concentration polarization regime [1]. Data from all the membranes extrapolate to the same retentate concentration at zero flux. This is the gel point or gel concentration. From the data in Fig. 4, this occurs at a CPC concentration of 530 mM CPC and 53 mM TBP [9]. This is all above the concentration at which rejections become poor (220 mM CPC), so the practical operating range of the process is limited by rejection considerations, not by the gel concentration. The flux rate increases as the transmembrane pressure increases, with corresponding small increases in rejection of pollutant or surfactant [9].

From considerations of flux and rejection, the membrane used should be 10,000 MWCO or larger. The optimum membrane will represent a trade-off between permeate purity and flux rate and will depend

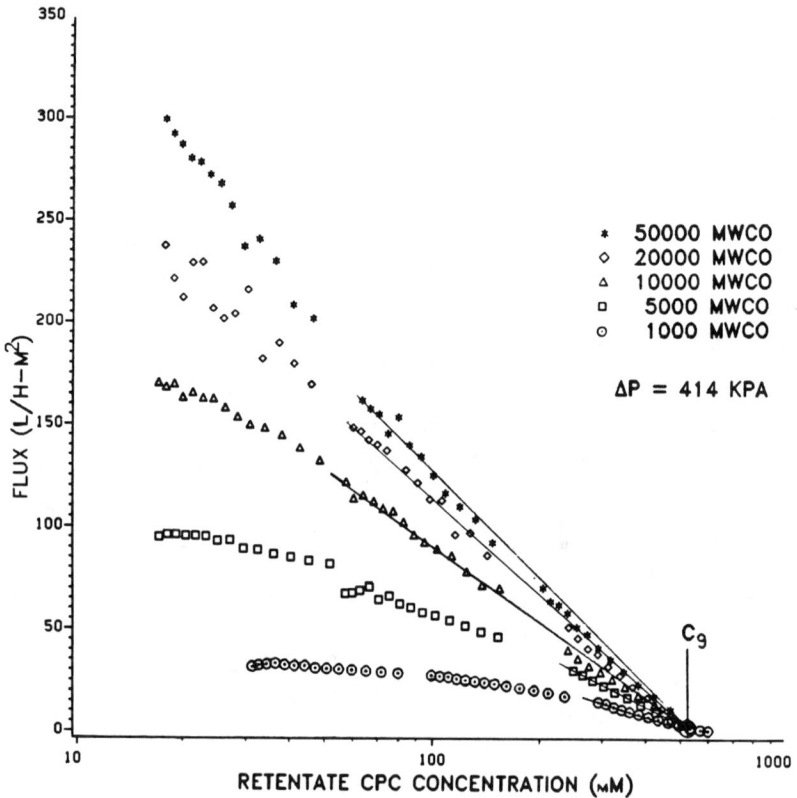

FIG. 4 Flux rates for various membrane pore sizes. (From Ref. 9.)

on the specific situation. However, for CPC as the surfactant, a pore size of 20,000 or 50,000 MWCO will probably be nearly optimum. The pressure should be as high as possible without membrane compression effects. This will depend on the specific membrane manufacturer. In this work, pressures up to 414 kPa were used without adverse effects.

5. Effect of Pollutant Type

The ability of the micelle to solubilize the organic solute depends on the structure of the solute. The pollutant concentration in the permeate and the rejection of this solute are shown in Table 1 for a variety of organics under the same conditions. These data correspond to a pollutant/surfactant ratio of 1:10. Membrane performance is often characterized in terms of rejection, defined here as

TABLE 1 Comparison of Rejection of Organic Pollutants of Various Structures in MEUF[a]

Pollutant	Pollutant concentration in permeate (mM)	Rejection of pollutant (%)
n-Hexanol	1.56	92.2
n-Heptanol	0.323	98.4
n-Octanol	0.141	99.3
Phenol	1.42	92.9
m-Cresol	0.526	97.4
4-tert-Butylphenol	0.0767	99.6
Benzene	2.33	88.4
Toluene	0.800	96.0
n-Hexane	0.0500	99.8
Cyclohexane	0.204	99.0

[a]Conditions: surfactant, CPC; retentate [CPC] = 0.2 M; retentate [pollutant] = 0.02 M; pressure = 414 kPa; membrane, 1000 to 20,000 MWCO; temperature, 30°C.

$$\text{rejection} = 1 - \frac{[\text{pollutant in permeate}]}{[\text{pollutant in retentate}]}$$

For all these compounds, rejection is greater than 88%, reaching as high as 99.8%. Therefore, MEUF can effectively remove a wide range of organics from water. In a homologous series (e.g., n-alcohols, alkylphenols, alkylbenzenes), as the size of the alkyl chain increases, the efficiency of MEUF increases. This effect occurs because the higher homologs are less water soluble, so they tend to solubilize to a greater extent at a given concentration in solution.

TBP has a higher rejection than n-octanol, even though both have about the same water solubility. This is due to the specific attractive interaction between the solute benzene ring and the positively charged surfactant hydrophilic groups in the micelle [11,12]. When the pollutant of interest is aromatic in nature, this is an additional incentive to the use of cationic surfactants in MEUF.

Phenol exhibits higher rejections than benzene, even though the former is much more water soluble. Clearly, the hydroxyl group in the phenol is enhancing the solubilization of this compound relative to the benzene.

Although water solubility gives a general guide to solubilization capacity of a solute in a micelle, specific interactions also need to be considered. A group contribution method to estimate solubilization of a solute from its chemical structure would clearly be of use in predicting the performance of MEUF for a solute without needing experimental data. Current work in our laboratory is aimed at this goal.

6. Staging of MEUF

As a chemical engineering unit operation, it is logical to consider staging as a way to improve MEUF performance. The permeate can be fed to another MEUF unit after additional surfactant has been added to it and further purified. This is illustrated for removal of TBP from a stream using CPC as the surfactant in Fig. 5. In that flow diagram, the total surfactant added in the process per unit volume original feed solution and the ultimate recycle ratio or output ratio (overall process permeate/feed) are constrained to be the same in the double-stage as in the single-stage unit. In the two-stage process, the surfactant is added to the initial feedstream as well as to the permeate from the first unit being fed to the second unit.

As seen in Fig. 5, the permeate from the two-stage unit has a lower pollutant concentration than that from the one-stage unit by a factor of 3.9 in this sample case. However, the membrane area required would be approximately twice as much in the two-stage process as in the one-stage process (i.e., the membrane area in *each* MEUF unit in the two-stage process would be equal to that in the single-stage process). Therefore, staging represents an opportunity to balance permeate purity against capital cost. However, if the constraint of a constant permeate purity, a constant membrane area, and a constant recycle ratio exist, staging provides no advantage over nonstaged processes.

7. Effect of Operating Variables on Performance

The permeate and retentate compositions, recycle ratio, surfactant requirements, and membrane area for the treatment of wastewater streams of various compositions under different operating conditions are shown in Table 2. The surfactant is assumed to be CPC in these calculations, and the retentate is assumed to be concentrated to 220 mM CPC, since it has already been shown that rejections are poor above this concentration.

The first case is a base case. The pollutant is TBP and the TBP/CPC ratio in the feed is 1:10. Note that the permeate TBP

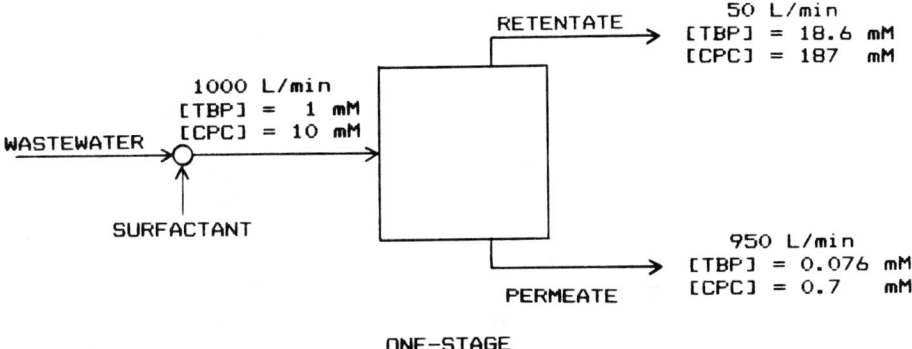

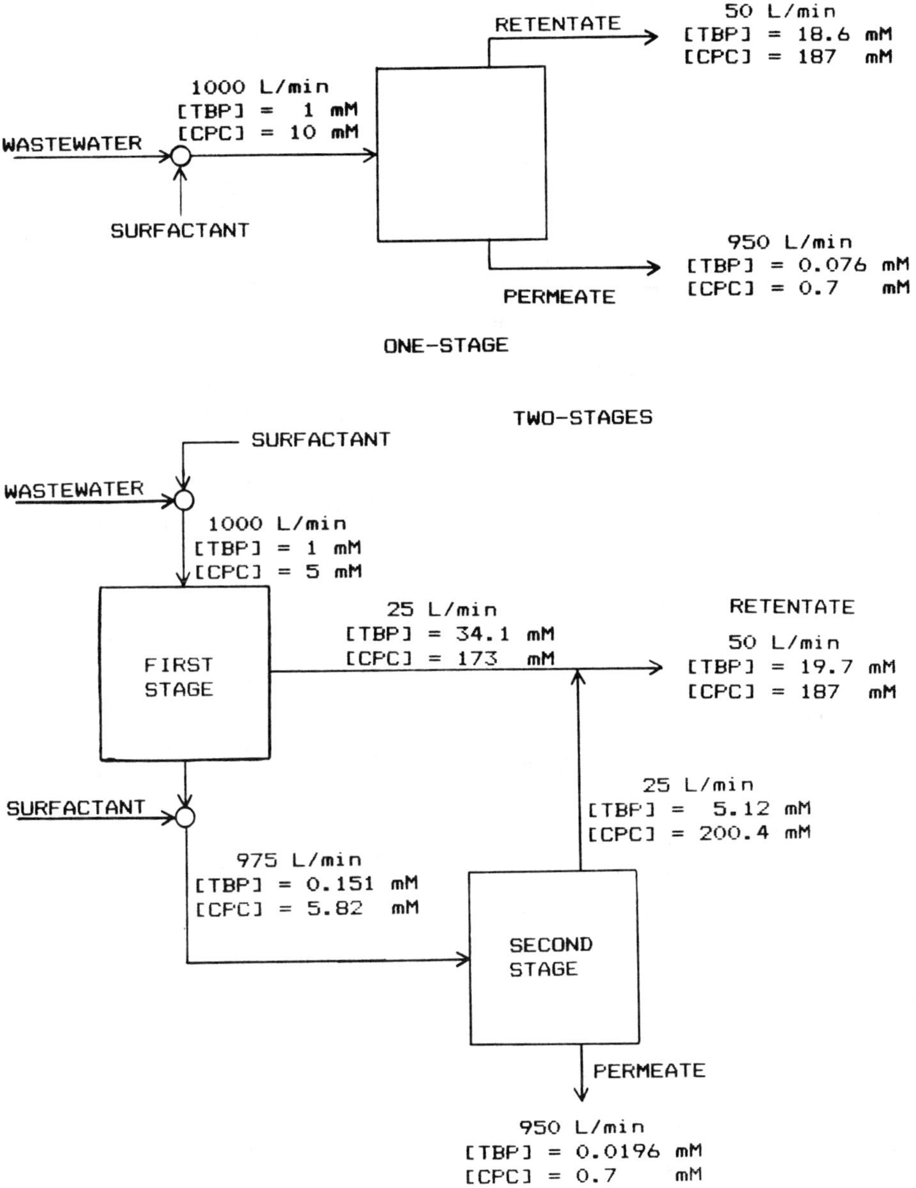

FIG. 5 Comparison of a one-stage process to a two-stage process for MEUF.

TABLE 2 Effect of Operating Variable on MEUF Performance[a]

Case	Pollutant	[Pollutant] in wastewater (mM)	CPC added to wastewater (kg/1000 L)	[CPC] in wastewater (mM)	Membrane MWCO	Permeate concentration (mM) Pollutant	Permeate concentration (mM) CPC	Membrane area [m^2/(1000 L/min)]	Permeate/Feed (L/L)	Retentate concentration (mM) Pollutant	Retentate concentration (mM) CPC
1	TBP	1	3.6	10	10 000	0.076	0.7	730	0.96	22	220
2	TBP	1	7.2	20	10 000	0.038	0.7	700	0.91	11	220
3	TBP	2	7.2	20	10 000	0.076	0.7	700	0.91	22	220
4	TBP	1	3.6	10	50 000	0.10	1.4	510	0.96	22	220
5	TBP	1	3.6	Two-stage process	10 000	0.020	0.7	1460	0.96	22	220
6	n-Octanol	1	3.6	10	10 000	0.14	0.7	730	0.96	22	220

[a]Pressure, 414 kPa; temperature, 30°C.

concentration is 7.6% of that in the feed and 0.34% of that in the retentate. The recycle ratio is 96%.

The effect of decreasing the pollutant/surfactant ratio is shown in case 2. As already discussed, the permeate pollutant concentration decreases. However, a negative effect is that the recycle ratio decreases. This is caused by the practical limitation on the maximum feasible retentate concentration.

The effect of treating a stream with a higher pollutant concentration at a constant pollutant/surfactant ratio is shown by case 3. The permeate composition remains unchanged, but the recycle ratio decreases because of the limit on maximum retentate concentration.
This might imply that if an extremely dilute wastewater stream were being treated, an extremely high recycle ratio would be attainable.
However, at a constant pollutant/surfactant ratio, as the surfactant concentration in the feed decreases, the monomer concentration becomes an increasing fraction of the surfactant until so little surfactant would be in micellar form that an effective separation could not be performed.

The effect of using a membrane with larger pore sizes is shown in case 4. The permeate pollutant concentration is increased by 31%, while the membrane area required for this example is reduced by 30%. Membrane pore size represents a compromise between permeate purity and membrane area.

The effect of staging the MEUF process (as demonstrated by the example in Fig. 5) is shown by case 5. A purer permeate, but a higher membrane area requirement results from staging, as already discussed.

Although most numerical examples have been based on TBP as the pollutant because we have done more studies on this compound than on any other, the effect of using a different solute is shown by case 6, with n-octanol as the pollutant. As demonstrated in Table 1, octanol does not solubilize as effectively into the micelles as does TBP. This results in a higher pollutant concentration in the permeate, but no other effects.

A summary of the effect of operating variables on performance characteristics of MEUF is shown in Table 3. Optimization of variables for an industrial application depends on the specific situation, but Table 3 shows directional effects to aid in variable manipulation.

8. Specific Application of MEUF

A nondestructive separation technique used to separate a dissolved material from a stream can be applied to a situation where the purified stream produced is the main goal, or to a situation where the goal is the concentration of some valuable product for further downstream purification.

TABLE 3 Qualitative Effect of Operating Variables on MEUF Performance

Variable whose value is increasing	Effect on			
	Surfactant required per unit volume	Permeate purity	Membrane area required	Recycle ratio
[Pollutant]/[surfactant] ratio	Decreases	Decreases	Increases	Increases
[Pollutant]	Increases	Unchanged	Decreases	Decreases
Membrane pore size	Unchanged	Decreases	Decreases	Unchanged
Number of stages	Unchanged	Increases	Increases	Unchanged
Tendency of pollutant to solubilize	Unchanged	Increases	Unchanged	Unchanged

Wastewater cleanup would be an example of the need to purify an aqueous stream for emission or reuse. The focus of this discussion of MEUF has been wastewater applications with pollutants as the species rejected.

The pollutants that can be most effectively removed from wastewater using MEUF are, in general, those which are least soluble in water. However, if the solubility is low enough, not enough pollutant could dissolve in the stream to reach toxic levels and cause a problem. Therefore, there is a range of solubilities where MEUF can be useful in pollution control. A compound such as 1-chlorooctadecane would probably be too insoluble to be of interest, whereas methanol would be too water soluble to solubilize well in micelles and be effectively removed by MEUF.

Most of the high-value products produced in dilute solution, which need to be concentrated to higher concentration before use, are biological or pharmaceutical in nature. For example, products of biotechnology are often produced as dilute components in "broths." If the components produced are high-molecular-weight proteins (e.g., MW > 20,000), ordinary ultrafiltration could separate them from the solution. MEUF would be advantageous to use only if the surfactant had a specific affinity for the compound of interest, so that it was separated not only from the bulk of the aqueous solution, but also from other by-products of the broth. Building both specificity and micellization properties into the surfactant molecule is an area for future research. For lower-molecular-weight compounds of value, MEUF may be economically superior to ordinary ultrafiltration since larger-pore-size membranes can be used.

9. Estimating Design Parameters for MEUF

In considering the use of MEUF in a specific application, the ability to estimate design parameters to perform economic evaluations would be useful.

In estimating the purity of the permeate, the pollutant can be assumed to be at its equilibrium unsolubilized concentration in the retentate. Of course, the solubilization will depend on the surfactant used. In the absence of other factors or special incentives, CPC has been shown to be effective in this application and should be considered for use. The solubilization value can be measured by a variety of techniques [5,13–15] or may be available from equilibrium solubilization constants in the literature. In an homologous series of solutes, the permeate concentration of pollutant decreases by approximately a factor of 3 for each additional methylene group on the alkyl chain (see Table 1). Therefore, crude estimates of permeate purity could be made for a series of homologous compounds if the solubilization characteristics of one member were known. For example, for phenolics or n-alcohols, the data in Table 1 could be used to estimate performance. Of course, a single experiment using a stirred cell ultrafiltration under expected conditions could always be used to measure permeate purity.

Once the permeate purity is defined for a given condition, the surfactant concentration needed in the feed to the MEUF can be calculated by assuming that the permeate pollutant concentration is inversely proportional to the surfactant concentration. The final retentate should not be concentrated over 220 mM in surfactant (for CPC as the surfactant). This dictates the permeate/feed ratio from a material balance.

As a worse case, except for membranes of high MWCO (e.g., >20 000 for CPC), the surfactant concentration can be assumed to be equal to its mean ionic molality in the permeate. This is approximately equal to the surfactant's critical micelle concentration (CMC), which is readily accessible [16] or easily measured. For example, for CPC, the CMC is equal to 0.88 mM.

The membrane area requirements, as dictated by the flux, are very dependent on the type of ultrafiltration device used and conditions under which the separation is run. It has been shown that the MEUF operates in the concentration polarization regime under expected operating conditions. Therefore, the flow patterns affect the boundary layer next to the membrane surface and affect the flux. As a result, the flux rates will be different in hollow fiber, spiral wound, stirred cell, and other membrane devices. To determine flux rates and membrane requirements accurately, experimental data on the actual type of device to be used are necessary. For a preliminary estimate, one could use the flux data in Fig. 4 or the membrane areas in Table 2.

B. Heavy Metal Removal

MEUF can be used selectively to remove multivalent ionic species from aqueous solution. Since the most prevalent pollutant of this type of current interest is heavy metals, this section will focus on these toxins. Heavy metals can be present at levels above permitted emission levels in wastewater from synfuels plants, metal plating plants, and chemical operations, among others. Heavy metals are also a problem in abandoned mines, the contaminated water from which can pollute groundwater supplies.

The principle of removing heavy metals from water using MEUF is illustrated in Fig. 6. Micelles composed of ionic surfactants have a high charge and counterions (ions of opposite charge to the surfactant) will bind to the surface of the charged micelle. Multivalent counterions will bind in higher proportion to the surfactant than will monovalent ions, relative to the respective concentrations in solution. The higher the valence state, the greater the binding. In this dis-

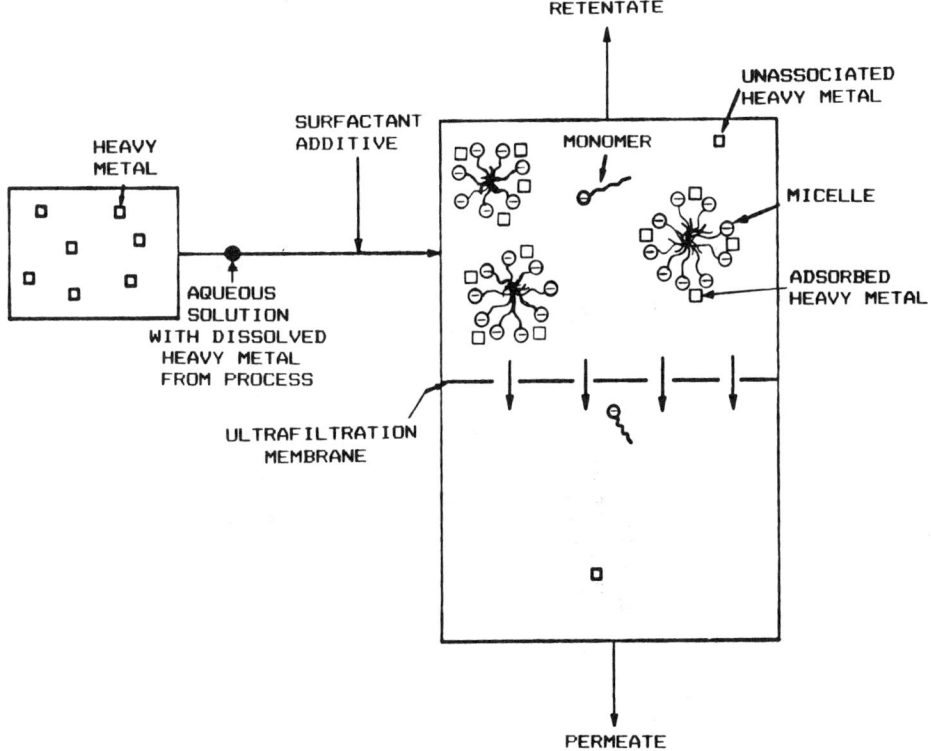

FIG. 6 Schematic of micellar-enhanced ultrafiltration process to remove dissolved multivalent heavy metals from water.

cussion, cationic counterions will be addressed; analogous arguments for anionic counterions could also be made.

If anionic surfactant is added to the stream, the multivalent heavy metal will bind preferentially on the micelle. As a result, the metal ion will be more dilute in the permeate relative to the feed and more concentrated in the retentate. Following the discussion in Section II.A.2, the anionic surfactant cannot have as long an alkyl chain as CPC for use at room temperature; otherwise, it will precipitate. As a consequence, a smaller-pore-size membrane must be used for this application of MEUF than in removal of dissolved organics because of smaller micelle size.

As an example of the effectiveness of MEUF in heavy metal removal, sodium dodecyl sulfate (SDS) was used as the surfactant in the removal of divalent zinc (added to solution as $ZnCl_2$). The results are shown in Table 4.

TABLE 4 Removal of Zn^{2+} from Wastewater Using MEUF[a]

Run	Retentate concentration (mM)		$[Zn^{2+}]/[SDS]$ in retentate	Permeate concentration (mM)		Zn^{2+} rejection (%)	Permeate $[Zn^{2+}]$ / Retentate $[Zn^{2+}]$
	Zn^{2+}	SDS		Zn^{2+}	SDS		
1A	0.12	49	0.0024	0.00097	7.2	99.2	0.0081
1B	0.55	190	0.0029	0.0052	8.9	99.1	0.0094
2A	0.66	41	0.016	0.0051	6.6	99.2	0.0077
2B	4.2	210	0.02	0.031	7.3	99.3	0.0074
3	10.0	95	0.105	0.084	4.1	99.2	0.0084

[a]Conditions of runs: surfactant, SDS; added electrolyte, $ZnCl_2$; pressure, 414 kPa; membrane, 1000 MWCO; temperature, 30°C.

The SDS concentration in the permeate is approximately equal to that of the CMC (6.6 mM), which approximates the mean ionic molality of the surfactant monomer in the retentate. Therefore, micelles are essentially completely rejected under the conditions studied.

Under the conditions used, the rejection of zinc is greater than 99% for all runs. Therefore, the removal of divalent ions from solution is feasible. Hirasaki and Lawson [17] demonstrated that divalent calcium could be removed from solution using this technique.

Comparing runs 1A and 1B and runs 2A and 2B, as the concentration of Zn^{2+} increases in the retentate at an approximately constant zinc/surfactant ratio, the permeate contains an increasing concentration of zinc. This is in contrast to the MEUF of organics, where the permeate would be of constant composition when this ratio is constant.

Comparing runs 1A and 2A and runs 1B and 2B, as the retentate zinc concentration increases at approximately constant surfactant concentration, the permeate zinc concentration increases. In fact, the permeate zinc concentration/retentate zinc concentration ratio is approximately constant over the entire set of conditions studied. This ratio varies by 27%, while the retentate surfactant concentration varies by a factor of 5 and the retentate zinc/surfactant ratio varies by a factor of 44. This would seem to imply that the separation of zinc is fairly independent of surfactant concentration, but the conclusion cannot be carried too far because the absence of surfactant would result in no separation. Rather, the reason probably has to do with the extremely high rejections; under the conditions studied, almost all the zinc is bound onto micelles. Additional micelles do not have much effect. If the surfactant concentration were lower, this would not be true.

The physics of counterion binding onto charged micelles is complex and models are becoming available to describe the phenomenon [17–19]. Such models will need to be applied to these systems to understand the MEUF process on other than an empirical basis.

Although insufficient data exist to outline a design parameter for this separation process, some observations may be made which should give guidance. MEUF gives extremely high separations in one pass for divalent heavy metals. Trivalent or even tetravalent counterions would be even easier to separate. When some surfactant concentration range is reached, additional surfactant in a one-pass operation does not improve separation performance perceptably. The surfactant concentrations in Table 4 are in this range and are too high for efficient commercial separation of Zn^{2+}.

Industrial waste streams rarely contain only one pollutant. The type of wastewater streams containing heavy metals would often also contain monovalent salts. These will compete for binding sites on the micelle and reduce the efficiency of the separation of the multivalent ion.

If only one or a few of the ionic species in the wastewater require removal, it seems wasteful to remove all the counterions. Since MEUF in this application depends only on electrostatic attraction, little selectivity occurs on other than a valence basis. If a chelating or complexing agent were added to solution (e.g., a crown ether), it could tie up the counterion of interest, then be solubilized into the micelle and the heavy metal of interest selectively removed from solution by normal MEUF [2].

III. FOAM SEPARATIONS

The basic idea in foam separations is simple: Air bubbles rise to the top of a liquid column. If a surfactant is in a column of liquid into which air is sparged, the rising bubbles form a foam above the surface of the liquid. Any materials in the liquid that can be made to attach to the surface of the rising bubbles will then be floated to the top of the liquid, where they will collect in the foam. If there is a mixture of materials in the liquid, and some of them are more prone than others to attach to the rising bubbles, these materials are also separated from one another by the flotation process.

It is probable that ore flotation is the most commonly known surfactant-based separation process. Ore flotation, however, is only one application of foam separation. Interestingly, given the current activity aimed at developing industrial-scale bioseparations processes, foam separation was used at least as early as the 1940s for the separation of proteins, including catalase, diastase, tyrosinase, and urease [20-23]. Such processes have much in common with ore flotation.

Foam separation processes can conveniently be divided into particulate flotation and colligend flotation processes [24]. Particulate flotation consists of processes in which solid particles, such as bacterial spores or mineral particles, are separated from an aqueous stream. Colligend flotation processes consist of the flotation of ionic and molecular substances. Put another way, particulate processes deal with flotation of undissolved materials; colligend processes, with flotation of dissolved materials. What both types of processes have in common is that the target material, whether dissolved or undissolved, must accumulate sufficiently on the surface of air bubbles sparged into the solution/suspension for it to be separated into the foam that accumulates above the liquid surface. We consider these processes as surfactant-based processes because, for the separation to be effective, either the target material must itself be surface active, or a surfactant must be added to the liquid to enhance its concentration at the gas/liquid interface.

A. Particulate Flotation

1. Types of Particles Successfully Floated

Ore or mineral flotation is currently the most industrially important example of a particulate flotation process and may be thought of as a model for other particulate processes. Particulates that have been successfully removed from suspension by flotation include bacterial spores [25], algae [26], clays [27], and colloidal precipitates [28–30]. Each of these has in common with ore flotation the need for addition of a suitably charged surfactant and either adjustment of the pH or addition of an ion that promotes adsorption of the surfactant on the surface of the particulate.

2. Flotation Mechanisms

The flotation mechanism is not at all well understood in most of these application. In each case there must be a driving force which causes the particle floated from the solution to adhere to the surface of the bubbles of air rising through the liquid column. Reasoning, though, from the more extensively studied ore flotation process, there is a very strong correlation between flotation efficiency and the extent of equilibrium surfactant adsorption on the particles to be floated. This is clearly shown to be true by numerous experiments, even though the flotation process itself occurs far from equilibrium and involves imposition of a third interface, the solid/gas interface [31]. An understanding of the mechanisms that result in surfactant adsorption at aqueous/solid interfaces is then important in applying foam separation to a particulate system. Continuing to use ore flotation as a model, the flotation mechanism is generally attributed to a change in the contact angle of the particle surface from water wetting to non-wetting. This reversal is widely believed to occur because of the formation of surfactant aggregates called hemimicelles on the particle surface in local monolayers, adsorbed with the hydrophilic groups next the the surface, presenting a now hydrophobic layer to the solution (Fig. 7). The primary difficulty with this explanation is that it requires the surfactant molecules to aggregate so that their hydrophobic groups are exposed to the aqueous phase. There is no other system in which such a structure is believed to exist. Surfactant aggregates, such as micelles, are generally believed to form so that their hydrophilic groups are exposed to the aqueous solution (see Fig. 1). An alternative, admittedly speculative mechanism which avoids this difficulty is illustrated in Fig. 8. In this explanation, at low equilibrium surface coverages, bubbles of air contacting the surface do not adhere because of the contact angle. When, however, the bubble contacts a surface with substantial concentrations of adsorbed surfactant, the second layer of the surfactant aggregate, which is now called an admicelle (for *adsorbed micelle*), rapidly diffuses

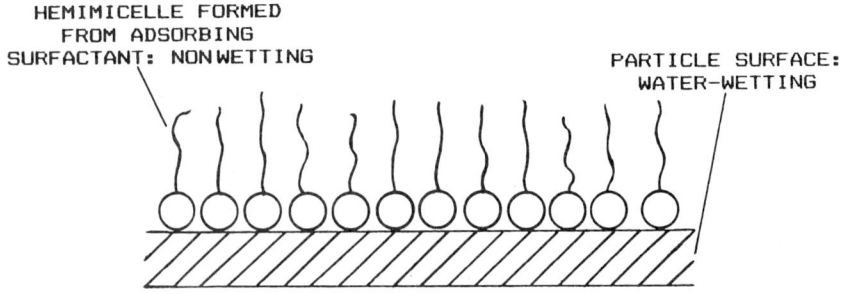

FIG. 7 Reversal of wettability of particle surface by hemimicelle formation.

along the solid-gas interface. The free energy of the system is at a minimum when the hydrocarbon tails of the surfactants are in the gas phase and the hydrophilic groups are next to the solid surface. The adsorbed surfactant thus acts to cause the particle to adhere to the bubble surface as the bubble rises through the solution. The particle is thus very accurately described as being floated out of the suspension by the bubble.

3. Process Variables

Whichever mechanism eventually proves to be more correct, for particulate flotation to occur, it is necessary for aggregates of surfactant to begin to form on the surface of the particle by adsorption from solution. A goal of any process design in foam particulate separation will of necessity be promotion of surfactant aggregate formation on the surface of the particles. Several parameters can be varied to cause an increase in admicelle formation. The first of these is surfactant concentration. As the surfactant monomer concentration increases, so does the surfactant adsorption. At the CMC, however, an adsorption pleteau is reached, and added surfactant only goes to form more micelles rather than more admicelles. At high surfactant concentrations, not only is micelle formation not helpful to particulate flotation, but the surface of a bubble may actually become coated with adsorbed surfactant before it contacts the surface of the particle. The head-head interaction between the surfactants on the particle surface and those on the bubble surface results in a mutual repulsion of the two surfaces, a decrease in occurence of adhesion, and thus a decrease in process efficiency. Commercial surfactants that contain a variety of molecular types may even exhibit a decrease in the equilibrium surfactant adsorption on the particle above the CMC. Even before the CMC is reached, surfactant adsorption at the air-

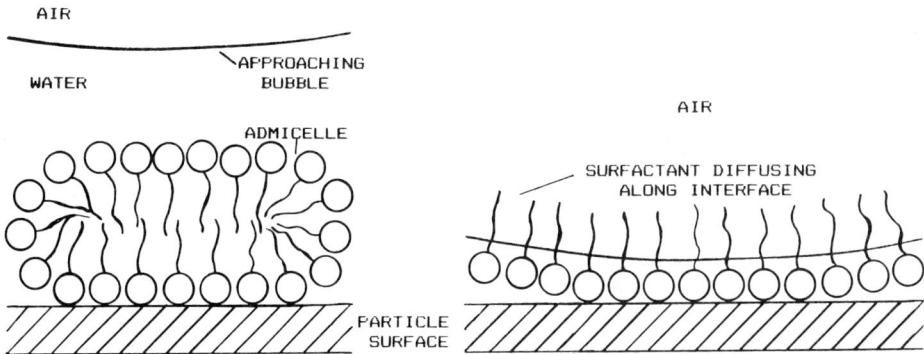

FIG. 8 Particle/bubble adhesion promoted by surfactant adsorption.

water interface may become large enough for the process to become less efficient. Although the surfactant concentration should generally be kept below the CMC, it may be better in some systems to keep it even lower, but never so low that no aggregates form on the surface. An efficient foam separation process for a particulate system will have to balance the effect of increasing surfactant adsorption at the solid/liquid interface to improve flotation efficiency with the effect of increasing surfactant adsorption at the air/water interface to decrease the efficiency.

Still using mineral flotation as a model particulate system, a typical surfactant adsorption isotherm is presented in Fig. 9. Following Fuerstenau and Somasundaran [31], most workers divide typical system isotherms into three or four regions, as illustrated by the schematic. Region I corresponds to low surface coverage by surfactants and an absence of indication of surfactant aggregate formation. It is sometimes termed the Henry's law region and has been modeled, for example, by the Stern-Grahame equation or an ion-exchange mechanism. It is generally accepted that the region I/region II transition corresponds to the onset of surfactant aggregate formation on the mineral oxide surface, as indicated by the sharp increase in the slope of the isotherm. There is very strong experimental and theoretical evidence to indicate that these aggregates form locally on the surface (i.e., by patchwise adsorption); they are referred to as hemimicelles or admicelles to emphasize the micelle-like aspects of their structure and behavior. The concentration at which the region I/region II transition occurs is widely referred to as the HMC (hemimicelle concentration) or the CAC (critical admicelle concentration). Because the efficiency of the flotation process is so dependent on the formation of aggregates on the surface of the solid, it is critical that

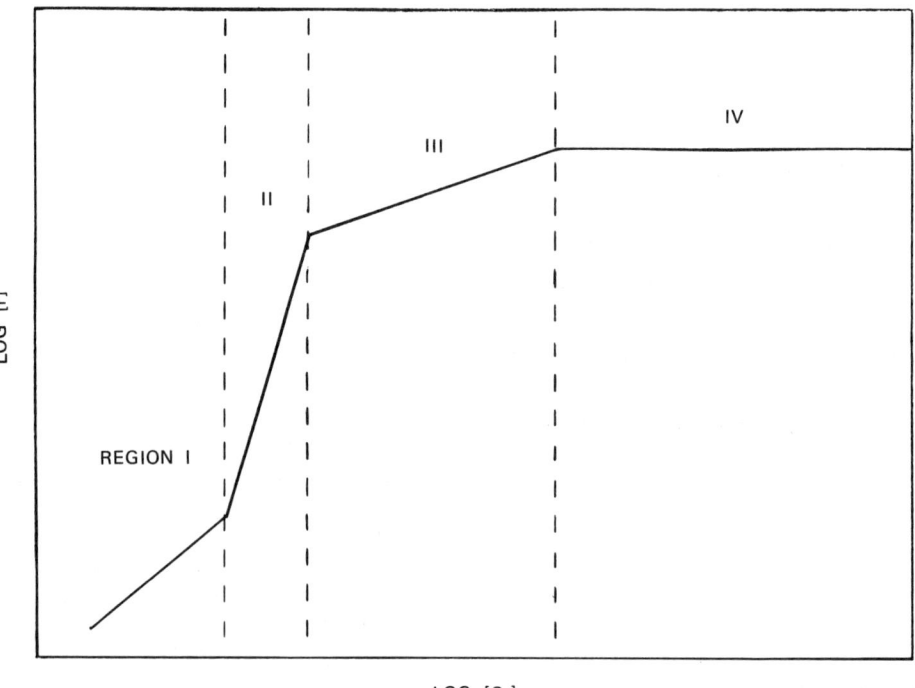

FIG. 9 Typical relationship between surfactant concentration and surfactant adsorption.

the system parameters be adjusted so that the region I/region II transition is reached. Region III begins where the slope of the adsorption isotherm starts to decrease, a phenomenon attributed by various workers to reversal of surface charge from superequivalent adsorption of surfactant ions or to interactions of surface heterogeneity and surface aggregate formation. The decrease in the rate of adsorption increase at the solid-liquid interface with surfactant concentration makes it more likely that increasing the surfactant concentration in region III may result in a decrease in flotation efficiency, as surfactant adsorption at the air-water interface may now increase more rapidly than that at the solid-liquid interface. The region III/region IV transition occurs either at the CMC or upon completion of bilayer coverage of the surface. Various surfactant/mineral oxide systems exhibit either an adsorption plateau or an adsorption maximum above the CMC. Workers who have observed adsorption plateaus above the CMC for monoisomeric systems generally attribute them to the micelles serving as chemical potential sinks for the system and attribute the maximum above the CMC to mixed micelle effects, which amounts to extending the pseudo-phase separation model of micelle formation to surfactant adsorption [32]. Other workers indicate that such an explanation is inadequate and that the electrostatic exclusion of micelles from the interfacial region is the most important mechanism at work in the production of the low concentration adsorption maximum [31]. In any event, the behavior of the adsorption isotherm indicates that in general there will be little or no advantage in increasing the equilibrium concentration of the surfactant beyond the CMC, and there will often be no net gain in efficiency once the region II/region III transition has been reached.

The reason that surfactant aggregates form on the particle surface before they form in the bulk seems clearly to be the presence of charge on the solid surface and the corresponding increased concentration of oppositely charged surfactant ions in the vicinity of this charge. Since charge is so important, the uppermost consideration in choosing a surfactant for a flotation process is that it be oppositely charged to the surface of the particle to be floated at the conditions of the process. Similarly, if the process stream contains a mixture of particles, those that have the greatest charge density at process conditions can be separated out most readily. When the sign of the surface charge of the particle is unknown, measurement of the particle's electrophoretic mobility is the surest, most straightforward way of determining it.

Once this overriding factor of surface charge at process conditions is used to decide on the use of an anionic or cationic surfactant, other factors may be considered. In general, factors that lower the CMC (i.e., favor micelle formation) will also increase the concentration of admicelles at any given concentration of surfactant. It is also most often found that factors that lower the CMC increase surfactant adsorption even more than they lower the CMC. Hence such factors as increasing the size of the hydrocarbon moiety, reducing the extent of branching in the tail, and increasing the counterion concentration will tend to increase the adsorption and improve flotation at a given molar concentration of surfactant. Since surfactants are generally sold on a per weight basis, however, the usefulness of increasing the molecular weight of the surfactant, from an economic point of view, needs to be evaluated in each situation.

The presence of the surface of the particle does introduce some new parameters which are less significant in affecting CMCs. The most important of these is pH. In most situations, lowering the pH will increase H^+ adsorption and make the surface more positive, thus increasing the adsorption of anionic surfactant and decreasing that of cationic surfactants. Raising the pH has just the opposite effect. The role of pH in mineral flotation is so critical that H^+ and OH^- are called potential determining ions for these systems [31-34]. Increasing the concentration of any ions that can adsorb between the admicelle and the particle surface will, generally, increase the adsorption of the surfactant. For anionic surfactants, it will often be the case that the counterions of choice for increasing the adsorption are Li^+ > Na^+ > K^+ > Cs^+, although this situation can be oversimplified [34]. This effect with cationic surfactants is less important. Multivalent ions, especially with small crystal radii, can even induce adsorption on what would otherwise be a like-charge surface [35]. In all these cases, one must be careful not to exceed the solubility product constant of the surfactant and the counterion. Mixed surfactant systems will generally have greater tolerance to electrolyte concentration.

B. Colligend Flotation

Colligend flotation, the foam separation of dissolved materials from an aqueous stream, is generally analogous to particulate flotation. In colligend flotation, however, the surfactant is not adsorbing on the surface of the target material. Rather, there is an electrical attraction of the charged target material to a surfactant layer adsorbed at the air-water interface of the bubble. For high surfactant concentrations at the air-water interface, the need to maintain electroneutrality in the bulk solution results in great enhancements of any oppositely charged ions next to the adsorbed surfactant layer. As the bubble rises through the solution, it drags the other materials along with it by means of an electrical attraction between the surfactant layer and other ionic species. Low ionic strengths are then more critical for colligend flotation than they are for particulate flotation, since high ionic strengths reduce long-range electrical attractions. This phenomenon is generally referred to as compression of the electrical double layer or Debye shielding. Similarly, surfactant structures that give high adsorption densities at the air-water interface, such as reduced branchings and multiple hydrophobic groups, are likely to give greater efficiencies of colligend flotation processes.

Ion flotation, a particular example of colligend flotation, is effective both for recovering ions, including metal ions, and for scrubbing a process stream for ionic pollutants. Strontium, for example, can be completely removed from solution at concentrations of < 6 × 10^{-5} M by ion flotation [36]. If complete recovery of the ion is critical rather than simply reducing its concentration to tolerable levels, a stoichiometric excess of surfactant may be required. This can greatly affect the economics of the process unless the surfactant can be recovered readily or the material is a high-value product.

Some proteins are sufficiently surface active to be foam separated without the addition of surfactant. Foam separation has been successfully used to concentrate albumin from dilute solutions containing organic and inorganic materials [37], to purify catalase from amalase [21], to concentrate diatase from lipase [22], and to separate urease from catalase [38].

Dyes and organic acids, including phenolics [39], humic acids [40], and fatty acids [41], have also been successfully separated from dilute aqueous solution by colligend flotation. In these cases the mechanism of flotation is probably adsolubilization. Micellar solubilization is a well-known phenomenon. Sparingly soluble compounds will partition into micelles, when they are present, enabling one to increase the solution concentration of a sparingly soluble compound greatly beyond its solubility limit in the absence of micelles. Adsolubilization is a similar phenomenon. In this system, the hydrocarbon region of the surfactant layer at the air-water interface is more attractive to sparingly soluble compounds than is the bulk aqueous solution. In colligend flotation, however, the adsolubilized region is attached to a bubble, which the bouyancy effect causes to rise through the liquid, floating the material out of solution.

C. Precipitate Flotation

Precipitate flotation falls somewhat between particulate and colligend flotation. It deals with the recovery of ions, but unlike flotation, the first step is to precipitate out of the solution the target ion. From here on the process is almost exactly like colloid flotation. The only significant difference is that the precipitating ion now acts as a potential determining ion to the resulting colloidal particle. If it is possible for the precipitating ion to be present in excess, without unduly increasing the ionic strength of the solution, it can be used to charge the surface of the precipitate particles and increase adsorption on them, allowing them to be floated at lower surfactant concentrations. This appears to be a potentially highly efficient technique for recovery of ions, including heavy metal ions, from very dilute aqueous solutions [24].

D. Continuous-Flow Foam Separation Process

Although foam separation is often used as a batch process for laboratory-scale separations, it is not limited to batch operations. The basic design of a continous-flow foam separation unit is rather straightforward [42], as shown in Fig. 10. If the process feed is sprayed into the column of foam, the unit is said to be operating in stripping mode. If liquid drainage from the foam is recycled back into the foam column, the unit is said to be in enriching mode. Graphical procedures for design calculations for such a unit have been developed. The design calculations include estimates of the number of transfer units (NTU) and the height of a transfer unit (HTU) [42-44].

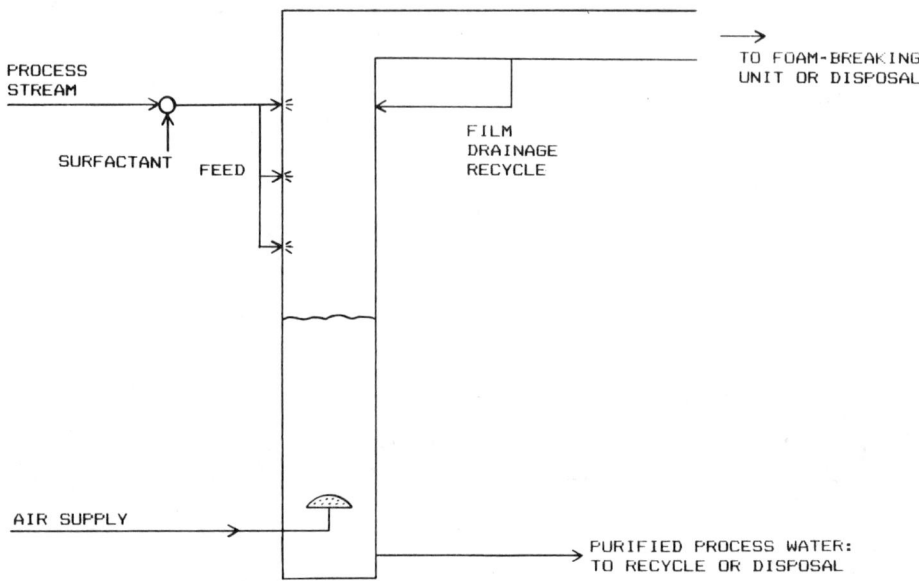

FIG. 10 Schematic of a typical foam separation unit.

It may well be the case that foam separation processes have not been more widely used because of the large volumes of liquid that units have to be designed to handle. Even though having some intrinsic advantages over, say, carbon adsorption, excessive capital cost and surfactant requirements are negative economic incentives. If, however, combined with MEUF or admicellar chromatography, which would greatly reduce the volume of liquid to be handled, these disincentives may largely dissappear. Note, for example, that if a MEUF unit achieved only a 10:1 product concentration, the volume of liquid in which the product was contained would be reduced by a factor of 10, so that a downstream foam separator would be reduced in size by a factor of 10. It will be interesting to see if such a combination of these processes achieves wide-scale industrial application in the future.

IV. SURFACTANT-ENHANCED CARBON REGENERATION

Adsorption beds containing activated carbon are widely used to remove organic pollutants from wastewater streams. When breakthrough occurs and the adsorber is no longer effective, the carbon must be regenerated. This involves removal of the adsorbed organic from the carbon surface. One traditional method is thermal regeneration [45], where the spent carbon is loaded into a furnace and the organic burned off. This is labor and energy intensive and some of the carbon is burned with the organic. An in situ method is hot-gas regeneration [45], where a hot gas such as steam [46] is injected into the bed and by a combination of heat-up and purging removes the organic. This method is effective only for low-molecular-weight, volatile, organic adsorbates.

Novel techniques for carbon regeneration have focused on in-situ methods because of the greater time and labor efficiency of such an operation. One such method is solvent regeneration [47], where a low-molecular-weight organic solvent is passed through the spent carbon in the bed. The original adsorbate desorbs into the solvent and is removed from the adsorber. The original organic solute and the solvent must then be separated in order to reuse the solvent. The carbon bed now has residual solvent and hot gas is passed through the bed to remove this volatile solvent. The solvent must be separated from the hot gas before it can be reused. Another class of in situ regeneration techniques involves biological regeneration methods [48]. In this case, microorganisms are injected into the bed and decompose the adsorbate into components with low adsorption. The process tends to be slow. Obviously, it can work only if the adsorbate is biodegradable.

A new in situ regeneration technique to be discussed here is surfactant-enhanced carbon regeneration (SECR), illustrated in Fig. 11. In SECR, a concentrated surfactant solution is passed through the adsorber containing the spent carbon. The adsorbate desorbs and is solubilized in the surfactant solution. This is illustrated in Fig. 12 at time = T. A concentrated solution containing organic adsorbate and surfactant results. The carbon (which is now free of the original solute) then has surfactant adsorbed on it. This is illustrated in Fig. 12 at time = T + ΔT. As shown in Fig. 11, water is then flushed through the carbon to remove the residual surfactant, after which the regeneration is complete. The wastewater stream is reintroduced to the adsorber, flowing countercurrent to the direction in which the regenerant and flush streams flowed through the bed.

Up to the point where the residual surfactant is flushed from the bed, this process is very similar to solvent regeneration. However, a principal advantage of using surfactants is that they are fairly inexpensive and innocuous to the environment, as evidenced by the large amount emitted to sewage systems each day from household usage (e.g., washing machines). Therefore, the residual surfactant is removed from the carbon by flushing with water and emitting directly to normal water treatment systems. This eliminates the complexities and expense of the closed purging system required for

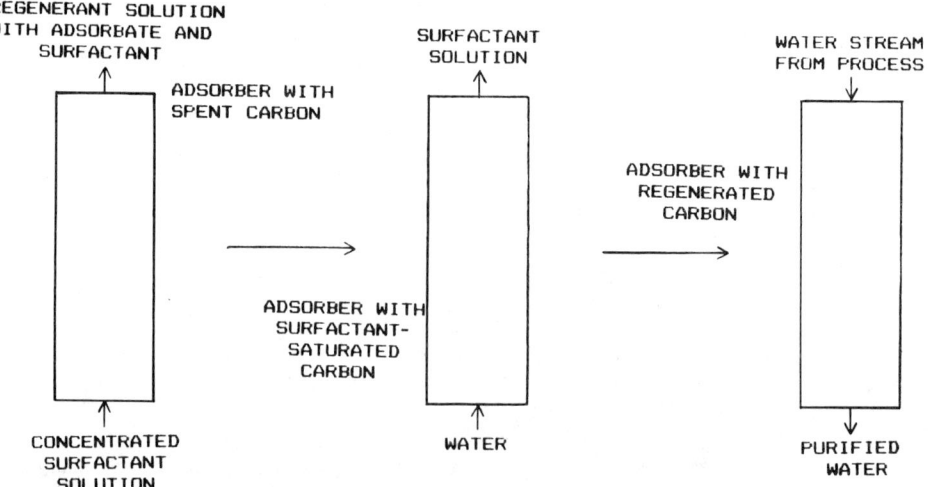

FIG. 11 Schematic of surfactant-enhanced carbon regeneration process.

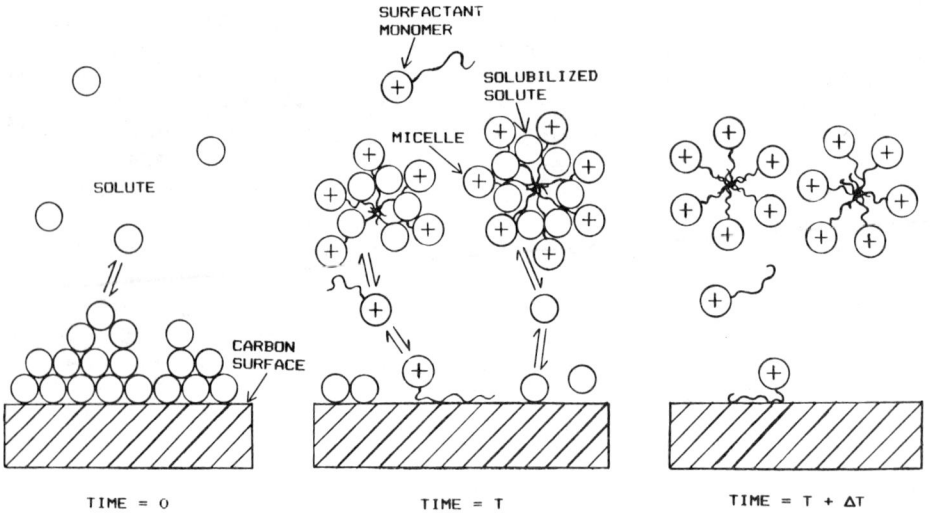

FIG. 12 Equilibria present at the various stages of a surfactant-enhanced carbon regeneration.

solvent regeneration where solvent such as acetone or methanol are used. The high energy costs of the hot-gas removal of residual solvent is also avoided.

A. Experimental Results

To test the feasibility of SECR, a model system was chosen for study [49]:

Adsorbate: 4-*tert*-butylphenol (TBP)
Adsorbate loading: 0.15 g of TBP per gram of carbon
Surfactant: cetylpyridinium chloride (CPC)
Granular activated carbon: Calgon Filtrasorb 300
Column: 2.5 cm in diameter, 100 cm in length
Temperature: 30°C

At the TBP concentrations present in the wastewater, 50,000 pore volumes (1 pore volume corresponds to the volume of liquid contained in a loaded adsorber bed) of wastewater could be treated before the carbon would become saturated. Of course, breakthrough would occur before complete saturation would be present, but for a well-designed adsorber, about 90% of saturation levels should be attained at breakthrough.

The fraction of TBP originally present on the carbon which is desorbed is shown in Fig. 13 as a function of the pore volumes of a 0.4 M solution of CPC run through the bed. Breakthrough occurred at about 1.4 pore volumes, indicating some adsorption of CPC on the carbon (otherwise, breakthrough would occur at 1 pore volume). After about 10 pore volumes of regenerant fluid had passed through the bed, about 50% of the TBP was desorbed. This corresponds to 33 h of regeneration at the flow rate used. After about 33 pore volumes of regenerant had passed through the bed, 70% of the TBP was removed. However, even after 160 pore volumes had been passed through, 20% of the TBP was still left on the carbon.

The column was then flushed with water to remove residual surfactant. After 10 pore volumes of water, about 30% of residual surfactant was removed. After 160 pore volumes, about 51% of the CPC was removed. TBP was observed to elute with the surfactant, confirming that not all of it was removed in the regeneration step.

B. Feasibility

In the model system studied above, when regenerant solution with a volume equal to 0.02% of the volume of the original stream treated had passed throught the carbon, 50% of the adsorbate had been removed. In this regeneration fluid, then, the original organic solute

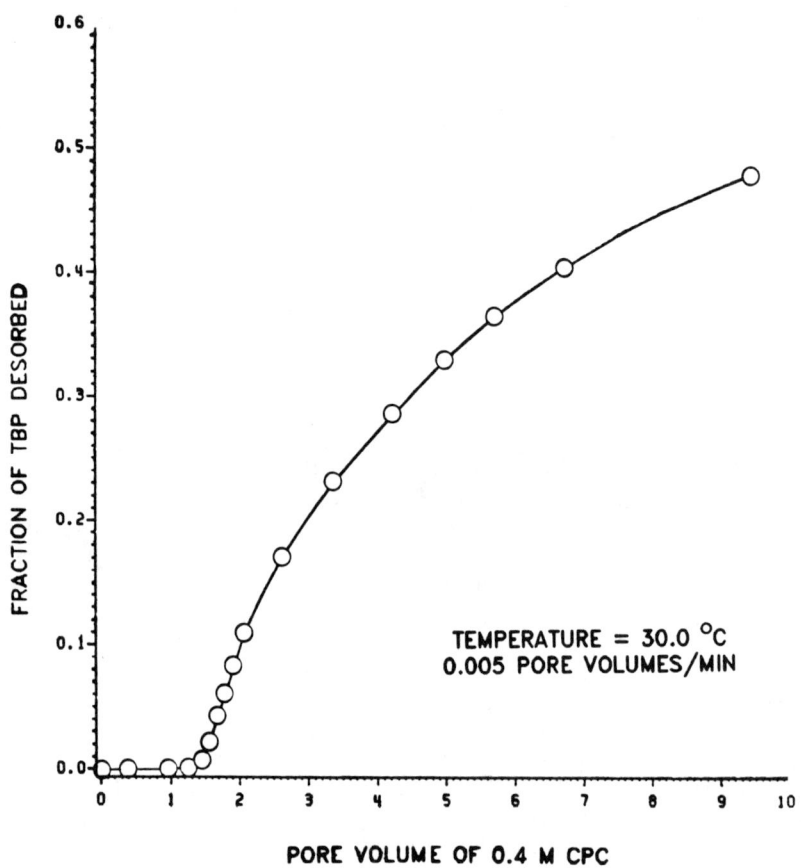

FIG. 13 Desorption of solute during a regeneration.

is concentrated by a factor of 2500 over its concentration in the stream originally being treated. When 70% of the adsorbate had been removed, the concentration ratio in regenerant to original feedstream is 1000. This indicates the tremendous potential of this technique for concentrating a dilute organic solute to high levels. However, removal of the last 20% of the organic from the carbon was difficult for the model system during regeneration.

It requires 0.32% of the volume of the original stream treated to flush 50% of the surfactant from the carbon after regeneration. As with the removal of the adsorbed organic solute, the majority of the adsorbed surfactant is easily removed, but almost complete removal requires large volumes of solution.

Conditions may be found which will result in more effective removal of adsorbed solute to near completion, and surfactants may be available which are easier to flush from the bed after regeneration. Also, by flowing the feedstream being treated countercurrent to the direction of flow of the regenerant solution and subsequent flush, the stream being treated would be exposed to the most purified carbon before being emitted from the bed. Even though the average removal of solute from the carbon may be, for example, 70%, that carbon which is farthest upstream during regeneration (farthest downstream during adsorption) may be substantially lower in residual adsorbate than the rest of the bed. This is illustrated in Fig. 11.

It is possible that SECR may be more effective at recovery of valuable dilute chemicals from aqueous solution (e.g., biochemicals produced in a fermentation broth). In this case, the ability to concentrate organic solutes to high levels by SECR is a great advantage. The residual solute left on the carbon after regeneration may be acceptable in that application.

Potential variations on application of SECR are many. For example, if a surfactant is used which selectively solubilizes a specific adsorbed molecule, relative to other adsorbed molecules, SECR would result in simultaneous concentration and separation of components from a solution. One can also envision combinations of SECR with MEUF or foam separations to yield effective new process schemes involving multiple surfactant-based separation processes.

In summary, SECR shows great promise as a new regeneration method for activated carbon. It may open up new applications for adsorption as a separation technique, such as recovery of valuable chemicals.

V. EXTRACTION INTO REVERSE MICELLES

Solvent extraction is a standard unit operation process in chemical engineering. It has been especially useful for scrubbing toxic materials in low concentrations from process streams. Especially valuable is that it is readily applied in a continuous rather than a batch mode. Unfortunately, such a process is not directly adaptable to bioseparations. Most proteins are relatively insoluble in apolar solvents, and those molecules that are dissolved in them frequently undergo irreversible denaturation.

It has been demonstrated, however, that a number of proteins can be solubilized in reverse micelles in apolar solvents, then recovered without irreversible denaturation or loss of activity. Proteins for which results have been reported include α-chymotrypsin, trypsin, pepsin, alcohol dehydrogenase, ribonuclease, lysozyme, peroxidase, α-amylase, and hydroxysteroid dehydrogenase [50–57]. Workers such as Goklen and Hatton [58] have recognized in this phenomenon a potentially important application of surfactant technology in the area of bioseparations.

A. Mechanism and Selectivity

When a dilute aqueous solution containing a protein is in contact with an organic solvent, under proper conditions, the protein will solubilize in reverse micelles in the organic phase (Fig. 14). It is known that parameters such as surfactant structure and concentration, nature of the organic phase, and the pH and electrolyte concentration of the aqueous phase are all important in determining the extent of solubilization. There have not been, however, systematic studies of the effects of these parameters, nor is it established what mechanisms are involved in causing these effects [58].

It is generally true, however, that transfer of the protein into the reverse micelle is favored by pH values at which the protein surface groups are either protonated or deprotonated, and by low electrolyte concentrations. The surfactant chosen should be oppositely charged to the protein at the pH of the aqueous phase. At the isoelectric point of the protein, and for increasing electrolyte concentration, the protein tends to transfer from the reverse micelle back into the aqueous phase. These observations suggest that minimization of the electrostatic contributions to the free energy of the system is the driving force for the solubilization step. When electrostatic interactions become less important, entropic considerations dominate, and the protein is desolubilized. Such considerations make it appear likely that system parameters can be adjusted to give preferential solubilization of one protein from a mixture of proteins, as well as preferential desolubilization of one protein over others.

B. Continuous-Flow Application to a Bioseparation

A preliminary process schematic for application of protein extraction by reverse micelles (PERM) is suggested in Fig. 15. Following primary separation of cell mass and cell debris (solid/liquid separation) from the output of a fermentation tank, the aqueous protein-rich stream is contacted with the organic phase in contactor 1. Surfactant is added and any needed adjustment to pH and electrolyte is made at this point. For proper choice of pH, electrolyte concentration, surfactant type, and organic solvent, the target protein will be transferred into reverse micelles in the organic phase. It is important that efficient contacting be accomplished but without generating shear rates that will damage the protein.

Two-phase flow from contactor 1 goes into separator 1. It is important that the organic solvent/surfactant/aqueous electrolyte system

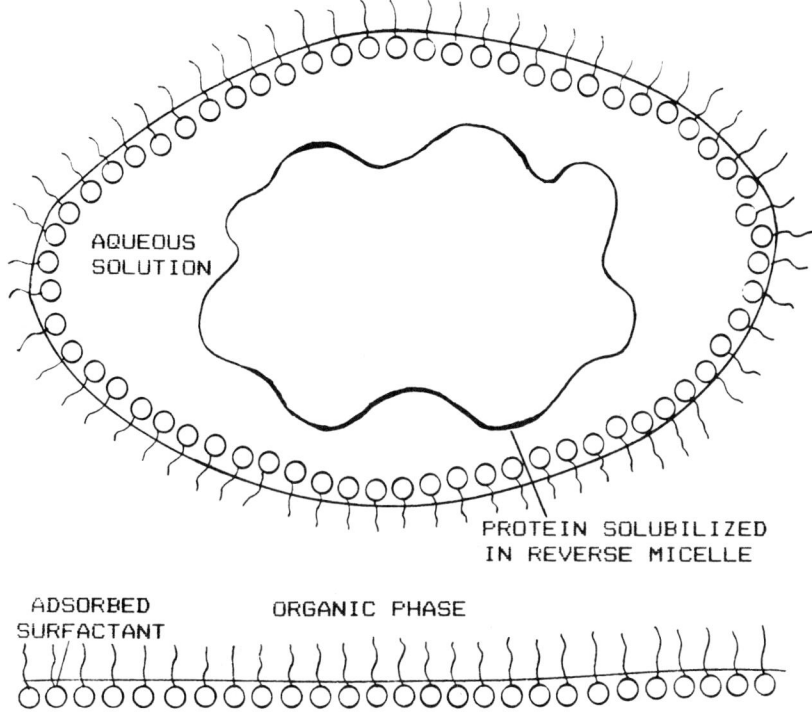
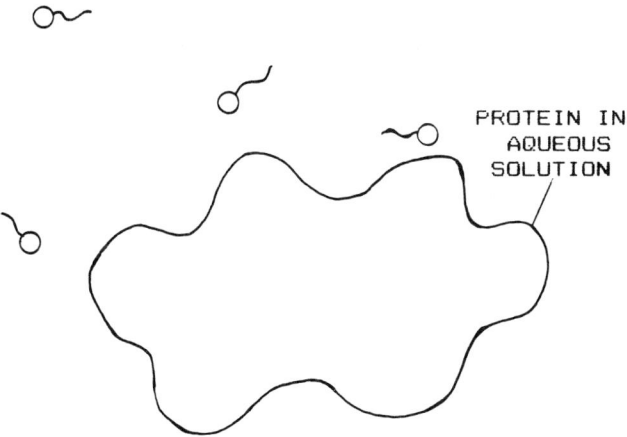

FIG. 14 Solubilization of protein in a reverse micelle.

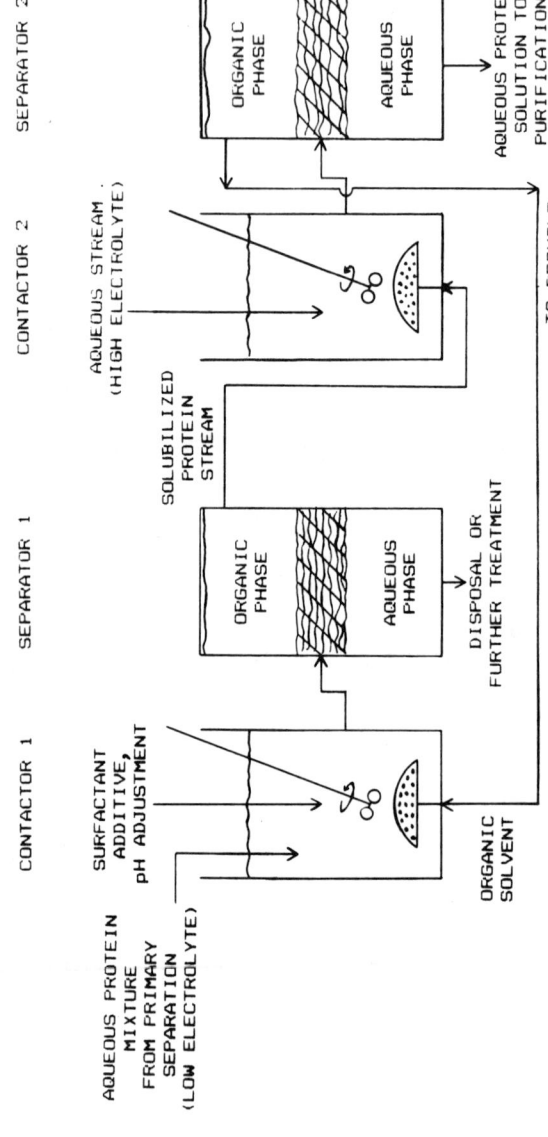

FIG. 15 Process schematic for protein extraction by reverse micelles (PERM).

not form an emulsion, so that the two phases can readily be separated. The aqueous bottoms, now depleted of the target protein, goes to disposal, or further treatment and eventual recycle. The organic stream from separator 1, containing the solubilized protein, goes into contactor 2. In the second contactor the pH and electrolyte of the aqueous feed are adjusted to induce transfer of the target protein back into the aqueous phase. Two-phase flow from contactor 2 goes into separator 2. The organic from separator 2 goes to recycle, perhaps with intermediate treatment for repurification. The aqueous stream from the second separator, now containing the target protein in aqueous solution goes to further purification. If the target protein exhibits sufficient surface activity, this stream may, for example, go to a foam separation unit. Although it is not evident from the diagram, the flow rate of the final aqueous stream need not be the same as that of the initial stream from primary separation. It would be a natural goal of the design to minimize the organic/aqueous ratio in contactor 1 while maximizing it in contactor 2. This allows the process not only to result in separation of the target protein from any others in the mixture, but also to result in concentration of the protein and a reduction in the volume of the stream that goes to the next purification step.

VI. ADMICELLAR CHROMATOGRAPHY

A. Adsolubilization

Many of the factors affecting surfactant adsorption on mineral oxides have been introduced in Section III. By proper manipulation of factors such as pH, surfactant structure, and counterion concentration, it is possible to obtain complete bilayer coverage of a mineral oxide surface with adsorbed surfactant [32,34]. Since some of these factors are reversible, it is also possible to manipulate them to obtain desorption of an already formed surfactant bilayer.

The phenomenon of adsolubilization has also been mentioned previously. A formal definition of adsolubilization suggested here is "the incorporation of compounds into surfactant surface aggregates, which compounds would not be in excess at the interface in the absence of the surfactant." This definition is modeled after definitions of solubilization widely used in the literature [5,59], and excludes simple coadsorption of, say, two surfactants. These definitions of solubilization are inappropriate for the phenomenon described here, although the physical mechanisms involved are nearly identical. Unfortunately, at the present time there are almost no data in the open literature concerning this phenomenon, beyond the fact that it has been observed to exist [60]. If the micelle-like behavior of admicelles extends into the area of adsolubilization, the phenomenon seems a promising basis for a new separations process which will be described below.

The promise of this phenomenon can be seen in the following example. Suppose that we wish to remove 1.6×10^{-4} M n-hexane from a process stream. The solubility of n-hexane in water at room temperature is 0.014 part per hundred on a weight basis, or 1.6×10^{-3} M. The maximum additive concentration of n-hexane in sodium oleate has been reported to be 0.46 molecule of n-hexane per molecule of surfactant [5]. These figures translate into a partition coefficient: $K = X/C = 228$ L/mol (K being the partition coefficient, X the mole fraction of n-hexane in the aggregate, and C the concentration of micelles). Using this micellar solubilization partition coefficient for partitioning into admicelles, we would predict adsolubilizations from our process stream into admicelles of only 0.05 mol of n-hexane adsolubilized per mole of adsorbed surfactant, since the alkane concentration in this hypothetical example is an order of magnitude below the saturation limit. If we were to use for the bed, however, an activated alumina such as Alcoa H-151, which has a surface area of 400 m^3/g and a bulk dry density of 50 lb/ft^3, and obtain bilayer coverage of oleate ions at 8 μmol of oleate per square meter of surface, we would obtain an adsolubilization of n-hexane per liter of bed of 0.13 molL. Since the bed void volume works out to be 0.8, the ratio of n-hexane in solution to n-hexane adsorbed is $0.8 \times 1.6 \times 10^{-3}/0.13 = 1/100$. In other words, we would achieve a 100-fold concentration of the n-hexane in the bed.

This degree of concentration of the adsolubilizate is not remarkable; rather, it is typical of any fixed-bed adsorption process, such as adsorption on activated carbon. This is the reason that fixed-bed adsorbers work so well at removing compounds from dilute aqueous solution. What is of interest in adsolubilization is that the surface bilayer, once saturated with adsolubilizate, is potentially removable by low-energy physical/chemical means, such as pH adjustment.

Another aspect of adsolubilization that makes it a potentially useful phenomenon is that micellar solubilizations can exhibit remarkable selectivity for one solubilizate over another. Table 5 contains maximum additive concentrations for estrone, estradiol, and three Δ^4-3-ketosteroids in sodium lauryl sulfate.

The main difference between the estrogens and the other steroid hormones in the table is that the estrogens carry a single aromatic ring, yet the solubilizations vary by an order of magnitude. The potential of using a surfactant's ability to solubilize certain compounds preferentially as a basis for a separation process has already been recognized [61]. The method proposed here has the additional advantage of being applicable in a fixed-bed mode and potentially, of producing much higher final solute concentration for downstream processing.

TABLE 5 Maximum Additive Concentrations in Sodium Lauryl Sulfate for Estrone, Estradiol, and Three Δ^4-3- Ketosteroids

Solubilizate	Moles of solubilizate per mole of surfactant
Estrone	0.0138
Estradiol	0.015
Progesterone	0.18
Testosterone	0.21
Deosycorticosterone	0.47

Source: Ref. 59.

B. Fixed-Bed Operation

A schematic of one possible scheme for implementing this technology for concentration, separation, or water purification is illustrated in Fig. 16. A process of this type has several potential advantages over conventional techniques. When compared to adsorption on activated carbon, the most obvious advantages are the reversibility of the bilayer formation and the resulting ease of bed regeneration. As the bilayer breaks up, the solute returns to the process stream at higher concentrations than the solutions from which it was adsorbed. There also appears to be a reasonable expectation of selective adsolubilization, so that separation and concentration steps can be combined. Even when multiple organic solutes are not separated from one another by this procedure, they are concentrated, making possible the application of the final purification steps on smaller volumes of solution.

When compared to membrane techniques, a major advantage is that adsolubilization techniques can be applied to low- and high-molecular-weight solutes with equal ease, as suggested by Table 6. The latter limitation of membrane techniques is also addressed by MEUF.

The implementation of adsolubilization as a separation technique suggested here is based on the use of inexpensive commercially available compounds. There is no reason at this stage of development of the process to suppose that expensive, specialized surfactants will be required.

If adsolubilization is implemented as a final treatment step for process waste steams before disposal, it is also advantageous that wide varieties of commercially available surfactants are biodegradable and acceptable for discharge into the environment. In all these applications the surfactant concentrations of interest are below the CMC of the surfactant, so that the concentrations are on the order of micromoles per liter.

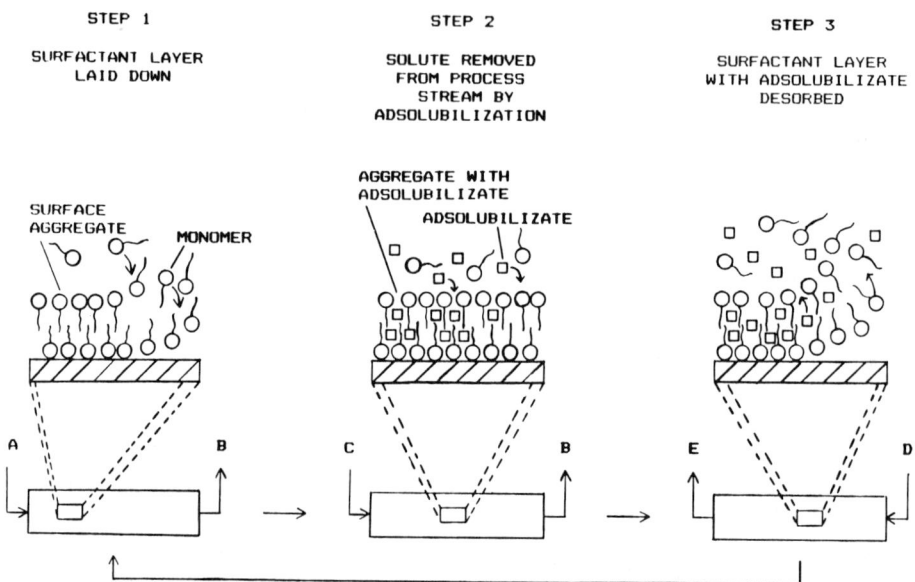

FIG. 16 Fixed-bed admicellar chromatography unit operation.

The proposed process can be viewed as having three steps. In the first step, a dilute surfactant solution (stream A) is equilibrated with a fixed bed of an adsorbate with an amphoteric surface, such as $\alpha\text{-}Al_2O_3$. Although the stream is dilute, it should be above the surfactant's CMC to minimize saturation time. This stream must be at a pH such that the surface will be oppositely charged relative to the surfactant. If the chosen surfactant needs to be nonionic in order to produce adequate adsolubilizations or to enhance selectivity, it will probably be necessary to mix it with an ionic surfactant and to choose an adsorbent on which the nonionic alone shows little adsorption. This is necessary so that the surfactant layer will readily desorb. During step 1, stream B will normally be free of surfactant and the target material, so that it may be recycled into the process or emitted.

After the surfactant layer has propagated sufficiently from the bed inlet, the process stream containing the adsolubilizate (stream C) is introduced into the column. Depending on the adsorption isotherm of the surfactant in the presence of the process stream solutes, it may be necessary to add low levels of surfactant to stream C. At most, the added surfactant will need to be at the surfactant's CMC, since the bilayer can be viewed as being in equilibrium with the surfactant monomer concentration. As the adsolubilizate wave propagates

TABLE 6 Maximum Additive Concentration of a Range of Solubilizates in 0.1 N Solutions of Dodecylamine Hydrocholride ($C_{12}HCl$), Sodium Oleate (NaC_{18}), and Potassium Laureate (KC_{12}), Respectively, at 25°C

Solubilizate	MAC (moles of solubilizate per mole of surfactant)		
	$C_{12}HCl$	NaC_{18}	KC_{12}
n-Hexane	0.75	0.46	0.18
n-Heptane	0.54	0.34	0.12
n-Octane	0.29	0.18	0.08
n-Decane	0.13	0.052	0.03
n-Dodecane	0.063	0.009	0.005
2,3-Dimethylpentane	0.62	0.35	0.11
3,3-Dimethylpentane	0.55	0.31	0.10
Diisobutylene	0.43	0.38	0.10
Methylcyclopentane	0.40	0.26	0.032
Cyclohexane	0.87	0.56	0.23
1,2,4-Trimethylcyclohexane	0.019	0.012	0.012
Benzene	0.65	0.76	0.29
Toluene	0.49	0.51	0.13
Ethylbenzene	0.38	0.40	0.20
p-Xylene	0.34	0.36	0.20
Methyl isobutyl ketone	1.78	1.82	1.20
Octylamine	0.13	0.07	0.07
n-Octanol	0.18	0.59	0.29
2-Ethylhexanol	0.36	0.47	0.064

Source: Ref. 59.

through the bed, adsolubilizate will be removed from the process stream. Initially, stream B will be devoid of surfactant and adsolubilizate. Sometime during step 2, surfactant will appear in stream B. Since the concentration will normally be at very low levels, it will often be possible to continue to recycle stream B without further treatment. In cases where the process is used solely for waste treatment, stream B may often be emitted to standard sewage systems because of the high environmental acceptability of most commercial surfactants. Alternatively, stream B may be used, after surfactant appears in it, to begin equilibrating a new bed with surfactant. The effluent from the new bed will again be free of surfactant.

Eventually, the bed will begin to approach saturation with the adsolubilizate. As the wave of adsolubilizate approaches the column exit, stream C must be switched to a new bed. Upon switching, step 3 is initiated. Although it may not be necessary to reverse the direction of flow, a new influent stream is now used. Stream D should be surfactant free and at a pH above (or below, depending on whether the surfactant is anionic or cationic) the pH where the surfactant begins to desorb. As the surfactant layer desorbs, the adsolubilizate is also desorbed from the surface, but now at a much higher concentration than before.

During step 3, effluent stream E will initially be at the same composition as stream C in step 2, and should be recycled. Eventually, a wave of high adsolubilizate concentration will appear in the effluent. Depending on the nature of the target molecule and the goal of the operation, this effluent may now be disposed of as is or further treated for removal or recovery of the adsolubilizate. This treatment may be by some conventional technique, perhaps one that would be used to treat stream C. Other times, the treatment method of choice may be another surfactant-based technique, such as MEUF or a foam separation. In some cases, stream E will have low-enough surfactant concentration to be used immediately. This obviously will depend on the application, but may be an element in surfactant selection. The key aspect to step 3 of the process is that if stream E must be treated further, it is now at a much higher adsolubilizate concentration. Any downstream equipment can now be smaller, and any high-energy processes can now be applied to much lower volumes of aqueous solution. This can have a major impact on process economics. It is easy to imagine the effect on overall project feasibility if use of an adsolubilization unit allows all downstream equipment to be downsized by a factor of 10 or 100. This might be critical in making a foam separation unit, for example, of manageable size.

When the concentration in effluent stream E falls below the value where it is economical to continue eluting the adsolubilizate, stream switching occurs again and the bed begins to be reequilibrated with surfactant.

VII. SUMMARY

In this chapter, five surfactant-based separation processes suitable for the treatment of aqueous process streams have been discussed. Of these five, only foam separations has been available for application until recently. Protein extraction by reverse micelles has only recently been proposed as a unit operation. Micellar-enhanced ultrafiltration has been discussed in only two previous publications. The concepts of admicellar chromatography and surfactant-enhanced carbon regeneration are discussed for the first time in the permanent literature in this chapter. Therefore, a spurt in development of surfactant-based separation is occurring. Because of the general advantages of being energy efficient and exposing fragile chemicals to mild conditions, with applications in biotechnology and pollution control, among other fields, surfactant-based separations to treat aqueous process streams are predicted to increase in importance and number in the future. In fact, these techniques may permit economical production of some new biochemically produced products, since downstream separation is a major limitation in their ultimate manufacture. These methods may also make economical the recovery of waste products in water effluent by permitting their efficient concentration.

One area that has not been developed at all is the combination of the various surfactant-based techniques presented here into a sequence of separation and purification steps. Use of an MEUF unit upstream of a foam separation unit could result in the foam separation unit being drastically downscaled, with concomitant decrease in capital, maintenance, and operating costs. Similarly, use of an admicellar chromatography unit upstream from a PERM unit could result in enhancement of the concentration downstream of the PERM unit, improving the purification of the target protein, and resulting in more efficient use of the surfactant involved. This combination of surfactant-based unit operations could then be followed by a foam separation unit.

Another possibility that has not been investigated is the incorporation of the surfactant into the final product. Because so many proteins are either insoluble in water or so sparingly soluble that they cannot be efficiently delivered in aqueous solution, solubilization is in common use in the pharmaceutical industry. There are no doubt many systems in which, after final purification of the target compound, surfactant is added as part of the final product which will actually be sold. By introducing this surfactant earlier in the process, in a surfactant-based separation process, it may well be possible to cut production costs drastically.

The hope of the authors is that this chapter will serve as a catalyst toward the implementation and realization of the promises of surfactant-based treatment of aqueous process streams.

ACKNOWLEDGMENT

Financial support for original results reported in this chapter was provided by the National Science Foundation (Grant CPE-8318864), Office of Basic Energy Sciences of the Department of Energy Grant DE-AS05-84ER13175, the Oklahoma Mining and Minerals Resources Research Institute, and the OU Energy Resources Institute. Colleagues and students who provided assistance were Lowry Blakeburn, Sherril Christian, Robert Dunn Jr., Rex Ellington, Lane Gibbs, Nancy Gullickson, Lisa Lyhane, and George Smith.

REFERENCES

1. M. C. Porter, in *Handbook of Separation Techniques for Chemical Engineers* (P. A. Schweitzer, ed.), McGraw-Hill, New York, 1979, Sec. 2.1.
2. P. S. Leung, in *Ultrafiltration Membranes and Applications* (A. R. Cooper, ed.), Plenum Press, New York, 1979, p. 415.
3. R. O. Dunn, Jr., J. F. Scamehorn, and S. D. Christian, Sep. Sci. Technol., 20:257 (1985).
4. P. Mukerjee, Adv. Colloid Interface Sci., 1:241 (1967).
5. P. H. Elworthy, A. T. Florence, and C. B. MacFarlane, *Solubilization by Surface Active Agents*, Chapman & Hall, London, 1968, Chap. 2.
6. M. J. Rosen, *Surfactants and Interfacial Phenomena*, Wiley, New York, 1978, Chap. 4.
7. P. Mukerjee, in *Solution Chemistry of Surfactants*, Vol. 1 (K. L. Mittal, ed.), Plenum Press, New York, 1979, p. 153.
8. R. S. Boethling, *Environmental Fate and Toxicity in Wastewater Treatment of Quaternary Ammonium Surfactants*, Environmental Protection Agency, Washington, D.C., 1984.
9. R. O. Dunn, Jr., J. F. Scamehorn, and S. D. Christian, Sep. Sci. Technol., 22:763 (1987).
10. E. Hutchinson, Z. Phys. Chem. N.F., 21:28 (1959).
11. R. Lianos, M. L. Viriat, and R. Zana, J. Phys. Chem., 88:1098 (1984).
12. M. Almgren, F. Grieser, and J. K. Thomas, J. Am. Chem. Soc., 101:279 (1979).

13. S. D. Christian, G. A. Smith, E. E. Tucker, and J. F. Scamehorn, Langmuir, 1:564 (1985).
14. E. E. Tucker and S. D. Christian, J. Colloid Interface Sci., 104:562 (1985).
15. A. E. Aboutaleb, A. M. Sakr, H. M. El-Sabbagh, and S. I. Abdelrahman, Pharm. Ind., 42:940 (1980).
16. P. Mukerjee and K. J. Mysels, *Critical Micelle Concentrations of Aqueous Surfactant Systems*, National Bureau of Standards, Washington, D.C., 1971.
17. G. J. Hirasaki and J. R. Lawson, SPE Reservoir Eng., 1:119 (1986).
18. J. A. Beunen and E. Ruckenstein, J. Colloid Interface Sci., 96:469 (1983).
19. J. F. Rathman and J. R. Scamehorn, J. Phys. Chem., 88:5807 (1984).
20. S. G. Davis, C. R. Fellers, and W. B. Esseler, Food Res., 14:417 (1949).
21. R. Bader and F. Schutz, Trans. Faraday Soc., 42:571 (1946).
22. W. Ostwald and W. Mischke, Kolloid Z., 90:205 (1940).
23. Y. Okamoto and E. J. Chou, in *Handbook of Separation Techniques for Chemical Engineers* (P. A. Schweitzer, ed.), McGraw-Hill, New York, 1979, Chap. 2.5.
24. E. J. Mahne, Chem. Can., 23:32 (1971).
25. W. A. Boyles and R. E. Lincoln, Appl Microbiol., 6:327 (1958).
26. A. J. Rubin, E. A. Cassel, O. Henderson, J. D. Johnson, and J. C. Lamb, Biotechnol. Bioeng., 8:135 (1966).
27. Y. Horikawa, Clay Sci., 4:281 (1975).
28. R. B. Grieves and D. Bhattacharyya, Nature, 207:476, (1965).
29. I. Sheiham and T. A. Pinfold, J. Appl. Chem., 18:217 (1968).
30. E. J. Makne and T. A. Pinfold, J. Appl. Chem., 19:57 (1969).
31. P. Somasundaran and D. W. Fuerstenau, J. Phys. Chem., 70:90 (1966).
32. J. F. Scamehorn, R. S. Schechter, and W. H. Wade, J. Colloid Interface Sci., 85:463 (1982).
33. J. A. Yopps and D. W. Fuerstenau, J. Colloid Sci., 19:61 (1964).
34. D. Bitting and J. H. Harwell, Langmuir, 3:500 (1987).

35. A. M. Gaudin and D. W. Fuerstenau, Trans. AIME, 202:958 (1955).
36. B. M. Davis and F. Sebba, J. Appl. Chem., 16:297, (1966).
37. S. I. Ahmad, Sep. Sci., 10:673 (1975).
38. M. London, M. Cohen, and P. B. Hudson, Biochim. Biophys. Acta, 13:111 (1954).
39. P. Moore and C. R. Phillips, Sep. Sci., 9:325 (1974).
40. L. D. Skrylev, A. N. Purich, and Y. S. Gorodentsev, Zh. Prikl. Khim. Leningrad, 48:685 (1975).
41. R. H. L. Howe, Prog. Hazardous Chem. Handl. Disposal Proc. Symp. 3rd, 1972, p. 130.
42. R. Lemlich, in *Adsorptive Bubble Separation Techniques* (R. Lemlich, ed.), Academic Press, New York, 1972.
43. P. A. Haas and H. F. Johnson, AICHE J., 11:319 (1965).
44. S. Fanlo and R. Lemlich, AICHE–I. Chem. E. Symp. Ser., (9):75 (1965).
45. R. A. Hutchins, in *Handbook of Separation Techniques for Chemical Engineers* (P. A. Schweitzer, ed.), McGraw-Hill, New York, 1979, Chap. 1.13.
46. J. F. Scamehorn, Ind. Eng. Chem. Process Des. Dev., 18:210 (1979).
47. T. Sutikno and K. J. Himmelstein, Ind. Eng. Chem. Fundam., 22:420 (1983).
48. W. A. Chudyk and V. L. Snoeyink, Environ. Sci. Technol., 18:1 (1984).
49. D. L. Blakeburn, M. S. thesis, University of Oklahoma, 1985.
50. F. J. Bonner, R. Wolf, and P. L. Luisi, J. Solid Phase Biochem., 5:225 (1980).
51. R. Hilhorst, C. Laane, and C. Veeger, FEBS Lett., 159:225 (1983).
52. P. L. Luisi, F. Henninger, M. Joppich, A. Dossena, and G. Casnati, Biochem. Biophys. Res. Commun., 74:1384 (1977).
53. P. L. Luisi, F. L. Bonner, A. Pellegrini, P. Wiget, and R. Wolf, Helv. Chim. Acta, 62:740 (1979).
54. P. L. Luisi, P. Meier, V. E. Inre, and A. Pande, in *Reverse Micelles* (P. L. Luisi and B. E. Straub, eds.), Plenum Press, London, 1984.

55. K. Martinek, A. V. Levashov, N. L. Klyachko, and I. V. Berezin, Dokl. Akad. Nauk SSSR, *236*:920 (1977).

56. P. Meier and P. L. Luisi, J. Solid-Phase Biochem., *5*:269 (1980).

57. R. Wolf and P. L. Luisi, Biochem. Biophys. Res. Commun., *89*:209 (1979).

58. K. E. Goklen and T. A. Hatton, Biotechnol. Prog., *1*:69 (1985).

59. J. W. McBain and E. Hutchinson, *Solubilization and Related Phenomena*, Academic Press, New York, 1955.

60. C. C. Nunn, Ph.D. dissertation, The University of Texas at Austin, 1981.

61. R. Nagarajan and E. Ruckenstein, Sep. Sci. Technol., *16*:1429 (1981).

4
Liquid Surfactant Membranes for Metal Extractions

Z. M. GU* and DARSH T. WASAN *Illinois Institute of Technology, Chicago, Illinois*

NORMAN N. LI *Allied-Signal, Inc., Des Plaines, Illinois*

I.	Introduction	128
II.	Principle of Metal Extraction by Liquid Membranes	129
	A. General Description of Liquid Surfactant Membranes	129
	B. Facilitated Transport	130
	C. Demulsification	132
III.	Metal Extraction Processes	133
	A. Published Systems of Metal Extraction by Liquid Membranes	134
	B. Typical Results of Laboratory and Pilot Tests	134
	C. Process Economics of Liquid Membrane Metal Extraction	141
IV.	Extractants	144
	A. Acidic Extractants	144
	B. Basic Extractants	144
	C. Neutral Extractants	150
	D. Synergistic Effect Between Extractant and Surfactants	150
V.	Enhancement of Facilitated Transport	151
VI.	Mathematical Modeling	154
VII.	Summary	160
	References	162

Present affiliation: Institute of Atomic Energy, Academia Sinica, Beijing, China.

I. INTRODUCTION

Microfiltration, ultrafiltration, reverse osmosis, dialysis and electrodialysis have been accepted as the most important polymeric membrane separation processes [1–6]. However, these polymeric membranes have generally suffered from low transmembrane flux and insufficient selectivity. These shortcomings limit the industrial applications of the polymeric membrane processes.

Attention has therefore focused on developing more efficient and more selective membrane separation processes. Liquid membranes have been recognized as a promising technology to overcome the inherent drawbacks of the polymeric membranes. The fluxes of liquid membranes are very high. This is because the liquid films offer much higher diffusivities than do the solid polymerical films and the liquid membranes can be made much thinner than the polymeric membranes. Furthermore, the liquid membranes can be "tailor made" by incorporating appropriate carriers in the membrane phase so that the separation properties of the membrane phase can be adjusted according to the specific separation needs [7].

It appears that liquid membrane technology is very promising for a variety of applications, including the separation of hydrocarbons [8–16], hydrometallurgy [17–37]. watewater treatment [38–51], biomedicine [52–58], and biochemical engineering, such as the encapsulation of enzymes and bacteria [59–67]. It is important to note that to date, liquid membrane technology has two commercial applications: one in metals extraction [68,69] and one in oil production as a wellcontrol fluid [70–73].

Liquid membranes can be divided into two types according to the configurations: polymer-supported liquid membranes and liquid surfactant membranes. Polymer-supported liquid membranes are made up of two forms. The first consists of a porous polymeric film with the liquid membrane materials held strongly in the pores. The membrane geometry can be either of plate or of hollow fiber type. Another form is a liquid film situated between two highly permeable thin polymeric films. The solute from the feed solution diffuses through one solid film, across the liquid membrane, and through another solid film to reach the receiving solution. Ward and Robb [74] demonstrated the marked increases in flux and selectivity for CO_2-O_2 separation by means of the facilitated transport in supported liquid membranes. Babcock et al. [76–78] and Baker et al. [75] discussed in detail the "coupled transport" of copper and uranium in supported liquid membranes.

One problem with supported liquid membranes is the insufficient area for mass transfer (unless hollow fibers are used). The liquid surfactant membrane concept, first proposed by Li [8], overcomes this problem by generating the necessary large surface area by way

Liquid Surfactant Membranes for Metal Extractions 129

of creating numerous emulsion-size spheres. The other problem that is unique to the supported liquid membrane is the slow loss from the membrane pores of the liquid containing expensive extractants, due to uneven pressures across the membrane.

The liquid surfactant membranes are a three-phase system: a water-oil-water (w-o-w) system or an oil-water-oil (o-w-o) system. For the w-o-w system, the oil phase that separates the two aqueous phases is the liquid membrane. For the o-w-o system, the liquid membrane is the water phase that separates the two oil phases.

This chapter will focus on liquid surfactant membranes. It should be pointed out, however, that the transport principles are the same for both the liquid surfactant membrane and the supported liquid membrane.

II. PRINCIPLE OF METAL EXTRACTION BY LIQUID MEMBRANES

A. General Description of Liquid Surfactant Membrane

In general, liquid surfactant membranes are made by forming an emulsion of two immiscible phases and then dispersing the emulsion into a third phase (the continuous phase). Usually, the encapsulated phase and the continuous phase are miscible. If it is to remain stable, the membrane phase must not be miscible with either phase. Therefore, the emulsion is of o/w type if the continuous phase is oil and of w/o if the continuous phase is water. To maintain the integrity of the emulsion during the separation process, the membrane phase usually contains certain surfactants, additives as stabilizing agents, and a base material which is a solvent for all the ingredients.

When the emulsion is dispersed by agitation in a continuous phase (the third phase), many small globules of emulsion are formed. Their size depends on the nature and concentration of the surfactants in the emulsion, the emulsion viscosity, and the mode and intensity of mixing. In general, the globule size is controlled in the range 0.1 to 1 mm in diameter. Thus a large number of globules of emulsion can easily be formed to produce a large membrane surface area for rapid mass transfer from either the continuous phase to the encapsulated phase, or vice versa. It should be noted that many such smaller droplets, typically 1 μm in diameter, are encapsulated within each globule.

Separation of mixtures can readily be achieved by selective diffusion of one component through the liquid membrane phase into the liquid of lower concentration. Once separation is effected, the three phases can be separated by first settling the emulsion and continuous phase and then breaking the used emulsion. For metal ion extraction, the emulsion is normally oil external since feedstreams are usually aqueous. Metal ions can be extracted through the oil membrane into the internal aqueous droplets of the emulsion. Individual chemical species can be trapped and concentrated in the internal phase for later disposal or recovery.

B. Facilitated Transport

The effectiveness of the liquid membrane process can be enhanced by utilizing a facilitated transport mechanism to maximize both the flux through the membrane phase and the capacity of the receiving phase for the diffusing species. There are two types of facilitated mechanisms that can be used to maximize the flux and capacity:

Type 1: Minimization of the concentration of the diffusing species in the receiving phase [43]. This is normally done by reacting the diffusing species with some other constituent in the receiving phase to form a product incapable of diffusing back through the membrane. This type of facilitated transport can be illustrated by the separation of phenol from wastewater as shown in Fig. 1. In this case, NaOH is encapsulated inside the emulsion droplets as the reagent which reacts with phenol to form sodium phenolate. Phenol is soluble in the oil-type liquid membrane and therefore can diffuse through it, whereas phenolate, in its ionic form, is insoluble in the membrane and therefore cannot diffuse back out into the treated water phase.

Type 2: Carrying the diffusing species across the membrane by incorporating "carrier" compounds in the membrane [20]. This type of carrier-mediated transport can be illustrated by the separation of metal ions from wastewater or mine leaching solutions as shown in Fig. 2. In these cases, a liquid ion-exchange reagent is incorporated in the membrane phase. Extraction of metal ions can then occur at the membrane/continuous phase interface, while stripping can occur at the membrane/internal phase interface under proper process conditions.

As an example of this type of facilitated transport, the extraction of cobalt is achieved according to the following chemical reactions [35]:

Extraction:

$$2HR + Co^{2+} \rightleftharpoons CoR_2 + 2H^+ \tag{1}$$

org.　　aq.　　org.　　aq.

Stripping:

$$2H^+ + CoR_2 \rightleftharpoons Co^{2+} + 2HR \tag{2}$$

aq.　　org.　　aq.　　org.

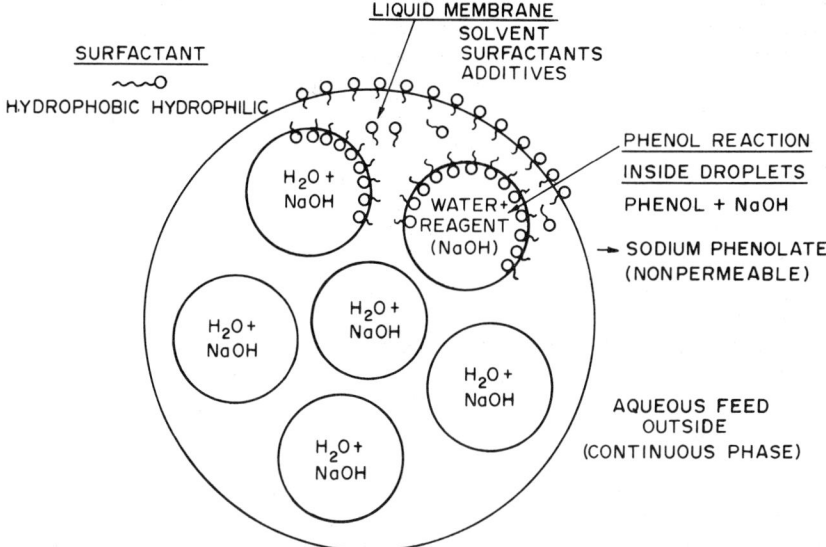

FIG. 1 Schematic diagram of liquid membrane system for phenol removal. (From Ref. 42.)

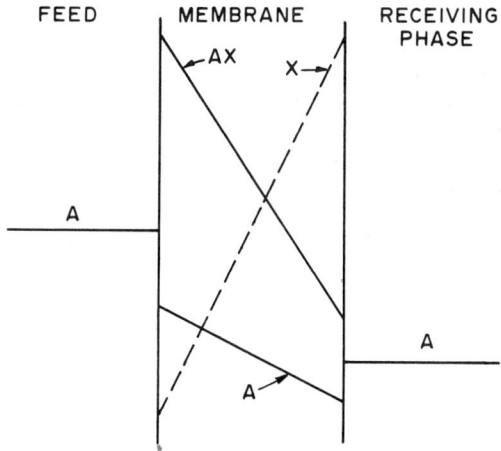

FIG. 2 The concept of facilitated transport. (From Ref. 20.)

where HR represents the protonated form of a liquid ion-exchange reagent which is used as the carrier or "transport facilitator." In this system, extraction [Eq. (1)] occurs at the membrane/external aqueous phase interface, while stripping [Eq. (2)] occurs at the membrane-internal aqueous phase interface. The cobalt is effectively concentrated in the encapsulated phase of the emulsion by the continuous permeation of hydrogen ions from the encapsulated phase to the external phase.

One of the important advantages of the liquid membrane process for metal extraction lies in the concurrent extraction and stripping in a single stage rather than in two separate stages as required by solvent extraction. In addition, by concurrently extracting and stripping, the liquid membrane process drives the extraction equilibrium, as shown in Eqs. (1) and (2), to the right by removal of the complexed ions as they are formed. This removes the equilibrium limitation inherent in solvent extraction. Li and Shrier [42] and Kitagawa et al. [45] have shown that they were able to remove chromium ion from a concentration of 100 ppm in a feed phase to a concentration of 182,000 ppm in an internal phase. Another benefit resulting from the nonequilibrium feature of a liquid membrane system is the significant reduction of the reagent inventory in the separation process. For an example, in copper extraction, the amount of the reagent required is only half that for solvent extraction [23].

C. Demulsification

Another important aspect of the liquid membrane process is the coalescence of the spent emulsion and the recovery of the membrane materials. There are two principal approaches for the demulsification of the spent emulsion: chemical and physical treatments. Chemical treatment, which involves the addition of demulsifier, is usually not suitable for the liquid membrane process, owing to contamination of the membrane phase by the demulsifier.

Physical methods include heating, centrifuging, solvent dissolution, high shear, and the use of high-volume electrostatic fields. Liquid membrane emulsions (both o/w and w/o types) can be effectively broken by the use of specially formulated solvent mixture [122] and high shear [123]. The specially formulated solvent mixture breaks the emulsion without damaging the surfactants. The solvents can easily be recovered later by evaporation, as they are low-boiling compounds. The method of demulsification by high shear includes the use of centrifugation as a first step. The high shear step can be achieved simply by pumping the half-broken emulsion after centrifugation through an emulsification machine.

The demulsification with electrostatic fields appears to be most efficient and convenient for w/o emulsions, which are used for metal extraction in a liquid membrane process [120–122]. Electrostatic coalescence is a technique widely used to separate dispersed aqueous droplets from nonconducting oils. The petroleum industry has been using it to separate brine emulsified in crude oil; the chemical industry uses it to resolve water-in-oil (w/o) emulsions generated during extraction (such as removal of naphthenic acids from gas-oils). Since this type of demulsification is strictly a physical coalescence process, the technique is most suitable for breaking liquid membranes to recover surfactants for reuse.

The use of electrostatic coalescence to separate liquid membrane emulsions is different from the conventional application of such coalescence. First, the liquid membrane emulsions contain much more aqueous phase (as high as 50% versus roughly 5%), and the recovered oil may contain more water (about 1%) than pipeline-quality dehydrated crude (about 0.3%). Second, since these emulsions are made intentionally, they contain high concentrations of potent emulsifying surfactants. Finally, the properties of the oil phase must be preserved for reemulsification.

Since the use of conventional electrostatic coalescers with bare metal electrodes for the separation of rich water-in-oil emulsion often causes sparking and the formation of a spongelike emulsion, Hsu et al. [120–122] invented insulated electrodes to be used in coalescers that allow the application of high electric fields at the oil-water interfaces without causing the above-mentioned problems. This results in a clean separation of oil from water.

An interesting aspect of the reuse of the spent emulsion for some cases is the regeneration of liquid membranes without demulsification. Dines [119] has developed a method to convert the trapped impermeable species in the internal phase into a species that is permeable, thereby removing this concentrated species from the internal phase of the emulsion. For the uranium extraction by liquid membranes, the permeable species to the internal phase is U^{6+}, which is trapped in this phase by reducing it to U^{4+}, an impermeable species, in the presence of hydroquinone and light. At the same time, hydroquinone is converted to quinone. After separation of the emulsion containing uranium in the +4 nonpermeable state from the external aqueous phase, the emulsion is subjected to agitation in the presence of a fresh aqueous solution and in the absence of light whereby the quinone oxidizes the U^{4+} to U^{6+}, which then permeates through the membrane into the fresh aqueous solution.

III. METAL EXTRACTION PROCESSES

The increased consumption of metals has resulted in a need to improve the existing methods and to explore new technology for the recovery, separation, concentration, and purification of metals. The outstanding features exhibited by the liquid surfactant membranes make this technology quite competitive with the conventional treatment methods for the recovery of valuable metals and the removal of toxic materials from dilute effluent streams.

A. Published Systems of Metal Extraction by Liquid Membranes

Extensive work has already been reported for liquid membrane metal extractions. A series of metal species including alkali, alkaline earth, transition, and heavy metals has been studied. Among these metals, copper is most widely used to study the liquid membrane formulation (especially extractants and surfactants), process kinetics, mass transfer model, and engineering evaluation [22,28,80-82].

Table 1 lists some of the liquid membrane processes published in the literature for the metal extractions. It should be pointed out that quite a few workers have studied the metal extraction by polymer-supported membranes, such as the extraction of copper [83,84], uranium [85], plutonium [86], americium [87], gold [88], and so on. These processes can in principle be adopted by the liquid surfactant membranes.

B. Typical Results of Laboratory and Pilot Tests

Laboratory work includes the screening of membrane materials for the specific separation problem, the study of mass transfer mechanisms and the factors that influence the membrane stability, and the liquid membrane mass transfer rates. The study of hydrodynamic conditions in the separation process is also important for achieving effective separations. On the basis of this type of work, the optimum membrane formulation and the operational conditions can be determined, and the basic data necessary for the scale-up of the separation process can be obtained.

Frankenfeld et al. [22] studied the influence on copper extraction of some important variables, such as membrane viscosity, treatment ratio, complexing agent concentration, internal droplet size, and copper extraction in the internal phase. They found that the magnitude of membrane viscosity represents a compromise of membrane leakage and mass transfer rate. They concluded that the more viscous membrane maintained its integrity considerably longer and gave better overall extraction. They also round that varying the concentration of extractant over a fairly wide range had only a minor effect on extraction rates. In the "loading" test for internal phase, they demonstrated that up to a loading of 50 g/L in the internal phase, the permeation rate was relatively unaffected by copper loading.

Martin and Davies [28] examined the possibility of using the liquid surfactant membranes to recover copper from very dilute solutions. A solution of 200 g of H_2SO_4 per liter containing additionally up to 20,000 ppm Cu^{2+} was used as a stripping solution. Under this condition, mass transfer was achieved with very large reverse concentration gradients across the membrane, from a feed solution at 120 ppm Cu^{2+} into a stripping solution containing 20,000 ppm Cu^{2+}.

Weiss et al. [89] studied the separation of mercury from wastewater by liquid surfactant membranes. Dibutylbenzoylthiourea was used as an extractant in the membrane phase. To test the enrichment of the internal phase, the emulsion was recycled to contact with fresh feed solution for 3 min each time. The initial feed solution contained 117 g Hg per cubic meter. With a 12-fold cycle, 25 cm^3 of water took up 286 mg of Hg, which corresponds to 11.44 kg of Hg per cubic meter and a 97.8-fold enrichment as compared with the initial solution.

Kitagawa et al. [45] studied the removal of toxic heavy metal ions as Cr^{6+}, Hg^{2+}, Cd^{2+}, and Cu^{2+} from wastewater. The results showed that liquid membranes were capable of reducing the levels of these heavy metal ions from several hundred ppm to less than 1 ppm under batch and continuous conditions. The results of laboratory tests for the extraction of Cr^{6+}, Hg^{2+}, Cd^{2+}, and Cu^{2+} were shown in Fig. 3.

Li and Shrier [42] have suggested a conceptual process diagram for water treatment by liquid surfactant membranes as shown in Fig. 4. Based on this flow diagram Kitagawa et al. [45] performed the experiments for wastewater treatment with a two-stage counterflow mixer-settler equipment as shown in Fig. 5. This equipment had a maximum treating capacity of 30 L/min of wastewater. The results of Cr^{6+} extraction with this pilot plant showed that the concentration of Cr^{6+} in the effluent water was continuously and steadily reduced from 100 to below 1 ppm. This extraction was accomplished with a small amount of emulsion, which was recycled and reused continuously for 31 h. After that time, the Cr^{6+} concentration in the internal stripping solution reached 182,000 ppm.

Column (differential) contactor was also used by some workers to perform pilot tests. Li and Chang [30] designed a continuous countercurrent extraction column system for removing Cr^{6+} from wastewater as shown in Fig. 6. They used a glass column with 6.2 cm in diameter and 140 cm in length. The emulsion in the column was dispersed by electromagnetic stirrer. Using 2% Alamine 336 as a carrier in the membrane phase, they achieved 99% removal of Cr^{6+} from wastewater containing 250 ppm Cr^{6+}.

Protsh and Marr [90] studied the separation of zinc from wastewater with a stirred countercurrent column. They achieved the enrichment factor of 228 in their pilot plant. With a 200-ppm zinc in

TABLE 1 Published Systems of Metal Extraction by Liquid Surfactant Membranes

Metal ion	Carrier	Solvent	Surfactant	Stripping agent
Li^+, Na^+, K^+, Rb^+, Cs^+, Tl^+, Ag^+	DC18C6	Toluene	Span 80	$Li_4P_2O_7$
Cs^+	Br_2DCC		Span 80/85	HNO_3
Sr^{2+}	$Br_2DCC/PEG300$		Span 80	HNO_3
Co^{2+}	D2EHPA	Cyclohexane/polybutadiene	Span 80	HNO_3
	LIX64N		ARIACEL80	DTPA
Ni^{2+}	LIX 64N		Span 80	HCl
Cu^{2+}	SME 529	Shell sol.	Span 80 or Span 20	H_2SO_4
	LIX 64N	S100N	ECA4360	H_2SO_4
	D2EHPA	Cyclohexane/polybutadiene	Span 80	HNO_3
	Benzoylacetone	Toluene	Span 80	HCl
Zn^{2+}	D2EHPA	Kerosene	Span 80	H_2SO_4
$Cd(CN)_4^{2-}$	Aliquat 336	S100N	Span 80	EDTA pH 4–6

Metal ion	Carrier	Diluent	Surfactant	Stripping agent
Hg^{2+}	Alamine 336	s100N	Span 80	NaOH or H_2SO_4
	Dibutylbenzoyl-thiourea	Decane/hexane	Rofetan OM, Rofetan OD, Span 80	$HCl/(NH_2)_2CS$
	Oleic acid/linolic acid	$C_{11}-C_{15}$ paraffins	Arlacel C (sorbitane sesquioleate)	H_2SO_4
Pb^{2+}	DC18C6	Toluene	Span 80	$Na_4P_2O_7$
$Cr_2O_7^{2-}$	Alamine 336	S100N	Span 80	NaOH or H_2SO_4
Ce^{3+}	TOPO		Span 80/20	HNO_3
La^{3+}	D2EHPA		S-205	HCl
TcO_4^-	TOA		Span 80	NH_4OH
UO_2^{2+}	Kelex 100		Span 80	H_2SO_4
	D2EHPA/TOPO	LOPS	ECA 4360	Reductant in H_3PO_4
U^{4+}	OPPA (Octylphenyl-phosphoric acid)			Oxidant

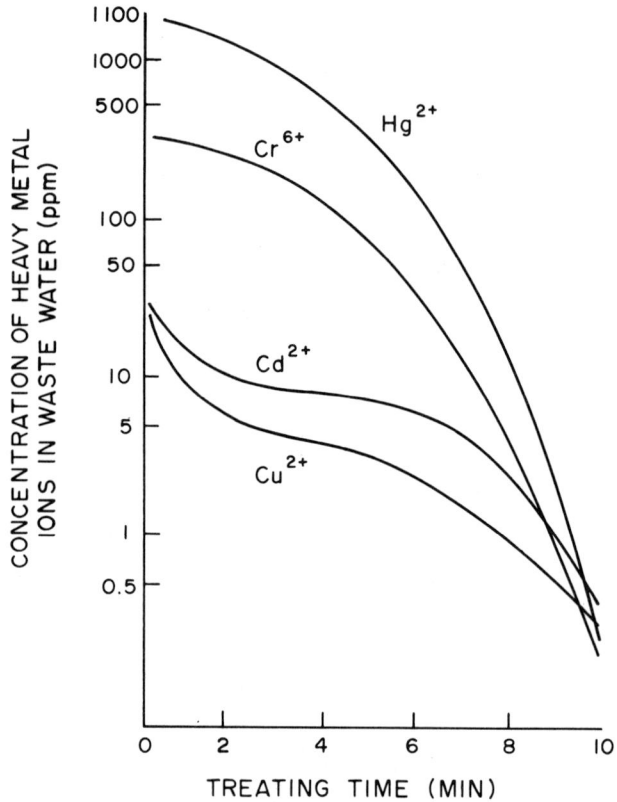

FIG. 3 Removal of Cr^{6+}, Hg^{2+}, Cd^{2+}, and Cu^{2+} from wastewater by liquid surfactant membranes. (From Ref. 45.)

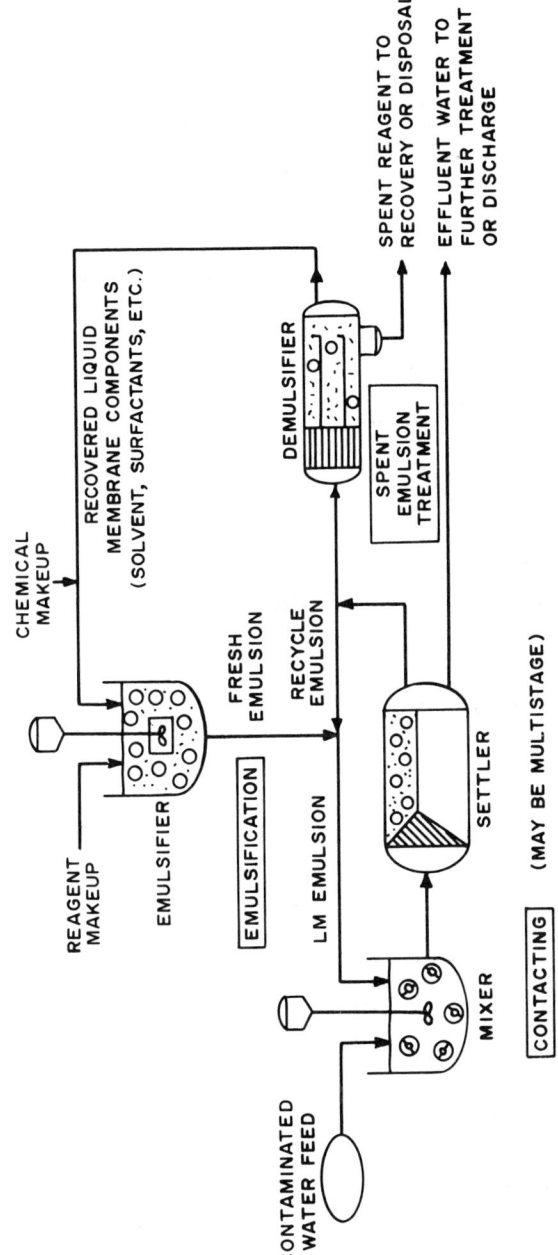

FIG. 4 Schematic diagram of conceptual liquid membrane water treatment process. (From Ref. 42.)

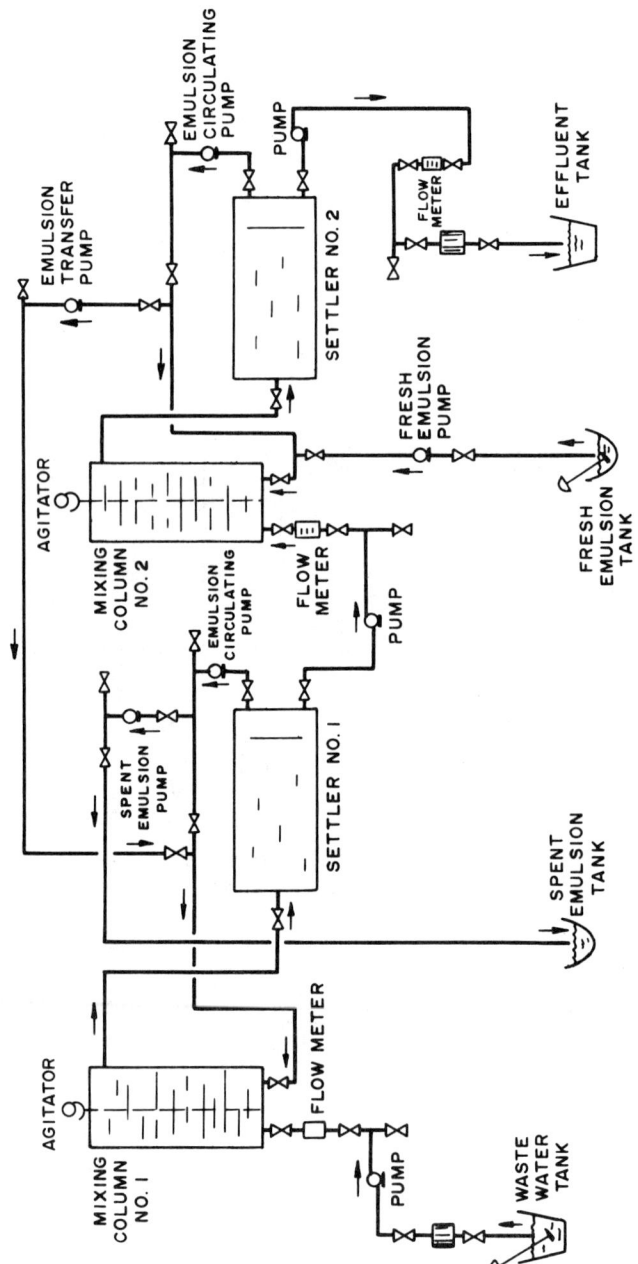

FIG. 5 Two-stage continuous countercurrent pilot plant.

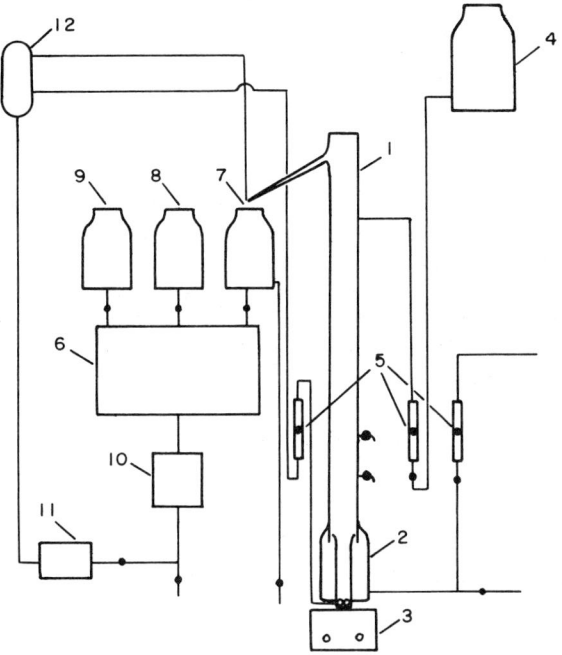

FIG. 6 Flow diagram of the continuous countercurrent liquid membrane extraction column. 1. Extraction column; 2. emulsion dispersion device; 3. electromagnetic stirrer; 4. feed tank; 5. flow meter; 6. emulsifier; 7. circulated emulsion reservoir; 8. oil tank; 9. NaOH solution tank; 10. emulsion tank; 11. pump; 12. potential tank. (From Ref. 30.)

the feed solution, an enrichment of 45,600 ppm in the resulting stripping solution was obtained within one stage.

C. Process Economics of Liquid Membrane Metal Extraction

Hayworth et al. [91] compared the liquid surfactant membranes with commercial solvent extraction technology for the extraction of uranium from wet process phosphoric acid. For 90% extraction of hexavalent uranium by D2EHPA/TOPO from the feed acid, only three liquid membrane stages were needed, while eight solvent extraction stages were usually needed for extraction and stripping. Liquid membranes, however, required a separate coalescer and emulsifier—offsetting to some extent the capital cost savings associated with the elimination of the stripping stages. Because simultaneous extraction and stripping took place in liquid membranes, the circulation rate of organic phase could be increased to a feed-to-organic ratios of 18:1 compared to the 1:1 normally found in solvent extraction. The economic comparison of liquid membranes and solvent extraction showed that the liquid membrane was about 20% cheaper in capital cost and 25% cheaper in operating cost as shown in Tables 2 and 3.

The estimation of economics was also made by Davy McKee Co. in the United Kingdom. Using a basis of plant production of 36,000 tons of copper per year from an acid leaching solution of 2.5 g of copper per liter, Davy McKee showed a 40% investment savings for liquid surfactant membranes as compared to solvent extraction. The operating cost for both processes is about the same. The estimates are summarized in Table 4. Part of the economic study was based on the results of a continuous run that lasted 9 days in which real leach liquor was used. Copper extraction of 92% was achieved after 10 min of residence time of the liquor in the liquid membrane extraction unit.

TABLE 2   Capitol Cost Estimates for Uranium Recovery[a]

	M dollars		
		SX	
	LM	Minimum pretreatment	Extensive pretreatment
On-site	14.0	15.4	19.4
Off-site	5.8	6.3	6.6
	19.8	21.7	26.0
Project contingency	4.8	5.4	6.5
Process allowance	2.7	—	—
	27.3	27.1	32.5
Solvent inventory	0.1	0.9	0.9
Total investment	27.4	28.0	33.4

[a]Basis: 400,000 tons/year P_2O_5 acid capacity, central Florida location, second quarter 1979 construction costs.

TABLE 3 Operating Cost Estimates for Uranium Recovery[a]

	Dollars per pound U_3O_8		
		SX	
	LM	Minimum pretreatment	Extensive pretreatment
Organic makeup			
Circulation loss	0.1	3.9	3.9
Raffinate loss	0.1	0.2	0.2
Crude loss and treatment	1.0	2.9	0.0
Chemicals and supplies	1.4	0.9	1.6
Utilities	0.7	0.9	1.0
Labor, maintenance, taxes, and insurance	7.1	7.1	8.6
Depreciation	4.6	4.7	6.6
Total operating costs	15.0	20.6	20.9

[a]Basis: 400,000 tons/year P_2O_5 acid capacity, 350,000 pounds/year U_3O_8 recovery, second quarter 1979 costs.

TABLE 4 Estimated Cost of Copper Recovery from Ore Leachers

	Solvent extraction	Liquid membranes
Copper recovered (K/metric tons/yr)	36	36
Stages	5	1
Plant investment (M$)[a] (major savings: reduction of stages)	13	8
Organic inventory (M$)	2	1
Direct operating cost (cents)	1.8	1.7

[a]Includes facilities to make and break emulsion in LM case. The investment in both cases includes only the facilities to extract the copper from the clarified leach liquor and to concentrate it into the electrolysis liquor. Not included are preparation and conditioning of the leach liquor; cost of the copper electrolysis plant.
Source: Refs. 31 and 92.

IV. EXTRACTANTS

The metal extraction by means of liquid surfactant membranes is based on the mechanism of carrier-mediated transport. Therefore, it is very important to incorporate a suitable extractant (carrier) into the liquid membrane phase to enhance the effective solubility of the metal species in the membrane phase, thus sharply increasing the metal flux.

According to the functional group, the extractants are generally divided into three classes: (1) acidic extractants, (2) basic extractants, and (3) neutral extractants. Tables 5 to 7 list some of the extractants which are suitable to be used as membrane carriers.

A. Acidic Extractants

In order to extract a positively charged metal ion from an aqueous solution, it is necessary to combine with an anionic species to form an uncharged species. Acidic extractants are most effective for extracting cations by exchanging the cations for their protons. Acidic extractants that are commonly used for liquid membrane cation extractions include chelating extractants, such as β-diketones, hydroxyoximes and 8-hydroxyquinolines, and alkylphosphorous compounds. In general, the coordination complex of the chelating extractant with the positively charged metal ion is very specific. Therefore, the chelating extractants can be used to achieve selective separation of metals. Although alkylphosphorous coumpounds, which are often called "liquid cation exchangers," are less selective, they are less expensive and their metal salts are more soluble in organic solvents than are metal chelates [93]. Therefore, they are also widely used in metal extraction processes.

In recent years a series of new extractants have been synthesized to improve the extraction process. For example, D2EHPA is a commonly used organophosphoric acid for Co/Ni separation in sulfate medium. However, the separation factor, $\beta_{Co/Ni}$, is very low ($\beta_{Co/Ni} = 2.2$). Chuei et al. [94] use a new acidic phosphonic extractant 5709 to replace D2EHPA. This new extractant exhibits excellent extraction performance. The value of $\beta_{Co/Ni}$ is 435.4, which is more than 100 times higher than that of D2EHPA system. Rikelton et al. [95] have found that the dialkylphosphinic acids, such as Cyanex 273, show even better Co/Ni separation factor.

B. Basic Extractants

Basic extractants, such as high-molecular-weight primary, secondary, and tertiary amines and the quarternary alkylammonium ions, are used for the extraction of anionic metal complexes (or neutral metal com-

TABLE 5 Acidic Extractants

Extractant	Structure
Chelating compounds	
β-Diketones	
TTA (2-ethenoyltrifluoroacetone)	$CF_3-\underset{OH}{C}=CH-\underset{\parallel}{C}-$ with S double bond
AA (acetylacetone)	$CH_3-\underset{OH}{C}=CH-\underset{\parallel}{C}-CH_3$ with O
BA (benzoylacetone)	$CH_3-\underset{OH}{C}=CH-\underset{\parallel}{C}-Ph$ with O
Hydroxyoximes	
LIX63 (General Mills)	$CH_3-(CH_2)_3-\underset{C_2H_5}{CH}-\underset{OH}{CH}-\underset{\parallel}{C}=\underset{NOH}{C}-\underset{C_2H_5}{CH}-(CH_2)_3-CH_3$
LIX65N (General Mills)	$CH_3-(CH_2)_8$-phenyl(OH)-C(=NOH)-phenyl

TABLE 5 (Continued)

Extractant	Structure
LIX64N (General Mills)	LIX65N + LIX63 (~1%)
LIX70 (General Mills)	C_9H_{19}–C$_6$H$_2$(OH)(Cl)–C(=NOH)–C$_6$H$_5$ + LIX63
SME529 (Shell)	CH_3–$(CH_2)_8$–C$_6$H$_3$(OH)–C(CH$_3$)=NOH
P17 (Acorga)	CH_3–$(CH_2)_8$–C$_6$H$_3$(OH)–C(CH$_2$C$_6$H$_5$)=NOH
P50 (Acorga)	CH_3–$(CH_2)_8$–C$_6$H$_3$(OH)–CH=NOH

β-Hydroxyquinolines (oxines):

Kelex 100
(Ashland Chemical)

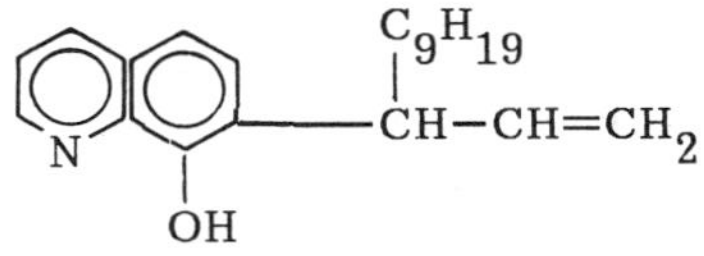

$$\underset{\text{OH}}{\underset{|}{\text{N}}}\!\!\!\diagup\!\!\!\diagdown\!\!\!\overset{C_9H_{19}}{\underset{|}{CH}}-CH=CH_2$$

Alkylphosphorous compounds
Organophosphoric acids

D2EHPA
[di-(2-ethylhexyl)-phosphoric acid]

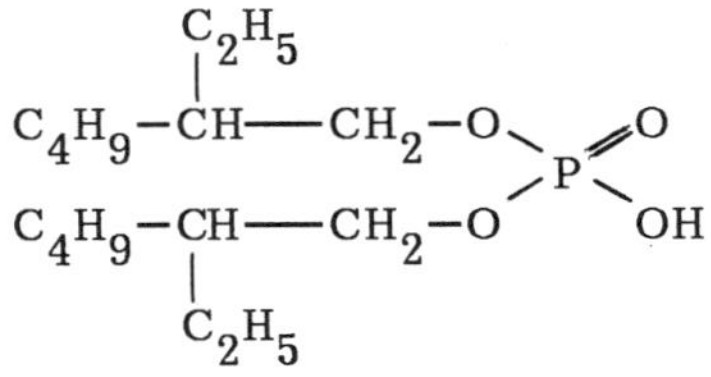

$$C_4H_9-\underset{\underset{C_2H_5}{|}}{CH}-CH_2-O\diagdown_{P}\diagup^{O}$$
$$C_4H_9-\underset{\underset{C_2H_5}{|}}{CH}-CH_2-O\diagup\ \ \diagdown OH$$

DBP
(dibutyl phosphoric acid)

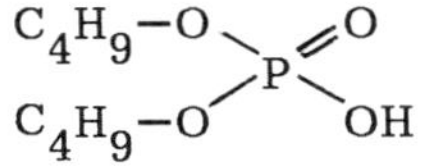

$$C_4H_9-O\diagdown_{P}\diagup^{O}$$
$$C_4H_9-O\diagup\ \ \diagdown OH$$

Organophosphonic acids:

PC 88A

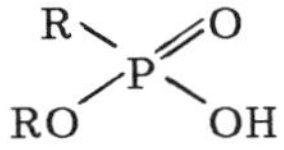

$$R\diagdown_{P}\diagup^{O}$$
$$RO\diagup\ \ \diagdown OH$$

Organophosphinic acids:

Cyanox 272
(American Cyanamid Co.)

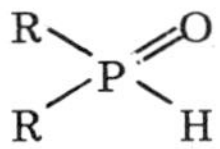

$$R\diagdown_{P}\diagup^{O}$$
$$R\diagup\ \ \diagdown H$$

plexes). Similar to the alkylphosphoric acids, these extractants can be regarded as "liquid anion exchangers."

In aqueous solutions many metal ions can form a variety of anionic complexes with sulfate, halide, cyanate, thiocyanate, and a number of other anionic ligands. Examples of anionic metal complex that commonly exist in the solutions of hydrometallurgical processes and electroplating processes are $Cd(CN)_4^{2-}$, $Cr_2O_7^{2-}$, $AuCl_4^-$, $UO_2(SO_4)_2^{2-}$, and $V_3O_9^{3-}$. The ease of formation of anionic complexes of different metal ions with the anion ligand in the aqueous solution varies greatly. For this reason, the selective extraction is obtained by using suitable amines. For the liquid membrane metal extraction, commercial products of amines such as Alamine 336 and Aliquat 336 are widely used.

TABLE 6 Basic Extractants

Extrahant	Structure
Primene JMT	$CH_3-\underset{\underset{CH_3}{\|}}{\overset{\overset{CH_3}{\|}}{C}}-(CH_2-\underset{\underset{CH_3}{\|}}{\overset{\overset{CH_3}{\|}}{C}})_4-NH_2$
Amberlite LA-2	$CH_3-(CH_2)_{11}-\underset{}{\overset{\overset{H}{\|}}{N}}-(\underset{\underset{CH_3}{\|}}{\overset{\overset{CH_3}{\|}}{C}}-CH_2)-\underset{\underset{CH_3}{\|}}{\overset{\overset{CH_3}{\|}}{C}}-CH_3$
TOA (trioctylamine)	$(C_8H_{17})_3N$
TNA (trinonylamine)	$(C_9H_{19})_3N$
TLA (trilaurilamine)	$(Lauryl)_3N$
Alamine 336	R_3N (R: C_8-C_{10})
Aliquat 336	$CH_3(C_8H_{17})_3N^+$

TABLE 7 Neutral Extractants

Organic phophoryl compounds

- TBP (tri-n-butyl phosphate): $(But-O)_3-PO_4$
- TBPO (tri-n-butyl phosphine oxide): $(Butyl)_3-P=O$
- TOPO (tri-n-octyl phosphine oxide): $(Octyl)_3-P=O$

Macrocyclic compounds

- Crown ethers
  - DB18C6 (di-benzo-18-crown-6)
  - DC18C6 (di-cyclohexane-18-crown-6)
  - 18C6 (18-crown-6)

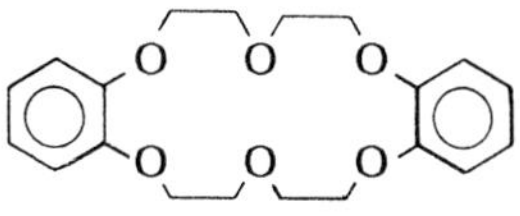

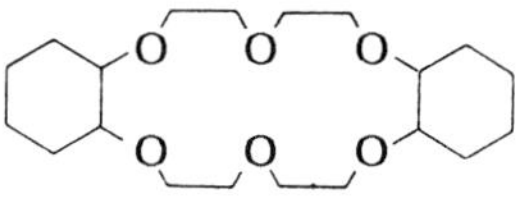

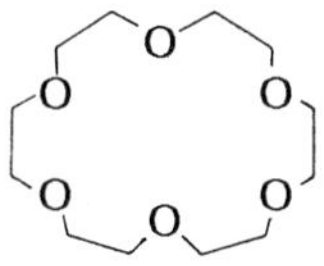

- Cryptands
  - Cryptand (2,2,1)
  - Cryptand (2,2,2)

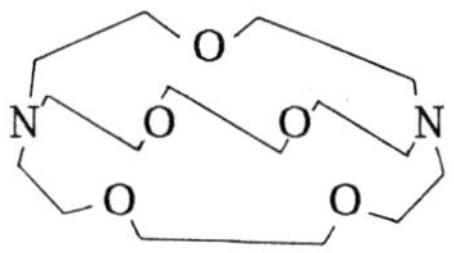

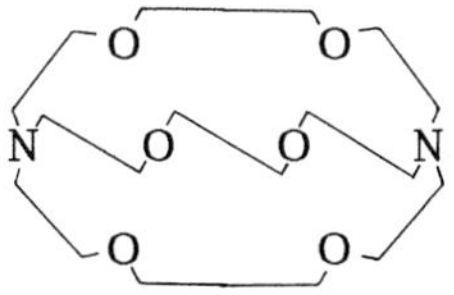

- Antibiotic microcycles
  - Monesin

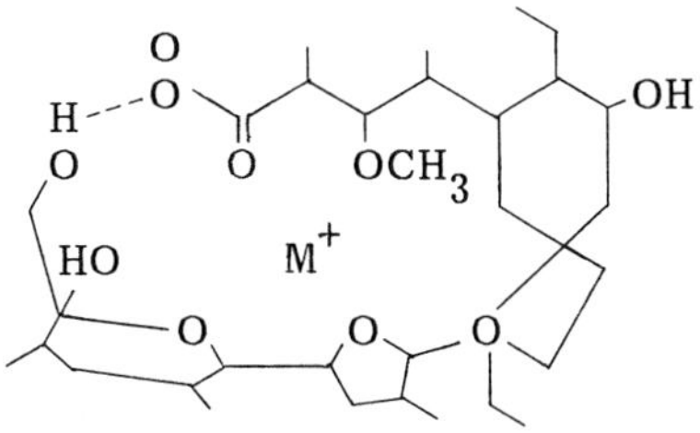

149

C. Neutral Extractants

Neutral extractants often extract uncharged metal complexes in aqueous solutions. In this type of extraction, the metal species is coordinated with two different types of ligands (i.e., a water-soluble anion and an organic-soluble electron-donating functional group). Most of the neutral extractants that were investigated in the liquid membrane studies are organic phosphoryl compounds and macrocyclic molecules.

Phosphoryl compounds (i.e., trialkyl phosphates and phosphine oxides) are used extensively as extractants for actinides and lanthanides. They are especially useful for the recovery of uranium and plutonium in the spent-fuel reprocessing of nuclear plant.

Macrocyclic ligands, such as crown ethers and their derivatives, and the naturally derived antiobiotic macrocycles, such as valinomycin and monesin, have the remarkable property of selectively complexing metal cations [96,97]. When these macrocyclic molecules are incorporated as carriers into liquid membranes, the fluxes of different metal ions can differ enormously. Reusch and Cussler [98] found that the flux of K^+ through a liquid membrane containing dibenzo-18-crown-6 ether was 4000 times larger than the flux of lithium ions.

Izatt, Christensen, and their co-workers studied extensively the incorporation of a number of macrocyclic carriers in liquid membrane systems to facilitate the transport of several alkali, alkaline earth, and transition metal cations [99]. They have undertaken systematic investigations of the influence on carrier-facilitated cation transport rates and selectivities of macrocyclic ligand structure [100], ratio of cation and macrocycle cavity diameters [101], and equilibrium constant for macrocycle-cation interaction [102]. One of the most attractive results in their studies is the remarkably high transport selectivity for Pb^{2+} over the other cations by the dicyclohexane-18-crown-6 in the membrane phase; the ratio of Pb^{2+} concentration to the concentration of another cation in the mixture is as low as 1:100 [103].

Since separations among the alkali, alkaline earth, and trivalent lanthanide cations have traditionally been difficult to achieve, membranes containing macrocyclic ligand carriers offer an inviting alternative for selective separations of these cations. At present, the initial cost of the macrocyclic compounds is still very high. However, compared with solvent extraction, the inventory of these expensive compounds in the process is much lower. This would make these extractants quite competitive in a liquid membrane process for separation of precious metals.

D. Synergistic Effect Between Extractants and Surfactants

For liquid membrane processes, the membrane composition is of great significance for obtaining high stability as well as selectivity and permeability. The membrane selectivity and permeability are dependent on the use of effective extractants. Although many surfactants have been studied, much work remains to be done to select or synthesize new surfactants for various separation applications. Yoshikazu Miyake et al. [117] have found that for the copper extraction by SME 529, the extraction rate is accelerated by anionic surfactants but retarded by cationic and nonionic surfactants. This synergistic effort between surfactants and carriers is not yet well understood and needs further study.

The liquid membrane composition can sometimes be simplified to achieve the separation task. Cui [118] has recently found that some compounds, such as polyethylene glycol, can be used both as an extractant and as surfactant for metal extraction by liquid surfactant membranes. With such compound incorporated in the membrane phase, a higher mass transfer rate for metal extraction may be obtained because the surfactant, being a membrane carrier, will not form a barrier to mass transfer of the metal species across the interfaces.

V. ENHANCEMENT OF FACILITATED TRANSPORT

The mass transfer process of liquid membrane metal extraction involves the diffusion processes of the metal species in the organic membrane phase and the two aqueous phases with simultaneous interfacial chemical reactions, that is, extraction reaction occurring at the interface between the membrane phase and the continuous aqueous phase and the stripping reaction occurring at the interface between the membrane phase and the internal aqueous phase. To achieve fast separations, it is necessary to reduce the mass transfer resistance to diffusion at these interfaces. This can be achieved by using membrane diluent to ensure low membrane viscosity and high diffusivity and by improving the physical conditions of w/o emulsion preparation and the hydrodynamic conditions of the liquid membrane system, such as the geometry of reactors and stirring speed of continuous aqueous phase.

Besides the physical means to increase the separation rate and selectivity, there exist several chemical ways to enhance the liquid membrane separations by changing the interfacial chemical reactions as discussed below.

Coextractants are quite useful for accelerating the heterogeneous chemical reactions. For example, it has long been discovered that the extraction rate of metal ions by LIX 65N alone is very low and LIX 63 itself is rather inert for metal extraction, but the metal extraction by LIX 65N is tremendously enhanced by the use of a coextractant of LIX 63. Therefore, LIX 63 is incorporated into LIX 65N to form a new effective extractant system, LIX 64N [104].

For the extraction of copper from aqueous solutions using a liquid membrane process, Martin and Davies found that with SME 529 as the extractant, the extraction rate of copper was reduced in the presence of Fe^{3+} in the external aqueous phase. By adding a "kinetic accelerator," RD 547, which was developed by Shell International Chemicals, an improvement in the extraction of copper was obtained. This "kinetic accelerator" is an α-hydroxyoxime having the functional group as follows:

$$-\underset{\underset{OH}{|}}{C}-\underset{\underset{NOH}{|}}{C}-H$$

Osseo-Asare and his co-worker [106,107] have demonstrated that in combination with appropriate extractant, sulfonic acids can be incorporated as catalysts into the membrane phase to enhance the metal extractants. For the extraction of Ni^{2+} and Co^{2+} by LIX 63, they observed the increased flux after the addition of dinonylnaphthalene sulfonic acid (HDNNS) to the membrane phase. They proposed that the reversed micelles formed by sulfonates enhanced the metal extraction by specific solubilization of both the metal and extractant, resulting in an increase in the interfacial concentration of the reacting species.

The extractant can also be incorporated into the aqueous phase. Gu et al. [36] analyzed the mass transfer resistances in a well-stirred system of liquid surfactant membranes in which carrier-mediated facilitation was utilized. They found that owing to the very high ratio of the interfacial area for stripping to that for extraction, the resistance of stripping could be neglected compared with that of extraction. Therefore, the extraction reaction would become the rate-controlling step if the reaction was very slow. Thus, accelerating the extraction reaction in this case would enhance significantly the overall mass transfer process of the liquid membrane system.

On the basis of this analysis, they have developed a method to enhance the metal extraction in a liquid membrane system [35,37]. This method consists of introducing certain anionic ligands, such as carboxylates, to continuous aqueous phase containing cations to be extracted to accelerate the extraction rate of cations, thus enhancing the overall mass transfer process. For example, the liquid membrane of Co^{2+} by D2EHPA is originally slow. It takes about 15 min for 80% Co^{2+} recovery. However, with the addition of 0.1 M acetate as a ligand in aqueous feed solution, only 2 min is needed for 95% recovery as shown in Fig. 7. This ligand effect can be explained by the replacement of coordinated water molecules in the hydrated metal ions by the ligand. The ligand-metal complex can then react quickly

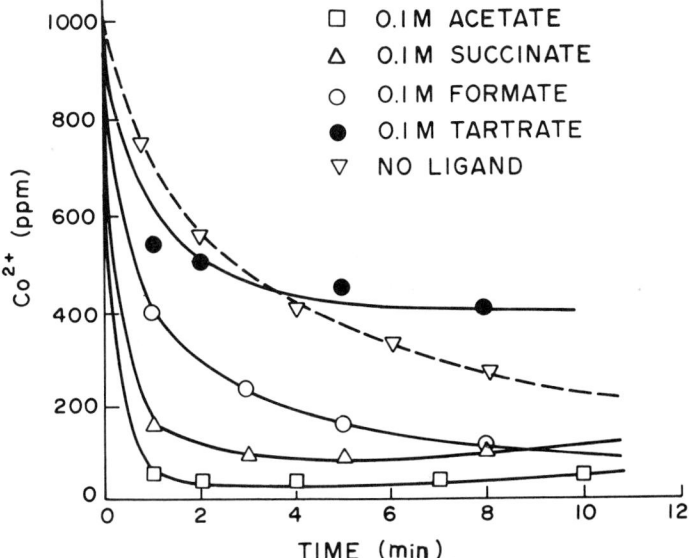

FIG. 7 Effect on kinetics of different ligands in external phase. (From Ref. 37.)

with the extractant in the membrane phase. In this way, the extraction rate is accelerated.

Since a number of metal extraction processes are coupled with the transport of hydrogen ions, the driving force across the membrane may also be maintained high by minimizing the concentration of diffusing species in the internal aqueous phase. This can be achieved by common-ion effect or by the irreversible reaction in the internal phase.

Li et al. [108] adopted the common-ion effect to improve the extraction processes of heavy metals. For the removal of heavy metals, such as Mn^{2+}, Cr^{3+} and Fe^{3+}, from wastewater by means of liquid membranes, the heavy metals can be trapped in the internal phase as sulfide or hydroxide precipitates. However, this process is restricted by the solubility of the precipitates in the internal aqueous phase.

An irreversible reaction in the internal phase can also be used to increase the driving force for mercury extraction. They used 1,1-di-n-butyl-3,3-benzoyl-thiiourea (DBBT) in the organic membrane phase as an extractant to transport mercury. Two reversible reactions with the extractant occur at the interfaces:

Extraction:

$$2R^2NCSNHCOR^1 + Hg^{2+} \rightleftarrows 2H^+ \, Hg(R^2NCSNCOR^1)_2 \quad (3)$$
   org.   aq.   aq.   org.

Stripping:

$$Hg(R^2NCSNCOR^1)_2 + 2H^+ \rightleftarrows 2R^2NCSNHCR^1 + Hg^{2+} \quad (4)$$
   org.   aq.   org.   aq.

$R^1 = C_6H_5$, $R^2 = (n\text{-}C_4H_9)_2$

To intensify the stripping reaction, an additional reagent, thiourea, is introduced in the internal phase together with the stripping reagent, HCl. In the internal phase, the following irreversible reaction takes place:

$$Hg^{2+} + 2HCl + 4(NH_2)_2CS \rightarrow Hg[(NH_2)_2CS]_4Cl_2 + 2H^+ \quad (5)$$

This reaction drives the stripping reaction [Eq. (4)] toward the right-hand side.

In the study of liquid membrane uranium extraction, Hayworth et al. [91] converted U^{6+}, which is a permeable species, into an impermeable species, U^{4+}, by a reductant. In this way, the uranium was efficiently trapped and concentrated in the internal phase.

VI. MATHEMATICAL MODELING

The purpose of modeling work is to make a meaningful interpretation of experimental extraction results, acquire a deeper understanding of the mechanism of mass transfer process and interactions among various process parameters, and obtain the basic data for the engineering design of the liquid surfactant membrane separation process. Up to the present time, several models have been proposed in literature and Kremesec [109], Stroever and Varamaso [110], and Borwankar et al. [123] have reviewed these models.

Generally speaking, the metal extraction by liquid surfactant is based on solute diffusion with simultaneous interfacial chemical reactions. In this system, the resistance to mass transfer comes from the following steps: (1) eddy diffusion of metal ions outside emulsion globules, (2) extraction reaction occurring at the interface between external aqueous phase and membrane phase, (3) molecular diffusion of the metal-extractant complex across the membrane (including interfacial surfactant layers) where eddy diffusion is assumed to be negligible, (4) stripping reaction occurring at the interface between membrane phase and internal aqueous phase, and (5) molecular diffusion of the stripped metal ions inside the internal droplets. The mathematical treatment of these steps can be quite complicated even when the chemical reactions are rather simple. For diffusional-chemical reaction systems, different mathematical treatments should be applied to different occasions.

In the case of large Damköler numbers (toward infinity), which correspond to very high reversible chemical reaction rates or very low diffusion rates (thick membranes), the chemical reactions can be assumed at equilibrium, and the concentration of the metal complex in the oil membrane phase can be expressed in terms of the metal concentration in the aqueous phase according to the equilibrium relationship since the reactions can be considered instantaneous. In this case, the resistance to mass transfer due to interfacial chemical reactions can be neglected compared with the resistance to diffusion through the two aqueous phases, the oil membrane phase, and the interfacial films. The liquid membrane mass transfer process can then be regarded as a diffusion-controlled process, and a simple diffusion-controlled mathematical model can be applied to describe this process. This situation is often encountered in liquid membrane metal extraction systems.

In the case of very small Damköler numbers (toward zero), which relate to very slow interfacial chemical reactions of very quick diffusion (thin membrane), the resistance of interfacial chemical reactions is much higher than the resistance of solute diffusion. In this case the interfacial chemical reactions control the overall extraction rate of the liquid membrane process. The liquid membrane mass transfer can thus be regarded as a kinetically controlled process. This situation is less common in liquid membrane systems because the extractants with slow kinetics are usually avoided when considering the liquid membrane formulations.

In the case of intermediate Damköler numbers, the chemical reaction rates are not many orders of magnitude larger than diffusion rates, and the reversible chemical reactions can no longer be assumed to be at equilibrium. In this case, the resistances due to diffusion and chemical reactions cannot be neglected with respect to each other. The mathematical treatment to this problem becomes very difficult. Stroever and Varanasi [110] suggested the "combined Damköler technique" as the promising approach to solve this problem. However, no paper has yet been published to model this situation. Most of the published papers are restricted to the study of the mass transfer processes in which the interfacial chemical reactions are assumed to be instantaneous. In this chapter we review briefly the development of some diffusion-controlled models applied to metal extractions.

There are two approaches to mathematical modeling of liquid surfactant membrane systems. The first approach is to establish rather simple mathematical equations to calculate the separation results for plant design; the other is to use a set of detailed mathematical equations to gain fundamental understanding of the separation process.

In dealing with the mass transfer in a liquid surfactant membrane system by the first approach, the double emulsion globules are often simplified to concentric spherical shells [20,111]; that is, all of the fine encapsulated aqueous droplets are assumed to be coalesced into a single large droplet, and the membrane phase (a network of stagnant oil interstices between the droplets) is viewed to form a spherical shell with a fixed thickness around the coalesced single large droplet. This concept is expressed in Fig. 8.

Cahn and Li [43] used the first approach to derive a simplified extraction rate equation for quick process design and scale-up calculations:

$$\frac{dC}{d\Theta} = (D)(A)\frac{\Delta C}{\Delta X} \qquad (6)$$

where C is the concentration of permeate, Θ the permeation time, D the diffusivity of the permeating species, A the permeation area, ΔC the concentration difference of the permeating species on either side of the membrane, and ΔX the membrane thickness.

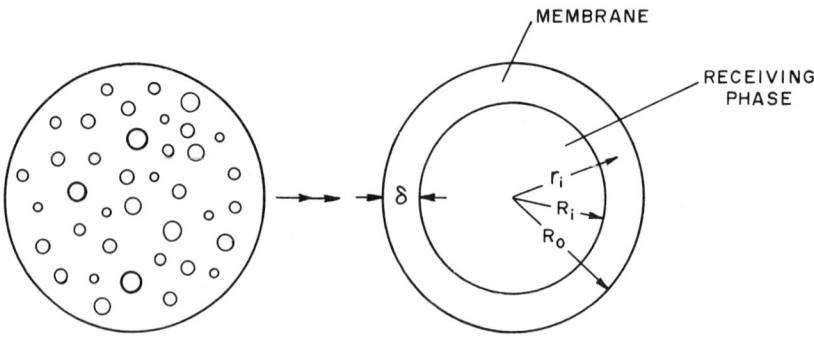

FIG. 8 Simplified model of an emulsion globule. (From Ref. 20.)

Since A and ΔX are difficult to measure for a liquid membrane system, the group $(D)(A)(\Delta X)$ is replaced by $D'(V_E/V_W)$. Essentially, ΔX has been combined into D' (permeation rate constant) and it has been assumed that when the emulsion breaks up into globules, the total surface area of the globules is proportional to the amount of emulsion used in the treatment per unit of external phase (V_E/V_W).

Equation (6) is easily rearranged and integrated to give

$$\ln \frac{C_{in}}{C_{out}} = D' \frac{V_E}{V_W} \theta \qquad (7)$$

D' can be estimated from batch experiments using Eq. (7) and employed in the design of continuous multistage mixers.

For an n-stage system, each stage being of equal volume, the appropriate volume per mixer will be

$$V_n = n \cdot K \left(\frac{C_{in}}{C_{out}} - 1 \right) \qquad (8)$$

where

$$K = \frac{V_W}{D'} \left(\frac{V_W}{V_e} + 1 \right) \qquad (9)$$

where v_W and v_e are the flow rates of feed and emulsion, respectively.

According to this simple model, the mass transfer rate is proportional to the average concentration difference between the external and internal phases. Although these equations are rough approximations of the process, the experimental results can be analyzed quickly to get some meaningful parameters. Therefore, this model is useful as the preliminary approach to the engineering design.

The second approach to the mathematical treatment for the liquid membrane system is somewhat complicated. In this case the mathematical equations should approach as far as possible the realistic physical situation. The concentric spherical shell model is obviously inadequate to describe the transport process in such a system. Instead, the model must reflect the emulsion heterogeneity that results from the presence of the encapsulated droplets dispersed in the emulsion globules.

Kopp et al. [112] have assumed that the encapsulated droplets are symmetrically distributed and fixed inside the emulsion globules.

They proposed a model to describe the mass transfer process in the emulsion globule for copper extraction in terms of a moving boundary at which a reaction between copper complex and the internal reagent occurs. This boundary moves toward the globule center as the internal reagent is consumed. The molar flux J of copper is given as

$$J = \alpha \sqrt{D \cdot t}$$

where α is a constant, D the diffusivity, and t the reaction time. They have found that as long as the carrier is saturated with copper at the surface of the emulsion globule, mass transfer is proportional to $\sqrt{t}$. The solution of this mass transfer model is for planar geometry and the copper concentration at the interface between the membrane and external phases is assumed to be constant. It should be pointed out that when the solution derived from planar geometry is used to represent the mass transfer in a spherical geometry, it is reasonable only for a very thin layer from the globule surface. This limits the range of applicability of this model.

Similar to the concept of Kopp et al., Ho et al. [113] have developed the advancing front model for a spherical geometry, as shown in Fig. 9. It is assumed that all emulsion globules in the system are uniform in size, and that mass transfer without the globules is by diffusion only. Coalescence of the globules is assumed to be negligible because the external phase is well stirred. The emulsion itself can be treated as a continuum and the encapsulated internal reagent is considered to be a uniformly distributed and immobile species within the continuum. Interfacial chemical reactions are considered to be at equilibrium. Hence the solute is unable to permeate into the globule beyond those droplets that are completely depleted of reagent. Thus there must exist a sharp boundary, or reaction front, at which the reaction takes place, and which separates the inner region containing no solute from the outer region which contains no reagent. As the reagent is consumed by the reaction, this reaction front advances into the globule. External phase mass transfer resistance is assumed to be negligible, owing to the sufficient agitation of this phase.

Besides the use of spherical coordination, this model takes care of the concentration changes at the globule surface and in the external phase. This is a significant improvement over the existing models. However, in this model the mass transfer resistance of the peripheral layer around the emulsion globule, which might be more important than the internal resistance, is neglected.

Casamatta et al. [114] have proposed a permeation model in an emulsion globule for spherical geometry. Although their model deals with the mass transfer of hydrocarbon (without carrier), its principle is also suitable for the carrier-mediated transport as long as the chemical reactions are very fast. In their mode, the o/w emulsion

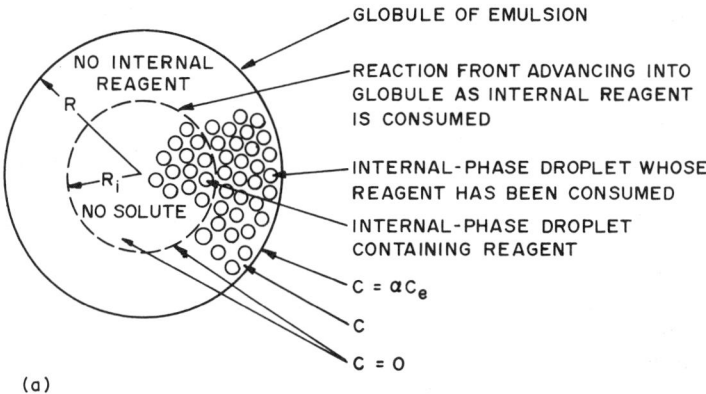

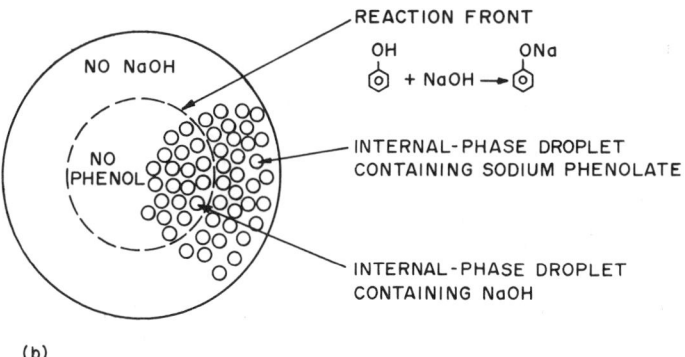

FIG. 9 Schematic diagram of the advancing-front model: (a) general case; (b) phenol example. (From Ref. 113.)

globule is assumed to be surrounded by a peripheral water layer and the fine droplets are fixed in the inner core of the globule. The hydrocarbon migration in an emulsion globule can be described as the unsteady-state diffusion into a composite sphere comprised of the inner core and the peripheral water layer. Its transfer in the inner core includes the diffusion through a network of stagnant aqueous interstices between the droplets in parallel with the permeation across each droplet. This model identifies the peripheral water layer of an emulsion globule as the controlling mass transfer resistance. The overall mass transfer coefficient K_i for an emulsion globule is given by

$$K_i = \frac{D_i m_i}{\delta} \qquad (11)$$

where D_i is the diffusivity, m_i the partition coefficient of component i between the aqueous and organic phases, and δ the thickness of the outer water layer.

This model gives a good understanding of the behavior of a liquid surfactant membrane system. The inadequacy of this model arises from the fact that the presence of the other water layer surrounding the emulsion globule is attributed solely to geometrical considerations. Also, the assumption of solute diffusion across each droplet may be untenable because the internal droplets are separated by the membrane phase.

Gu et al. [33,36] have noticed the effect of surfactant on interfacial mass transfer rates in a liquid surfactant membrane system. They have found that this interfacial resistance to molecular diffusion due to the presence of surfactant at the interface may reach as high as 2500 s/cm [35]. Obviously, such a high resistance to mass transfer should not be neglected when a mass transfer process is discussed. Based on this experimental finding, they developed a model of diffusion-controlled mass transfer in liquid surfactant membranes for fast chemical reactions. This model takes into account the diffusion both across the outer surfactant layer and through the inner core of the emulsion globules. The concentration profile inside an emulsion globule is shown in Fig. 10. Their results indicate that the thin layer of surfactant existing at the interface between emulsion globules and continuous aqueous phase offers the predominant barrier to mass transfer in a liquid surfactant membrane system.

Another phenomenon which is often omitted in the mathematical modeling is the leakage of liquid membranes. However, leakage can become significant in the case of weak membranes and stringent mixing conditions. Boyadzhiev et al. [115] have taken leakage into consideration in their modeling of hydrocarbon extractions by liquid membranes. Frankenfeld et al. [22], Ho and Li [116], and Borwankar et al. [123,124] have discussed the leakage mechanisms comprehensively and developed a general mathematical model containing overall mass transfer coefficients for extraction and leakage in a liquid membrane extraction process.

In general, more fundamental studies are needed to determine the interfacial kinetics and interfacial rheological properties, such as interfacial elasticity, interfacial shear viscosity, and interfacial tension, and to model or correlate these parameters with interfacial mass transfer parameters and membrane stability.

VII. SUMMARY

Liquid surfactant membranes are effective and versatile tools for performing a large variety of separations. They are particularly attrac-

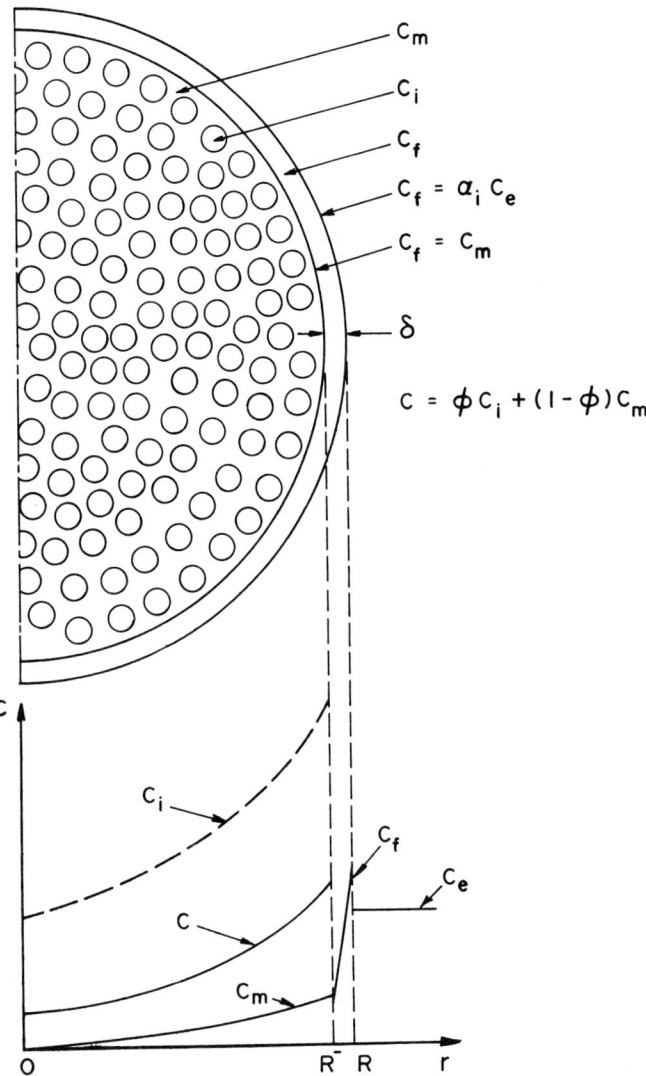

FIG. 10 Concentration profile inside an emulsion globule. (From Ref. 33.)

tive for metals extractions, where equilibrium considerations make solvent extraction methods less efficient.

When the process of extracting metals by liquid membranes was first developed by Li and his associates some time ago, the timing was unfortunate because of the recession in the United States and thus the poor metals market around that time. The metals market is now improving and the first commercial process of extracting zinc by liquid membrane has been achieved [68]. Further growth of the metals market would certainly lead to more commercialization of metals extraction applications of the liquid surfactant membrane process.

ACKNOWLEDGMENT

This work was partially supported by the Industrial Waste Elimination Research Center established by EPA at Illinois Institute of Technology.

REFERENCES

1. N. N. Li, R. B. Long, and E. J. Henley, Ind. Eng. Chem., 57:18 (1965).
2. C. E. Rogers, M. Fels, and N. N. Li, in *Recent Developments in Separation Science*, Vol. 2 (N. N. Li, ed.), CRC Press, Boca Raton, Fla., 1972, p. 107.
3. D. Spriggs and N. N. Li, in *Membrane Separation Processes* (P. Mears, ed.), Elsevier, Amsterdam, 1976, p. 39.
4. N. N. Li and Winston Ho, in *Perry's Chemical Engineers' Handbook*, 6th ed., McGraw-Hill, New York, 1984, pp. 17–24.
5. S, S, Kulkarni and N. N. Li, in *Encyclopedia of Science and Technology*, 6th ed., McGraw-Hill, New York, 1984, pp. 17–14.
6. H. K. Lonsdale, J. Membr. Sci., 10:81 (1982).
7. N. N. Li and J. W. Frankenfeld, in *Handbook of Separation Process Technology*, Wiley, New York, 1987, pp. 840–861.
8. N. N. Li, U. S. Patent 3,410,794 (1968).
9. N. N. Li, U. S. Patent 3,696,028 (1972).
10. N. N. Li, U. S. Patent 4,056,462 (1977).
11. N. N. Li, U. S. Patent 4,155,844 (1979).
12. N. N. Li, AIChE J., 17(2):495, 1971.

13. N. N. Li, Ind. Eng. Chem. Process Des. Dev., *10*:215 (1971).
14. N. N. Li, in *Membrane Processes in Industry and Biomedicine* (M. Bier, ed.), Plenum Press, New York, 1971, p. 175.
15. R. P. Cahn and N. N. Li, in *Membrane Separation Processes* (P. Mears, ed.), Elsevier, Amsterdam, 1976, p. 327.
16. R. P. Cahn and N. N. Li, J. Membr. Sci., *1*(2):129 (1976).
17. N. N. Li, U. S. Patent 3,779,907 (1973).
18. N. N. Li, U. S. Patent 4,086,163 (1978).
19. N. N. Li, U. S. Patent 4,292,181 (1981).
20. E. S. Matulevicius and N. N. Li, Sep. Purif. Methods, *4*(1):73 (1975).
21. N. N. Li, J. Membr. Sci., *3*:265 (1978).
22. J. W. Frankenfeld, R. P. Cahn, and N. N. Li, Sep. Sci. Technol., *16*(4) (1981).
23. R. P. Cahn, J. W. Frankenfeld, N. N. Li, D. Naden, and K. N. Subramanian, in *Recent Developments in Separation Science*, Vol. VI (N. N. Li, ed.), CRC Press, Boca Raton, Fla., 1981, p. 51.
24. E. L. Cussler and D. F. Evans, J. Membr. Sci., *6*:113 (1980).
25. E. L. Cussler, AIChE J., 17405, (1971).
26. R. M. Isatt, M. P. Biehl, J. D. Lamb, and J. J. Christensen, Sep. Sci. Technol., *17*:1351 (1982).
27. K. H. Lee, D. F. Evans, and E. L. Cussler, AIChE J., *24*:860 (1978).
28. T. P. Martin and G. A. Davies, Hydrometallurgy, *2*:315 (1977).
29. A. Hochhauser and E. L. Cussler, AIChE Symp. Ser., *71*:136 (1975).
30. S. P. Li and Y. C. Chang, Proc. vol. (II), *CIES/AIChE Joint Meet. Chem. Eng.*, Beijing, China, Sept. 19–22, 1982, Chemical Industrial Press, pp. 571–582.
31. N. N. Li, R. P. Cahn, D. Naden, and R. W. M. Lai, Hydrometallurgy, *9*:277 (1983).
32. H. C. Hayworth, W. S. Ho, W. A. Burns, and N. N. Li, Sep. Sci. Technol., *18*(6):493 (1983).
33. Z. M. Gu, H. F. Zhang, D. T. Wasan, and N. N. Li, paper presented at the National AIChE Meeting, Denver, Colo., Aug. 30, 1983.

34. E. J. Fuller and N. N. Li, J. Memb. Sci., 18:251 (1984).
35. D. T. Wasan, Z. M. Gu, and N. N. Li, "Faraday Disc. Chem. Soc.," 77:67 (1984).
36. Z. Gu, D. T. Wasan, and N. N. Li, Sep. Sci. Technol., 20:599 (1985).
37. Z. Gu, D. T. Wasan, and N. N. Li, J. Membr. Sci., 26:129 (1986).
38. N. N. Li, U. S. Patent 3,617,546 (1971).
39. N. N. Li, U. S. Patent 3,647,488 (1972).
40. N. N. Li, U. S. Patent 4,029,744 (1977).
41. N. N. Li, U. S. Patent 4,064,040 (1977).
42. N. N. Li and A. L. Shrier, in *Recent Developments in Separation Science*, Vol. I (N. N. Li, ed.), CRC Press, Boca Raton, Fla., 1972, p. 163.
43. R. P. Cahn and N. N. Li, Sep. Sci., 9(6):505 (1974).
44. J. W. Frankenfeld and N. N. Li, in *Recent Developments in Separation Science*, Vol. III, Part B (N. N. Li, ed.), CRC Press, Boca Raton, Fla., 1977.
45. T. Kitagawa, Y. Nishikawa, J. W. Frankenfeld, and N. N. Li, Environ. Sci. Technol., 11:602 (1977).
46. R. P. Cahn, N. N. Li, and R. M. Minday, Environ. Sci. Technol., 12(9):1051 (1978).
47. H. H. Downs and N. N. Li, J. Sep. Process Technol., U. K., (1982).
48. R. E. Terry and N. N. Li, J. Membr. Sci., 10:305 (1982).
49. Q. J. Qi, M. X. Xu, P. L. Tang, Z. M. Gu, L. Y. Zhu, and F. J. Cui, paper presented at the Symposium of LM Separations, Dalian, China, Sept. 24–28, 1982.
50. Y. C. Chang and S. P. Li, Desalination, 47:351 (1983).
51. R. M. Kurzeja, D. T. Wasan, and N. N. Li, submitted to Sep. Sci. Technol.
52. N. N. Li, U. S. Patent 3,942,527 (1976).
53. N. N. Li, U. S. Patent 4,183,918 (1980).
54. N. N. Li and W. J. Asher, in *Chemical Engineering in Medicine*, ACS Advances in Chemistry Series, American Chemical Society, Washington, D. C., 1973, p. 1.

55. H. W. Wallace, W. J. Asher, and N. N. Li, Trans. Am. Soc., Artif. Intern. Organs, 21:80 (1973).

56. W. J. Asher, K. C. Bovee, J. W. Frankenfeld, R. W. Hamiton, L. W. Henderson, P. C. Holtzapple, and N. N. Li, *Liquid Membrane System Directed Toward Chronic Uremia*, Kidney International Volume, Vol. 7, 1975, p. S-409.

57. J. W. Frankenfeld, W. J. Asher, and N. N. Li, in *Recent Developments in Separation Science*, Vol. IV (N. N. Li, ed.), CRC Press, Boca Raton, Fla., 1978, p. 39.

58. J. W. Frankenfeld and N. N. Li, in *Treatise of Analytical Chemistry*, Part I, *Theory and Practice*, Wiley, New York, 1982, pp. 251–280.

59. N. N. Li, U. S. Patent 3,740,315 (1973).

60. N. N. Li, U. S. Patent 3,897,308 (1975).

61. N. N. Li, U. S. Patent 4,098,736 (1978).

62. S. W. May and N. N. Li, Biochem. Biophys. Res. Commun., 47(5):1179 (1972).

63. R. R. Mohan and N. N. Li, Biotechnol. Bioeng., 16:513 (1974).

64. S. W. May and N. N. Li, in *Enzyme Engineering*, Vol. 2 (E. K. Pye and L. B. Wingard, eds.), Plenum Press, New York, 1974, p. 77.

65. R. R. Mohan and N. N. Li, Biotechnol. Bioeng., 17(8):1137 (1975).

66. S. W. May and N. N. Li, in *Biomedical Applications of Immobilized Enzymes and Proteins* (T. M. S. Chang, ed.), Plenum Press, New York, 1977.

67. N. N. Li, in *Treatise of Analytical Chemistry*, Part I, *Theory and Practice*, Wiley, New York, 1982, pp. 251–280.

68. *Zinc Recovery*, Lenzing AG, Lenzing, Austria, 1986.

69. R. Marr presentation given at Engineering Foundation Conference on Separation Technology, Scholoss Elman, Germany, Apr. 26–May 1, 1987.

70. "How Exxon Broke New Ground in Complex Fluids to Solve a Serious Drilling Problem," Chem. Eng. News, Dec. 9, 1985; see also "A Cure for Lost Circulation," Lamp Mag., 17 (1985).

71. N. N. Li, U. S. Patent 4,359,391 (1982).

72. C. R. Dawson, N. N. Li, and D. E. O'Brien, U. S. Patent 4,397,354 (1983).

73. C. R. Dawson, N. N. Li, and D. E. O'Brien, U. S. Patent 4,568,392 (1986).
74. W. J. Ward and W. L. Robb, Science, 156:1481 (1967).
75. R. W. Baker, M. E. Tuttle, D. J. Kelly, and H. K. Lonsdale, J. Membr. Sci., 2:213 (1977).
76. W. C. Babcock, R. W. Baker, E. D. LaChapelle, and K. L. Smith, J. Membr. Sci., 7:71 (1980).
77. W. C. Babcock, R. W. Baker, I. D. LaChapelle, and K. L. Smith, J. Membr. Sci., 7:89 (1980).
78. W. C. Babcock, R. W. Baker, E. D. LaChapelle, and K. L. Smith, Proc. Hydromet. 1981, Society of Chemical Industry, Manchester, England, 1981.
79. E. L. Cussler and D. F. Evans, Sep. Purif. Methods, 3(2):399 (1974).
80. J. Strzelbicki and W. Charewicz, J. Inorg. Nucl. Chem., 40:1415 (1978).
81. F. Nakashio and K. Kondo, Sep. Sci. Technol., 15(4):1171 (1980).
82. W. Volkel, W. Halwachs, and K. Schügerl, J. Membr. Sci., 6:19 (1980).
83. Toshihiko Imato, Hosei Ogawa, Shigeharu Morooka, and Yasuo Kato, J. Chem. Eng. Jpn., 14(4):289 (1981).
84. R. Chlarizia, A. Castagnola, P. R. Danesi, and E. P. Horwitz, J. Membr. Sci., 14:1 (1983).
85. Stylianos Sifniades, Theodore Largman, Allen A. Tunick, and Fred W. Koff, Hydrometallurgy, 7:201 (1981).
86. Giulio M. Gasparini, Giuseppe Grossi, Maurizio Casarci, and Armando D'Andrea, paper presented at the International Solvent Extraction Conference, Denver, Colo., Aug. 26–Sept. 2, 1983, p. 395.
87. R. P. Danesi, E. P. Horwitz, R. Chiarizia, and P. Rickert, paper presented at the 3rd Symposium on Separation Science and Technology for Energy Applications, Gatlinburg, Tenn., June 28–July 1, 1983.
88. M. Burgard, P. Elisoamiadana, M. J. F. Leroy, and H. S. Park, paper presented at the International Solvent Extraction Conference, Denver, Colo., Aug. 26–Sept. 2, 1983.
89. Siegfried Weiss, Valerie Grigoriev, and Peter Mühl, J. Membr. Sci., 12:119 (1982).

90. M. Protsch and R. Marr, paper presented at the International Solvent Extraction Conference, Denver, Colo., Aug. 26—Sept. 2, 1983, p. 66.

91. H. C. Hayworth, W. S. Ho, W. A. Burns, Jr., and N. N. Li, Sep. Sci. Technol., 18(6):493 (1983).

92. W. A. Burns, in *Liquid Membrane Seminar*, UMIST/University of Manchester, Manchester, England, 1980.

93. Tatsuya Sekine and Yuko Hasegawa, *Solvent Extraction Chemistry, Fundamentals and Applications*, Marcel Dekker, New York, 1977.

94. Chuei Bing-yi, Yu Jin-fen, and Zhang Liang-ping, paper presented at the International Solvent Extraction Conference, Denver, Colo., Aug. 26—Sept. 2, 1983, p. 1983.

95. W. A. Rikelton, D. S. Flett, and D. W. West, paper presented at the International Solvent Extraction Conference, Denver, Colo., Aug. 26—Sept. 2, 1983, pp. 195—1986.

96. J. J. Christensen, D. J. Eatough, and R. M. Izatt, Chem. Rev., 74:351 (1984).

97. J. D. Lamb, P. R. Brown, J. J. Christensen, I. S. Bradshaw, D. G. Garrick, and R. M. Izatt, J. Membr. Sci., 13:89 (1983).

98. C. F. Reusch and E. L. Cussler, AIChE J., 19:736 (1973).

99. R. M. Izatt, J. D. Lamb, J. L. Oacarson, and J. J. Christensen, paper presented at the 3rd Symposium on Separation Science and Technology for Applications, Gatlinburg, Tenn., June 28—July 1, 1983.

100. J. D. Lamb, R. M. Izatt, D. G. Garrick, J. S. Bradshaw, and J. J. Christensen, J. Membr. Sci., 9:83 (1981).

101. R. M. Izatt, D. V. Dearden, E. R. Witt, D. W. McBride, and J. J. Christensen, Solvent Extr. Ion Exch., 2:459 (1984).

102. J. M. Lamb, J. J. Christensen, J. L. Oscarson, B. L. Nielsen, B. W. Asay, and R. M. Izatt, J. Am. Chem. Soc., 102:6820 (1980).

103. J. M. Lamb, R. M. Izatt, P. A. Robertson, and J. J. Christensen, J. Am. Chem. Soc., 102:2452 (1980).

104. R. R. Swanson, Int. Solvent Extr. Conf., 1977.

105. N. N. Li and J. W. Frankenfeld, in *Handbook of Separation Process Technology*, Wiley, New York, 1987.

106. K. Osseo-Asare, K. L. Lin, and D. J. Chaiko, Proc. Int. Solvent Extr. Conf., Denver, Colo., Aug. 26—Sept. 2, 1983, p.313.

107. K. Osseo-Asare and M. E. Keeney, Sep. Sci Technol., 15(4): 999 (1980).
108. N. N. Li, R. P. Cahn, and A. L. Shier, U. S. Patent 4,081,369 (1978).
109. V. J. Kremesec, Sep. Purif. Methods, 10(2):319 (1981).
110. Pierter Stroever and Padma Prabodh Varahasi, Sep. Purif. Method, 11(1):29 (1982).
111. Kazuo Kondo, Katsuhide Kita, Isao Koida, Jin Irie, and Fumiyuki Nahashio, J. Chem. Eng. Jpn., 12(3) 203 (1979).
112. A. G. Kopp, R. T. Marr, and F. E. Moser, Inst. Chem. Eng. Symp. Ser., 54:279 (1978).
113. W. S. Ho, T. A. Hatton, E. N. Lightfoot, and N. N. Li, AIChE J., 28(4):662 (1982).
114. Gilbert Casamatta, Claude Chavarie, and Henri Angelino, AIChE J., 24(6):945 (1978).
115. L. Boyadzhiev, T. Sapundzhiev, and E. Bezenshek, Sep. Sci., 12(5):541 (1977).
116. W. S. Ho and N. N. Li, paper presented at the ACS Meeting, New York, Aug. 23—28, 1981.
117. Yoshikazu Miyake, Yoichiro Takenoshita, and Masaaki Teramoto, paper presented at the International Solvent Conference, Denver, Colo., Aug. 26—Sept. 2, 1983, p. 301.
118. F. J. Cui, paper presented at the Symposium on Inorganic Chemistry sponsored by Chinese Association of Chemistry, September 1983.
119. M. B. Dines, Regeneration of Liquid Membrane without Breaking Emulsion, U. S. Patent 4,337,225 (1982).
120. E. C. Hsu, N. N. Li, and T. Hucal, U. S. Patent 4,415,426 (1983).
121. E. C. Hsu, N. N. Li, and T. Hucal, U. S. Patent 4,419,200 (1983).
122. E. C. Hsu and N. N. Li, Sci. Technol., 20(243):115 (1985).
123. R. P. Borwankar, C. C. Chan, D. T. Wasan, Z. M. Gu, and N. N. Li, to be published in AIChE J., (1988).
124. R. P. Borwanker, R. M. Kurzeja, and D. T. Wasan, submitted to Environ. Sci. Technol. (1987).

5
Surfactants in Aqueous Emulsification Separation of Oleic and Stearic Acids

NORMAN O. V. SONNTAG *Consultant, Red Oak, Texas*

I.	Introduction	169
II.	Practical Aspects of Emulsion Separation Technology	173
III.	Theoretical Aspects of Emulsion Separation Technology	175
IV.	Optimization of Theoretical and Practical Aspects of Oleic/ Stearic Acid Separation by Aqueous Solutions of Surfactants	186
	References	188

I. INTRODUCTION

Perhaps the first report of surfactant aqueous solutions for the separation of liquid oleic from solid stearic acid occurred in 1905 with the issuance of a German patent to Fratelli Lanza [1]. This pioneering process involved the use of sulfooleic acid, a Twitchell fat-splitting catalyst and one of the earliest anionic surfactants. The patent carefully pointed out that it was necessary to stir the fatty acid mixture with the aqueous surfactant. Use of 1°Be sulfuric acid along with the surfactant was also involved. The fatty acid mixture was added to the bath of water/sulfooleic acid containing sulfuric acid. The upper layer was finally removed by decantation, although filtration was said to be used also to separate the solid fatty acids from the aqueous liquid.

In 1909, Twitchell, who had invented sulfonated fatty acids as fat-splitting catalysts [2], patented the use of these surfactants in water for the separation of oleic and stearic acids [3]. Twitchell's process advocated treating the mixed liquid-solid fatty acids with a relatively small percentage of a sulfofatty acid, then washing out the liquid fatty acid with water from the solid and decanting the liquid oil from the wash water. Neither Fratelli Lanza or Twitchell's processes were ever developed commercially. Evidently, the methods of filtering the liquid acids from the solid acids presented difficulties. The decantation consumed excessively long cycle times. At least 10 to 15 years would expire before the development of centrifugal equipment to efficiently separate layers by specific gravity.

Hapgood [4], in 1921, was the first worker to attempt an aqueous separation of olein and stearin from fats without the use of surfactant. His process, assigned to De Laval Separator Co., involved the use of centrifuges first to separate the lighter olein layer from the lower layer at 90° F, and then to separate the aqueous phase of the lower layer from the stearin at above the melting point of the stearin. Presumably, Hapgood reasoned that the use of water as a separative agent in the first step was needed to provide a higher specific gravity and accordingly, to facilitate the centrifugal separation of the heavier stearin-aqueous suspension from the lighter olein layer. The difficulties in the centrifugal separation, resulting from the intransient nonflowable stearin cake in the centrifuges, made Hapgood's process difficult to operate. No translation to industrial scale was ever attempted.

During the period 1939–1950, Schutte investigated and patented in detail the aqueous separation of partially solidified fat and wax mixtures. His initial U.S. Patents [5–12] involved the use of water as a vehicle with air as an emulsifying agent for the separation of liquid and solid components of waxes by filtration. Subsequently, he extended this technique to other fat or wax separations [7], to fatty acids or glycerides [8–10], and to improvements in the designs of the centrifuges [11,12], which he preferred for separation. This processing approach had advantages over prior technology. However, the use of air or inert gases as emulsifiers did not afford optimum surface properties in the water phase to afford an efficient, trouble-free separation. Schutte's technology was never commercially developed.

It was not until the development of commercial centrifuges that a large impetus was given to the aqueous surfactant separation of oleic and stearic acids. This occurred in Italy and West Germany in the early 1950s. The patent literature substantiates the fact that concurrent work on the separation of natural fat triglycerides, fatty alcohols, and even methyl esters was also carried out. Much of the early development work was concerned with a variety of fatlike materials. Apparently, the first disclosure of success in the centrifugal separation was revealed in a British patent in 1953 [13], assigned to Henkel & Cie, Dusseldorf, West Germany. A large number of international patents were issued between 1955 and 1976. These patents were assigned to Henkel & Cie and covered various separations of

Surfactants in Separation of Acids

oleic and stearic acids, many triglycerides, fatty alcohols, and other esters [14-30].

As far as separation of oleic and stearic acids is concerned, the Henkel separative technique, which has become known today as the Henkel or "hydrophilization" process [31,32], involves the centrifugal separation of a lighter oleic acid layer from a heavier loosely emulsified aqueous stearic acid layer. The presence of inorganic salts in the aqueous phase increases the specific gravity of the heavier lower layer.

Mixed anionic emulsifiers and inorganic salts are used in the aqueous phase to enhance the separation. More recent modification of the separation includes the withdrawal of a portion of the aqueous phase from the stearic acid water emulsion, a melting of the solid, and washing of the melted stearic acid in order to conserve the use of surfactants and salt [33].

In 1956, an Italian patent [34] assigned to SABO-S.A. revealed a process for the separation of mixtures of long-chain fatty alcohols of differing melting points. The process comprised the addition of water, then stirring and chilling to form a mass of two parts. One part of the mass consisted of liquid alcohols of lower melting point. The other was a nonhomogeneous emulsion of solid, higher-melting alcohols in water. The two parts were then separated.

In 1960, the Tai-Yo Fishery Co. Ltd. of Japan patented [35] an aqueous surface-active agent separation in which, for example, soybean fatty acids were separated with the aid of pressure. Sulfuric acid was also present in the aqueous phase.

In 1967, VEB Deutsche Hydrierwerk, Rodleben, East Germany, reported [36] that animal fatty acids could be separated by emulsification with an aqueous solution of anionic detergent. The lighter upper layer containing the unsaturated acids could be removed by centrifugation. At this time, the advantages of this separative technology over solvent separation methods were clearly appreciated.

It was not until 1970 that successful filtration processing involving water surfactant emulsification of oleic and stearic acids was disclosed. An initial Canadian patent [37], followed shortly by an American patent [38], described the Kraft process. This process involved evaporation under vacuum of a liquid such as water to provide the cooling effect. The semisolidified mixture of liquid and solid fatty acids so obtained could then be treated with an aqueous surfactant salt mixture to afford an aqueous oily "loose" emulsion that could readily be separated. The separation was also shown to be possible by means other than centrifugation. In ordinary filtration, both liquefied oleic acid and an aqueous phase containing salt and surfactant could be allowed to pass through a solid cake on a filter surface. Other modifications and innovations of the water separation of oleic and stearic acid quickly followed. Today, the rotary vacuum belt filtration apparatus is entirely adequate for this separation. Alternatively, plate and frame filters and pressure filters such as the Carver Press or the Eimco Burwell Press are also suitable.

One significant modification of the technology is exemplified in a Unilever N.V. patent [39]. Triglycerides were shown to be separable according to melting points by the water emulsification procedure. The separation depended on the addition of surfactants of the nonionic monoglyceride type along with other water-soluble surfactants, such as potassium oleate, sodium decyl sulfate, or sodium dodecyl sulfate. One example involved the use of 2% of *Myverol 1898*, a 90% type of safflower monoglyceride, and 0.6% potassium oleate dispersed in peanut oil at high temperatures. After cooling the dispersion 18 h to 0°C and stirring 1 h with a 5% aqueous solution of sodium sulfate (at a dispersion ratio 0.6:1), 75% of the olein was recovered by centrifugation. Under comparable conditions, the use of 0.5% sodium decyl sulfate and 2% $MgSO_4 \cdot 7H_2O$ (without the potassium oleate) was said to afford an olein in only 55% yield. The use of potassium soaps of unsaturated fatty was claimed to give significantly higher yields than the use of relatively higher priced water-soluble anionic surfactants (e.g., sodium alkyl sulfates).

A Romanian patent assigned to Intreprinderea de Sapun "Stela" [40] indicates how animal fatty acids of acid number 185 to 200, saponification number 190 to 200, and iodine value 55 to 80 may be separated into an unsaturated fatty acid of iodine value 95. The fatty acid mixture was crystallized at 10°C, then was treated with an aqueous solution containing 0.5% sodium lauryl sulfate, C-10 alkyl sulfate, and 2% magnesium sulfate and separated in a centrifuge at 2 atm to give liquid fatty acids. The liquid fatty acids had an acid number of 195 to 205, a saponification number of 195 to 205, and a "congealing point" of 8°C. After splitting, distillation, and separation, 3600 kg of liquid fatty acids was obtained from 4500 kg of animal fat plus 4500 kg of refined fatty acids. Use of refined fatty acids in the feedstock afforded an increase in yield.

It is obvious from what has been developed that a great many innovations and modifications of water emulsification are possible for the separation of fats according to melting point. In the past, the preferred technology has been the use of highly water soluble anionic surfactants augmented with the use of water-soluble salts such as magnesium sulfate, sodium sulfate, or sodium chloride. In the 1950s to 1970s centrifugal separation of an aqueous liquid/solid mixture of fat appeared to be preferable. But at about 1970 the use of other separative equipment, such as filters, presses, and so on, was shown to be suitable. Faced with a choice of separative processing during the period 1972–1979, two American companies (Darling and Co., Chicago, Illinois, and A. Gross and Co. Millmaster Onyx Group, Newark, New Jersey, now Witco-HumKo Corp.), and one Canadian organization (Canada Packers Ltd., Toronto, Ontario), selected the Henkel process to separate stearic and oleic acids with centrifuges. Since that time a number of other American companies have adopted water surfactant separation with the use of various filtration equipment. Apparently, we can conclude from these trends that the economics for water emulsification separation does compete satisfactorily with that for low-temperature solvent crystallization [41]. It is more than likely that water emulsification separation will have displaced solvent separation for oleic acid production in the near future.

II. PRACTICAL ASPECTS OF EMULSION SEPARATION TECHNOLOGY

Prior to about 1950 there were no practical methods for the efficient separation of fatty mixtures by melting point. Attempts were unsuccessful because filtration of fat mixtures or fatty acid ester mixtures, as such or with water alone, were difficult to accomplish. In 1961, Debrus [42] provided a review of the various methods that had been suggested prior to 1960. At or about 1955, it became evident that water emulsification with surfactants could be applied successfully to the separation of either oleic acid from stearic acid or of other fat components, such as triglycerides, fatty alcohols, and methyl esters. Stein in 1960 provided a practical discussion of the methods and procedures for separating fatty substance by the emulsification process [43]. He characterized the separative technique in the following words

> The mixtures of solid and liquid fat substances which is to be separated (in the following for simplicity's sake we generally refer to fatty acids) is treated with an aqueous solution of a surface active substance. As a result of the surface tension activity of these solutions the liquid fatty acids are dissolved ("fluidized") from the outer layers of the crystalline portions, are washed away and are emulsified in the aqueous solution. The outer layer of the crystal, which, in the original material, understandably, was wetted with oil, is now wetted with water, and the crystals move over into the water phase. Similar experiments were conducted by Twitchell about 50 years ago. However, he did not find a suitable method for separating these dispersions. A sedimentation of the dispersion by means of a centrifuge, results in layers of the kind in which the oily portion has separated and is found above the aqueous phase. The water phase contains the total crystalline component. The separation between oil and water is very sharp. The crystalline fat portions are completely contained in the aqueous layer in spite of the high content of solid components. This procedure, with

the aid of a fully jacketed centrifuge, permits separation of these dispersions. This was surprising since the suspension of stearin contained substantial solid material and actually must be designated as sludge. By continuous flow, the centrifuge delivers the liquid fat portion and the aqueous flow separately. The aqueous phase contains the suspended crystalline fat components. The separation of these components from water is accomplished by melting the fat and then using a separatory box or a separator.

Clearly, Stein envisioned that the separation procedure with a centrifuge became a separation of liquid from liquid rather than, as was the case in the past, a filtration separating liquid from solid. Consequently, most of the difficulties which were related in the past to the filtration of fatty materials were eliminated. It is, of course, obvious that the mixture must contain solid and liquid components. Stein continues his summary of the process:

The raw material is continuously crystallized with the aid of a so-called scraper cooler apparatus. The fatty acid mixture which now has been brought to separation temperature is immediately brought in contact with the aqueous solution of surface active substance at which time the rewetting occurs. The dispersion then enters the fully enclosed centrifuge wherein the separation takes place. The liquid fatty acids leave the centrifuge on one side while the dispersion of the solid portion in water leaves from the other side. The solid portions are separated and the aqueous phase is returned for further use in the process.

Braae [44] about 1961 presented a practical discussion of the application of centrifugal separators for the separation of liquid and solid fats with detergent solutions. In 1961, Lange [45] discussed aspects of the phenomena of wetting, spreading, and rewetting as they were appreciated at that time. Practically speaking, the function of inorganic salts in aqueous surfactant separation may be a twofold one where centrifugation is used, and a single function with respect to separation by filtration. The inorganic salts increase the specific gravity of the aqueous phase, thus affording more convenient centrifugation of the lower aqueous layer, and also exerting a desirable effect on limitation of overemulsification of the solid stearic acid. When filtration is applied to the separation, there is no gravity advantage from dissolved salt content, but there is a desirable effect on the surfactants properties. In summary, it may be said that the emulsion is a loose and easily formed one. As a matter of fact, high agitation under shearing conditions overemulsifies the stearic acid in the water phase and renders its separation by a centrifuge or a filter much more difficult.

III. THEORETICAL ASPECTS OF EMULSION SEPARATION TECHNOLOGY

The chemical literature on the physical and colloidal chemistry of surfactants, in both theory and practice, is voluminous. Any attempt to cover completely the theoretical aspects of emulsification, wetting, rewetting, and the other aspects involved in the aqueous emulsification separation of oleic and stearic acid is obviously beyond the scope of the present chapter. However, we attempt a brief treatment of the theoretical principles that underlie the effective separation.

At about the time of the development of the Henkel process, the theoretical background for surface-active agents was adequately covered in several important treatises. In 1949, Schwartz and Perry [46] provided the original edition of their monumental treatise *Surface Active Agents*. This was followed in 1958 by a second volume [47]. The value of both volumes was so important that they were reprinted in 1977—1978 [48]. Becher's original edition of *Emulsions: Theory and Practice* appeared in 1957 [49], with a second edition in 1966 [50]. The second edition was also reprinted in 1977 [51]. Since that time the development of surface-active theory has been enormous. Today (1986) the physical chemistry of surfactant action has been comprehensively condensed in Rosen's *Surfactants and Interfacial Phenomena* [52]. Becher has also provided an updated three-volume treatise titled *Encyclopedia of Emulsion Technology* [53].

One of the earliest discoveries in the physical chemistry of ionizable surface-active agents was the decrease in surface and interfacial tensions in the presence of inorganic salts. The effect of surface tension lowering has been correctly ascribed to ions of appropriate charge which lower the critical concentration of micelle formation. In effect, the salt cations lower the repulsion between surface-active ions. Consequently, more of the surface-active ions can be packed into the surface film. Surfact tension is lowered, correspondingly. Aickin [54] found K^+ to be most effective — with $K^+ > NH_4 > Na^+ > Li^+$ in effectiveness. The least hydrated ions are apparently the most effective in reducing surface tension. In the case of cationic surface-active agents, the added ions were also effective in increasing surface activity.

There is a time lag in attaining an equilibrium value of surface tension. This has been appreciated for many years and has been studied by many investigators. Sodium cetyl sulfate solutions, for example, attain surface equilibrium very slowly (taking several days) in concentrations below the critical concentration for micelle formation. On the other hand, with the bulk solution at higher concentrations, the solution consists largely of micelles. At higher concentrations, equilibrium is attained in a matter of minutes or seconds [55]. In the case of very dilute solutions, however, the rate of fall after the first few hours is very small, and for all practical purposes the value obtained after such a time interval may be regarded as final [56].

Rosen [57] points out that although the question of micelle shape has still not been settled completely, considerable data are available on micelle aggregation numbers and the factors that govern them. In aqueous solution, the aggregation number appears to increase with increase in hydrophobic group, and with increase in the binding of the counterions to the micelle in ionics. Binding of the counterion to the micelle in aqueous solution appears to decrease with an increase in the polarizability and charge. Thus, for alkyl sulfate micelles, the order of decreasing binding is $Cs^+ > Rb^+ > Na^+ > Li^+$ [58]. In Table 1 are summarized the aggregation numbers of a series of sodium alkyl sulfates as determined by Tartar and Lelong [59]. Such numbers could be pertinent in this application because, although the data are too sparse to permit generalization, there could be an insidious relationship between aggregation numbers and performance of sodium alkyl sulfates.

The critical micelle concentrations of several sodium alkyl sulfates, as determined by several workers, are shown in Table 2. Here in an aqueous medium the CMC decreases as the number of carbon atoms in the hydrophobic group increases to about 16. One generalization is that the CMC is halfed by the addition of one methylene group to a straight-chain hydrophobic group attached to a single-terminal hydrophilic group.

Klevens in 1953 [69] developed an empirical relationship between CMC, C_{CMC}, and the number of carbon atoms N in the hydrophobic chain of water-soluble anionic surfactants:

$$\log C_{CMC} = A - BN$$

where A is the constant for a particular ionic head at given temperature and B is a constant 0.3 (= log 2) at 35°C. The relationship

TABLE 1 Aggregation Numbers of Sodium Alkyl Sulfates in Water

Compound	Temperature	Aggregation number
$C_{10}H_{21}SO_4^- Na^+$	23	50
$C_{12}H_{25}SO_4^- Na^+$	23	71

Source: Ref. 59.

TABLE 2 Critical Micelle Concentrations of Sodium, Potassium, and Lithium Alkyl Sulfates in Water

Compound	Temperature	CMC (M)	Reference
$C_8H_{17}SO_4^-Na^+$	40	1.6×10^{-1}	60
$C_{10}H_{17}SO_4^-Na^+$	40	3.3×10^{-2}	61
$C_{11}H_{23}SO_4^-Na^+$	21	1.6×10^{-2}	62
$C_{12}H_{25}SO_4^-Na^+$	40	8.6×10^{3}	63
$C_{13}H_{25}SO_4^-Na^+$	40	4.3×10^{-3}	64
$C_{14}H_{29}SO_4^-Na^+$	40	2.2×10^{-3}	63
$C_{15}H_{31}SO_4^-Na^+$	40	1.2×10^{-3}	64
$C_{16}H_{33}SO_4^-Na^+$	40	5.8×10^{-4}	61
$C_{18}H_{37}SO_4^-Na^+$	50	2.3×10^{-4}	65
$C_{12}H_{25}SO_4^-Na^+$	25	8.2×10^{-3}	66
$C_{12}H_{25}SO_4^-Li^+$	25	8.9×10^{-3}	67
$C_{12}H_{25}SO_4^-K^+$	40	7.8×10^{-3}	68
$C_{12}H_{25}SO_4^-Na^+$ (0.01 M NaCl)	21	5.6×10^{-3}	62
$C_{12}H_{25}SO_4^-Na^+$ (0.03 M NaCl)	21	3.2×10^{-3}	62
$C_{12}H_{25}SO_4^-Na^+$ (0.1 M NaCl)	21	1.5×10^{-3}	62
$C_{16}H_{33}SO_4^-Na^+$	40	5.9×10^{-4}	61

was applicable to a correlation of structural units in anionic surface-active agents such as soaps, alkanesulfonates, and alkyl sulfates in aqueous medium. The relationship makes it apparent that the statement of the CMC being halfed for each increase in the hydrophobic chain of one carbon is a valid one. This relation is of interest in the context of this chapter because it leads to a practical determination of optimum chain length in the anionic surfactant in the absence of dissolved salts. Table 3 lists constants for the relationship

$$\log C_{CMC} = A - BN$$

which have been determined for n-alkyl sulfates.

The order of effectiveness for ions in electrolytes in decreasing the CMC of nonionic surfactants is $\frac{1}{2}SO_4^{2-} > F^- > BrO_3^- > Cl^- > Br^- > NO_3^- > I^- > CNS^-$ [72] and $NH_4^+ > K^+ > Na^+ > Li^+ > \frac{1}{2}Ca^{2+}$ [74]. On the other hand, the order of decreasing effectiveness of the anion in depressing the CMC for sodium laurate and sodium napthenate solutions is $PO_4^{3-} > B_4O_7^{-2} > OH^- > CO_3^{2-} > HCO_3^- > NO_3^- > Cl^-$ [73].

Anionic lauryl sulfates exhibit a decrease of CMC with added electrolyte in the order $Li^+ > Na^+ Cs^+ > N(CH_3)_4^+ > N(C_2H_5)_4^+ > Ca^{2+}, Mg^{2+}$ [74]. When electrolytes are added to aqueous solutions of ionic surfactants, the aggregation number is increased. Possibly, the ionic heads have an electrical double layer; this is compressed with the result that there is a mutual repulsion in the micelle. More ions are present in the micelle with no increase in free energy of the system.

Corrin's equation [75] relates the concentration of electrolyte with the CMC, as follows:

$$\log C_{CMC} = -a \log C_i + b$$

TABLE 3 Constants for the Relationship $\log C_{CMC} = A - BN$

Surfactant	Temperature (C°)	A	B	Reference
Sodium n-alkyl-1-sulfates	45	1.4_2	0.30	69
Sodium n-alkyl-1-sulfates	60	1.3_5	0.28	70
Sodium n-alkyl-2-sulfates	55	1.2_8	0.27	71

where a and b are constants for a given ionic head at a certain temperature and C_i is the total (monovalent) counterion concentration in moles per liter. Thus in the separation systems described, the relevance is that a delineation of the magnitude of the decrease in CMC, at least in the presence of monovalent salts, is possible.

In addition to considerations of the emulsification performance of the surfactant in the oleic-stearic acid separation (as reflected by surface tension reductions), and the various phenomena related to critical micelle concentration, a review of the fundamental adsorption and wetting theories that relate to the separation is pertinent. The separation is extremely complex because there is the dual adsorption of surfactant from the aqueous phase onto the solid hydrophobic surface of the stearic acid and a simultaneous adsorption of surfactant onto the surface of the liquid hydrophobic oleic acid.

Rosen [76,77] points out that there are six mechanisms to consider at a solid/liquid interface: ion exchange [78–80], ion pairing [79,80], hydrogen bonding [79–81], adsorption by polarization of electrons [81], adsorption by dispersion forces [80,82], and hydrophobic bonding. The latter mechanism occurs to a certain extent when similar hydrophobic groups in the surfactant are attracted sufficiently to enable them to escape from the aqueous environment and to be adsorbed onto the solid stearic acid by chain aggregation. It is even possible for surfactant molecules to be adsorbed out of the aqueous phase onto or adjacent to other surfactant molecules already adsorbed [78,83,84].

Aveyard and Haydon [85] give the fundamental equation for the calculation of the "adsorbed" solvate (component 2) from a binary solution onto a solid adsorbent:

$$\frac{n_0 \Delta x_2}{m} = n_2^s - n_1^s x_2$$

where n_0 is the total number of moles of solution before adsorption, $\Delta x_2 = x_{2,0} - x_2$, $x_{2,0}$ is the mole fraction of component 2 before adsorption, x_1 and x_2 are the mole fractions of components 1 and 2 at adsorption equilibrium, m is the mass of adsorbent in grams, and n_1^s and n_2^s are the number of moles of components 1 and 2 adsorbed per gram of adsorbent at adsorption equilibrium. In the case where component 2 is much more strongly adsorbed onto the solid adsorbent than the solvent (component 1) (undoubtedly the case with sodium decyl sulfate from water onto stearic acid) and $n_0 \Delta x_2 \approx n_2$, where n_2 is the change in the number of moles of component 2 in solution, $n_1^s \approx 0$, and $x \approx 1$; therefore

$$n_2^s = \frac{n_2}{m} = \frac{C_2 V}{m}$$

where $C_2 = C_{2,0} - C_2$, $C_{2,0}$ is the molar concentration (in moles per liter) of component 2 before adsorption, C_2 the molar concentration of component 2 (the surfactant) at adsorption equilibrium, and V the volume of the liquid phase, in liters.

Rosen also points out how the adsorption isotherm may be determined. For dilute solutions of surfactants, the number of moles of surface-active solute absorbed per unit mass of the solid substrate may be calculated from the concentration of the solute in the liquid phase before and after the solution is mixed with the solid adsorbent and equilibrium reached. A number of analytical techniques are currently available for determining a change in the concentration of the surfactant [86]. The surface concentration, C_2^s, in mol/cm^2, of the surfactant may be calculated if a_s, the surface area per unit mass of the solid adsorbent, in cm^2/g (the specific surface area) is known.

$$C_2^s = \frac{\Delta C_2 \times V}{a_s \times m}$$

The adsorption isotherm can then be plotted in terms of C_2^s as a function of C_2. The surface area per adsorbate molecule on the adsorbent a_2 in square angstroms (for values in nm^2, divide by 10^2) is

$$a_2 = \frac{10^{16}}{N C_2^s}$$

where N is Avogadro's number.

Langmuir's adsorption isotherm [87] was developed to explain adsorption under strict conditions in which (1) the adsorbent is homogenous, (2) both solute and solvent has equal molar surface areas, (3) both surface and bulk phases exhibit ideal behavior, and (4) the adsorption film is monomolecular. It is surprising how closely many surfactant solutions follow the relationship despite the obvious nonadherence to the conditions specified.

$$C_2^s = \frac{C_m^s C_2}{C_2 + a}$$

where C_2^s is the surface concentration of the surfactant, in mol/cm^2; C_m^s the surface concentration of the surfactant, in mol/cm^2, at monolayer adsorption; C_2 the concentration of the surfactant in the liquid phase at adsorption equilibrium, in mol/l; and a a constant [= 55.5 exp ($\Delta G^\circ/RT$)], in mol/l, at a given temperature T, where ΔG° is the free energy of adsorption at infinite dilution. Unfortunately, much of the adsorption data in the literature has been obtained on materials with strongly charged sites such as quartz, α-alumina, silver iodide, and the like. Relatively few studies have been concerned with adsorption from aqueous solutions onto nonpolar, hydrophobic adsorbents.

Somewhat related to the case of adsorption of homologous sodium alkyl sulfates onto stearic acid is that prevailing in the case of the adsorption of well-purified monofunctional anionic and cationic surfactants onto polyethylene or polypropylene [88]. Here the adsorption behavior is in accord with the Langmuir type. The adsorbate is perpendicular to the substrate and the phenomenon shows surface saturation in the range of the critical micelle concentration of the adsorbate. Adsorption onto these substrates may be primarily by dispersion forces.

Electrolyte addition causes an increase in both the efficiency of adsorption of ionic surfactants and the effectiveness of adsorption. In the former case, the neutral electrolyte decreases the electrical repulsion between similarly charged adsorbed ions, permitting close packing. These tendencies relate to the separation process by tending to show how inorganic salts, while they decrease emulsification of the liquid and solid phases, increase the absorption of the surfactant on both phases, thus increasing the wetting performance of the system. In the case of adsorption from one liquid surface to another, the use of adsorption isotherms is experimentally difficult; consequently, the uses of surface (or interfacial) tensions at the surface boundaries are employed to describe adsorption phenomena.

The fundamental adsorption equation of Gibbs [89], in its most general form, describes liquid/liquid phenomena adequately:

$$d\gamma = -\sum_i \Gamma_i d\mu_i$$

Where $d\gamma$ is the change in surface or interfacial tension of the solvent, Γ_i the surface excess concentration of any component of the system, and $d\mu_i$ the change in chemical potential of any component of the system.

Generally, for dilute aqueous solutions of ionic surfactants in the absence of other electrolyte, the Gibbs equation is used in the form

$$d\gamma = -2.303 RT \Gamma_2 \, d \log C_2$$

For dilute solutions containing electrolytes (most commercial surfactants contain free sodium sulfate or other salts), the activity coefficient is given by the Debye-Hückel theory as

$$\ln f_i = -K |Z_+ Z_-| \sqrt{1/2 \Sigma_i c_i z_i^2}$$

where Z_i is the valence of any ion in the system and K is a constant. Then the solute is a completely dissociated 1:1 electrolyte,

$$\ln f_2 = -K \sqrt{C_2}$$

and

$$d \ln f_2 = \frac{-K dC_2}{2 \sqrt{C_2}}$$

For solutions of 10^{-2} M or less, certain assumptions can be made and the Debye-Hückel equation may be applied in simple form without significant error as

$$d\gamma = -4.606 RT \Gamma_2 d \log C_2$$

where activity coefficients are expected to deviate significantly from unity, for example, when divalent or multivalent ions are present in the solution or concentrations of the surfactant exceed 10^{-2} M, d log C_2 in the appropriate equation may be replaced by $d(\log C_2 + \log f_2)$ and log f_2 calculated in water at 25°C, from the Debye-Hückel equation,

$$\log f_2 \frac{-0.509 |Z_+ Z_-| \sqrt{I}}{1 + 0.33a \sqrt{I}}$$

where the total ionic strength of the solution

$$I = 1/2 \Sigma_i C_i z_i^2$$

and a is the mean distance of approach of the ions, in angstroms [90].

Rosen [91] presents an excellent treatment of the calculation of surface concentrations and area per molecule at the interface by the use of the Gibbs equation, and also discusses the effectiveness of adsorption at liquid-liquid interfaces. Table 4 lists some effectiveness of adsorption $\Gamma m°$, area per molecule at surface saturation a_m, and efficiency of adsorption pC_{20} data of sodium alkyl sulfates at the aqueous solution/heptane interface.

Several generalizations result from physicochemical data, which have implications for the separation of oleic and stearic acids:

1. Change in the length of the hydrophobic group of ionic surfactants beyond 10 carbon atoms appears to have almost no effect on the effectiveness of adsorption at an aqueous solution/heptane interface.
2. Branching of the alkyl chain in the hydrophobic group also appears to have an unusually small effect — a small increase in in the area per molecule at the interface. However, when the number of carbon atoms in a straight-chain hydrophobic group exceed 16 at the aqueous solution/heptane interface, there is a significant decrease in the effectiveness of adsorption, which has been attributed [96] to coiling of the long chain, with a consequent increase in the cross-sectional area of the molecule at the interface. In its effect on adsorption, a carbon atom on a branch of the hydrophobic group appears to be equivalent to about one-half of a carbon atom in a straight chain. A phenyl group that is part of a hydrophobic group has the effect of about three and one half $-CH_2-$ groups in a straight hydrophobic chain. This suggests that straight-chain-medium-length (C_8 to C_{10}) sodium alkyl sulfates might exhibit optimum absorption properties.
3. In ionic surfactants, those with more tightly bound counterions [ions with small hydrated radii (e.g., Cs^+, K^+, NH_4^+)] appear to be more effectively adsorbed than those with less tightly bound ones (Na^+, Li^+, F^-), although the effect, except for tetraalkylammonium salts [97], is rather small. On the other hand, the nature of the hydrophilic group has a major effect on the effectiveness of adsorption. In general, the area per molecule at aqueous solution interfaces appears to be determined by the cross-sectional area of the hydrated hydrophilic group at the interface. Carboxylates generally give higher saturation values than sulfonates or organic sulfates [93].
4. Factors producing a significant change in Γm are the nature of the nonaqueous phase in adsorption at the liquid-liquid interface. It has been found that saturation adsorption increases with an increase in the interfacial tension between the two phases [93]. Temperature changes from 20 to 85°C have little effect on saturation adsorption. A very convenient measure of the efficiency of adsorption at a liquid-liquid interface is the negative logarithm of the concentration of surfactant in the bulk phase required to produce a 20-dyn/cm reduction in the surface or interfacial tension of the solvent, $-\log C\ (-\Delta y = 20) \equiv pC_{20}$.

Following are some conclusions relating efficiency of adsorption to the structure of sodium alkyl sulfates:

1. Efficiency of adsorption at liquid-liquid interfaces increases linearly with the number of carbon atoms in the straight-chain hydro-

TABLE 4 Effectiveness of Adsorption, Γ_m; Area/Molecule at Surface Saturation, a_m; and Efficiency of Adsorption, pC_{20}, of Sodium Alkyl Sulfates at the Aqueous Solution/Heptane Interface

Compound	Temperature (°C)	Γ_m (moles/cm^2 × 10^{10})	a_m (Å^2)	pC_{20}	References
$n\text{-}C_8H_{17}SO_4^-Na^+$	50	2.3	72	1.61	92
$n\text{-}C_9H_{19}SO_4^-Na^+$	20	3.0	56±2	—	93
$n\text{-}C_{10}H_{21}SO_4^-Na^+$	50	3.0	54	2.11	92, 93
$n\text{-}C_{10}H_{21}SO_4^-Na^+$	50	3.2	52	—	94
$n\text{-}C_{12}H_{25}SO_4^-Na^+$	20	3.1	53	—	93

$n\text{-}C_{12}H_{25}SO_4^-Na^+$	50	2.0_5	56	2.72	92, 93
$n\text{-}C_{12}H_{25}SO_4^-Na^+$	50	3.2	52	—	94
$n\text{-}C_{14}H_{29}SO_4^-Na^+$	50	3.2	52	3.31	92, 93
$n\text{-}C_{14}H_{29}SO_4^-Na^+$	50	3.2_5	51	—	94
$n\text{-}C_{16}H_{33}SO_4^-Na^+$	50	3.0_5	54	3.89	92, 93
$(C_{10}H_{21})(C_7H_{15})CHSO_4^-Na^+$	20	2.8_5	58	—	95
$n\text{-}C_{18}H_{37}SO_4^-Na^+$	50	2.3	72	4.42	92

phobic group, suggesting, perhaps, that available and economical saturated C_{20} to C_{24} homologs from fish oils be incorporated in suitable anionic detergents.

2. pC_{20} is increased by 0.56 to 0.6 for each increase in the chain by two $-CH_2-$ groups (a surface concentration close to saturation value can be obtained with only 25 to 30% of the bulk-phase surfactant concentration). (This suggests that an optimum choice of chain-length selection of surfactant would result in more economical and effective surfactant usage.) In contrast to the effectiveness of adsorption, where increases beyond C_{16} chain lengths decreases effectiveness, efficiency of adsorption appears to increase steadily with an increase in chain length of the hydrophobic group up to at least 20 carbon atoms (suggesting, in contrast to conclusion 1, that a compromise between effectiveness and efficiency of adsorption must always be made for the choice of optimum surfactant chain length).

Theoretical treatments of the phenomenon of wetting are usually limited to those involving the ability of water or an aqueous solution to displace air from a liquid or solid surface (e.g., the wetting of cotton by surfactant solutions as in the Draves test) [98]. The separation of oleic-stearic acids by aqueous surfactant solutions, as it is generally carried out, is probably a wetting phenomenon of the immersion type in which the aqueous surfactant phase displaces liquid oleic acid from the surface of solid stearic acid. It is well known that the addition of electrolytes to aqueous surfactants increases the wetting power along with the decrease in surface tension [99].

The theory is quite sparse in the area of explaining the advantageous effect of bivalent salts such as magnesium sulfate in the aqueous phase on the mechanism of rewetting. Suffice it to say that there is as yet no complete theoretical explanation for the superior performance of magnesium compared to sodium sulfate. Presumably, the known decrease in the CMC or the decrease of surface tension has, at best, a secondary effect on the ability of magnesium sulfate to function as a rewetting additive.

IV. OPTIMIZATION OF THEORETICAL AND PRACTICAL ASPECTS OF OLEIC/STEARIC ACID SEPARATION BY AQUEOUS SOLUTIONS OF SURFACTANTS

The developed industrial processing for most triglyceride, fatty alcohol, and fatty acid aqueous surfactant separations calls for the use of about 0.5% (by weight) of sodium lauryl sulfate (sodium decyl sulfate appears to be preferred in Henkel separations) and about 2% (by weight) of magnesium sulfate solutions. A number of other conditions are important and are necessary if optimum separation is to be achieved. These include the requirement to crystallize the liquid fatty acid raw material slowly so that relatively large size crystals of solid fat are produced before the treatment with the aqueous surfactant, the need for minimum agitation during the emulsification-wetting step to avoid size reduction of the solid crystals, a preference for carrying out the emulsification-wetting step by introducing the wetting agent first, followed by the inorganic salt, and an avoidance of any conditions that will produce "overemulsification."

Stein [43] appreciated that his developed Henkel or "hydrophilization" process was more a wetting phenomenon than an emulsification. In the early days he referred to his process as the "rewetting" process. It is now apparent that all these separations are examples primarily of a rewetting and adsorption phenomenon in which simultaneous emulsification of the solid and liquid components is purposely severely restricted. The adsorption of the sodium alkyl sulfate surfactant on the stearic acid solid phase is maximized. The decyl homolog is preferred because this appears to be the optimum chain length among sodium alkyl sulfates for highest effectiveness of adsorption. However, the cost of sodium lauryl sulfate from whole coconut oil raw material is substantially less than that of sodium decyl sulfate, which requires costly fractional distillation of the methyl "cocoate" intermediates. Sodium lauryl sulfate surfactant corresponds to an average chain length of about 12.8. However, the use of this material in oleic-stearic aqueous surfactant separations is a compromise between its price and its efficiency. During the past five years in the United States, the price of sodium lauryl sulfate, on a salt-free 100% solids basis, has ranged from 30 to 40 cents/lb less than that of sodium decyl sulfate as a 90% C_{10} chain-length composition. Considering the efficiency involved in the use of the decyl homolog and the relatively small quantities required, its preferential use now appears to be well established.

The use of magnesium sulfate as the preferred inorganic salt is based primarily on its moderating effect on the overemulsification of solid stearic acid — and less so on the liquid oleic acid phase. The bivalent ionic salt, although it decreases surface tension somewhat and increases the aggregation number (micelles are larger), lowers the critical micelle concentration, thus lessening the overall tendency to emulsify the solid. The effectiveness of the SO_4^{2-} ion in accomplishing this has probably been largely underestimated in the past. Note that bivalent cations such as Ca^{2+} and Mg^{2+} depress the CMC less than monovalent salts such as Na^+, K^+, or Li^+ [74].

Magnesium sulfate has a secondary advantage where centrifugation is employed in that the separation step is made more efficient or speeded up because of the higher specific gravity of the aqueous salt/stearic phase, but this is minor. The use of purified magnesium sulfate is worth the increased cost over sodium chloride. Another advantage is the fact that magnesium decyl sulfate and magnesium sulfate through ionic interchange has increased solubility over the sodium salt which is of value in the use of hard water aqueous solutions. Thus magnesium sulfate, although higher in cost than sodium sulfate, is preferred.

An important economic consideration in aqueous surfactant separations is the recovery and reuse of both surfactants and salts. Usually, loss of these additives can amount to as much as 2.2 kg/ ton of wetting agent solution per mass if nondistilled fatty acid feedstocks are employed. By a "withdrawal" and recycle technique [33], the loss of surfactant could be lowered to an 0.8-kg/ton solution. The monitoring and process control of both surfactant and salt content can be accomplished very satisfactorily by in line computer-controlled analysis and self-controlled adjustments of the concentrations.

One area in which additional improvement in the ease and efficiency of separation of oleic-stearic acids by aqueous surfactants can be achieved is in the use of additional substances such as nonionic surfactants and certain polymeric colloid materials. Our own experience is that higher recoveries can be achieved in the separation of tallow fatty acids through the presence of 0.5% sodium decyl sulfate, 2% magnesium sulfate, and 1.5% of nonhydrogenated tallow monoglyceride of the 90% type. A larger recovery of unsaturated fatty acid components from the solid acids is accomplished by filtration through a conventional rotary vacuum filter. The improvement in the separation is of the order of 9.3% [100].

REFERENCES

1. Fratelli Lanza, Ger. Patent 191,238 (1905).
2. E. Twitchell, U.S. Patent 601,603 (1898); JACS, 22:22—26 (1900).
3. E. Twitchell, U.S. Patent 918,612 (1909).
4. C. H. Hapgood, U.S. Patent 1,381,705 (1921).
5. A. H. Schutte, U.S. Patent 2,168,140 (1939).
6. A. H. Schutte, U.S. Patent 2,168,306 (1939).
7. A. H. Schutte, U.S. Patent 2,296,456 (1942).
8. A. H. Schutte, U.S. Patent 2,296,457 (1942).
9. A. H. Schutte, U.S. Patent 2,296,458 (1942).
10. A. H. Schutte, U.S. Patent 2,370,999 (1945).
11. A. H. Schutte, U.S. Patent 2,431,142 (1947).
12. A. H. Schutte, U.S. Patent 2,534,210 (1950).
13. Henkel & Cie, Brit. Patent 702,354 (1953).

14. Henkel & Cie, Australian Patent 158,773 (1954).
15. Henkel & Cie, Ger. Patent 925,674 (1955).
16. Henkel & Cie, Brit. Patent 724,222 (1955).
17. Henkel & Cie, Brit. Patent 743,166 (1956).
18. Henkel & Cie, Ger. Patent 1,010,062 (1957).
19. W. Stein and H. Hartmann, U.S. Patent 2,800,493 (to Henkel & Cie) (1957).
20. Henkel & Cie, Ger. Patent 970,292 (1958).
21. Henkel & Cie, Ger. Patent 1,088,490 (1960).
22. H. Hartmann and W. Stein, U.S. Patent 2,972,636 (to Henkel & Cie) (1961).
23. Henkel & Cie, Ger. Patent 1,107,659 (1961).
24. Henkel & Cie, Ger. Patent 1,102,739 (1962).
25. H. Waldmann and W. Stein, U.S. Patent 3,052,700 (to Henkel & Cie) (1962).
26. Henkel & Cie, Ger. Patent 1,136,786 (1962).
27. Henkel & Cie, Ger. Patent 1,256,645 (1967).
28. H. Hartmann and W. Stein, U.S. Patent 3,733,343 (to Henkel & Cie) (1973); Ger. Patent 2,155,988 (1973).
29. W. Stein and H. Hartmann, U.S. Patent 3,737,444 (to Henkel & Cie) (1973); Ger. Patent 2,030,569 (1971).
30. W. Stein and H. Hartmann, U.S. Patent 3,953,485 (to Henkel & Cie) (1976).
31. Chem. Eng. News, May 15, 1967, pp. 62–63; J. Am. Oil Chem. Soc., 45(10):569A (1968) (Lurgi); Chem. Week, June 7, 1978, p. 42 (Rust Eng. Co.).
32. W. Stein, J. Am. Oil Chem. Soc., 45(6):471 (1968).
33. W. Stein and H. Hartmann, U.S. Patent 4,009,213 (to Henkel & Cie) (1977).
34. S. A. Prodotti Chimica Botozzi & Co. a. Bergamo, Ital. Patent 542,115 (1956).
35. K. Azuma, I. Koshiro, and S. Tsukamoto, Jap. Patent 2685 (to Tai-Yo Fishery Inc.) (1960).
36. G. Lauermann (VEB Deut. Hydrierwerk, Rodleben, Germany), Abh. Dtsch. Akad. Wiss. Berlin Kl. Chem. Geol. Biol. 1966, 6:502 (1967).

37. G. R. Payne, W. B. Campbell, and N. S. Yanick, Can. Patent 837,647 (to National Dairy Prod. Corp.) (1970).
38. G. R. Payne, W. B. Campbell, and N. S. Yanick, U.S. Patent 3,541,122 (to Kraftco Corp.) (1970).
39. J. H. M. Rek, U.S. Patent 4,049,687 (to Unilever N.V.) (1977).
40. D. Longu, I. Floree, and I. Mitrea, Rom. Patent 82,244 to Intreprinderea de Sapun "Stela" (1983).
41. N. O. V. Sonntag, J. Am. Oil Chem. Soc., 56(11):861A (1979).
42. O. J. Debrus, Seifen Oele Fette Wachse, 87(12):344 (1961).
43. W. Stein, *Proc. Orig. Res. 3rd Int. Congr. Surf. Act. Subs.*, Vol. 4, Sec. D, pp. 120—125, Cologne, 1960.
44. B. Braae, *Application of Centrifugal Separators for the Fractionation of Liquid and Solid Fats by Means of Detergent Solution*, A B Separator, Stockholm, Sweden, about 1961.
45. H. Lange, *Proc. 3rd Int. Cong. Surf. Act. Subst.*, Vol. II, Sec. B, pp. 415—425, Cologne, 1961.
46. A. M. Schwartz and J. W. Perry, *Surface Active Agents, Their Chemistry and Technology*, Interscience, New York, 1949.
47. A. M. Schwartz, J. W. Perry, and J. Berch, *Surface Active Agents and Detergents*, Vol. II, Interscience, New York, 1958.
48. A. M. Schwartz and J. W. Perry, *Surface Active Agents, Their Chemistry and Technology*, R. E. Krieger, Melbourne, Fla., 1978; A. M. Schwartz, J. W. Perry, and J. Berch, *Surface Active Agents and Detergents*, Vol. II, R. E. Krieger, Melbourne, Fla., 1977.
49. P. Becher, *Emulsions: Theory and Practice*, Reinhold, New York, 1957.
50. P. Becher, *Emulsions: Theory and Practice*, 2nd ed., ACS Monograph Series 162, Reinhold, New York, 1966.
51. P. Becher, *Emulsions: Theory and Practice*, 2nd ed., ACS Monograph Series 162, R. E. Krieger, Melbourne, Fla., 1977.
52. M. J. Rosen, *Surfactants and Interfacial Phenomena*, Wiley, New York, 1978.
53. P. Becher, ed., *Encyclopedia of Emulsion Technology*, Vol. 1, *Basic Theory*, Vol. 2, *Application*, Vol. 3, *Basic Theory/Measurements/Applications*, Marcel Dekker, New York, 1983, 1985, 1987.
54. R. G. Aickin, J. Soc. Dyers Colour., 60:36 (1944).
55. A. Lottermoser and B. Baumgurtel, Kolloid-Beih., 41:73 (1935); Trans. Faraday Soc., 31:200 (1935); N. K. Adam and H. L.

Shute, Trans. Faraday Soc., 34:758 (1938); G. C. Nutting, F. A. Long, and W. D. Harkins, J. Am. Chem. Soc., 62:1496 (1940); 63:84 (1941).

56. E. Dreger, G. I. Keim, G. D. Miles, L. Shedlovsky, and J. Ross, Ind. Eng. Chem., 36:610 (1944); G. D. Miles and L. Shedlovsky, J. Phys. Chem., 48:57 (1944).

57. M. J. Rosen, *Surfactants and Interfacial Phenomena*, Wiley, New York, 1978, pp. 83—93.

58. I. D. Robb and R. Smith, J. Chem. Soc. Faraday Trans. 1, 70:287 (1974).

59. H. V. Tartar and A. Lelong, J. Phys. Chem., 59:1185 (1955).

60. H. B. Klevens, J. Phys. Colloid Chem., 52:130 (1948).

61. H. C. Evans, J. Chem. Soc., 1956:579 (1956).

62. H. F. Huisman, Proc. K. Ned. Akad. Wet., Ser. B, 67:388 (1964).

63. B. D. Flockhart, J. Colloid Sci., 16:484 (1961).

64. E. Götte and M. J. Schwuger, Tenside, 3:131 (1969).

65. E. Götte, *Proc. 3rd Int. Cong. Surface Activity, Cologne, 1:* 331 (1966).

66. P. H. Elworthy and K. J. Mysels, J. Colloid Sci., 21:331 (1966).

67. K. J. Mysels and L. H. Princen, J. Phys. Chem., 63:1696 (1959).

68. K. Meguro and T. Kondo, J. Chem. Soc. Japan. Pure Chem. Sec., 77:1236 (1956).

69. H. B. Klevens, J. Am. Oil. Chem. Soc., 30(2):74 (1953).

70. M. J. Rosen, J. Colloid Interface Sci., 56:320 (1976).

71. M. J. Schick and F. M. Fowkes, J. Phys. Chem., 61:1062 (1957).

72. N. H. Schick, J. Colloid Sci., 17:801 (1962).

73. P. A. Demchenko, N. N. Zakharova, and L. G. Demchenko, Ukr. Khim. Zh., 28:611 (1962).

74. M. J. Rosen, *Surfactants and Interfacial Phenomena*, Wiley-Interscience, New York, 1978, p. 102.

75. M. L. Corrin and W. D. Hawkins, J. Am. Chem. Soc., 69:683 (1947).

76. M. Rosen, *Surfactants and Interfacial Phenomena*, Wiley, New York, 1978, pp. 26—34.

77. M. Rosen, J. Am. Oil Chem. Soc., 52(11):431 (1975).

78. T. Wakamatsu and D. W. Fuersfenau, in *Adsorption from Aqueous Solution* (W. J. Wever, Jr., and E. Natijeric, eds.), American Chemical Society, Washington, D.C., 1968, pp. 161—172.

79. H. Rupprecht and H. Liebl, Kolloid Z. Z. Polym., 250:719 (1972).

80. J. P. Law, Jr., and G. W. Kunze, Soil Sci. Soc. Am. Proc., 30:321 (1966).

81. L. R. Snyder, J. Phys. Chem., 72:489 (1968).

82. H. Kölbel and K. Hoerig, Angew. Chem., 71:691 (1959); H. Kölbel and P. Kuehn, Angew. Chem., 71:211 (1959).

83. S. G. Dick, D. W. Fuerstenau, and T. W. Healy, J. Colloid Interface Sci., 37:595 (1971).

84. C. H. Giles, A. P. D'Silva, and I. A. Easton, J. Colloid Interface Sci., 47:766 (1974).

85. R. Aveyard and D. A. Haydon, *An Introduction to the Principles of Surface Chemistry*, Cambridge University Press, Cambridge, 1973.

86. M. J. Rosen and H. A. Goldsmith, *Systematic Analysis of Surface Active Agents*, 2nd ed., Wiley-Interscience, New York, 1972.

87. L. Langmuir, J. Am. Chem Soc., 40:1361 (1918).

88. F. G. Greenwood, G. D. Parfitt, N. H. Picton, and D. G. Wharton, in *Adsorption from Aqueous Solution*, (W. J. Weber, Jr., and E. Matijevic, eds.), Advances in American Chemistry Series 79, American Chemistry Society, Washington, D.C., 1968, pp. 135—144.

89. J. W. Gibbs, *The Collected Works of J. W. Gibbs*, Vol. I, Longman, Green and Co., London, 1928, p. 119.

90. E. A. Boucher, T. M. Grinchuk, and A. C. Zettlemoyer, J. Am. Oil Chem. Soc., 45:49 (1968).

91. M. J. Rosen, *Surfactants and Interfacial Phenomena*, Wiley, New York, 1978, pp. 59—71.

92. W. Kling and H. Lange, Proc. 2nd Int. Congr. Surf. Act., London, 1951, Vol. I, p. 295.

93. F. van Voorst Vader, Trans. Faraday Soc., 56;1067 (1960).

94. H. Lange, Kolloid Z., *152*:155 (1957).
95. J. R. Livingston and R. Drogin, J. Am. Oil Chem. Soc., *42*: 720 (1965).
96. P. Mukerjee, Adv. Colloid Interface Sci., *1*(3):241 (1967).
97. K. Tamki, Bull. Chem. Soc. Jpn., *40*:38 (1967).
98. H. Draves, Am. Dyest. Rep., *28*:425 (1939).
99. M. J. Rosen, *Surfactants and Interfacial Phenomena*, Wiley, New York, 1978, p. 197.
100. N. O. V. Sonntag, unpublished data.

6
Surfactants in Flotation

P. SOMASUNDARAN and R. RAMACHANDRAN *Henry Krumb School of Mines, Columbia University, New York, New York*

I.	Introduction	195
II.	Flotation Technique	197
	A. Thermodynamic Aspects	198
	B. Surfactants for Flotation	200
III.	Surfactant Adsorption	201
	A. Electrostatic Factors	201
	B. Chain-Chain Interactions	209
	C. Chemical Forces	210
	D. Surface Precipitation	212
	E. Xanthate Interactions	215
	F. Surface Chelation	219
	G. Surfactant Solution Chemistry	222
IV.	Modifying Agents	225
References		230

I. INTRODUCTION

Flotation is an important separation technique today, with its applications ranging from selective separation of minerals to microorganisms and even ions. The flotation process utilizes the differences in the surface properties of the particulates, normally with the addition of reagents to achieve the separation. Surfactants play a critical role in the flotation process since their interaction with the

TABLE 1 Flotation Techniques Classified on the Basis of Mechanism of Separation and Size of Material Separated

Mechanism	Size range		
	Molecular	Microscopic	Macroscopic
Natural surface activity	Foam fractionation; example: detergents from aqueous solutions	Foam flotation; examples: microorganisms, proteins	Froth flotation of non-polar minerals; example: sulfur
In association with surface-active agents	Ion flotation, molecular flotation, adsorbing colloid flotation; examples: Sr^{2+}, Pb^{2+}, Hg^{2+}, cyanides	Microflotation, colloid flotation, ultraflotation; examples: particulates in wastewater, clay, microorganisms	Froth flotation; example: minerals such as silica Precipitate flotation (first and second kind); example: ferric hydroxide

Source: Ref. 1.

particle surface essentially determines the hydrophobicity of the particles and their probability of attachment to bubbles during collisions in the flotation cell. A large number of surfactants and auxiliary reagents are used in flotation process depending on the properties of the particles to be processed and the ultimate aim of the process. In this chapter these surfactants are discussed with emphasis on the mechanisms by which they act in various flotation processes. Many types of flotation processes exist today, and these are classified on the basis of size and the mechanism of flotation in Table 1: foam fractionation for the separation of surface-active species such as detergents in aqueous solutions, foam flotation for naturally surface-active organisms and proteins, ion flotation for the separation of ions, micro and ultra flotation for separation of very fine particles, and froth flotation for the separation of minerals. In addition, there are nonfoaming techniques such as bubble fractionation, solvent sublation, and oil flotation [2—4] in which the floated material is collected selectively in a liquid phase that is not compatible with the bulk aqueous media. Among these, froth flotation is the only technique with wide industrial application. Here the role of surfactants in froth flotation of minerals is therefore dealt with in detail.

II. FLOTATION TECHNIQUE

In froth flotation, mineral particles are initially stirred, often intensely with the reagent solution in a mixer, and then transferred into a specifically designed cell for the actual flotation process (Fig. 1). Air is sucked or pushed into the cell and dispersed by a rotating impeller. Hydrophobic particles attach to the air bubbles and levitate to the top of the cell, where they are skimmed off and hydrophilic material is collected at the bottom of the cell. In froth flotation, surfactants have to impart hydrophobicity selectively, to the desired particles, for successful separation. Selective adsorption of the surfactant on the mineral surface itself is a function of a number of parameters such as surface composition of the solid, mineral-solution equilibria and solution chemistry of the surfactant [5—14]. The principles involved in flotation and the complex role of surfactants in different mineral systems are discussed in the following sections.

Mineral particles can be separated by flotation only if hydrophobicity can be selectively imparted to those particles that are to be floated. Since most minerals are naturally hydrophilic (sulfur, talc, molybdenite, graphite, and coal are hydrophobic), they are treated with a surfactant that adsorbs on the mineral surface,

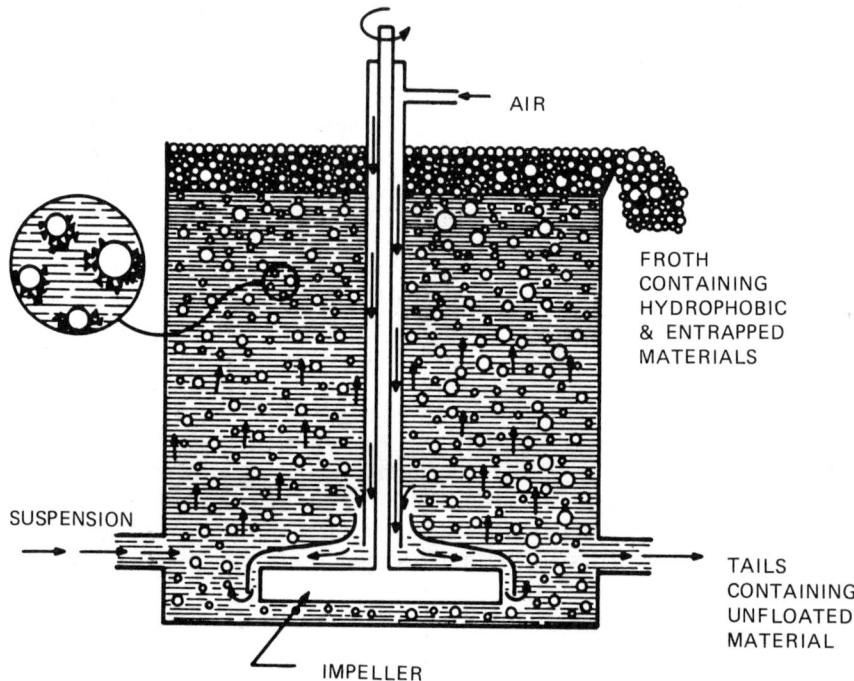

FIG. 1 Schematic representation of a flotation cell.

rendering them hydrophobic. These surfactants adsorb with their polar head oriented toward the mineral surface and the hydrophobic tail pointing toward the solution, thus making the mineral surface hydrophobic.

A. Thermodynamic Aspects

At equilibrium, the forces at the point of contact between an air bubble and a solid in a liquid is defined by the Young-Dupre equation,

$$\gamma_{sg} = \gamma_{sl} + \gamma_{lg} \cos \theta \tag{1}$$

where γ_{sg}, γ_{sl}, and γ_{lg} are the surface tensions at the solid-gas, solid-liquid, and liquid-gas interface, respectively, and θ is the contact angle. For flotation to occur, $\cos \theta$ has to be a minimum or the contact angle has to be a maximum. This condition is met if

$$\cos\theta = \frac{\gamma_{sg} - \gamma_{sl}}{\gamma_{lg}} \qquad (2)$$

is also minimum or if γ_{sg} is as small as possible, keeping γ_{sl} and γ_{lg} at maximum values. It can be seen that the surfactant must adsorb, if possible with maximum effect at the solid-gas interface.

Another way to look at the process of particle-bubble attachment is as a destruction of a solid-liquid and liquid-gas interface and a creation of a solid-gas interface. The free-energy change for this process is given by Dupre's equation,

$$\Delta G = \gamma_{sg} - (\gamma_{sl} + \gamma_{lg}) \qquad (3)$$

Combining Eqs. (1) and (2), we get

$$\Delta G = \gamma_{lg}(\cos\theta - 1) \qquad (4)$$

It is clear from Eq. (4) that for all finite values of θ there is a net decrease in ΔG upon particle bubble attachment, which leads to the condition that

$$\gamma_{sg} - \gamma_{sl} < \gamma_{lg} \qquad (5)$$

for particle bubble attachment. In practice, a minimum contact angle is required for flotation; for example, a 10° angle is reported for the quartz dodecylamine system [15]. The application of Eqs. (1) to (5) to an actual flotation system is limited because these equations are valid only under equilibrium conditions. In an actual system, dynamic conditions prevail and equations based on dynamic factors such as minimum particle-bubble contact time would be more relevant [16–18].

Another approach in this regard is that of Derjaguin [19], who used the concept of disjoining pressure inside the liquid film between the particle and the bubble to examine the attachment process. Disjoining pressure is defined as the derivative of the free energy of the system with respect to the interlayer thickness per unit surface area.

$$P = \frac{\partial G}{\partial A} \qquad (6)$$

As flotation is directly dependent on film rupture, the value of disjoining pressure must be negative for particle-bubble attachment. Three factors contribute to P:

$$P = P_{vw} + P_{edl} + P_s \qquad (7)$$

where P_{vw} is the van der Waals contribution, P_{edl} the electrical double-layer contribution, and P_s the surface hydration contribution. The values of these components are directly dependent on the nature and orientation of the adsorbed surfactant species and can be estimated as described in many works [20–25]. It is to be noted that theoretical determination of the values in both Young's and Derjaguin's treatment requires prior knowledge of the orientation of the surfactant at the solid/liquid interface. Recent experiments using fluorescence and electron spin resonance spectroscopy [26] have yielded valuable information on the structure and orientation of adsorbed alkyl sulfates on alumina. Similar data on other mineral-surfactant systems would prove very useful to yield a better insight into the flotation process.

B. Surfactants for Flotation

Surfactants with a large variety of polar and nonpolar groups are available for flotation and they can be classified based on their function and on their chemical composition. A surfactant whose primary role is to render the solid surface hydrophobic is called a collector. Surfactants whose primary role is to provide the required stability to the top froth layer in the flotation cell and to influence the kinetics of particle-bubble attachment are referred to as frothers. These are usually nonionic surfactants such as cresols which can enhance the rate of film thinning and contribute to the stability of the particle-bubble aggregates. The exact mechanism of frothers in flotation is not clearly established. The role of the frothers in particle bubble attachment is supposedly due to the reduction in the repulsion forces between the bubble and the particle by suitable alignment of the surfactant dipoles at the three-phase point of contact. It is the ability of the nonionic surfactants to align their dipoles that make them good frothers. Experimental evidence to confirm this theory based on the relaxation time of the frother is, however, lacking. Lekki and Laskowski [27] showed the induction time to decrease markedly in chalcocite flotation using ethyl xanthate in the presence of α-terpineol at pH 9.7. Surface tension, zeta potential, and infrared spectroscopic data in the literature indicate significant interaction between frothers and collectors in flotation systems. Commercial frothing agents and their utilization have been reviewed comprehensively by Booth and Freyberger [28]. Table 2 shows frother usage in different flotation systems. Surfactants can also be classified on the basis of (1) the charge of the polar group: cationic, anionic, or nonionic; (2) the nature of the hydrocarbon chain: alkyl, aryl and so on; or (3) functionality of the polar group: thio compounds, oximes, amines, carboxylates, and so on. Typical flotation surfactants with their ionization characteristics are listed in Table 3.

III. SURFACTANT ADSORPTION

Adsorption of the surfactant on mineral surfaces is dependent on a number of factors, such as the surface charge and surface chemical composition of the mineral, and the solution chemistry of the mineral and the surfactant. Major forces involved in causing surfactant adsorption are electrostatic attraction, covalent bonding, hydrogen bonding, and nonpolar bonding between the surfactant and the mineral, as well as the lateral attraction between the adsorbed species. In the following sections the various surfactant adsorption mechanisms are discussed with specific examples from the literature.

A. Electrostatic Factors

Electrostatic forces will play a significant role in systems where the surfactant and the mineral are charged. For example, in the system alumina/dodecyl sulfonate, significant adsorption was observed only below pH 9, where the mineral is positively charged (Fig. 2). Similar correlations on the electrostatic dependence of surfactant adsorption have been observed for calcite [39,40], apatite [41], corundum [31,32], quartz [47], monazite [34], zircon [35], magnetite [36], and tricalcium phosphate [37].

It is to be noted that the mineral surface charge itself is mainly dependent on the mineral solution equilibria operating in the system under the processing conditions. Several mechanisms [38] have been proposed for the surface charge generation of various systems. Oxides such as hematite, silica, and alumina are considered to acquire their charge due to hydrolysis and pH-dependent dissolution. This process can be represented by the following reactions:

$$MOH = MO^- + H^+$$

$$H^+ + MOH = MOH_2^+$$

TABLE 2 Frother Usage

Frother	Gold ores,[a] 52 mills	Simple copper ores,[b] 66 mills	Complex copper ores,[c] 35 mills	Lead ores,[d] 95 mills
Pine oil	15	17	23	4
Cresylic acid	8	11	8	27
Pine oil and cresylic acid	19	5	3	2
Methylisobutylcarbinol (MIBC)	—	14	26	29
Other alcohols	4	3	3	—
Polyglycol types	13	21	8	9
Triethoxybutane (TEB)	15	8	3	—
None	—	—	—	9
Combinations (other than pine oil and cresylic acid)				
Polyglycols and				
(a) Pine oil	4	5	8	2
(b) Cresylic acid	8	1	3	2
(c) Pine oil and cresylic acid	6	—	—	—
MIBC and				
(a) Pine oil	2	5	6	2
(b) Cresylic acid	4	3	6	12
(c) Pine oil and cresylic acid	—	—	—	—
Miscellaneous	2	7	3	2

[a] Ores from which gold is the only important product recovered by flotation, probably in most cases in pyrite.
[b] Ores from which copper is the product of major importance. 49 ores classed as containing copper only, 17 ores classed as containing some gold and/or silver.
[c] Ores from which two important products are obtained, generally by selective flotation. 14 Cu-MoS$_2$ ores; 13 Cu-Zn ores; 2 Cu-Co ores; 6 Cu-Ni ores.

Percentage of total mills					
Zinc ores,[e] 104 mills	Pyrite ores 14 mills	Bulk sulfide ores,[f] 23 mills	Coal,[g] 15 mills	Amine flotation, 21 mills	Fatty acid, soap, sulfonate flotation, 61 mills
13	50	17	27	30 (mica; feldspar)	8
17	22	13	—	—	5
6	—	9	—	—	—
33	—	25	53	25 (KC1)	11
4	7	4	7	5 (KC1)	5
16	7	9	13	5 (feldspar)	7
—	7	—	—	—	—
—	—	—	—	30 (quartz)	64
2	—	—	—	—	—
2	—	—	—	—	—
—	—	—	—	—	—
1	—	9	—	—	—
3	—	5	—	—	—
—	7	—	—	—	—
3	—	9	—	5	—

[d]4 lead ores; 71 Pb-Zn ores; 15 Pb-Cu-Zn ores. In the last named subgroup it was assumed that a lead, copper float was aimed primarily at recovering the lead.
[e]8 zinc ores; 70 Pb-Zn ores, 13 Pb-Cu-Zn ores; 13 Cu-Zn ores. (Zinc frother unidentified in one Pb-Zn ore and two Pb-Cu-Zn ores.)
[f]Bulk sulfide flotation from ore, not followed by differential flotation. Eight instances of flotation of sulfides from tungsten ores.
[g]Survey limited to domestic mills in this group.
Source: Ref. 28.

TABLE 3 Characteristics of the Major Types of Collectors

Type	Formula[a]	Ion	Ionization
Fatty acid soap	$R-C{\overset{O}{\underset{O^-\cdots Na^+}{\diagup}}}$	$RCOO^-$	Weak
Alkylphosphate	$R-O-P(=O)(O^-\cdots Na^+)(O^-\cdots Na^+)$	RPO_4^-	Weak
Alkylsulfate	$R-O-S(=O)(=O)-O^-\cdots Na^+$	RSO_4^-	Strong
Alkylsulfonate	$R-S(=O)(=O)-O^-\cdots Na^+$	RSO_3^-	Weak
Primary amine salt	$R-\overset{H}{\underset{H}{N}}-H^+\cdots Cl^-$	RNH_3^+	Weak

Surfactants in Flotation

Secondary amine salt	$\begin{array}{c}R'\\\|\\R-N-H^+\cdots Cl^-\\\|\\H\end{array}$	$RNR'H_3^+$	Weak
Tertiary amine salt	$\begin{array}{c}R'\\\|\\R-N-H^+\cdots Cl^-\\\|\\R'\end{array}$	$RN(R')_2H^+$	Weak
Quaternary amine salt	$\begin{array}{c}R'\\\|\\R-N-R'^+\cdots Cl^-\\\|\\R'\end{array}$	$RN(R')_3^+$	Strong
n-Alkyl xanthates	$R''-O-\underset{\underset{S}{\|\|}}{C}-S^-\cdots Na^+$	$ROCS_2^-$	Strong

[a] R, long-chain alkyl group containing eight or more carbon atoms; R', a short alkyl chain, usually a methyl group; R´, normally ethyl or amyl group.

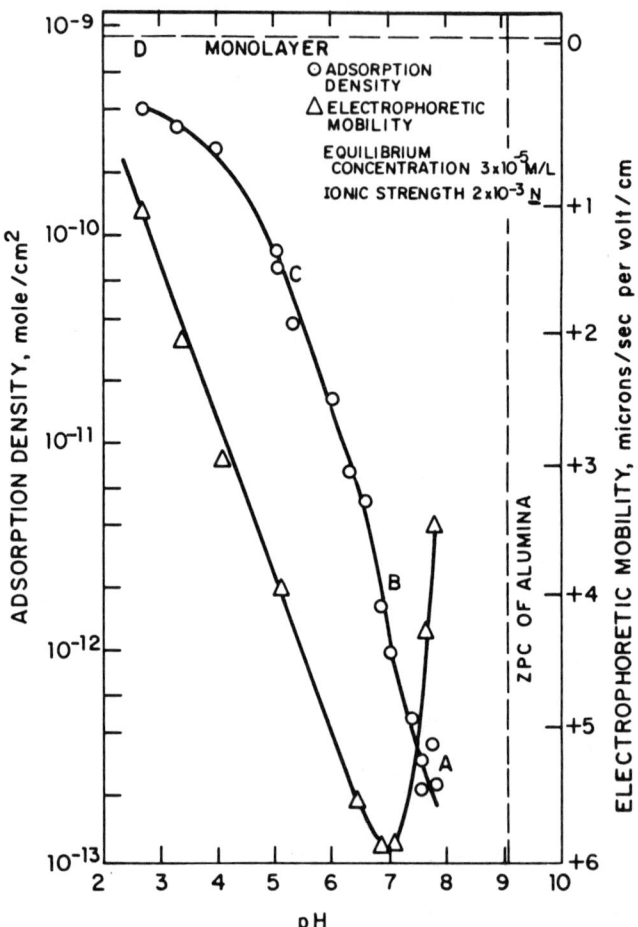

FIG. 2 Adsorption of dodecyl sulfate on alumina as a function of pH.

where M represents the metal atom. It is clear from these equations that the oxide surface will be positive at low pH and negative at high pH values. The pH at which the surface is completely neutral is referred to as the point of zero charge or PZC of the mineral. PZCs of several oxides are listed in Table 4. For salt-type minerals such as calcite, apatite, and dolomite, the preferential dissolution of ions, their reactions with the solution constituents, and subsequent adsorption-precipitation on the surface are mainly responsible for the generation of surface charge. For example, apatite equilibria in water can be represented by the following reactions:

TABLE 4 PZCs of Several Oxides

Mineral	PZC	References
Anatase	5.9	5
Barite	9.5	71
Calcite	8 – 10.8	72
Cassiterite	4.5	73
Chromite	5.6 – 7.2	90
Corundum	9 – 9.4	74
Cuprite	7 – 9.5	91
Dolomite	7.0	76
Fluoroapatite	4 – 6	92
Fluoroapatite (syn)	5.2	76
Goethite	6.7	77
Hematite	4.8 – 6.7	90
Kaolinite	5 – 6	88, 89
Quartz	2.3 – 3.7	83
Rutile	6.0	83
Talc	3.5	84
Tenorite	9.5	85
Zircon	5.8	86

$$Ca_{10}(PO_4)_6(F,OH)_2 \rightleftarrows 10Ca^{2+} + 6PO_4^{3-} + 2(F^-,OH^-) \quad K_{sp}(F) = 10^{-118}$$

$$PO_4^{3-} + H^+ \rightleftarrows HPO_4^{2-} \quad 10^{12.3}$$

$$HPO_4^{2-} + H^+ \rightleftarrows H_2PO_4^- \quad 10^{7.2}$$

$$H_2PO_4^- + H^+ \rightleftarrows H_3PO_4 \qquad 10^{2.2}$$

$$Ca^{2+} + H_2O \rightleftarrows CaOH^+ + H^+ \qquad 10^{-12.9}$$

$$Ca^{2+} + 2H_2O \rightleftarrows Ca(OH)_2(s) + 2H^+ \qquad 10^{-22.8}$$

$$F^- + H^+ \rightleftarrows HF \qquad 10^{3.1}$$

$$Ca^{2+} + HPO_4^{2-} \rightleftarrows CaHPO_4(aq) \qquad 10^{2.7}$$

$$CaHPO_4(aq) \rightleftarrows CaHPO_4(s) \qquad 10^{4.3}$$

$$Ca^{2+} + H_2PO_4^- \rightleftarrows CaH_2PO_4^+ \qquad 10^{1.1}$$

$$Ca^{2+} + 2F^- \rightleftarrows CaF_2(s) \qquad 10^{10.4}$$

$$Ca^{2+} + F^- \rightleftarrows CaF^+ \qquad 10^{1.0}$$

The distribution of the activities of the dissolved species can be calculated using the information above and is shown in Fig. 3 as a function of pH. It is clear that in the acidic pH range, Ca^{2+} activity governs the behavior of hydroxyapatite, whereas in the alkaline pH range CO_3^{2-} and HCO_3^- activities will predominate. Zeta-potential results on apatite [39,40] support the calculations above (Fig. 4). At pH 7.2, where the zeta potential is zero, the activities of the positive and negative species counteract on another. Similar correlations between zeta-potential and species-distribution diagrams have been obtained for other minerals, such as calcite [41], dolomite [42], and magnesite [43].

Electrostatic adsorption of surfactants can also be influenced by the other charged species present in the system [44–46]. For example, multivalent cations enhance the flotation of quartz using fatty acids due to the uptake of these ions bearing a charge that is opposite to that of the surfactant [45]. In contrast, amine flotation of quartz is depressed by KNO_3 due to the competition between K^+ and the aminium ions for adsorption on the negative sites of the quartz surface (Fig. 5).

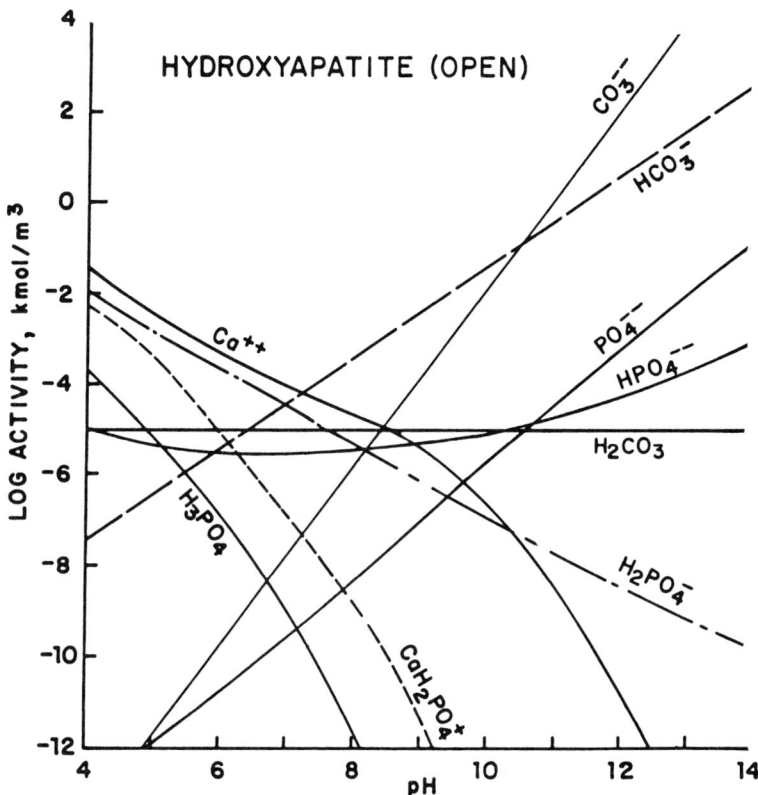

FIG. 3 Species distribution diagram of hydroxyapatite.

B. Chain-Chain Interactions

In the case of flotation of non-sulfide minerals, a major force of collector adsorption results from the association of hydrocarbon chains of adsorbed surfactant at the mineral surface to form two-dimensional aggregates called hemimicelles. Figure 6 illustrates this phenomenon for the alumina dodecylbenzene sulfonate system. Hemimicellization occurs due to the favorable energetics of the partial removal of hydrocarbon chains from the aqueous environment. Recent work by Chandar et al. using fluorescence and electron spin resonance spectroscopy [26] has shown the evolution of two-dimensional aggregates of dodecylsulfate at the alumina surface, supporting the mechanism of associative interactions of the surfactant. These authors further probed the microfluidity of the adsorbed layer and found it to be more viscous than micelles. The evolution of these aggregates

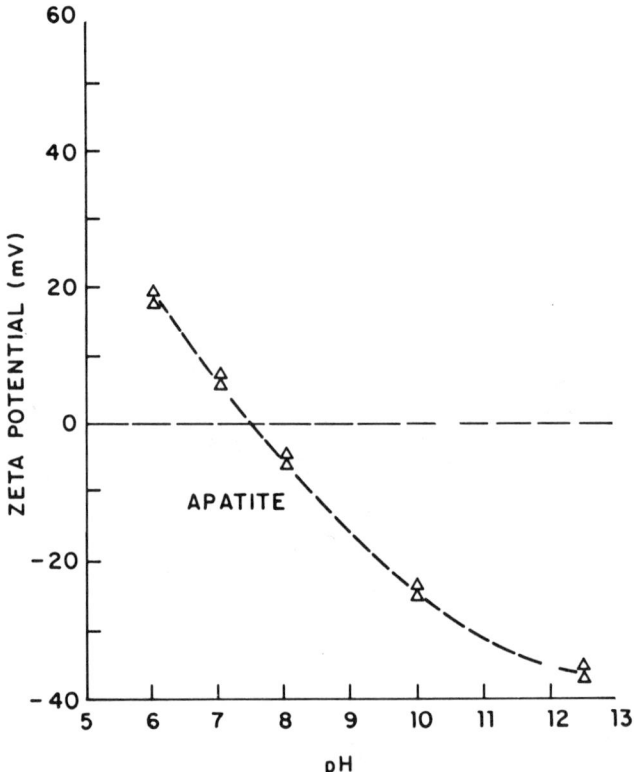

FIG. 4 Zeta potential of apatite as a function of pH.

at the solid-liquid interface is illustrated schematically in Fig. 6b. It is to be noted that the extent of association and hence adsorption will depend on the chain length and branching of the surfactant, the size of the ionic head group, and other features, including the presence of double bonds or perfluoro groups.

C. Chemical Forces

Adsorption of surfactant due to specific covalent bond interaction with the mineral surface is termed chemisorption. The chemical bond between the surfactant and the solid is more system specific than are the other bonds. The chemisorption phenomenon is illustrated here with examples of different mineral-surfactant systems.

Fatty acids adsorption on calcite, barite, and fluorite has been proposed by several investigators to occur by chemisorption [48—52].

Surfactants in Flotation 211

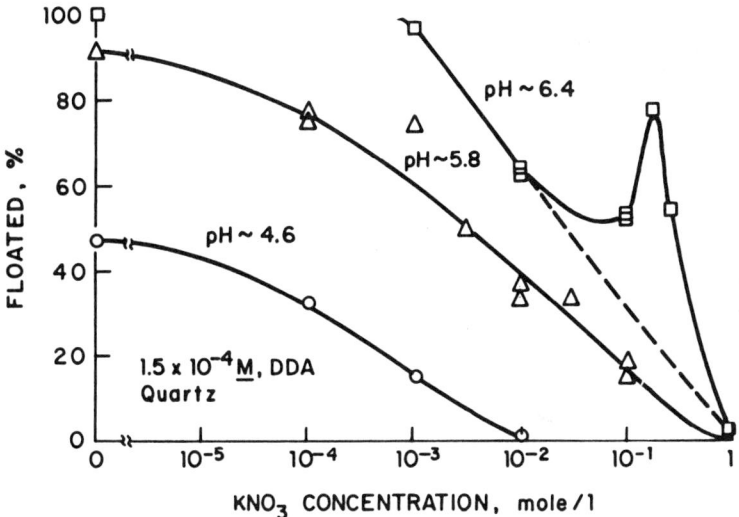

FIG. 5 Flotation recovery of quartz as function of KNO_3 concentration at pH 4.6, 5.8, and 6.4. 1.5×10^{-4} M/liter of dodecylammonium acetate was used as collector. (From Ref. 47.)

Chemisorption in the fluorite-oleate system is thought to take place by an ion-exchange mechanism in which a stoichiometric amount of lattice ions such as F^- is released into the solution during oleate adsorption. Experimental evidence of chemisorption is mainly from infrared spectra of the solid after adsorption. Thus Miller et al. [53] observed that the intensity of the =CH bond of adsorbed oleate decreased at high temperature in the fluorite-oleate system and attributed this phenomenon to an oxidation-polymerization reaction between the oleyl groups at the fluorite surface. However, the FTIR technique suffers from a major drawback, owing to the possibility of alterations in the chemical state of the adsorbed surfactant during the preparation of the sample for spectroscopic investigation.

In examining the chemical forces involved, a major factor that is often ignored is the change in the active chemical form of the surfactant with change in solution pH. This has often forced researchers to adopt the alternative of chemisorption as the mechanism of adsorption. For example, flotation maximum at pH 8 for the hematite-oleate system has been proposed to be due to covalent bonding between the surfactant species and the mineral surface (Fig. 7). Oleate itself, however, has been found to exhibit maximum surface tension lowering at pH 8, suggesting formation of highly surface

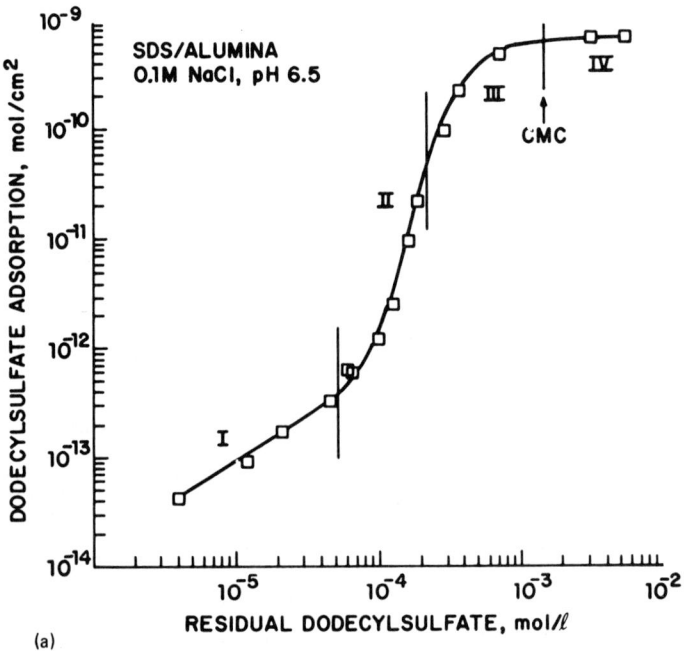

FIG. 6 (a) Adsorption isotherm of dodceyl sulfate on alumina; (b) schematic representation of evolution of surface aggregates at different regions of the adsorption isotherm. (From Ref. 29.)

active complexes in this pH region [54,55]. Ionomolecular complexes that can form between oleic acid and oleate in this pH range can be expected to give rise to a flotation maximum. Indeed, the surface activity of the acid-soap oleate complex hematite flotation exhibit a maximum at the same pH (Fig. 7).

D. Surface Precipitation

An important factor that has to be taken into account in this regard is the possibility of the precipitation of the surfactant on the particle surface. Since the concentration of an ionic surfactant at an oppositely charge surface is higher than that in bulk, precipitation of it can occur at the interface even below concentrations required for precipitation in the bulk. Such surface precipitation can indeed be expected to lead to a sharp dependence of adsorption on solution conditions. Adsorption of multivalent species on oxide surfaces exhibits a sharp pH dependence; in the past this has been attributed

Surfactants in Flotation

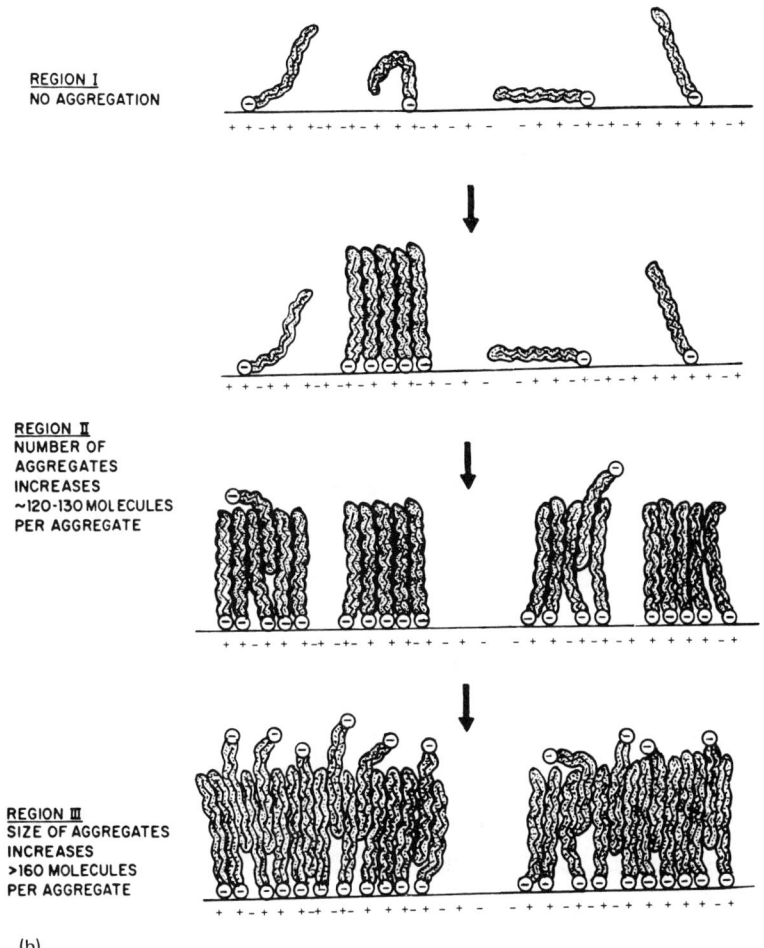

REGION I
NO AGGREGATION

REGION II
NUMBER OF
AGGREGATES
INCREASES
~120-130 MOLECULES
PER AGGREGATE

REGION III
SIZE OF AGGREGATES
INCREASES
>160 MOLECULES
PER AGGREGATE

(b)

FIG. 6 (Continued)

to chemisorption of the metal hydroxy species. For example, the adsorption of Ca species on silica (Fig. 8) shows a sharp increase around pH 11 which was proposed to be the result of chemisorption of $CaOH^+$ since $Ca(OH)^+$ forms in measurable amounts in the bulk around pH 10 to 11. However, if the accumulation of Ca^{2+} ions (as measured by the adsorption) is taken into account, it becomes clear that surface precipitation of $Ca(OH)_2$ can take place around pH 11. This correlates with the adsorption isotherm, suggesting

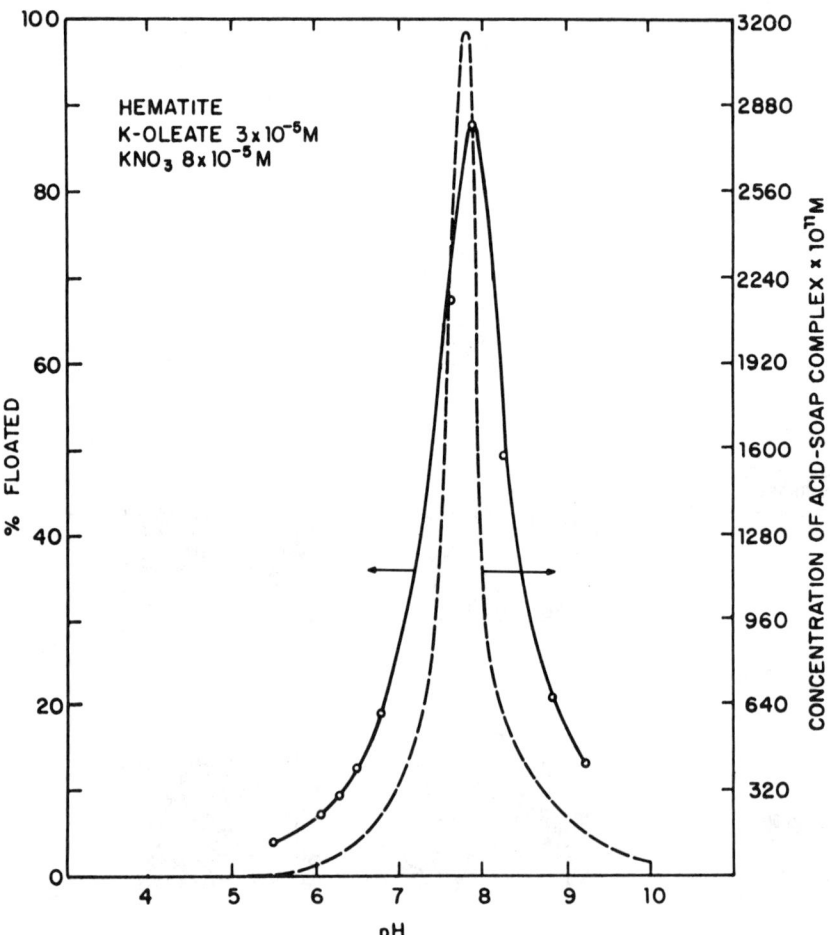

FIG. 7 Correlation of hematite flotation and formation of acid soap complex.

that surface precipitation rather than chemisorption is responsible for the steep rise of the isotherm. Similar correlations have been obtained for the cobalt-silica and the alumina-dodecylsulfonate systems [56].

Adsorption of surfactants on semisoluble minerals has also been considered in the past to be due to chemisorption. Figure 9 shows flotation behavior of dolomite using oleate. By determining the Ca^{2+} activity in equilibrium with the mineral and the surfactant required

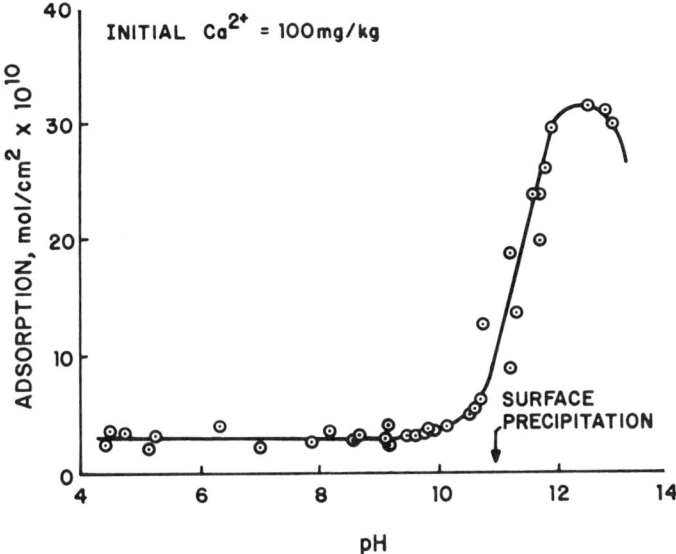

FIG. 8 pH dependence of Ca adsorption on quartz. K_{sp} Ca(OH) = $10^{-5.2}$. pH of surface precipitation for different assumed values of sdsorbed layer thichness (x): x=35 °A, pH 11.4; x=10 °A, pH 10.96.

to precipitate the Ca-surfactant salt, Ananthapadmanabhan and Somasundaran [56] clearly showed the flotation maximum in this system to correlate with the onset of surface precipitation of the calcium salt. Evidently, a careful distinction between chemisorption and surface precipitation has to be made based on the chemical equilibria involved in each system.

E. Xanthate Interactions

Sulfide mineral flotation is an important example where chemical interactions between the surfactant and the mineral plays a governing role. The most commonly used collector for sulfides is xanthate, which is essentially a derivative of carbonic acid (H_2CO_3) in which the two oxygens are replaced by sulfurs and one hydrogen by an alkyl or aryl group. Different xanthates used in sulfide flotation are listed in Table 5 along with their solubility products. The exact nature and mechanism of interactions in the xanthate-sulfide system are not established precisely despite many decades of research partly because of the wide and complex variations in the surface properties

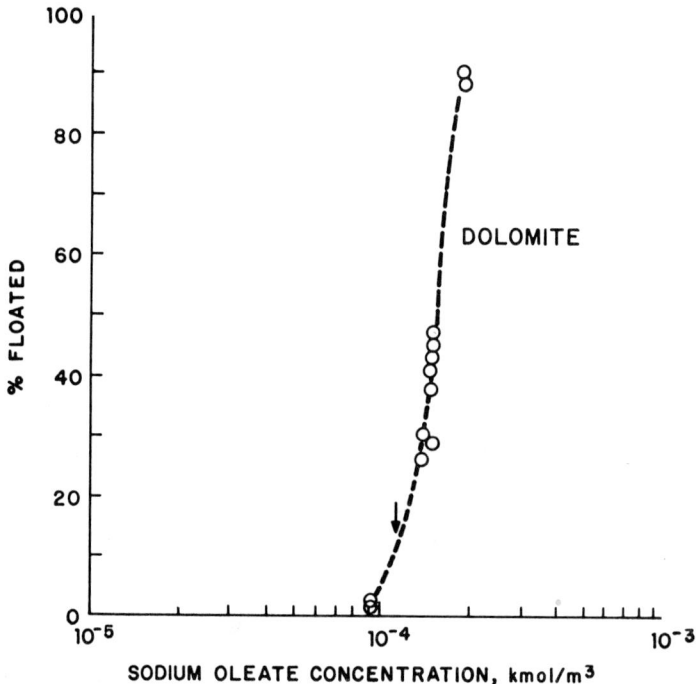

FIG. 9 Flotation of dolomite as a function of oleate concentration. Arrow indicates onset of surface precipitation of Ca oleate (pH 10). Ca^{2+} (dolomite) = 2.8 x 10^{-3} kmol m^3, K_{sp} (Ca oleate) = 4 x 10^{-16}. (From Ref. 93.)

of sulfides. Thus while it is known that oxygen is necessary for xanthate adsorption [57-65], the role of oxygen in critical reactions involving xanthate oxidation to dixanthogen,

$$2ROCS_2 = (ROCS_2)_2 + e$$

$$ROCS_2 + \tfrac{1}{2}O_2 + H_2O = (ROCS_2)_2 + 2OH$$

and consequent hydrophobization of the solid is not clearly established. Finkelstein [59] proposed a two-site electrochemical mechanism involving oxidation of the lattice sulfide ions to explain sulfide flotation. Xanthate adsorption was attributed to ion exchange

TABLE 5 Solubility Products for Metal Xanthates at 20°C[a]

		Ag^+	Pb^{2+}	Zn^{3+}	Cu^{2+}	Ni^{2+}	Co^{2+}	Fe^{2+}	Mu^{2+}
Ethyl xanthate	$C_2H_5OCSS^-$	18.6	16.7	8.2	24.2	12.5			
Isopropyl xanthate	$C_3H_7OCSS^-$	18.6	17.8		24.7	13.4			
Butyl xanthate	$C_4H_9OCSS^-$	19.5	18.0		26.2				
Butyl (i−) xanthate	$C_4H_9OCSS^-$	19.2	17.3		26.3				
Amyl (i−) xanthate	$C_5H_{11}OCSS^-$	19.7	17.6		27.0	14.5			
Hexyl (n−) xanthate	$C_6H_{12}OCSS^-$	20.8	20.3		29.0	16.5	14.3		
Octyl (i−) xanthate	$C_8H_{17}OCSS^-$	20.4	21.3			17.7			
Nonyl (n−) xanthate	$C_9H_{19}OCSS^-$	22.6	24.0	16.2	30.0	22.3	21.3	11.0	9.9
Nonyl (i−) xanthate	$C_9H_{19}OCSS^-$	21.3				21.7			
Lauryl xanthate	$C_{12}H_{25}OCSS^-$	23.8	26.3	19.5	37.0	23.0			

[a]All values are given as negative logarithm of the solubility product extrapolated to infinite dilution.

between the oxidation products and the xanthate species. This mechanism was analogous to that of metal corrosion and it was expected that the rest potential of the mineral surface would be between the reversible potentials of the anodic and cathodic reactions. Experimental data on the rest potentials of the galena-xanthate-oxygen system, however, show significant variations possibly due to the mineralogical heterogeneities leading to different extent of oxidation at the interface. Poling [65] reported adjacent localities of the same particulate sample to show variations of several hundred millivolts in potentials with the more anodic areas adsorbing higher amounts of xanthate. The complex nature of xanthate adsorption and hydrophobisation of sulfides is still not adequately explained, mainly because of the surface chemical heterogeneities of different sulfide minerals that result in different mechanisms dominating in different systems.

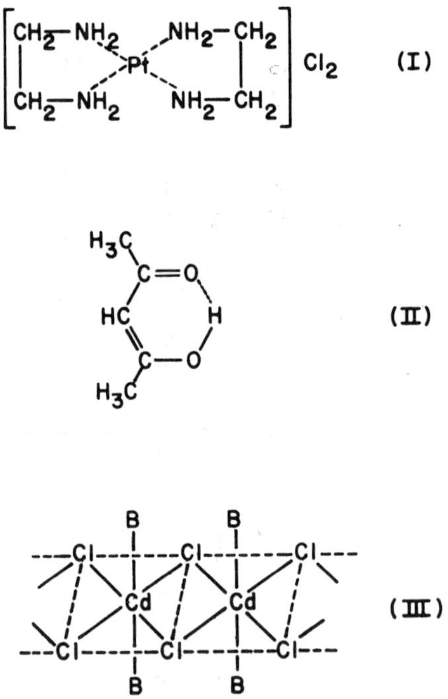

FIG. 10 General structure of chelating agents.

TABLE 6 Examples of Chelating Agents as Collectors

Reagent	Minerals
O-O type	
Cupferron	Cassiterite, uraninite, hematite
Salicylaldehyde	Cassiterite
α-Nitroso β-naphthol	Cobaltite
Acetylacetone	Malachite, chrysocolla
Alkylhydroxamic acid (IM50)	Chrysocolla, hematite and minerals containing Ti, Y, La, Nb, Sn, and W
Phosphonic acids	Cassiterite
N-O type	
β-Hydroxyoximes	Cu oxide minerals
α-Hydroxyoximes	Cu oxide minerals
8-Hydroxyquinoline	Cerussite, pyrochlore, chrysocolla
N-N type	
Diphenylguanidine	Cu minerals
Dimethylglyoxime	Ni minerals
Benzotriazole	Cu minerals
S-S type	
Xanthates	
Dithiophosphates	All sulfide minerals
Dithiocarbamates	
N-S type	
Mercaptobenzothiazole	Sulfide and tarnished sulfides
Dithizone	Sulfides
S-O type	
N-benzoyl-O-alkyl thionocarbamate	Sulfides

F. Surface Chelation

Another group of collectors used for the flotation of copper, nickel, and other metal sulfides is chelating agents that adsorb by forming metal hydrophobic complexes characterized by ring structures. Figure 10 shows examples of three types of chelating agents: I, where the metal is coordinated to the four nitrogens of two molecules of ethylene diamine; type II, showing an intramolecular hydrogen bridge; and type III, involving polynuclear halogen bridges. The two basic requirements for a molecule to form metal chelates are (1) it should

TABLE 7 Structure and Water Solubility of Various Hydroxyoximes

Name	Mol. wt.	Water solubility M	Stock solution ($\underline{M}$)
SALO (salicylaldoxime)	137.1	2×10^{-1}	10^{-2}
OHAPO (O-hydroxy acetophenone oxime)	151.2	4.5×10^{-3}	2×10^{-3}
OHBuPO (O-hydroxy butyrophenone oxime)	179.2	5.0×10^{-4}	3×10^{-4} and 5×10^{-4}
OHBePO (O-hydroxy benzophenone oxime)	213.2	1.5×10^{-4}	1.5×10^{-4} and 2×10^{-4} (1.2% acetone)

Surfactants in Flotation

Structure	Name	MW		
	2H5MeApo (2-hydroxy-5-methyl acetophenone oxime)	166.2	3×10^{-4}	3×10^{-4}
	2H5BAO (2-hydroxy-5-methoxy benzaldoxime)	168.2	5×10^{-3}	5×10^{-3}
	2HNAO (2-hydroxy-1-naphthaldoxime)	189.2	1×10^{-4}	1.25×10^{-4}
	OHCHO (O-hydroxy cyclohexanone oxime)	129.0	~1.0	5×10^{-1}

Source: Ref. 19.

have suitable functional groups, and (2) the functional groups must be situated so as to permit ring formation with the metal as the closing member.

In general, it has been observed that a chelate should preferably be neutral if it should function as a collector and be charged to function as a depressant. Selectivity in chelation is achieved by making use of the differences in stability constants of the metal chelates and by optimizing solution properties such as pH and ionic strength [66,67]. Examples of common chelating agents and their structures are given in Tables 6 and 7. It is important to distinguish between the role of surface and bulk chelates in flotation systems. Thus flotation has been shown to correlate with surface chelation rather than bulk chelation in the tenorite-salicylaldoxime system (Fig. 11).

G. Surfactant Solution Chemistry

In addition to the aqueous chemistry of the minerals and the interactions between surfactant and dissolved inorganic species, the solution chemistry of the surfactant itself, plays a significant role in flotation. As mentioned earlier, hydrolyzable surfactants such as fatty acids and amines can undergo associative reactions to form highly surface active ionomolecular complexes leading to enhanced flotation. For example, dodecylamine reactions in aqueous media can be represented by the following reactions [69]:

$$RNH_3^+ \rightarrow RNH_2 + H^+ \qquad pK_a = 10.63$$

$$2RNH_3^+ \rightarrow (RNH_3)_2^{2+} \qquad pK_d = -2.08$$

$$RNH_3^+ + RNH_2 \rightarrow (RNH_2 \cdot RNH_3)^+ \qquad pK_{ad} = -3.12$$

$$RNH_2(1) \rightarrow RNH_2(aq) \qquad pK'sol = 4.69$$

$$C_T = C_{RNH_2} + C_{RNH_3^+} + 2C_{(RNH_3)_2^{2+}} + 2C_{(RNH_2 \cdot RNH_3^+)} \qquad \text{(mass balance)}$$

The dimerization constant for dodecylamine hydrochloride is assumed here to be the same as that for dodecylamine thiosulfate [68]. Species distribution diagrams of dodecylamine hydrochloride at two

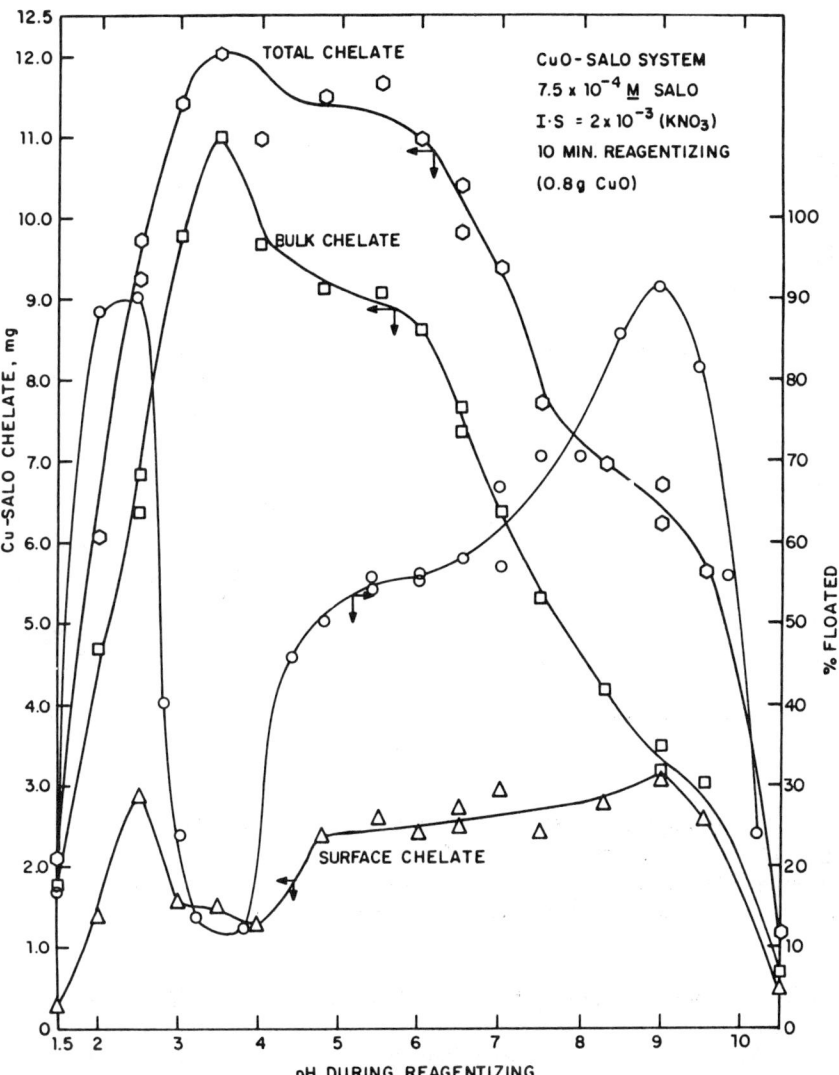

FIG. 11 Correlation of bulk and surface chelation of salicylaldoxime with tenorite and its species with flotation in the tenorite-salicylaldoxime system. (From Ref. 67.)

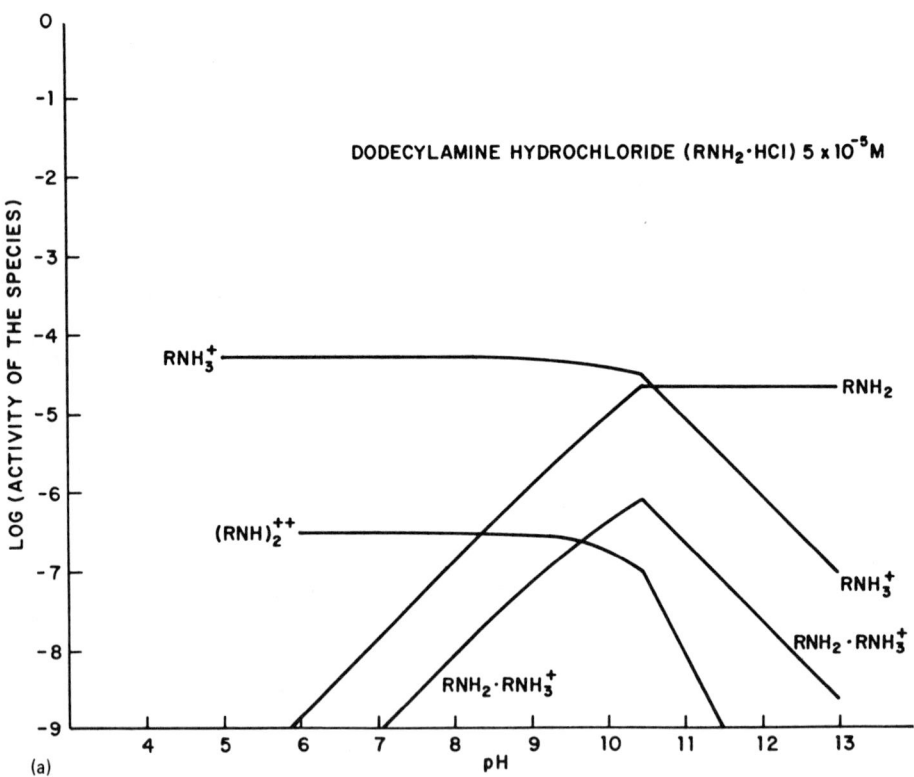

FIG. 12 (a) Species distribution diagram of dodecylamine as a function of pH, total amine = 10^{-5} M, below the solubility limit; (b) species distribution diagram of dodecylamine as a function of pH, total amine = 5×10^{-5} M, above the solubility limit.

different concentrations, 5×10^{-5} M (above the precipitation limit of neutral dodecyl amine) and 10^{-5} M (below the precipitation limit) are shown in Fig. 12. Two important aspects of these calculations are: (1) the pH of maximum amine-aminium complex has shifted from a higher value (10.9) to a lower value (10.4) with increasing total dodcylamine concentration, and (2) the pH of precipitation coincides with the pH of maximum complex formation. Figure 13 shows flotation of quartz using dodecylamine at 10^{-5} M and 5×10^{-5} M concentrations with a maximum at pH 10 to 11. It is to be noted that at 5×10^{-5} M amine concentration, the pH of flotation maximum coincides with the pH of maximum activity of aminium dimer in both

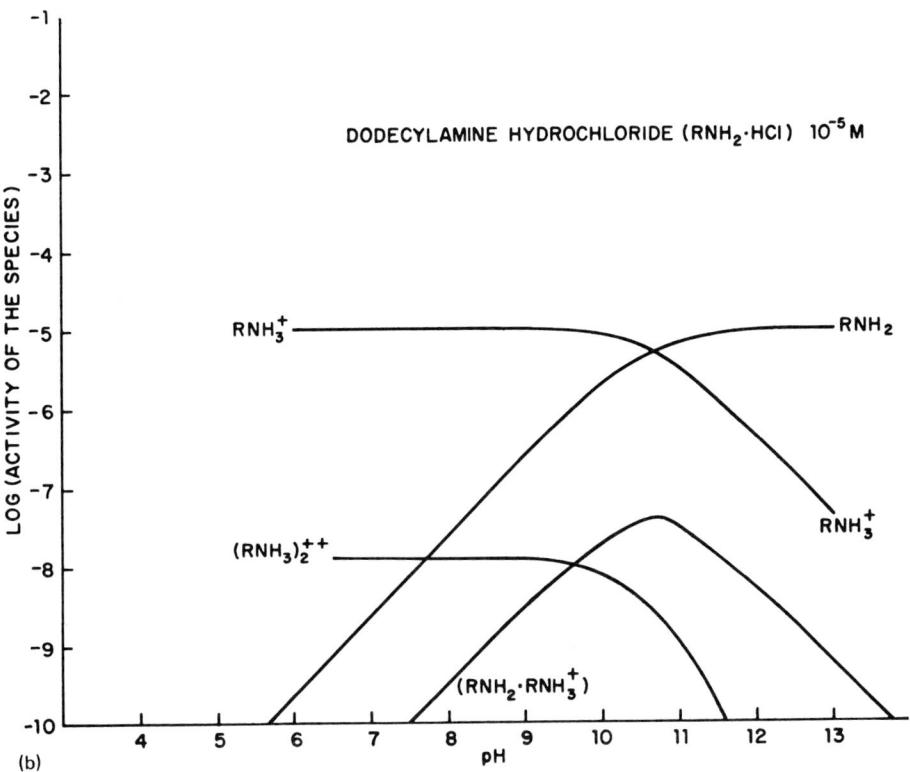

FIG. 12 (Continued)

the amine aminium complex and in amine precipitation. However, at 10^{-5} M amine concentration, there is *no* precipitation, and flotation maximum can be correlated solely to the formation of the iono-molecular complex. Similar correlations between collector association and flotation have also been observed for the oleate-hematite system [69].

IV. MODIFYING AGENTS

In addition to surfactants that function primarily as collectors and frothers, other chemicals are added as depressants, deactivators, activators, pH modifiers, and dispersants to most flotation systems. Depressants act by retarding or inhibiting the flotation of the solid selected. Polymers such as starch and tannin are common depressants. Interestingly, depression of flotation does not always take

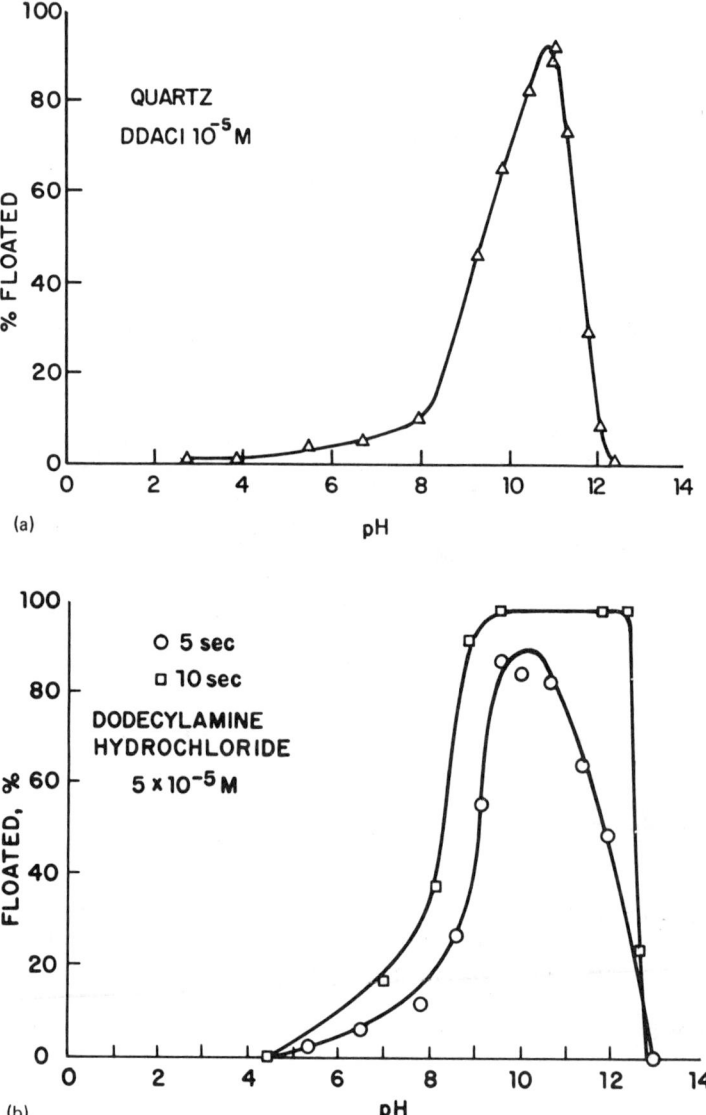

FIG. 13 Hallimond cell flotation of quartz as a function of pH at (a) 5×10^{-5} M amine and (b) 10^{-5} M amine.

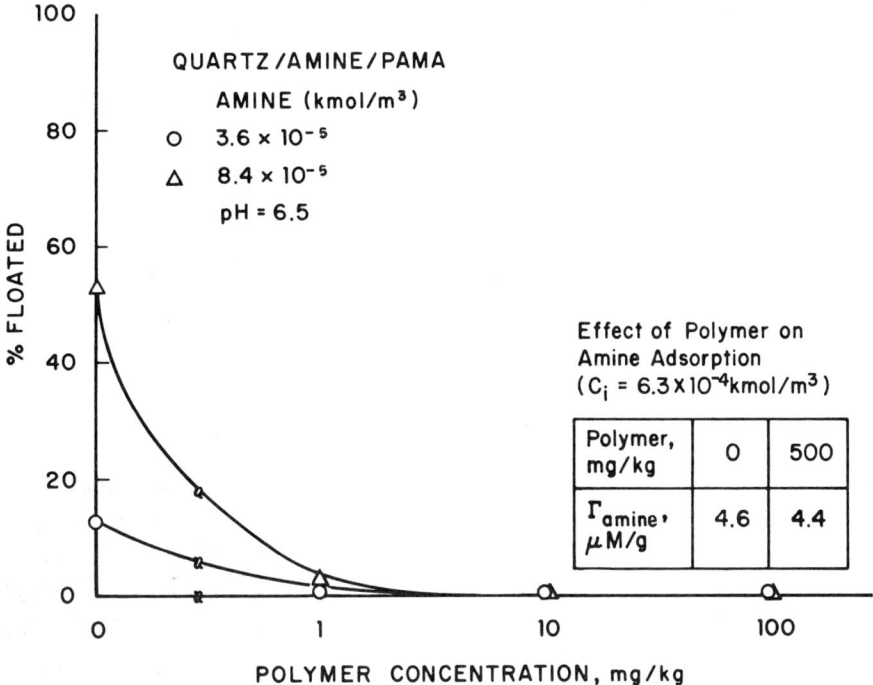

FIG. 14 Depression of flotation of quartz using dodecylamine by the cationic polymer PAMA at natural pH. Adsorption of sulfonate in the presence and absence of polymer is shown in the inset.

place, due to inhibition of collector adsorption. For example, flotation of quartz using amine is depressed by a cationic polymer even though the adsorption is totally unaffected [70] (Fig. 14). Based on information obtained from electrophoretic and surface tension tests, flotation depression has been attributed to masking of the adsorbed amine by the adsorbed massive polymer species. Indeed, such polymers can also be expected to activate flotation: for example, in this case by an anionic surfactant such as alkyl sulfonate (Fig. 15). Figure 16 schematically illustrates the masking phenomenon and the consequent effect on flotation due to variations in surface characteristics after adsorption of polymer and surfactant. Activators thus enhance the adsorption of the surfactant and thereby flotation. Sphalerite activation by copper and quartz activation by multivalent cations are additional examples. Deactivators are reagents that usually react with activators to retard flotation selectively for example,

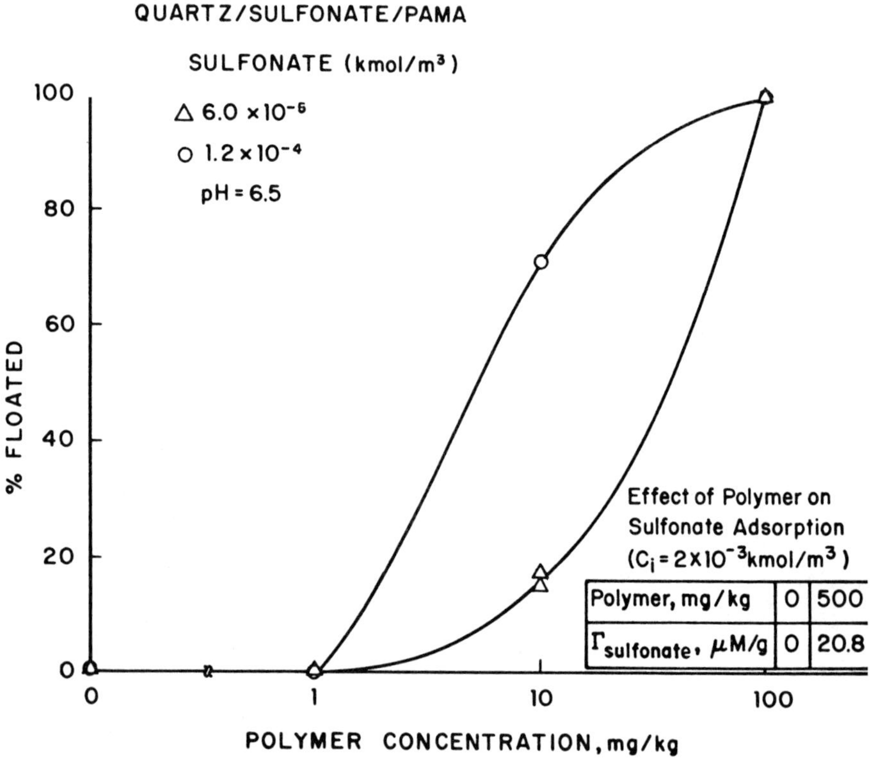

FIG. 15 Activation of flotation of quartz dodecylsulfonate by the cationic polymer PAMA at natural pH. Adsorption of sulfonate in the presence and absence of polymer is shown in the inset.

FIG. 16 Schematic representations of (a) cationic polymer PAMA and dodecylamine coadsorption on quartz particles resulting in flotation depression, (b) adsorption in the quartz/dodecylsulfonate system, and (c) cationic polymer PAMA and dodecylsulfonate coadsorption on quartz particles resulting in flotation activation.

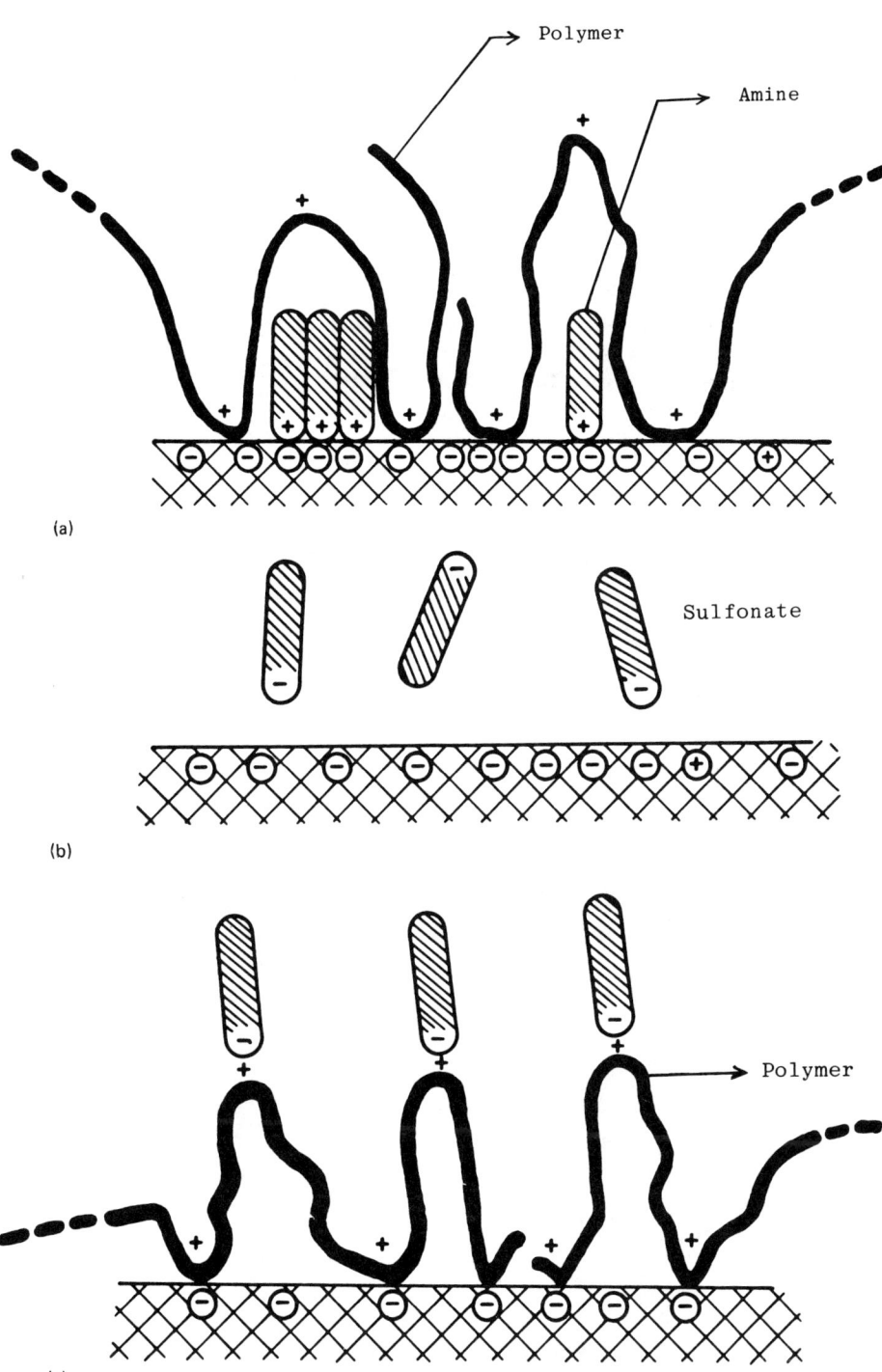

FIG. 16

addition of cyanide in xanthate flotation of zinc sulfides causes deactivation of copper flotation. In addition to these chemicals, dispersants such as sodium silicate and modifiers such as oxalic acid, tartaric acid, and ethylenediaminetetraacetic acid (EDTA) are often used in flotation systems either to complex with the deleterious chemical species or colloids that can coat the particles to be floated.

It is clear from the examples above that the role of the surfactant in flotation is influenced by a number of factors. For any flotation system, selection of the best surfactant is dependent on the identification of the relevant interactions between many species that are released into the system on the dissociation of the surfactant as well as the dissolution of the mineral.

ACKNOWLEDGMENTS

The authors acknowledge the support of the National Science Foundation and the Union Carbide Corporation for support of this work.

REFERENCES

1. P. Somasundaran, Sep. Purif. Methods, 1:117 (1972).
2. R. M. W. Lai, and D. W. Fuerstenau, Trans. AIME, 241:549 (1968): see also S. Raghavan, and D. W. Furstenau, AIChE Symp. Ser., 71 (1975).
3. G. N. Shah, and R. Lemlich, Ind. Eng. Chem. Fundam., 9:250 (1970).
4. I. Sheiham, and T. A. Pinfold, Sep. Sci., 7:43 (1972).
5. A. M. Gaudin, *Flotation*, 2nd ed, McGraw-Hill, New York, 1957.
6. M. C. Fuerstenau, *Froth Flotation*, 50th Ann. Vol., AIME, New York, 1962.
7. K. L. Sutherland, and I. W. Wark, *Principles of Flotation*, Australian Institute of Minerals and Metals, Melbourne, 1955.
8. V. A. Glembotskii, V. I. Klassen, and I. N. Plaskin, *Flotation*, translated from Russian by R. E. Hammond, Primary Sources, New York, 1972.
9. M. C. Fuerstenau, *Flotation*, Vols. I and II, A. M. Gaudin Memorial International Flotation Symposium, 1976.
10. P. Somasundaran, and R. B. Grieves, AIChE Symp. Ser., 160 (1975).

11. A. S. Joy, and A. J. Robinson, in *Recent Progress in Surface Science* (J. F. Danielli, K. G. Pankhurst, and A. C. Riddiford, eds.), Academic Press, 1964, p. 169.
12. P. Somasundaran, in *Separation and Purification Methods,* Vol. 1 (E. S. Perry and C. J. Vannoss, eds.), Marcel Dekker, New York, 1972, p. 117.
13. P. Somasundaran, Sep. Sci., *10* (1):93 (1975).
14. P. Somasundaran, Trans. Soc. Min. Eng. AIME, *241*:105 (1968).
15. D. W. Fuerstenau, and S. Raghavan, *Freiberger Forschunschefte*, Deutscher Verlag für Grundstoffindustrie, Leipzig, 1978, pp. 75–109.
16. D. Reay, and G. A. Ratcliff, Can. J. Chem. Eng., *51*:178 (1973).
17. J. P. Anfruns, and J. A. Kitchener, in *Flotation*, A. M. Gaudin Mem. Vol. 2 (M. C. Fuerstenau, ed.), AIME, New York, 1976, pp. 625–637.
18. G. J. Jameson, S. Nam, and M. M. Young, Miner. Sci. Eng., *9*:103 (1977).
19. B. V. Derjaguin, and S. S. Dukhin, Trans. IMM, *70*:221 (1960).
20. J. Laskowski, Miner. Sci. Eng., *6*(4):223 (1974).
21. A. Frumkin, and Gorodetzkaya, Acta Physicochim., URSS, *9*:327.
22. L. R. Flint, and W. J. Howarth, J. Chem. Eng. Sci., *26*:1155 (1971).
23. H. J. Schulze, *Physico-Chemical Elementary Processes in Flotation*, Elsevier, Amsterdam, 1984.
24. A. Sheludko, Coll. Polym. J., *191*:52 (1963).
25. J. Laskowski, *Advances in Mineral Processing* (P. Somasundaran, ed.), Society of Mining Engineers, Colorado, 1986.
26. P. Chandar, P. Somasundaran, Kenneth Waterman, and N. J. Turro, accepted for publication in J. Colloid and Interface Sci., and J. Phys. Chem.
27. Lekki and J. Laskowski, Trans. IMM, *80*:174 (1971).
28. R. B. Booth, and W. L. Freyberger, in *Froth Flotation* (D. W. Fuerstenau, ed.), 50th anniversary vol., AIME, New York, 1962, pp. 258–276.

29. P. Somasundaran, and D. W. Fuerstenau, J. Phys. Chem., 70:90 (1966).
30. I. Iwasaki, S. R. B. Cooke, and Y. S. Kim, Trans. AIME, 223:113 (1962).
31. H. J. Modi, and D. W. Fuerstenau, Trans. AIME, 217:381 (1960).
32. H. J. Modi, and D. W. Fuerstenau, J. Phys. Chem., 61:640 (1957).
33. I. Iwasaki, S. R. B. Cooke, and A. F. Colombo, *Flotation Characteristics of Goethite*, Report of Invest. 5593, U. S. Bureau of Mines, Washington, D.C., 1960.
34. H. S. Choi, and K. U. Whang, Korean Chem. Soc., 7:91 (1963).
35. H. S. Choi, and K. U. Whang, Trans. Can. Ins. Min. Metall., 66:242 (1963).
36. I. Iwasaki, S. R. B. Cooke, and Y. S. Kim, Trans. AIME, 223:113 (1962).
37. H. S. Hanna, Ph.D. thesis, Ain Shamms University, Cairo, 1968.
38. G. A. Parks, in *Chemical Oceanography* (S. P. Riley and G. Skirrow, eds.), Academic Press, 1975.
39. O. J. Amankonah, P. Somasundaran, and K. P. Ananthapadmanabhan, paper presented at the 114th Annual AIME Meeting, New York, 1985.
40. O. J. Amankonah, and P. Somasundaran, Colloids Surf., 15:335–353 (1985).
41. O. J. Amankonah, P. Somasundaran, and K. P. Ananthapadmanabhan, Colloids Surf., 15:295–307 (1985).
42. P. Somasundaran, K. P. Ananthapadmanabhan, and O. J. Amankonah, in Proc. 15th Int. Min. Proc. Congr., Cannes, France, 1985, Vol. II, pp. 244–254.
43. O. J. Amankonah, Ph. D. thesis, Columbia University, 1985.
44. M. C. Fuerstenau, D. A. Rice, P. Somasundaran, and D. W. Fuerstenau, Trans. IMM (London), 73:381 (1965).
45. M. C. Fuerstenau, C. C. Martin, and R. B. Bhappu, Trans. AIME, 226:449 (1963).
46. F. Z. Saleeb, and H. S. Hanna, J. Chem. UAR, 12(2):237 (1969).

47. P. Somasundaran, Trans. AIME, 255:64 (1974).
48. R. O. French, J. Phys. Chem., 58:80J (1954).
49. A. S. Peck, and M. E. Wadsworth, Proc. 7th Int. Min. Proc. Cong. (N. Arbiter, ed.), Gordon and Breach, New York, 1965, p. 259.
50. A. Bhar, M. Clement, and H. Surmatz, Proc. 8th Int. Minl. Proc. Congr., Leningrad, Paper S-11, 1968.
51. U. Blissing, Dissertation, Bergakademie Frieberg, 1969, Ger-Text.
52. H. L. Shergold, Trans. IMM, 81(3):C148 (1972).
53. J. D. Miller, J. S. Hu, and M. Mishra, paper presented at the 116th annual AIME meeting, New Orleans, La., March 2–6, 1986 (Preprint 86-14).
54. P. Somasundaran, K. P. Ananthapadmanabhan, and I. B. Ivanov, J. Colloid Interface Sci., 99(1):128 (1984).
55. K. P. Ananthapadmanabhan, Ph.D. thesis, Columbia University, 1981.
56. K. P. Ananthapadmanabhan, and P. Somasundaran, Colloids Surf., 13:65–72.
57. A. Granville, N. P. Finkelstein, and S. A. Allision, Trans. IMM, 81:C1 (1972).
58. A. M. Gaudin, and N. P. Finkelstein, Nature, London, 207:389 (1965).
59. N. P. Finkelstein, Sep. Sci., 6:227 (1970).
60. J. Leja, L. H. Little, and G. W. Poling, Trans. IMM, 72:407 (1963).
61. G. W. Poling, and J. Leja, J. Phys. Chem., 67:2121 (1963).
62. N. P. Finkelstein, Trans. IMM, 76:C51 (1967).
63. A. Pomianowski, and J. Leja, Can. J. Chem., 41:2219 (1963).
64. S. A. Allision, L. A. Goold, M. J. Nicol, and A. Granville, Metall. Trans., 3:2613 (1972).
65. G. W. Poling, in *Flotation*, A. M. Gaudin Mem. Vol. 1 (M. C. Fuerstenau, ed.), AIME, New York, 1967.
66. D. R. Nagaraj, and P. Somasundaran, Min. Eng., 33:1351 (1981).

67. P. Somasundaran, and D. R. Nagaraj, *Reagents in Mineral Industry*, Inst. of Min. & Met., Lon., 1984, pp. 209—219.
68. P. Mukerjee, J. Phys. Chem., *62*:1404 (1958).
69. P. Somasundaran, and K. P. Ananthapadmanabhan, in *Solution Chemistry of Surfactants* (K. L. Mittal, ed.), Plenum Press, New York, 1979.
70. P. Somasundaran, and J. Cleverdon, Colloids Surf., *13*(1): 73—85 (1985).
71. P. G. Johansen, and A. S. Buchanan, Aust. J. Chem., *10*:398 (1957).
72. P. Somasundaran, and G. E. Agar, J. Colloid Interface Sci., *24*:433 (1967).
73. J. Laskowski and S. Sobieraj, Inst. of Min. & Met., Lon., *28*: C163 (1969).
74. G. A. Parks, Chem. Rev., *65*:177 (1965).
75. M. Robinson, J. A. Pask, and D. W. Fuerstenau, J. Am. Chem. Soc., *47*:516 (1964).
76. W. Stumm and J. J. Morgan, *Aquatic Chemistry*, Wiley, New York, 1970.
77. G. A. Parks, Adv. Chem. Ser., *6*(67):121 (1967).
78. M. C. Fuerstenau, G. Gutierrez, and D. A. Elgillani, Trans. AIME, *241*:319 (1968).
79. P. Somasundaran, J. Colloid Interface Sci., 27(4):659 (1968).
80. F. Z. Saleeb and P. L. de Bruyn, Electroanal. Chem. Interface Electrochem., *37*:99 (1972).
81. I. Iwasaki, S. R. B. Cooke, D. H. Harraway, and H. S. Choi, Trans. AIME, *223*:97 (1962).
82. R. D. Kulkarni and P. Somasundaran, Am. Electrochem. Soc., *31* (1972).
83. A. M. Gaudin and D. W. Ferstenau, Trans. AIME, *202*:66 (1955).
84. I. Iwasaki, S. R. B. Cooke, and H. S. Choi, Trans. AIME, *220*:394 (1961).
85. O. Huber and J. Weigl, Wochenbl. Papierfabr., *97*(10):359 (1964).

86. J. M. Cases, Trans. AIME, 247:123 (1970).
87. K. P. Ananthapadmanabhan and P. Somasundaran, Trans. Indian Inst. Met., 32(2):177–194 (1979).
88. R. Ramachandran and P. Somasundaran, accepted for publication in Colloids Surf. (1986).
89. P. B. Lorenz, Clays Clay Miner., 17:223 (1969).
90. M. C. Fuerstenau, D. A. Elgilliani, and J. D. Miller, Trans. AIME, 247:11 (1970).
91. D. W. Fuerstenau, Pure Appl. Chem., 24:135 (1970).
92. P. G. Berube, and P. L. de Bruyn, Electrochemistry, 37:99 (1972).
93. B. M. Moudgil and R. Chanchani, Flotation of Calcite and Apatite Using Sodium Oleate as Collector, SME-AIME preprint 83-160.

7
Surfactants in Herbicide Dispersions

NORMAN O. V. SONNTAG *Consultant, Red Oak, Texas*

I.	Introduction	237
II.	Surface Properties of Importance in Herbicide Dispersion Applications	238
III.	Historical Development	241
IV.	Surfactant Choice	242
	A. Surface Activity	242
	B. Relationship to Vehicle	250
	C. Other Factors	251
V.	New Horizons	252
	References	258

I. INTRODUCTION

If the present-day success of weed control technology in agriculture is attributable to the development and effective use of organic herbicides, then, certainly, the use of herbicide adjuvants, particularly surfactants, can be assigned a major share of the credit. Few, if any, herbicides would perform at higher than a small fraction of their potential herbicidal activity were it not for the homogeneous dispersion of the herbicide concentrate in an application vehicle. An effective spray application vehicle permits even distribution of herbicide over the surface of the foliage or soil. Even distribution and penetration of the outer lipid layers of the weed leaf are required to kill the weed selectively.

Hodgson [1] estimated that U.S. farmers spent $3.6 billion on
401.1 million pounds of herbicides in 1976 for chemical weed control.
At a 5% level, Hodgson estimated that this corresponded to the use
of 19.8 million pounds of assorted surfactants in these herbicide
products. Based on detailed estimates for industrial surfactant annual
growth [2,3], including applications for agricultural chemicals (insecticides, herbicides, fungicides, and rodenticides), it is quite likely
that the 1987 annual American use of surfactants in herbicide dispersion is 27 million pounds. Although this estimate is necessarily imprecise, it does indicate the relative economic importance of use of surfactants in herbicide applications. Kapusta [4] points out that during
1985, approximately 25% of the soybean acreage in Illinois (i.e., 1
million hectares) was sprayed with postemergence herbicides. Almost
all of these applications included the addition of an enhancing agent
such as a phytobland petroleum concentrate at a dosage rate of about
2.5 L/ha. The petroleum oil may typically contain emulsifier or surfactant at levels of 15% or more.

Environmental and economic considerations in the last decade have
been largely responsible for the trend toward lowering the amount of
paraffinic hydrocarbon vehicles in herbicide applications. Work on
vehicle development has continued intensively with aqueous surfactant
systems and on systems containing certain vegetable oils [e.g., once
refined or degummed soybean oil (with and without surfactants)].
These trends can be expected to substantially decrease but probably
not eliminate the amount of surfactants employed in herbicide applications in the next decade. The alternative approach involving the use
of vegetable oils such as soybean oil as a carrier or vehicle, in which
some herbicides are soluble, could eliminate the need for surfactant
in some cases. In such instances, the use of water as a carrier may
also decrease because the total volume of spray required would be
lowered substantially [4].

In this chapter we attempt a brief review of the use of conventional
surfactants in aqueous application of herbicides to weeds. The potential shift in surfactant use with the employment of vegetable oils as
carriers for herbicide sprays is also monitored.

II. SURFACE PROPERTIES OF IMPORTANCE IN HERBICIDE DISPERSION APPLICATIONS

Surfactants perform a number of different functions in herbicide dispersions. Surfactants are used primarily in aqueous dispersions,
where they reduce the surface tension and consequently increase the
spreading and adhesion wetting of the weed surface. This results in
uniform coverage of the weed surface, greater absorption, reduction
in the rate of evaporation, and other desirable effects. Related to
this is the lowering of the contact angle of the spray on the plant

surface. Some surfactants modify the phytotoxicity of the herbicides. When phytotoxicity is enhanced, this can be a significant advantage, provided that the full selectivity between the weed and the protected crop is not adversely effected.

Van Valkenburg [5] has attempted to classify herbicide adjuvants with respect to their specific mode of action. He makes two general classifications: spray modifier adjuvants and activator adjuvants. The former relate to the nature and efficiency of the spray operation; the latter relate to the more efficient distribution of the biologically active herbicide and the influence of the adjuvant on the activity and/or selectivity of the herbicide. "Spreaders," usually nonionic surfactants, applied at 13 to 52ml per 100L of spray, cause spray droplets to adhere and spread on foliage. "Sticker/spreaders" are similar but usually have an added agent to aid in the retention of the toxicant on the foliage to improve adhesion during wet weather. These adjuvants can be anionic materials such as fatty acids, polymerized fatty acids, or synthetic polymer latexes. "Stickers" are generally added agents designed to cause agricultural chemicals to adhere better during the spraying operation and resist "wash-off" during rainy weather. They are ordinarily insoluble in water, are polymeric, and possess a finite degree of tackiness to enhance adhesion. One can appreciate that there is a considerable overlapping among spreaders, spreader/stickers, and stickers in their mode of action. In many instances, a single or blended surfactant system provides all three types of action in the dispersion operation.

Van Valkenburg also characterizes two other types of spray modifier adjuvants: *spray thickeners for reducing spray drift* and *foaming agents for reducing drift*. Depending on the type of spray equipment used, various sizes of spray droplets may be produced. Usually, the smaller droplets, say below 10 μm, adhere well but unfortunately are more easily carried by wind or air currents several kilometers from the target. Spray additives that produce a larger proportion of droplets in the 50 μm range or larger are usually based on water-souble or swellable hydrophilic polymers. Resulting sprays have a substantially reduced tendency to be caught by the wind and accordingly drift from the target. *Foaming agents for reducing drift* generate larger droplets based on the use of air rather than water. The technique is relatively new. More data are needed to establish the effects of these additives on biological activity. Since a high surface area is generated in the liquid there could also be a higher rate of liquid evaporation. Such an increase in evaporation rate would result in increased drift and diminished penetration of foliage because of the decreased time the liquid phase is in contact with the foliage.

Depending on the herbicide employed and, of course, the vehicle used in its dispersion as spray, a great many surfactants have found use in herbicide dispersion. The preference today for water dispersion appears to be toward the universal use of nonionic structural types in order to optimize the inertness between surfactant and herbicide, and to minimize the effect of varying water hardness. Table 1 identifies the seven surfactants which were selected for evaluation in a recent University of Florida experimental program [6] for evaluation of atrazine and ametryne herbicides. These surfactants possessed HLB (hydrophilic lipophilic balance) values that varied between 8.6 and 16.7. Evaluations at Southern Illinois University using soybean oil, with and without surfactants [4], indicated the importance of the choice of emulsifiers with certain herbicides. For example, the herbicide metribuzin is insoluble in soybean oil. A 5:1 ratio of metribuzin in water allowed it to be sprayed while suspended in soybean oil. However, it was pointed out that effective surfactant formulations would have to be developed to permit greater uniformity of application.

TABLE 1 Typical Nonionic Surfactants Evaluated in Atrazine and Ametryne Aqueous Dispersions

Chemical name	Trade name	HLB	Source
Alkylaryl polyoxyethylene glycol (AAPOEG)	Ortho X-77	13.5	Chevron Chem. Co. Richmond, Calif.
Sorbitan monolaurate (SML)	Span 20	8.6	ICI America Inc. Wilmington, Del.
Nonylphenyl polyoxyethylene ethanol (NP + EO)	Sterox-NJ	13.0	Monsanto St. Louis, Mo.
Trimethyl nonylpolyethoxyethanol (TMNPEOE)	Surfactant WK	11.7	E.I. DuPont Co. Wilmington, Del.
Octylphenoxy polyethoxyethanol (OP + EO)	Triton X-114	12.4	Rohm & Haas Co. Philadelphia, Pa.
Modified Phthalic glycerol alkyl resin (MPGAR)	Triton B-1956	—	Rohm & Haas Co. Philadelphia, Pa.
POE (20 sorbitan monolaurate) (POESML)	Tween-20	16.7	ICI America, Inc. Wilmington, Del.

Source: Ref. 6.

III. HISTORICAL DEVELOPMENT

McWhorter [7] points out that surfactant interactions with herbicides were recognized from the earliest days of experimentation with 2,4-D [8]. Sometime later it was noted that increases in activity of herbicides and growth regulators resulted from the addition of surfactants [9-13]. By 1950, the advantages of surfactants in spray solutions were generally appreciated. However, agricultural adjuvants were not available for sale at that time, so use of household detergents was frequently recommended [14 -17]. Before 1960, the use of surfactants with herbicides undoubtedly was greater in industrial weed control than in agriculture.

Although use by farmers was quite limited, it was readily apparent that combinations of surfactants with herbicides in agriculture were quite advantageous. This was obvious in the 1950's [18-26]. By 1950, s-triazine or subsituted urea herbicides were found to be effective with the addition of oil or surfactant in the spray operation. McWhorter [27] showed that postemergence applications of s-triazine to corn, soybeans, and cotton were successful with paraffinic oil used as vehicles. Later evaluations in Mississippi showed an increased effectiveness of ureas and s-triazine for postemergence weed control when applied in oil or in oil-water emulsions [28-30]. Holstun and Bingham [28] referred to the paraffinic oils as "phytobland oils." By the mid-1960's, American [31-39] and Canadian research [40,41] reached its most concentrated peak in studying the effectiveness of oil-water emulsions. Atrazine was accepted by American farmers on a very wide scale and was applied in oil-water emulsions. In 1978, probably more than 8000 ha of corn were treated annually with postemergence applications of atrazine in oil-water emulsions.

Diuron and other substituted urea herbicides were used until 1960 for preplanting and preemergence weed control at planting. McWhorter determined that these herbicides were highly effective as postemergence weed control when surfactant was included in spray mixtures [42,43]. Cotton farmers enthusiastically adopted this technique. In Mississippi by 1961, weed in 170,000 ha of cotton were treated with postemergence applications of Diuron surfactant solution. By 1971, 1,600,000 ha of cotton were treated with herbicide surfactant solutions. By 1978, mixtures of MSMA (monosodium methanearsonate) plus substituted urea or s-triazine herbicides in water surfactant mixtures were commonly used throughout the American cotton belt.

In the literature, vegetable oils such as soybean oil are commonly referred to as "pesticide dispersants." Obviously, this also includes use of vegetable oils as herbicide vehicles or carriers. The burgeoning uses of soybean oil (SBO) in agricultural herbicides was estimated to be at 25 millon pounds in 1983 [44]. By 1990, projected SBO use will be approximately 500 million pounds [45,46].

Considering surfactants, it was estimated that 25 million pounds of total surfactants was used in 1985 for the dispersion of herbicides. Accordingly, we can logically expect that as an increasingly greater proportion of water and vegetable oils is used in preference to paraffinic hydrocarbons for this purpose, a different type of surfactant system will be optimum for each oil vehicle employed. In the decade 1990–2000, if vegetable oils do supplant hydrocarbons and water for this purpose, surfactants such as monoglycerides of octanoic, decanoic, dodecanoic (lauric), or tetradecanoic (myristic) acids, or mixtures of these, may be the preferred surfactants. Use of surfactants from vegetable oils does assume a suitable cost position for these materials.

IV. SURFACTANT CHOICE

A. Surface Activity

Surfactants provide a variety of functionalities in herbicidal formulations. They may function as an activator adjuvant, a sprayer modifier adjuvant, a utility modifier or an emulsifying agent, or as a dispersing agent for wettable powders. Regardless of the usage, the surface activity of the surfactant is the primary basis for selection. Surface-active properties that are determinative in herbicide dispersion choice are surface tension, wetting properties, contact angle, micelle formation, and hydrophilic-lipophilic balance (HLB). Van Valkenburg [47] gives an adequate discussion of the theoretical background of surface properties as they relate to herbicide dispersion. In that discussion it was found necessary to lower the surface tension of a liquid by means of surfactant addition so that the liquid could be emulsified with an ordinarily insoluble second liquid. Such surface tension lowering was also effective in permitting the dispersion of particulate solids into liquid phases. The binary systems that are required to be emulsified consist of paraffin oil-water and vegetable oil-water. It is obvious that not every surfactant would be expected to perform equivalently in both media. In the past, many oil dispersions incorporated the use of salts of sulfonic acids as surfactants. These are relatively inexpensive and readily available paraffin oil-water emulsifiers. When water was introduced as the vehicle, nonionic rather than anionic emulsifiers were utilized.

In the early development of vegetable oils as vehicles, water was used as the spray vehicle. Where water-vegetable oil systems are employed, the surfactants generally found to be useful for paraffin oil-water emulsions have been used, sometimes quite successfully. Of course, they would not necessarily be optimum in performance. In the case in which vegetable oils are used without the use of water, surfactants may still be required. Surfactants are needed because whereas some herbicides are soluble in vegetable oils, many are not.

TABLE 2 Effect of Various Surfactants and Their Concentrations on Surface Tensions

Surfactant concentration (% v/v)	Surface tension (dyn/cm)[a]					
	AAPOEG (Ortho X-77)	NP + EO (Sterox NJ)	MPGAR (Triton B-1956)	OP + EO (Triton X-114)	SML/POESML (Span-20 + Tween 20)	Sun Oil 11E[b]
0	72.8	72.8	72.8	72.8	72.8	72.8
10^{-5}	69.5	68.4	71.8	67.4	71.4	57.5
10^{-4}	58.5	64.6	72.2	62.9	65.7	52.2
10^{-3}	37.7	39.6	45.1	48.2	51.5	47.4
10^{-2}	32.2	32.2	31.0	32.4	32.6	36.6
10^{-1}	32.0	32.4	30.3	30.3	30.4	33.0
0.125	31.6	32.4	30.1	30.1	30.1	32.4
0.25	31.4	32.6	29.7	29.5	29.8	32.6
0.50	31.6	32.4	29.9	29.7	30.0	32.4
1.0	31.6	32.2	29.7	29.7	28.9	32.6
2.0	31.2	32.2	29.9	29.1	29.8	32.4

[a]See Table 1 for identification of the abbreviations.
[b]Sun Oil 11E is a phytobland hydrocarbon oil containing about 2% nonionic emulsifier.
Source: Ref. 6.

TABLE 3 Effect of Various Surfactants on Surface Tension of Two Herbicides in Various Concentrations

Herbi-cide	Rate (kg/ha)	Surface tension (dyn/cm)[a]									
		No surfac-tant	AAPOEG	SML	NP + EO	TMNPEOE	MPGAR	OP + EO	POESML	SML POESML	Sun Oil 11E[b]
Atrazine	1.0	57.0	31.9	29.0	32.1	28.6	33.0	30.4	39.3	34.8	34.0
	2.0	51.2	31.7	29.4	32.0	28.7	33.1	30.8	39.5	34.7	36.9
	4.0	41.7	31.9	30.6	32.1	28.7	33.1	30.6	39.2	34.0	37.6
Ametryne	1.0	35.5	32.2	30.4	31.9	28.7	32.6	30.6	38.4	35.2	34.6
	2.0	33.3	31.9	31.0	31.9	28.9	32.6	30.7	38.4	34.4	34.7
	4.0	33.5	32.1	31.3	31.8	28.8	32.7	30.8	38.3	35.2	34.4

[a]See Table 1 for identification of the abreviations.
[b]Sun Oil 11E is a phytobland hydrocarbon oil containing about 2% nonionic emulsifier.
Source: Ref. 6.

An emulsifier is therefore, required to yield uniform distribution. This area of herbicide emulsification represents an important field with more effort needed for development of satisfactory surfactant formulations.

Table 2 shows the effect of six nonionic surfactants, including mixtures, at various concentrations, on surface tension of soft water. Table 3 summarizes the effect of various concentrations of surfactants on the surface tension levels of two herbicide solutions. It can be seen from Tables 2 and 3 that the use of surfactants reduced the surface tension in most instances. The steepest diminishing of surface tension levels occurred at lower surfactant concentrations. The surfactant concentrations ranging from 10^{-5} to 10^{-2}% were most effective in reducing surface tension. In the range 10^{-2} to 2% the further reduction of surface tension was slight or did not exist.

The critical micelle concentration (CMC) is defined as the concentration of surfactant above which further reduction of surface tension does not occur or occurs to a relatively low extent. This, of course, varies from surfactant to surfactant depending on chemical composition and structure. In the University of Florida evaluation program the CMC levels varied as follows: AAPOEG (Ortho X-77), SML (Span-20), NP + EO (Sterox-NJ), MPGAR (Triton-B 1956), POESML (Tween 20), and SML/POESML (Span-20 + Tween 20) had CMC levels of approximately 10^{-2}%. For TMNPEOE (Surfactant-WK), OP + EO (Triton X-114), and Sun Oil 11E, CMC levels were approximately 10^{-1}% (Table 2). Water hardness levels, from 250 to 1000 ppm as calcium ion, did not affect surface tension appreciably. Up to the limits tested, water hardness would not be expected to affect the results for herbicide solutions when nonionic surfactants are added to these solutions. The surfactants studied can be arranged in the following order as to effect in reducing surface tension: TMNPEOE > OP + EO > SML > SML/POESML > MPGAR > AAPOEG > NP + EO > Sun Oil 11E > POESML.

The surface tension values of 0.5, 1, and 2% (v/v) concentrations in suspensions of commercial formulations of atrazine and ametryne were determined to simulate application rates of 1, 2, and 4 kg/ha in 400 L/ha [6]. These values are given in Table 3. Singh et al. [6] noted that atrazine solutions at 1.0, 2.0, and 4.0 kg/ha had substantially higher surface tension values than similar dosage rates of ametryne in the absence of surfactant. An increase in the rates of atrazine from 1 kg to 4 kg per hectare reduced the surface tension from 57 dyn/cm to 41.7 dyn/cm, while the ametryne rate increases did not affect the surface tension. The authors attributed these results to the fact that the commercial ametryne concentrate probably contained sufficient surfactant to provide optimum CMC. The addition of surfactants to atrazine solutions reduced surface tension, whereas the addition of surfactants to ametryne solutions did not affect the surface tension except for a reduction of about 5 dyn/cm observed for TMNPOE (Surfactant-WK) (compared to no surfactant addition). The addition of POESML (Tween 20) to ametryne increased surface tension, probably caused by an interaction between POESML and the unknown surfactant(s) already present in the commercial ametryne formulation. The work of Singh et al. [6] clearly points out that surface tension measurements should be carried out in laboratory herbicide-surfactant formulation work. The results should also be applied to those cases where the herbicide manufacturer has already incorporated some surfactant into the herbicide concentrate.

The wetting properties of the surfactant are related to, but can be treated as distinct from, the effects on surface tension. Wetting activity is of paramount importance in determining the uniform distribution of the herbicide on the leaf surface and its absorption thereon. Consequently, the ability of the surfactant dissolved in the vehicle to wet the target leaf surface is of vital importance.

Wetting is frequently expressed in terms of the contact angle that a drop of liquid makes on a solid surface. It is obvious that the lower the contact angle (and the lower the surface tension of the liquid), the more extensively the solid surface becomes wetted by the liquid. Laboratory determinations of contact angles are not easy to carry out. Great care must be taken to assure that the measurement surfaces are clean and the actual measurements are usually the statistical average of a number of separate determinations. It is further understood that the contact angle determined in the laboratory on clean glass or Teflon surfaces is not identical to the angle of the herbicide-surfactant vehicle on a leaf surface.

Contact angles of six nonionic surfactants in water at various concentrations from 2% (v/v) down to 10^{-3}% were determined by Singh et al. [6]. A contact angle gonimeter was employed to measure angles of contact on a clean Teflon surface. The data are given in Table 4. The effect of several nonionic surfactants on the contact angles of two herbicide water solutions at 1.0, 2.0, and 4.0 kg/ha rates are shown in Table 5.

It can be observed that the use of surfactant reduced the contact angle. The increase in surfactant concentration from 10^{-5} to 10^{-2}% decreased the contact angle sharply until the CMC point was reached. Of the six combinations evaluated, NP + EO (Sterox NJ) showed the greatest contact angle decrease down to the 46 to 48° range (10^{-1} to 2%). SML/POESML (Span 20 + Tween 20) showed the least reduction, down to the 52 to 65° range (10^{-1} to 2%). The authors [6] determined that the contact angle of 1000 ppm hard water (as calcium ions) was slightly lower than distilled water, but that water hardness levels of 250, 500, 750, and 1000 ppm did not affect the contact angle. Accordingly, the data of Table 4 relate to soft water only.

TABLE 4 Effect of Various Surfactants and Their Concentrations on Contact Angle

Surfactant Concentration (% v/v)	Contact angle (deg)[a]					
	AAPOEG (Ortho X-77)	NP + EO (Sterox NJ)	MPGAR (Triton V-1956)	OP + EO (Triton X-114)	SML/POESML (Span-20 + Tween 20)	Sun Oil 11E[b]
0	110	110	110	110	110	110
10^{-5}	90	96	97	96	102	75
10^{-4}	78	91	99	91	94	73
10^{-3}	67	70	72	78	79	71
10^{-2}	63	48	61	57	72	53
10^{-1}	61	48	60	54	65	51
0.125	60	48	58	53	64	50
0.25	58	49	56	52	58	49
0.50	51	48	57	51	55	47
1.0	51	47	56	50	54	48
2.0	50	46	55	49	52	48

[a]See Table 1 for identification of the abbreviations.
[b]Sun Oil 11E is a phytobland hydrocarbon oil containing about 2% nonionic emulsifier.
Source: Ref. 6.

TABLE 5 Effect of Various Surfactants on Contact Angle of Herbicide Solutions

Herbicide	Rate (kg/ha)	Contact angle (deg)[a]								
		No surfactant	AAPOEG (Ortho X-77)	SML (Span-20)	NP + EO (Sterox NJ)	TMNPEOE (Surfactant WK)	MPGAR (Triton B-1956)	OP + EO (Triton X-114)	POESML (Tween 20)	SML/POESML (Span-20 + Tween 20)	Sun Oil 11E[b]
Atrazine	1.0	59	54	65	59	35	36	47	76	69	46
	2.0	83	53	64	59	35	36	50	76	68	48
	4.0	93	55	64	59	35	38	50	76	69	51
Ametryne	1.0	58	56	62	59	36	35	50	75	65	53
	2.0	62	55	62	60	36	37	51	77	66	53
	4.0	65	55	63	58	35	36	50	74	66	53

[a]See Table 1 for identification of the abbreviations.
[b]Sun Oil 11E is a phytobland hydrocarbon oil containing about 2% nonionic emulsifier.

Table 5 shows that in the absence of surfactant, atrazine solutions had higher contact angles than ametryne solution at the two highest rates. The contact angle increased with increased rates of application of atrazine and ametryne. The increases were greater with atrazine than with ametryne. The addition of surfactant to both atrazine and ametryne solutions lowered the contact angle considerably. Increasing concentrations of surfactant from 0.5 to 2% had no effect on the contact angle except for the case of MPGAR (Triton B-1956), where lower contact angle values were noted at 2% compared with 0.5 and 1%. In the presence of surfactants, no difference was found between contact angles of atrazine and ametryne solutions at any concentration of surfactant. Minimum contact angles were noted with the use of TMNPEOE (Surfactant-WK) and maximum angles with POESML (Tween-20).

In a practical sense, the determination of surface tension and contact angle on commercial samples of herbicides with surfactants is indefinite because of the presence of unidentified amounts of unspecified surfactants. On the other hand, technical evaluation laboratories in the herbicide industry and academic organizations are increasingly classifying surfactants into strong, weak, and intermediate surfactants, respectively, based on their surface tension and contact angle reduction ability.

Micelle formation of a surfactant undoubtedly plays a complex role in herbicide dispersion. When the surfactant molecules are dispersed in water, they orient according to lipophilic and hydrophilic segments with the former positioned away from the water and the latter toward the water. As the concentration of surfactant is increased, there is an increased tendency to form aggregates called micelles.

In the monomer state there is a tendency for the surfactant to be loosely linked with the herbicide through hydrogen bonding. This complex may or may not penetrate foliage as well as the herbicide itself. The overall lipid solubility determines the degree of penetration of the complex as compared to the herbicide. With increased surface-active concentration, more micelles are formed, and there is a decreased tendency for herbicide complex formation. As a result, herbicide molecules are solubilized by the micelles because the lipid-like organic herbicide is attracted to the center lipophilic segment of the surfactant. This defeats at least one function in herbicide dispersion because it has been shown [48] that herbicides solubilized by micelles cannot be transported through biological membranes. Desorption of solubilized herbicides from micelles is very slow. Thus, a surfactant concentration at somewhat below extensive micelle formation would appear to be advantageous. In an organic solvent, the opposite orientation of surfactant micelles would prevail. Here the micelle has the lipophilic segment on the outside and the hydrophilic segment faced to the inside.

The hydrophilic-lipophilic balance is an empirical scale developed by W. C. Griffin [49] in which a number of from 1 (oleic acid) to 20 (potassium oleate) represents the degree of hydrophobic (waterlike) (high numbers) and lipophilic (also called hydrophobic (low numbers) character of a particular surfactant. Each component structural feature can be assigned a value, and the total numerical value of HLB can be obtained as a weighted summation. The HLB system has been valuable in estimating the optimum surfactant required for an emulsification and, in particular, for selecting emulsifiers for herbicide formulations. However, valuable though it has been, it is only applicable to nonionic surfactants. Furthermore, in the presence of inorganic salts, usually present as an integral part of commercial emulsifiers, there is a decrease in the hydration of surfactants, and this shifts the apparent HLB to a lower scale or greater lipophilicity. Despite these shortcomings, the HLB continues to serve as a very useful tool. Becher has provided a superb discussion of HLB in his 1966 treatise [50] and has also provided an updated (1985) bibliography [51].

B. Relationship to Vehicle

Historically, at least, the development in the field of herbicide dispersions has transgressed through the use of paraffin oils, paraffin oils/water, to water used alone. The use of paraffin oils/water represents the predominating bulk of past and present usage. In the decade beginning with 1980 we observed the introduction of vegetable oils such as soybean oil for this purpose. Thus, vegetable oil and vegetable oil/water can now be added to the list of spray media for herbicide dispersion. It almost goes without saying that the choice of optimum surfactant depends on the media in which it is used. Although there is a plethora of background and experience with paraffin oils/water systems, there is considerably less information available in the literature concerning all-water systems. On the other hand, the background and data on vegetable oil and vegetable oil/water systems is extremely sparse. Most of the introductory work that has been undertaken thus far has been based primarily on the use of partially degummed soybean oil or vegetable oils fortified with unknown amounts of unidentified surfactants. Although this has afforded an appreciation of the advantageous overall prospects, it has thus far provided only little idea of the optimum conditions required and the specific surfactants needed for each herbicide.

Water exhibits a high surface tension of 72.8 dyn/cm at 20°C, but very little surfactant concentration is required to drop this value to the 30's, which is usually sufficient for proper emulsification of most oils. This fact minimizes the cost of conventional surfactants for this operation. Emulsions of the oil-in-water (o/w) type can usually be achieved with surfactants having HLB values ranging roughly from 8 to 18, and there would be some overlap in the utility of any surfactant for the paraffin oil or vegetable oil component of a o/w emulsion. On the other hand, to this point, the use of randomly chosen surfactants for soybean oil dispersion can hardly be accidentally "ideal" in surface-active properties. More evaluation will be required to explore this area thoroughly.

One problem with vegetable oils is the fact that the natural emulsifiers present in these oils, soybean oil particularly, are gum-like phospholipids. Although these are known as effective o/w emulsifiers, they may cause mechanical problems in the application and transfer of the herbicide emulsion. As gums, these components would have a tendency to plug spray nozzles. Crude soybean oil contains $1\frac{1}{2}$ to 3% phospholipids, about 25% of which is lecithin, and even so-called degummed soybean oil may contain as much as 200 ppm (as phosphorus) of these natural phospholipid emulsifiers. (This is why ASA has recommended once-refined or fully refined oil for this purpose.) Even oxidation of the linolenic component of the soybean oil might be expected to cause mechanical spray problems not usually encountered with paraffinic oil/water systems. Since both paraffin oils and vegetable oils exhibit small solubility for most of the nonionic surfactants, these surfactant materials are largely suspended in these vehicles with only a minor proportion dissolved. Fortunately, the amount of surfactant required to achieve a satisfactory emulsification is below the solubility limit in most cases. As a matter of fact, we have indicated that it is preferred to operate below a concentration where substantial micelles are generated. Thus considerations for optimum solubility of surfactants in many vehicles are relatively unimportant.

C. Other Factors

Such small quantities of surfactant are required in water, oil, or vegetable oil that considerations of oxidation stability of the surfactant are not pertinent. Still, in many cases, ester surfactants are structured from fatty acid components which are devoid of easy oxidizable linoleate and linolenate homologs. Where oleoesters are chosen, freedom from polyunsaturated components is desirable; generally, this figure is restricted to a level of polyunsaturates below 5%. As more and more vegetable oils are utilized in herbicide dispersion, the importance of oxidation stability in the nonionic esters will be apparent. For example, if soybean oil monoglycerides were chosen as a surfactant candidate, it would be advisable for it to be selectively hydrogenated to destroy the about 7% linolenic component to afford optimum oxidation stability in the surfactant. The difficulties with oxidized vegetable oils either as such, or in combination with water, are tendencies toward thickening and air oxidation. Such tendencies result in spraying difficulties and nozzle plugging. These problems are, of course, not encountered with polyoxglycol esters where the fatty acid component is saturated.

The requirement for surfactant "inertness" is manifested in two directions. Obviously, the surfactant must be chemically inert to the herbicide. With respect to the biological function of the herbicide, it is preferable that the surgactant enhance this functionality in any manner possible provided that the selectivity between target and protected plant is not adversely affected. We have already pointed out the disadvantageous effect of high micelle concentration with the use of larger than necessary concentrations of surfactant. Limitation to the required low concentrations affords both an economic and a biological functional advantage. Compared to the cost of the herbicide itself, the cost of most surfactants in herbicide dispersion is minimal. However, all new surfactants developed for this application must be priced in the range of the conventional nonionic surfactants already employed.

V. NEW HORIZONS

During the period 1968—1972, the use of emulsifiable crop origin oil (oil extracted from crops such as soybeans, cotton, flax, etc.) at 2.3 L/ha was found to be as effective as 9.3 L/ha of petroleum oil in enhancing weed control in corn with atrazine and cyanazine [52—54].

To date, the major carrier for the application of herbicides has been water, followed by paraffin oils as a distant second. The hydraulic sprayers employed have changed little in over 50 years. When a volume as high as 200 L/ha is used routinely, the spray operation may become inefficient, and there is a frequent need for refilling the spray tank. Recently, herbicide applications with rotary atomizers (controlled droplet applicators) were commercialized. This depends on the use of a spinning cone which discharges the spray solution from the rim of the cone as a series of very uniform droplets. Accordingly, it is possible to apply herbicides uniformly in carrier volumes as low as 2 to 3 liters/ha. Because the cost of spraying is thus reduced several dollars per hectare, this makes the use of certain crop origin oils attractive as a carrier for herbicides. Whitney et al. [55] have supplied data on flow rates of several vegetable oil/pesticide blank formulation mixtures through various agricultural spray nozzles.

Kapusta [4] described a series of field studies conducted in 1982 and 1983 at Southern Illinois University. This group evaluated soybean oil as a carrier for preemergence and postemergence herbicide and as an enhancing agent for postemergence herbicides (Tables 6 and 7). Preemergence and postemergence herbicides were applied with rotary nozzles employing a soybean oil carrier volume as low as 5 liters/ha. Afforded weed control was equal to that achieved when the herbicides were applied with flat fan nozzles in 187 L of water per hectare. Soybean oil employed as an enhancing agent for four postemergence soybean herbicides was equal to petroleum crop oil concentrate with three of the herbicides, but distinctly less effective with a fourth herbicide. In this work, unidentified emulsifiers were used in both the petroleum oils and soybean oil concentrates. In the case of the soybean oil, from 7 to 15% levels of emulsifier were incorporated directly in the oil before use. In the case of the petroleum oil concentrate, the unidentified surfactant amounted to 17% of the mixture. Table 8 shows an evaluation of soybean oil concentrate versus petroleum oil concentrate as enhancing agents with postemergence herbicides. This study involved control of Johnson grass in soybeans at Murphysboro, Illinois. Reducing the carrier volume to extremely low amounts was a major benefit observed for the use of soybean oil with rotary nozzles. This greatly increases the efficiency of spray operation because frequent refilling associated with current applications methods is eliminated. The cost of spraying can be decreased substantially. Carrier hauling is decreased, although there is an initial greater investment associated with rotary nozzles compared with convention nozzles. This investment would be recovered in several years.

Soybean oil can be adapted as a postemergence herbicide carrier applied with conventional sprayers because investment in new equipment is not needed. Results from the Johnson grass study (Table 8) and more recent research indicate that it is equal to PCOC in enhancing the activity of essentially all postemergence herbicides.

Growers using soybean oil as a carrier would need to clean their equipment frequently to eliminate films of oils and deposits on it during spraying operations. The compatability between the soybean oil and each herbicide should be determined in advance to preclude problems in the field. The use of personnel wearing protective clothing more routinely is recommended because the oil may increase absorption of the herbicide through the skin. (Such protection is also recommended with use of petroleum oils.) Studies continue at Southern Illinois University and at other universities to further evaluate the potential of soybean oil as a carrier and additive for herbicides.

In summary, it is perhaps pertinent to outline a number of observations that appear worthy of emphasis:

1. There is considerable ambiguity in herbicide dispersion terminology (i.e., terms such as adjuvant, carrier, concentrate, vehicle, and others are frequently misused). For this reason the American

TABLE 6 Evaluation of Soybean Oil as a Carrier for Preemergence Soybean Herbicides Applied with Rotary Nozzles, Carbondale, Ill., 1982

Treatment	Rate (kg/ha)	Application details				Percent control, rated June 20[c]							Soybean yield (kg/ha)
		Nozzle type[a]	Carrier type[b]	Carrier total (L/ha)	Volume oil (L/ha)	Gift	Yens	Colq	Corw	Jiwe	Ilmg		
Nontreated						0	0	0	0	0	0		1537
Alachlor	2.24	RN	SBO	7		98	97	97 ab	96	96 ab	68		2196
	2.24	RN	SBO	9		98	97	98 a	97	97 ab	84		2133
Metribuzin	0.42	RN	SBO	7		96	94	98 a	96	97 ab	80		2007
	0.42	RN	SBO	9		90	98	98 a	94	94 ab	85		2133
Alachlor + metribuzin	2.24 +0.42	RN	SBO	7		98	98	98 a	98	97 ab	76		1819
	2.24 +0.42	RN	SBO	9		98	98	98 a	98	98 a	67		2133
Alachlor	2.24	RN	SBO		5	98	97	97 ab	96	94 ab	70		2133
	2.24	RN	SBO		9	98	91	98 a	89	72 c	83		2070
Metribuzin	0.42	RN	SBO		5	96	94	98 a	97	96 ab	87		2196
	0.42	RN	SBO		9	95	94	98 a	98	97 ab	82		2258
Alachlor + metribuzin	2.24 +0.42	RN	SBO		5	98	78	96 bc	96	96 ab	72		2196
	2.24 +0.42	RN	SBO		9	98	90	98 a	98	97 ab	67		2007

Herbicide	Rate	Nozzle	Carrier	Volume							
Alachlor	2.24	RN	Water	9	97	96	98 a	98	97 ab	79	2133
Alachlor	2.24	RN	Water	19	98	93	98 a	98	96 ab	83	2133
Metribuzin	0.42	RN	Water	9	97	94	97 ab	97	96 ab	69	2007
Metribuzin	0.42	RN	Water	19	92	92	98 a	98	91 ab	71	2007
Alachlor + metribuzin	2.24 +0.42	RN	Water	9	98	98	98 a	98	98 a	86	1631
Alachlor + metribuzin	2.24 +0.42	RN	Water	19	98	93	98 a	98	98 a	62	2258
Alachlor	2.24	FF	Water	187	98	96	95 c	96	89 ab	73	2070
Metribuzin	0.42	FF	Water	187	98	89	98 a	98	98 a	76	1945
Alachlor + metribuzin	2.24 +0.42	FF	Water	187	98	98	98 a	98	98 a	88	1756

[a] RN, Micromax rotary nozzles at 2000 and 3500 rpm with water and soybean oil, respectively; FF, 8002 flat fan nozzles.
[b] SBO, Stoller Chemical Co. Natur'l Oil (93% once-refined soybean oil plus 7% emulsifier).
[c] Means within columns followed by one or more like letters are not different at the 5% level according to Duncan's multiple range test (values within a column not followed by letters are not significantly different).
Gift, giant foxtail; yens, yellow nutsedge; colq, common lambsquarters; corw, common ragweed; jiwe, jimsonweed; ilmg, ivyleaf morning glory.
Source: Ref. 4.

TABLE 7 Evaluation of Soybean Oil as a Carrier for Postemergence Soybean Herbicides Applied with Rotary Nozzles, Carbondale, Ill., 1982

Treatment	Application details				% Control, rated June 20[c]						Soybean yield (kg/ha)
	Rate (kg/ha)	Nozzle type[a]	Carrier volume (L/ha)	Carrier type[b]	Colq	Corw	Pesw	Jiwe	Vele	Ilmg	
Nontreated					0	0	0	0	0	0	2196
Bentazon	0.84	RN	5	SBO	97	92	95	93	98	38	2258
	0.84	RN	7	SBO	97	92	95	91	91	26	2196
	0.84	RN	9	SBO	97	94	97	89	98	40	2321
	0.84	RN	9	Water	98	97	97	98	98	41	2384
	0.84	RN	19	Water	98	97	97	98	98	62	2509
	0.84	RN	37	Water	98	98	98	98	98	48	2321
	0.84	FF	187	Water	98	98	98	98	98	52	2448

[a]RN, Micromax rotary nozzle at 3500 rpm; FF, 8002 flat fan nozzles.
[b]SBO, Stoller Chemical Co. Natur'l Oil (93% once-refined soybean oil plus 7% emulsifier).
[c]Colq, common lambsquarters; Corw, common ragweed; Pesw, Pa. smartweed; Jiwe, jimsonweed; Vele, velvetleaf; Ilmg, ivyleaf morning glory.
Source: Ref. 4.

TABLE 8 Evaluation of Soybean Oil Concentrate versus Petroleum Oil Concentrate as Enhancing Agents with Postemergence Herbicides for Johnsongrass Control in Soybeans, Murphysboro, Ill., 1983

		% Control,[a] Aug. 10 Additive, 2.3 L/ha		
Herbicide	Rate (g/ha)	None	PCOC[b]	SOC[c]
Nontreated				
Sethoxydim	84	8 d	18 d	28 cd
	140	18 d	37 cd	37 cd
	210	33 cd	45 c	47 c
Fluazifop-butyl	84	35 cd	30 cd	37 cd
	140	62 bc	80 ab	55 bc
	210	75 b	83 ab	77 b
Hoe 581	70	37 cd	52 c	32 cd
	140	62 bc	67 bc	57 c
	175	80 ab	—	—
DPX-Y6202	18	8 d	40 cd	5 d
	35	38 cd	73 b	27 cd
	53	67 bc	90 a	40 cd
	70	77 b	—	—

[a]Values within and between columns followed by one or more like letters are not different at the 5% level according to Duncan's multiple range test.
[b]Petroleum oil 83% plus emulsifiers 17%.
[c]Fully refined soybean oil 85% plus emulsifiers 15%.
Source: Ref. 4.

Soybean Association has published a brochure titled *Crop Oils Glossary* [56], in which many terms are defined.
2. Conventional surface tension measurements (such as Wilhemy's detachment method using a torsion balance [57]) and determination of contact angle with a goniometer on a Teflon surface provide much practical information pertaining to the performance of commercial herbicides and surfactants in dispersions.
3. There is a tendency on the part of herbicide manufacturers to incorporate unspecified quantities of unidentified surfactants in herbicide concentrates.

4. Commercial herbicides usually contain small quantities of inorganic salts and these effect the determination of surface tension and contact angle appreciably.
5. Surfactant quantities for optimum herbicide dispersion are well below the critical micelle concentration in water. This is a boon for both the biological and economical utilization of surfactants in herbicide dispersion.
6. The contact angle for a herbicide surfactant water dispersion determined on a glass or Teflon surface is proportional to but not identical to that on the target leaf surface.
7. As the trend to displace hydrocarbon phytobland oils with vegetable oils develops, we may see a trend in the use of surfactants for herbicide dispersion from nonionic ethoxylates to nonionic monoglycerides.

REFERENCES

1. R. H. Hogson, *Adjuvants for Herbicides*, Weed Science Society of America, Champaign, Ill., 1982, Preface, p. i
2. Colin A. Houston & Associates, Inc., Mamaroneck, N.Y., cited in P. L. Layman, *Chem. Eng. News*, Jan. 21, 1985, pp. 23–48, as quoting 2.9% annual growth in 1983 for industrial surfactants until 1990 (*Chem. Week*, Aug. 3, 1983, p. 25).
3. Chemark, a chemical market research firm, surfactant survey 1984, cited in *Chem. Week*, Dec. 12, 1984, SAS Section.
4. G. Kapusta, *J. Am. Oil Chem. Soc.*, 62(5):923 (1985).
5. J. W. Van Valkenburg, in *Adjuvants for Herbicides*, Weed Science Society of America, Champaign, Ill., 1982, pp. 1–9.
6. M. Singh, J. R. Orsenigo, and D. O. Shah, *J. Am. Oil Chem. Soc.*, 61(3):596 (1984).
7. G. C. McWhorter, in *Adjuvants for Herbicides*, Weed Science Society of America, Champaign, Ill., 1982, pp. 10–25.
8. P. W. Zimmerman and A. E. Hitchcock, *Contrib. Boyce Thompson Inst.*, 12:321 (1942).
9. W. B. Ennis, Jr., R. E. Williamson, and K. P. Dorchner, *Weeds*, 1:274 (1952).
10. A. E. Hitchcock and P. W. Zimmerman, *Contrib. Boyce Thompson Inst.*, 15:173 (1948).
11. H. M. Hull, *Weeds*, 4:22 (1956).
12. P. C. Marth, F. F. Davis, and J. W. Mitchell, *Bot. Gaz. Chicago*, 107:131 (1945).

13. J. W. Mitchell and C. G. Hamner, *Bot. Gaz. Chicago,* 105:474 (1944).

14. F. E. Edwards, W. B. Ennis, Jr., V. C. Harris, E. B. Hollingsworth, C. G. McWhorter, and O. B. Wooten, Jr., *Miss. Agric. Exp. Stn. Circ.,* 195 (1955).

15. F. E. Edwards, W. B. Ennis, Jr., V. C. Harris, J. T. Holstun, Jr., C. G. McWhorter, and O. B. Wooten, *Miss. Agric. Exp. Stn. Circ.,* 177 (1953).

16. H. E. Rea, *Tex. Agric. Exp. Stn.,* Bull. 902 (1958).

17. A. F. Wiese and H. E. Rea, *Tex. Agric. Exp. Stn. Res. Rep.,* 1962, 3 p. (1956).

18. G. E. Blackman, *Proc. 4th Int. Cong. Crop Protection,* Hamburg, 1957, Vol. I, Braunschweig, W. Germany, pp. 481–487 (1959).

19. A. S. Crafts, H. B. Currier, and H. R. Drever, *Hilgardia,* 27: 723 (1958).

20. A. B. Currier and C. D. Dybing, *Weeds,* 7:195 (1959).

21. C. D. Dybing and H. B. Currier, *Weeds,* 7:214 (1959).

22. V. H. Freed and M. Montgomery, *Weeds,* 6:386 (1958).

23. C. G. L. Furmidge, *J. Sci. Food Agric.,* 10:267 (1959).

24. W. C. Orgell and R. L. Weintraub, *Bot. Gaz. Chicago,* 119:88 (1957).

25. J. W. Mitchell and P. J. Linder, *Science,* 112:54 (1980).

26. A. E. Smith, J. W. Zukel, G. M. Stone, and J. A. Ridell, *J. Agric. Food Chem.,* 7:341 (1959).

27. C. G. McWhorter, Stoneville, MS, Table 2, pp. 295–296, *Rep. Res. Comm.* (W. B. Albert, Chairman), Proc. 9th South Weed Conf., 1956.

28. J. T. Holstun, Jr. and S. W. Bingham, *Weeds,* 8:187 (1960).

29. J. T. Holstun, Jr. and C. G. McWhorter, Stoneville, MS., Table 2, pp. 249–251, *Rep. Res. Comm.* (E. G. Rodgers, Chairman), Proc. 10th South. Weed Conf., 1957.

30. C. G. McWhorter and E. E. Schweizer, *Proc. Northeast, Weed Control Conf.,* 18:6 (1964).

31. O. C. Burnside, Res. Rep. North Cent. Weed Control Conf., 23: 77 (1966).

32. R. S. Colby, *Proc. Northeast, Weed Control Conf.,* 19:312 (1965).

33. W. L. Currey and R. H. Cole, *Proc. Northeast, Weed Control Conf.*, 20:297 (1966).

34. A. G. Dexter, O. C. Burnside, and T. L. Lavy, *Weeds*, 14:222 (1966).

35. L. C. Liu, R. D. Ilnicki, J. B. Regan, and E. J. Visinski, *Proc. Northeast. Weed Control Conf.*, 20:309 (1966).

36. G. Miller, *Minn. Proc. North Cent. Weed Control Conf.*, 24:57 (1967).

37. A. R. Putnam and S. K. Ries, *Proc. Northeast. Weed Control Conf.*, 19:300 (1965).

38. G. A. Wicks, O. C. Burnside, and C. R. Fenster, *Neb. Proc. North Cent. Weed Contr. Conf.*, 22:58 (1967).

39. A. F. Wiese and D. Owens, *Tex. Agric. Exp. Stn. Res. Rep.*, 2437:6 pages (1967).

40. G. W. Anderson and G. E. Jones, *Res. Rep. East. Sect. Natl. Weed Comm. Can.*, p. 16 (1964).

41. G. W. Anderson and G. E. Jones, *Res. Rep. East. Sect. Natl. Weed Comm. Can.*, p. 17 (1964).

42. C. G. McWhorter, *Weeds*, 11:265 (1963).

43. C. G. McWhorter and T. J. Sheets, *Proc. South. Weed Conf.*, 14:54 (1961).

44. *Fact: Number One Choice for Crop Oil*, American Soybean Association, St. Louis, Mo., 1982.

45. N. O. V. Sonntag, *J. Am. Oil Chem. Soc.*, 62(5):928 (1985).

46. *Chem. Mark. Rep.*, March 28, 1983, p. 1.

47. J. Wade Van Valkenburg, in *Adjuvants for Herbicides*, Weed Science Society of America, Champaign, Ill., 1982, pp. 3–5.

48. K. J. Mysels, in *Pesticide Formulation Research* (W. Van Valkenburg, ed.), Advances in Chemistry Series, American Chemical Society, Washington, D.C.,1969, pp. 24–38.

49. W. C. Griffin, *J. Soc. Cosmet. Chem.*, 5249 (1954).

50. ACS Monograph Series, American Chemical Society, Washington, D.C., P. Becher, *Emulsions: Theory and Practice*, 1966, pp. 231–252.

51. P. Becher ed., *Encyclopedia of Emulsion Technology*, Vol. 2, *Applications*, Marcel Dekker, New York, 1985, pp. 425–512.

52. G. R. Miller and W. W. Nelson, *Res. Rep. North. Cent. Weed Control Conf.*, *28*:123 (1971).
53. J. D. Nalewaja, *Proc. North. Cent. Weed Control Conf.*, *23*:12 (1968).
54. J. D. Nalewaja, *Res. Rep. North Cent. Weed Control Conf.*, *29*:294 (1972).
55. R. W. Whitney, L. O. Roth, and T. L. Underwood, *J. Am. Oil Chem. Soc.*, *63*(3):340 (1986).
56. *Crop Oils Glossary*, American Soybean Association, St. Louis, Mo., 1984.
57. L. S. Osipow, *Surface Chemistry, Theory and Industrial Applications*, R. F. Krieger, Melbourne, Fla., 1972, p. 473.

8
The Role of Surfactants in Emulsion Polymerization

BENJAMIN B. KINE and GEORGE H. REDLICH *Rohm and Haas Company, Spring House, Pennsylvania*

I.	Introduction	264
II.	Historical Perspective	266
III.	The Polymerization Process	267
	A. Interval I	268
	B. Interval II	268
	C. Interval III	269
	D. The Chemical Reactions	269
IV.	Growth of Particles	274
	A. Heterogenous Nucleation	274
	B. Homogenous Nucleation	285
V.	Colloidal Stability of Latex Particles	294
	A. The Basis of Stability	294
	B. The Effect of Added Surfactants	297
	C. Effect of Soap on Dispersion Stability	302
VI.	Practical Considerations for Surfactant Selection	306
VII.	Examples of Emulsion Polymerization	308
VIII.	An Emulsion Polymerization Plant	311

I. INTRODUCTION

Emulsion polymers are a major factor in the manufacture and formulation of products used in many raw material and applications areas. Among the major examples are such diverse products as paints, coatings, floor polishes, adhesives, textile finishes, carpet and textile backings, rubber goods, automotive and aircraft tires, and others. An understanding of emulsion polymerization and the subsequent behavior of the dispersion that results is necessary in order to design a successful commercial product. As a class, emulsion polymerization processes probably contain more "ingredients" than other industrial processes employed for the synthesis of a composition. In many ways the most critical ingredient, in the emulsion polymerization process, is the surfactant, or soap, both in the polymer initiation and growth steps and in the subsequent stabilization of the emulsion polymer particle, as well as in the maintenance of the final emulsion stability and properties throughout storage, formulation, and use conditions.

The essential components of an emulsion polymer system are monomer, dispersant, catalyst, and water. Prior to polymerization, water, which is the continuous phase, has dispersed in it droplets of monomer and surfactant micelles swollen with monomer. The dispersed components are in equilibrium with monomer, surfactant, and catalyst, which are soluble in the water phase. The monomer droplets are stabilized by surfactant which is positioned at the droplet-water interface. After polymerization the system consists of particles of polymer whose surfaces are stabilized by surfactant, which is in equilibrium with the surfactant dissolved in the water phase and the surfactant at the air-water interface, and all of the other interfaces that the dispersion contacts. The polymerization reaction, if properly conducted, can be almost quantitative in converting monomer to polymer—and if the reaction is conducted properly, very little monomer, if any, remains in the final product.

Classical emulsion polymerization processes comprise a series of steps that take place in a dispersed system in which there are several coexisting phases. In the batch-type polymerization, commonly termed the one-shot type, all the ingredients are charged to the kettle and the reaction initiated. The solids content of the final product is limited by the heat of polymerization that is generated and the ability to remove this heat. At this stage the key ingredients and their location in the system are:

	Aqueous phase	Micellar phase	Monomer droplet
Surfactant	= CMC	Micelles	Minor amount
Monomer	= Saturation	Swollen	Major amount
Initiator	"All"		

Surfactants in Emulsion Polymerization

At the other extreme, we have what is termed the gradual addition type of emulsion polymerization, in which the monomer is gradually pumped into a kettle that contains surfactant and initiator dissolved in water. Monomer is added at a rate that is commensurate with the rate at which heat can be removed. Intermediate to these are processes in which part of the monomer is put into the kettle initially, the remainder gradually fed in, or those in which initiator is fed in parallel with the monomer. The process has direct implications on the morphology of the polymer particle and invariably is reflected in the application and stability properties.

The initiation step in emulsion polymerization is generally believed to occur in the aqueous phase; the initiator decomposes, either thermally or via a redox process (a one-electron transfer reaction), to a free radical.

$$I \longrightarrow 2Rc^*$$

The following step has been proposed to occur in either of two general ways, the extreme depiction of each being:

1. The radicals enter a monomer-swollen emulsifier micelle where polymerization of monomer takes place, to form a monomer-swollen polymer particle.
2. The radicals initiate monomer polymerization in the aqueous phase, forming an oligomeric radical. The oligomeric radical precipitates when it grows long enough to exceed its solubility in the aqueous phase, thus forming a stable primary particle or coagulate with a previously formed latex particle.

In the first instance the surfactant supplies a micelle in which monomer dissolves, and to which the initiator radical migrates to start the actual polymerization. The second mechanism represents another role that the surfactant plays, that of stabilizing the growing oligomer via adsorption. In either case the first polymer molecule results in the formation of a polymer particle, which then continues to absorb monomer and to serve as a locus for further polymerization. As the polymer particle grows in volume and surface area, the role of the surfactant in maintaining the colloidal stability of the particle becomes a critical factor.

The colloidal stability of such latex particles can be described via the DLVO theory [42,43]. In general terms, the stability is a function of the colloidal charge on the particle, which defines a distance from the particle surface at which energy minimum occurs and the distance between particles where coagulation/flocculation would occur. The role of the surfactant following the particle formation stage is to modify the surface characteristics of the particle so that coagulation does not occur over a reasonable time period (which can range from hours to decades). The surfactant accomplishes this via adsorption to the surface of the latex particle, thereby either increasing the charge on the surface (if anionic or cationic soap) or by increasing the hydrophilicity of the surface with polyethylene oxide functionality of nonionic soaps, or by a combination of both of the foregoing mechanisms when the surfactant is a nonionic type which has been terminated with a charged moiety such as a sulfate, sulfonate, phosphate, or other charged groups. There is also a large favorable entropic effect due to the reduction in the number of molecules in solution which are now bound to the latex surface, and thereby more immobilized.

Regardless of the exact mechanism by which polymerization actually occurs, the role of soap is crucial, and its effect on the polymerization process and the final product depends very much on the nature of the soap (i.e., its micelle-forming properties and its related adsorption properties). It is taken as evident that the same hydrophobe-hydrophobe interactions that drive micelle formation can also drive the adsorption of the soap molecule to the particle surface.

II. HISTORICAL PERSPECTIVE

An emulsion is defined as "a substantially permanent heterogeneous liquid mixture of two or more liquids which do not normally dissolve in each other but are held in suspension one in the other, by mechanical agitation, or more frequently, by small amounts of additional substances known as emulsifiers" [1]. Unfortunately, the appellations "emulsion polymerization" and the product formed as a consequence of the reaction, the "emulsion polymer," is an unfortunate one since, with possible exceptions of rare circumstance, polymer does not form by a mechanism that involves the emulsion droplet.

An early German patent of 1909 described emulsion polymerization (German patent 250,690, 1909, Farbenfabrik Bayer A.G.). However, the misnomer probably is a result of the experiments conducted by R. P. Dinsmore in 1929 (U.S. Patent 1,732,795) which historically was one of the first reported English language works on emulsion polymerization. Dinsmore used the primitive emulsifiers oleic acid salts and casein for the "emulsion polymerization" of butadienes. It is interesting to note that he patiently conducted the reaction for a period of 6 months at reaction temperatures of 50 to 70°C. The misnomer "emulsion polymerization" still continues today and is used to describe the process that begins with an emulsion of monomer and ends as a dispersion of polymer.

In 1932, both Harkins and McBain, independently, observed that polymer particles form even from monomers of low water solubility in the absence of emulsifier, and that the formation of particles does not necessarily require the presence of monomer droplets [2].

They observed the size of the polymer particle to be much smaller than that of the droplets and thus concluded that the locus of polymerization must be other than in the droplets. Fikentscher by 1937 noted that even in the presence of micellar emulsifier, the "aqueous phase," not the monomer droplet, was the probable locus of polymerization [3]. The quantitative work of Heller and Klevens from 1943 to 1945 clearly showed the relationship between the surfactant or the emulsifier concentration and the number of particles formed both above and below the critical micellization concentration [4]. Harkins, in 1947, published a series of quantitative studies on both styrene and isoprene, both in the presence and absence of micelles, wherein it was observed that the rate of polymerization was much greater when micelles were present [5]. Based on this evidence it was proposed that the locus of polymerization was the micelle. The quantitative theory of Smith and Ewart was presented to the world the next year [6]. From this theory one was able to predict the absolute rate of the polymerization and the absolute particle concentration. As noted above, the early work was limited to monomers of very low water solubility. Work in the mid-1940s on monomers such as methyl methacrylate, which were more water soluble, raised doubts about assumptions that micelles were the exclusive initial locus of polymerization. Baxendale and co-workers, in 1946, published a definitive study showing that methyl methacrylate (MMA) polymerized, in the presence or absence of soap, via a homogeneous nucleation process in which soap micelles played no role [7].

Because of the importance of synthetic rubber to our economy and way of life, the bulk of the work reported in the early years was in terms of highly water insoluble monomers such as styrene and butadiene and therefore work, and theories on heterogeneous or micellar polymerization prevailed in the literature. Following the initial period, reports about homogeneous nucleation began to appear in much greater number; however, the complexities of the chemistry and exceptions to the rule or theory also increased. Controversies still rage. For the purposes of achieving a qualitative and semiquantitative discussion and understanding of the role of the surfactant in this chapter, we will initially use Smith-Ewart theory for illustrative purposes to avoid confusions, and then discuss more recent theories.

III. THE POLYMERIZATION PROCESS

Emulsion polymerization is commonly divided into three phases or intervals, during which the processes of initiation, particle growth, and "solid-phase" polymerization takes place.

A. Interval I

In this first stage initiation of the free-radical polymerization takes place and is characterized by particle nucleation. This stage is generally considered to be completed when the emulsifier micelles have disappeared. In a typical emulsion polymerization micellar disappearance usually occurs during the first 10 to 20% conversion.

B. Interval II

During this second stage, particle nucleation has ceased and particles grow in size. The source of monomer for this growth is from dispersed monomer droplets and that monomer which is dissolved in the continuous phase. In effect, an equilibrium is set up in which aqueous phase concentration of monomer remains relatively constant; monomer in the latex particle polymerizes, drawing monomer from the aqueous phase, which in turn is replenished from the monomer droplet. Therefore, as the latex particle grows, monomer droplets shrink and are consumed.

Monomer is usually more soluble in its polymer than in water, so that a driving force is provided for equilibrium. Regardless, though, depletion of monomer in the monomer/polymer particle by polymerization results in depletion of the monomer droplet. This also results in a monomer/polymer particle which is quite fluid and allows for two possible mechanisms of particle growth. In competitive growth described above, large particles can absorb more monomer than small particles. Although this results in an increase in volume, an increase in diameter is not relatively as great since particle diameter increases as the one-third power of the volume of added monomer (polymer). Therefore, diameters at the high and low end of the particle size distribution tend to become closer, thus resulting in a narrower particle size distribution. An added factor is that since the total surface area of the smaller particles is larger, the smaller particles tend to absorb more monomer; thus they grow faster, and there tends to be an increase in the monodispersity of the system. As polymerization proceeds, another pathway for particle growth is via coalescence of already existing particles. If two particles in close contact are insufficiently stabilized, they may collide. If they coalesce, since they are fluid particles of monomer-swollen polymer, this can result in a large particle of lower surface area than the two particles from which it was created. But since the number of stabilizing surface molecules is constant and since there is lower total particle surface area, the number of soap molecules per unit surface area is greater and the coalesced entities tend to be more stable than the smaller particles from which they were formed.

C. Interval III

Interval III begins when the monomer droplets disappear. It occurs usually at 60 to 90% conversion and is characterized by a rapid decrease in the aqueous-phase concentration of the monomer, and polymerization of the monomer in the swollen latex particle. During this phase, 30 to 60% of the polymerization can take place, yet the particle number and size remain fairly constant.

The question of why polymerization occurs in the nucleated particle and not in the monomer droplet can be answered by considering the relative surface areas and the number of each species involved. Since the usual particle size of a monomer droplet is on the order of 50 to 5000 µm, whereas that of the particle at this nucleated stage is about 0.025 µm, for each monomer droplet there are 1 million to ten million nucleated particles. The total number and surface area of nucleated particles, then, greatly exceed those of the monomer droplets; thus the probability of a free radical generated in the aqueous phase striking a nucleated particle far exceeds the probability of striking a monomer droplet. Therefore, the nucleated particle is the preferred site of polymerization. A typical reaction would have the relations shown in Table 1.

D. The Chemical Reactions

Emulsion polymers are prepared from monomers that can be polymerized by a free-radical mechanism. Free-radical polymerizations are rapid and may be characterized by the generally accepted chain-reaction steps of initiation, propogation, and termination. The termination reaction occurs by two mechanisms: termination by combination or termination by chain transfer. From an overall view it can be summarized that during a polymerization reaction a number (n) of vinyl-type monomer units polymerize to form a polymer chain n units in length.

$$nCH_2=CH{-}X \longrightarrow (CH_2-CH(X)-)_n \tag{1}$$

TABLE 1 Relation Between Micelles and Droplets

	Number (no./g emulsion)	Total surface area (cm^2/g emulsion)
Monomer droplet	$\sim 10^{6-9}$	$3 \times 10^{2-3}$
Micelle	$\sim 10^{17}$	3×10^{6}

When separated into its component steps the process by which the polymer chain is formed may be represented by the following simplified scheme:

1. The initiator (I), usually a peroxy compound, thermally or chemically induced, decomposes to yield two catalyst radicals (R_c^*). Some chemically induced decomposition reactions yield only one free radical. In practice the number is lower than one or two because a proportion of the radicals are removed by side reactions.

$$I \xrightarrow{k_d} 2R_c^* \tag{2}$$

2. The primary radical (R_c^*) reacts with a monomer molecule (M) to yield a propagating free radical.

$$R_c^* + M \xrightarrow{k_i} R_cM^* \tag{3}$$

3. In turn, the propagating free radical (R_cM^*) reacts with another monomer molecule to form the growing chain ($R_cM_2^*$):

$$R_cM^* + M \xrightarrow{k_p} R_cM_2^* \tag{4}$$

Chain growth continues:

$$R_cM_2^* + nM \xrightarrow{k_p} R_cM_{(n+2)}^* \tag{5}$$

until the reaction is terminated.

4. Termination may occur by two primary mechanisms—one is a bimolecular reaction between two polymer radicals. This is usually called termination by recombination and results in only one polymer molecule being formed.

$$R_cM_m^* + R_cM_n^* \xrightarrow{k_{tc}} R_cM_{m+n} \tag{6}$$

Or termination may occur by the abstraction of an atom or group and the formation of a terminated chain and a new free-radical species.

$$R_cM_m^* + HSR \xrightarrow{k_{tt}} R_cM_mH + RS^* \tag{7}$$

The reaction is called termination by chain transfer. Chain transfer reactions can and do occur with initiator or the initiator radical, monomer, solvent, polymer, surface-active agents, chain transfer agents, modifiers, and other components of the system. Termination by chain transfer leads to lower-molecular-weight polymers.

Kinetics offers a powerful tool for the quantitative understanding of a chemical system. By use of this tool we gain mechanistic insights into very rapid and complex polymerization reactions which cannot be studied effectively by other means. As an example, the total time required for the formation of polymer chains that have a molecular weight of about 1 million may be of the order of 0.1 to 1 s. In this time interval about 10,000 monomer molecules have reacted. Through the use of kinetics we are able to describe the various interdependent steps of the reaction and to quantitatively evaluate the influence of each of the reactions.

The kinetic equations that describe initiation, propagation, and termination use the events depicted in Eqs. (2) to (7). Two quantities that are relatively easy to measure and provide both mechanistic insight and practical usefulness are the rate of polymerization and the molecular weight of the polymer that is formed.

The derivations of the rate of polymerization (R_p) equation and that for the kinetic chain length (the number of monomer units per polymer chain) are as follows. Formation of initiator radicals from unimolecular decomposition of a catalyst molecule is given by Eq. (8), where R_d is the rate of catalytic decomposition and (I_0) the initial concentration of initiator (catalyst). Both I_0 and R_d may be measured directly.

$$R_d = \frac{d(Rc^*)}{dt} = 2k_d f(I_0) \tag{8}$$

Usually, side reactions also compete for the consumption of free radicals; thus, in the real world, fewer than the two radicals formed by thermal decomposition of the peroxide effectively initiate a growing chain. To correct for this the factor (f) is introduced into equation (8) and represents the fraction of radicals that effectively initiate a growing chain. For most free-radical reactions the value of f is between 0.5 and 1.0.

The rate of chain initiation, R_i, described in Eq. (3), is given by the expression

$$R_i = \left| \frac{d(M^*)}{dt} \right|_i = k_i (R_c^*)(M) \tag{9}$$

The units of k_i are gmol/L·s.

The rate at which the growing chain adds monomer units, the rate of polymerization, R_p, is given in Eq. (4) and is kinetically expressed by

$$R_p = -\frac{d(M)}{dt} = k_p(M)(M^*) \tag{10}$$

where (M) is the monomer concentration and (M*) the concentration of growing chains. The units of the specific reaction rate constant for propagation, K_p, are L/gmol·s. Under steady-state conditions the radical concentration in the reaction must be essentially constant; therefore, the rate of the formation of the growing chains, R_i, must equal the rate of termination of the growing chain, R_t. Thus

$$R_i = R_t = \left|\frac{d(M^*)}{dt}\right|_i = \left|\frac{d(M^*)}{dt}\right|_t \tag{11}$$

Bimolecular termination, via Eq. (6), leads to the expression for the rate of termination by combination, R_{tc}.

$$R_{tc} = \left|-\frac{d(M)^*}{dt}\right|_t = 2k_{tc}(M^*)^2 \tag{12}$$

Equation (12) allows the algebraic solution for the instantaneous concentration of free radicals:

$$(M^*) = \left|\frac{R}{2k_{tc}}\right|^{1/2} \tag{13}$$

Since $R_i = R_t$, [Eq. (11)], the instantaneous concentration of free radicals can be calculated in terms of constants and concentrations that are measurable.

$$(M^*) = \left|\frac{R}{2k_{tc}}\right|^{1/2} = \left|\frac{k_d f(I_0)}{k_{tc}}\right|^{1/2} \tag{14}$$

In bulk polymerization the reaction is carried out in a medium that is primarily monomer; in solution polymerization there is an additional solvent in which both the monomer and polymer are soluble.

Substituting Eq. (14) into Eq. (10) leads to the rate of polymerization for bulk polymerization or solution polymerization.

$$R_p = -\frac{d(M)}{dt} = k_p \left|\frac{k_d f(I_0)}{k_{tc}}\right|^{1/2}(M) \tag{15}$$

In the Smith-Ewart theory of emulsion polymerization, case 2, the number of free radicals per polymer particle approximately equals 0.5 [6]. The free-radical concentration in emulsion polymerization thus is

$$(M^*) = \frac{N}{2} \tag{16}$$

where N is the number of particles per unit volume.

Substituting Eq. (16) in Eq. (10) the rate of polymerization for emulsion polymerization accordingly may be written

$$R_p = -\frac{d(M)}{dt} = k_p(M)\frac{N}{2} \tag{17}$$

From Eq. (17) it is noted that the rate of polymerization can be affected by the choice of monomer since the monomer dictates the value of k_p, the monomer concentration, and the number of particles per unit volume. The rate of polymerization is not affected by the initiator concentration, the type of initiator, or the termination constant, which is contrary to the factors influencing bulk and solution polymerization [Eq. (15)], where the rate is dependent on

$$\left|\frac{k_d}{k_{tc}}\right|^{1/2}, \; I_0^{1/2}, \; k_p, \; \text{and} \; (M)$$

Since the application for which the polymer will be used very frequently dictates the monomer type and the dispersion particle size, the only variable that can be used to control the rate of polymerization is the monomer concentration. For monodisperse emulsion polymers (dispersions where all the particles are of the same size) the number of particles per unit volume of dispersion (N) is related to the particle size by the equation

$$N = \frac{(6)(w/cm^3)}{\pi \rho d^3} \tag{18}$$

where w/cm^3 is the weight of polymer per cubic centimeter, ρ the polymer density, and d the particle diameter. Thus, for a fixed rate of monomer feed or instantaneous monomer concentration, the rate of polymerization is dependent on control of the particle diameter which in turn is controlled primarily by the surfactant and the surfactant concentration. The inability to control particle size accurately in industrial practice leads to nonreproducible products.

Further, the application for which the polymer is designed also dictates the molecular weight of the polymer. Equation (19) relates the degree of polymerization or kinetic chain length, the number of monomer units per chain, with the number of particles per unit volume. The number of monomer units polymerizing per unit time, $\bar{X}$, is defined by R_p and the number of terminations per unit time R_t; then the ratio of the two must be the number of monomer units per chain. Thus for emulsion polymerization,

$$\bar{X} = \frac{R_p}{R_i} = \frac{R_p}{R_t} = \frac{k_p(M)N}{\Sigma} \qquad (19)$$

where Σ is the radical flux. Again assuming that the other factors in Eq. (19) are controlled by the application for which the dispersion is intended and by the process, variation in particle size, and therefore N, will cause a variation in molecular weight. Thus since N must be kept constant, the surfactant is the prime factor for controlling the performance of an emulsion polymerization and the product that results from it.

IV. GROWTH OF PARTICLES

A. Heterogeneous Nucleation

Unless influenced by a destabilizing or transient event (the agglomeration of growing particles to form much larger particles), particles grow in accordance with a one-third relationship of rate of growth of size (diameter) to rate of growth of volume. If d_0 is the diameter of a particle of volume V_i prior to the addition of an incremental volume a, the final diameter of the particle, d_f, is [8],

$$d_f = d_0 \left(\frac{V_i + a}{V_i} \right)^{1/3} \qquad (20)$$

The rate at which a particle can be grown is thus dependent on the rate of polymerization R_p. In industrial practice the rate at which monomer can be added to a reaction is usually limited by the rate at which the heat of polymerization can be removed from the reaction. Vinyl polymerizations are highly exothermic (see Table 2); heat removal usually is the limiting factor, not the intrinsic rate of polymerization.

To assure better control of the many concurrent reactions that occur, emulsion polymerizations frequently are conducted isothermally. The maximum rate of polymerization is limited by Eq. (21). R_p is

TABLE 2 Heat of Polymerization of Common Monomers

Monomer	$-\Delta H$ (kcal/mol)	T(°C)
Ethylene	24.2	25
Styrene	16.5	74.5
Vinyl Acetate	21.3	76.8
Vinyl Chloride	26.7	25
Vinylidene Chloride	17.5	74.5
Acrylic Acid	18.5	20
Methyl Acrylate	18.7	76.8
Ethyl Acrylate	18.6	74.5
Butyl Acrylate	18.6	74.5
Acrylonitrile	18.3	74.5
Methacrylic Acid	13.6	74.5
Methyl Methacrylate	13.6	25
Ethyl Methacrylate	14.2	74.5
Butyl Methacrylate	13.5	76.8
1,3-Butadiene	17.4	25
Chloroprene	16.2	61.3

Source: Ref. 44.

the rate of polymerization required for an isothermal reaction (6) (the rate at which polymerization may be conducted so that the reaction temperature will remain constant), A_c the available heat transfer of the reactor, U the overall heat transfer coefficient, T_r the reaction temperature, T_c the coolant temperature, V the reaction volume, $-\Delta H_p$ the heat of polymerization, and C the heat removed by other sources, such as condensers and heat exchangers.

$$R_p = \frac{A_c U(T_r - T_c) + C}{-\Delta H_p V} \tag{21}$$

The cooling surface is fixed by the geometry of the reactor and the level of the material in the reactor which is in contact with the surface.

The overall heat transfer coefficient (U) takes into account the condition of the surface (clean or fouled), the rate of transfer of material over the surface, and the heat transfer characteristics of the surface. The reaction and coolant temperatures fix the value of $T_r - T_c$; $-\Delta H_p$ is fixed by the volume of monomer fed into the reactor. Returning to Eq. (17), it is readily observed that R_p can be regulated only by the monomer concentration M since the composition determines the value of k_p, and the application for which the polymer dispersion was designed, the dispersion concentration and stability requirements, fix the value of N, the number of particles per unit volume. Since reactors are usually operated under conditions of maximum cooling, the rate of polymerization R_p can be regulated only by the rate of monomer addition, which in turn is limited by the rate of heat removal. From the above it is evident that N, the number of particles per unit volume, cannot be allowed to vary. Variation in the value of N will cause a change in the rate of polymerization, the molecular weight of the polymer, the stability of the dispersion, and not infrequently failure in the application for which the product is intended since the total surface area of the dispersion affects formulation and stability parameters.

A change in the diameter of the particles will change the total surface area per unit volume of dispersion by the inverse diameter of the dispersion [Eq. (22)]. The effect is quantitatively shown in Table 3.

$$S_{total} = \frac{6w}{\rho d} \tag{22}$$

where S is the total surface area of unit volume of unimodal dispersion, w the weight of polymer per unit volume, ρ the polymer density, and d the diameter of the particle in centimeters. Formulated materials such as paints are developed based on a specific dispersion of a given particle size. Because of the many interdependent variables in the system, paint manufacturers cannot make adjustments in formulation without subjecting themselves to the intolerable position of manufacturing products of variable quality. The effect of particle size variability is illustrated in the following examples. Increases in particle diameter will reduce the total total surface area of the particles; thus excess surfactant will be present, making the emulsion polymer product more likely to foam and the formulated product more prone to foaming during shipment, formulation, and application. The excess of surfactant per unit surface will also tend to increase the water sensitivity of the formulated product. A decrease in particle size increases the total surface area, thus creating excess particle surface. Systems with insufficient surfactant to stabilize the surface tend to be unstable and will tend to agglomerate during manufacture, storage, shipping, and formulation. Since fre-

TABLE 3 Relationship of Number of Particles and Surface Area to Diameter[a]

Diameter (μm)	Number of particles/cm^3 (× 10^{-11})	Surface area (cm^2) of N particles (× 10^{-4})	Nd/Nd 0.05	SAd/SAd 0.05
0.05	76,000	60	1	1
0.1	9,600	30	1/8	1/2
0.2	1,200	15	1/64	1/4
0.4	150	7.5	1/512	1/8
0.8	19	3.8	1/4096	1/16
1.2	5.5	2.5	1/13,824	1/24

[a]Particle density, 1; solids, 50%.

quently it is not realized that unaccounted for excess surface is present in the product, the formulated end product will also tend to be unstable.

The prime factors affecting N, the number of particles per unit volume, are the surfactant type and concentration and the temperature at which the reaction is conducted. Tables 4 and 5 show the relative effect of particle diameter change on the rate of polymerization and the molecular weight of product. A decrease in particle size increases the rate of polymerization. Since reactors usually are operated under conditions of maximum cooling, an increase in the rate of polymerization, if there is excess monomer present, will cause a temperature rise; R_p will then exceed that permitted by Eq. (21). For example, a rise in reaction temperature will cause conditions to exist which are outside the parameters under which the product was designed. As a consequence, the various competing reactions will no longer be in balance; thus the final product can differ significantly from the product that was intended. In addition, a decrease in particle size can account for a sometimes desirable and uncontrolled increase in product molecular weight.

An increase in particle size causes a decrease in the rate of polymerization R_p and a decrease in molecular weight. A decrease in rate of polymerization will cause the reaction temperature to drop; the danger therein is that excess monomer can build up in the reactor. Excess monomer increases the rate of polymerization and thus, since the reactor is operating at the maximum rate of heat transfer, can lead to a dangerous runaway reaction where the temperature can rise above the boiling point of one or more of the batch components or azeotrope thereof; such an event can cause excessive pressure to build up within the reactor, causing uncontrolled and dangerous discharge of the reaction components from the reactor. For the foregoing reasons it is important to control particle diameter. Surfactant is the prime variable in the control of particle diameter.

TABLE 4 Rate of Polymerization of Styrene Versus Particle Size

Particle size (μm)	Relative rate of polymerization (D0.4 = 1)
0.05	500
0.1	65
0.2	8
0.4	1
0.8	0.1

TABLE 5 Degree of Polymerization of Styrene Versus Particle Size

Particle size (μm)	Degree of polymerization (0.4 μm = 1)
0.05	500
0.1	65
0.2	8
0.4	1
0.8	0.1

In Smith-Ewart theory particles from via the micelle. At the start of a polymerization, monomer swells the micelle; a radical enters the micelle, it reacts with monomer, forming a monomer radical, which in turn reacts with more monomer, causing the polymer chain to grow. For each particle that forms and grows to the final particle there are about 1000 micelles initially present. The number of particles per unit volume, N, frequently is set during the initial period of growth. From that point, through the growth and final periods, the formation of new particles is avoided, if possible.

If one compares the size and number of micelles, monomer droplets, and final particles of a 50% solids, 0.2-μm, unit density unimodal dispersion, the relationship would be as follows: Micelles are small, of the order of 0.005 μm (50 Å) in diameter, and the initial micelle concentration would be about 10^{17} micelles/cm^3. The monomer droplets vary in diameter from about 5 to 5000 μm and would vary from about 10^6 to 10^9 per cubic centimeter. The final model dispersion has a diameter of about 0.2 μm and a particle number of about 1.2×10^{14} particles per cubic centimeter.

The aggregation number, A_n, is the number of soap molecules per micelle. The number of micelles per unit volume, N_m, depends on the aggregation number. In equation (23) Av is Avogadro's number, M_s the molecular weight of the soap, and W_s the weight of soap per unit volume.

$$N_m = \frac{W_s (Av)}{M_s A_n} \tag{23}$$

For a given weight of surfactant the number of micelles is inversely proportional to the molecular weight of the surfactant and the aggregation number. The aggregation number of a given surfactant composition is affected by temperature, electrolyte type and concentration, organic solvents or cosolvents, and perhaps agitation. Because of the multitude of interactions under reaction conditions, at this time it is impossible to define precisely the interactions of a complex and dynamic emulsion polymerization system. Compensation must be made for the effect of electrolyte, organic solvents, temperature, and so on, on the aggregation number and the CMC. In Table 6 and Fig. 1 the interaction of electrolyte with sodium dodecyl sulfate, a common anionic emulsifier used in emulsion polymerization, can be noted.

The addition of electrolyte tends to increase the aggregation number, while soluble organic material tends to decrease the aggregation number. Increasing aggregation number implies decreasing critical micelle concentration (CMC) (Table 6). CMC is the concentration of surfactant at which micelles first appear. As more surfactant is added to a solution, above the CMC, more micelles form. The surfactant that is present below the CMC is in equilibrium with the surfactant in the micelle but in terms of a particle formation can be considered wasted since it does not contribute to the number of micelles.

The presence of micelles permits the use of sparingly soluble monomers. Styrene monomer solubility in water varies with temperatures as shown in Table 7. Other monomers vary considerably in water solubility, as evidenced by Table 8.

At a reaction temperature of 85°C, the styrene solubility is 0.008 M or ~4.8 $\times$ 10^{18} molecules per cubic centimeter. In order to achieve meaningful rates of polymerization, the monomer concentration term in Eq. (17) must be increased two to three orders of magnitude. Solubilization of the monomer by the micelle enables this to

TABLE 6 Effect of Electrolyte on Micellar Parameters of Sodium Dodecyl Sulfate

NaCl normality	Aggregation number	CMC (mol/L)	References[a]
0	62	0.0093	1, 3
0.02	66	0.006	2, 3
0.03	72	0.005	1, 3
0.20	101	0.0021	1, 3
0.50	142	0.001	1, 3

[a](1) K. J. Mysels and L. H. Princen, J. Phys. Chem., 63:1696 (1959); (2) L. M. Kushner and W. D. Hubbard, J. Colloid Sci., 10:428 (1955); (3) M. L. Corrin and W. D. Harkins, J. Am. Chem. Soc., 69:683 (1947).

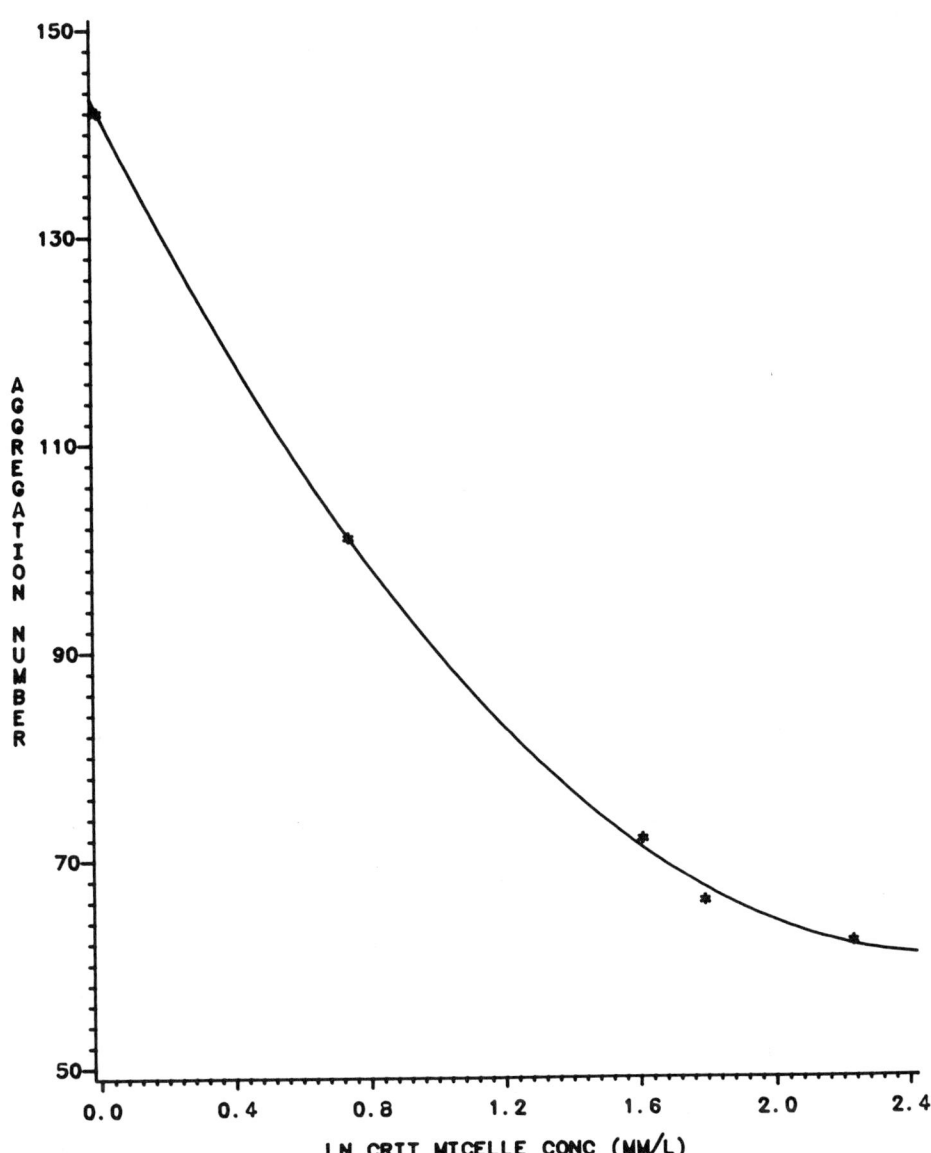

FIG. 1 Relationship between aggregation number and CMC (parameters for sodium dodecyl sulfate at different salt levels).

TABLE 7 Solubility of Styrene in Water Versus Temperature

Temperature (°C)	g/cm^3	Molecules/cm^3 ($\times 10^{-18}$)	Increase in solubility
0	9×10^{-6}	0.052	1
10	3.5×10^{-5}	0.2	4
30	2.35×10^{-4}	1.36	26
60	5.3×10^{-4}	3.06	59
85	8.25×10^{-4}	4.8	92

Source: *Physical Properties of Styrene Monomer*, Dow Chemical Co.

happen. Monomer enters and swells the hydrophobic portion of the micelle. Each micelle imbibes between 30 and 200 hydrocarbon molecules. Diffusion of the monomer into the center of the micelle occurs because the environment at the center of the micelle is very similar to that of a hydrocarbon fluid. Styrene monomer is much more soluble in the hydrocarbon than it is in water. Assuming that there are 10^{17} micelles present per cubic centimeter, the concentration of styrene in the micellar phase would rise to ~10^{20} molecules of styrene solubilized per cubic centimeter. Styrene concentration in the water phase is ~0.0089 M, whereas in the soap phase it will range from 1 to 5 M, which increases the monomer concentration term in Eq. (17) by a factor of 100 to 500 times, thus allowing for a rapid reaction.

The usual strategy for growing particles is to establish a fixed number of particles at the beginning of the polymerization and to grow particles without starting the growth of new particles. The particle nucleation stage is often difficult to reproduce in consecutive experiments when different batches of ingredients are used [9]. The required number of particles for a unimodal dispersion of a given particle size and solids level may be calculated by Eq. (18). Once N is established, the number of micelles required to start the growth of N particles will range between 100 and 1000 times N [9]. For example, to prepare a dispersion of 0.2-μm-diameter particles of polymer of density $\rho = 1.0$ g/cm^3, the particle number N may be calculated as follows:

$$N = \frac{6w}{\pi \rho d^3} = \frac{6 \times 0.5 \text{ g/cm}^3}{\pi (1.0) (0.2 \times 10^{-4} \text{ cm})} = 1.19 \times 10^{14}$$

TABLE 8 Water Solubility of Vinyl Monomers

Monomer	Water solubility (at 25–50°C; mM)
n-Octyl acrylate	0.34
Dimethylstyrene	0.45
Vinyltoluene	1.0
n-Hexyl acrylate	1.2
Styrene	3.5
n-Butyl acrylate	11
Chloroprene	13
Butadiene	15
Vinylidene chloride	66
Ethyl acrylate	150
Methyl methacrylate	150
Vinyl chloride	170
Vinyl acetate	290
Ethylene	200–600
Methyl acrylate	650
Acrylonitrile	1600
Acrolein	3100

Source: Reproduced from Ref. 9 by permission of John Wiley & Sons, Inc.

The number of micelles required to generate that number of particles ranges from 10^{16} to 10^{17} micelles per cubic centimeter. The exact number will depend on the aggregation number of the surfactant under the conditions of polymerization. At the start of the reaction, when only micelles swollen with monomer, water-soluble monomer, and monomer droplets are present, the number of styrene molecules soluble in the water phase is relatively close to the number of micelles. However, the number of styrene molecules in the micellar phase exceeds those in the water phase by two to three orders of magnitude. In addition, the number of micelles and the total micellar surface area can exceed that of the monomer droplets by several orders of magnitude (Table 1). Thus with the number of micelles exceeding the number of droplets by a factor of 10^8 or 100 million micelles for each droplet and the surface area of the micellar phase exceeding that of the droplets by a factor of 10^3 to 10^{-4}, the catalyst radicals, which are being generated at a rate of 10^{12} to 10^{14} radicals per cubic centimeter per second, are more likely to enter the micelle than the droplet. Following initiation, growth occurs for the same reasons in the growing monomer-polymer particle, which is now swollen with monomer that has diffused through the aqueous phase from the droplet.

Once sufficient polymer has been formed, the micelle vanishes; it has been converted into the polymer particle, which is now swollen with monomer and stabilized by surfactant. Again the monomer-polymer particles are more numerous than the droplets (by $\sim 10^5$ to 10^7); accordingly, it is statistically much more probable for a radical to find a monomer-polymer particle than a monomer droplet. The radicals find, and react in, monomer-polymer particles rather than in monomer droplets. Approaching the final stage of reaction, the monomer droplets have shrunken and begin to approach the size of the growing monomer-polymer particle. Even in this stage the growing particle still continues to be five to eight orders of magnitude more plentiful; thus reaction in the monomer-polymer particle still continues to predominate. Practioners try to avoid polymerization in the droplet since this leads to the uncontrolled formation of large particles, and/or also particles that are microscopic or nearly microscopic in size [10].

Particles do not contain a single polymer chain. During the growth of particles many polymer chains are formed in each particle. The diameter of the particle, however, can limit the number of polymer chains than can fit into it. Table 9 gives the number of polymer chains that can fit into one particle of a given volume and assumes that all polymer chains are composed of polyethyl acrylate (density 1.17 g/cm^3), molecular weight 1 million (kinetic chain length of 10^4).

TABLE 9 Number of Polyethylacrylate Chains per Particle

Particle diameter (μm)	Chains/particle (MW $\times 10^{-6}$)
0.01	0.37
0.014	1
0.05	46
0.1	369
0.5	46,100

B. Homogeneous Nucleation

The importance of homogeneous nucleation was not recognized at first because most of the early studies involved highly water insoluble monomers (e.g., styrene). In 1946, Baxendale and co-workers [7] published work with the more water soluble monomer methyl methacrylate (MMA). In this study, the kinetics of polymerization were studied in the presence and absence of the cationic surfactant cetyl trimethyl ammonium bromide (CTAB). The kinetics of the polymerization were similar to those of homogeneous solution kinetics, where mutual termination of free radicals occurs despite the precipitation of the polymer. Upon increasing the CTAB level, the rate of polymerization increased, which was attributed to inhibition of the rate of particle coagulation, a reflection of the reduction in the effective rate constant for mutual termination.

A subsequent paper by Priest [11] demonstrated that the polymerization of vinyl acetate (VAc) initially takes place as a solution process. He postulated that the polymer grows until it reaches the molecular weight at which its solubility in the aqueous phase is exceeded, thus creating the nucleation site or primary particle. On this basis, the number of particles should be equal to the number of chains initiated in solution; however, this number is reduced by two processes: (1) adsorption by existing primary particles of chains which had not completely polymerized to precipitation length, and (2) coalescence of particles in order to reduce the interfacial surface area to maintain the colloidal stability of the dispersion.

Ultimate particle size is then proposed to be a function of anionic sulfate group surface density on the particle, from the initiator radical, and on the level of surfactant or colloidal stabilizer present in the system. For example, Priest showed that increasing levels of the surfactant sodium lauryl sulfate (SLS) results in lower particle size (Table 10). Priest found that at 3.7×10^{-3} mol/L of SLS no micelles were formed, but that at 11.5×10^{-3} mol/L there are micelles present. On the basis of these results, Priest proposed the curve shown in Fig. 2 to describe the variation in particle number with degree of conversion.

During the initial stage of polymerization, particles are formed at a rapid uniform rate, represented by the ascending left-hand portion of the curve. As the concentration of nucleation centers increases, particle coalescence becomes more important, so that at the peak of the curve the rate of new particle formation and the rate of particle reduction via coalescence are balanced. Beyond this point coalescence predominates. The role of the surfactant in this scheme is to shift the curve to result in greater particle number and earlier conversion by stabilizing particles at a smaller size so that degree of coalescence is reduced.

Further evidence for homogeneous polymerization with vinyl acetate was generated by Napper's group [12,13] in studies of the poly-

TABLE 10 Effect of Surfactant on Particle Size[a]

SLS (mol/L X 10^{-3})	PS (nm)
0	410
1.36	107
3.7	70
11.5	39
46	34

[a]$t = 51°C$; K persulfate 2.5 X 10^3 mol/L; vinyl acetate/H20 = 0.058.
Source: Ref. 11.

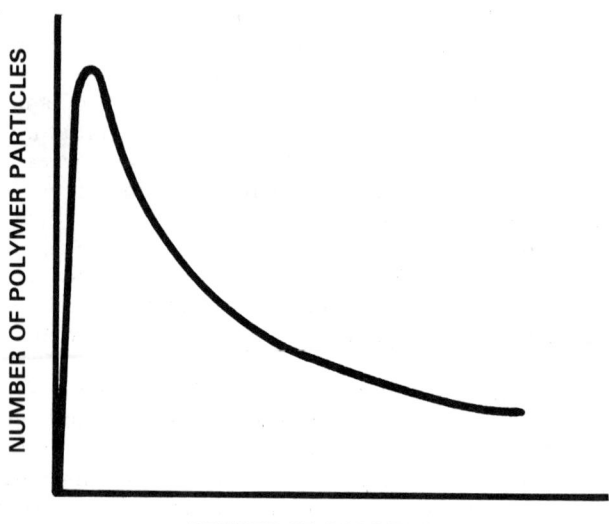

DEGREE OF CONVERSION

FIG. 2 Form of curve showing change in number of particles with degree of conversion in a typical persulfate-initiated vinyl acetate polymerization conducted in the absence of stabilizer. (From Ref. 11.)

merization process with anionic, nonionic, cationic, and no surfactants. This is summarized in Fig. 3, where it is seen that the effect of the anionic soap is to increase the rate of polymerization, the effect of the nonionic soap is neutral, while cationic soap reduces the conversion rate. When replotted as a function of the square root of the conversion, the conversion root is linear up to 50% conversion for anionics, nonionics, and no soap. However, linearity is not achieved until almost 20% conversion in the presence of the cationic soap, and then the slope becomes equal to that of the soapless polymerization. Particle number varies with the soap, as expected:

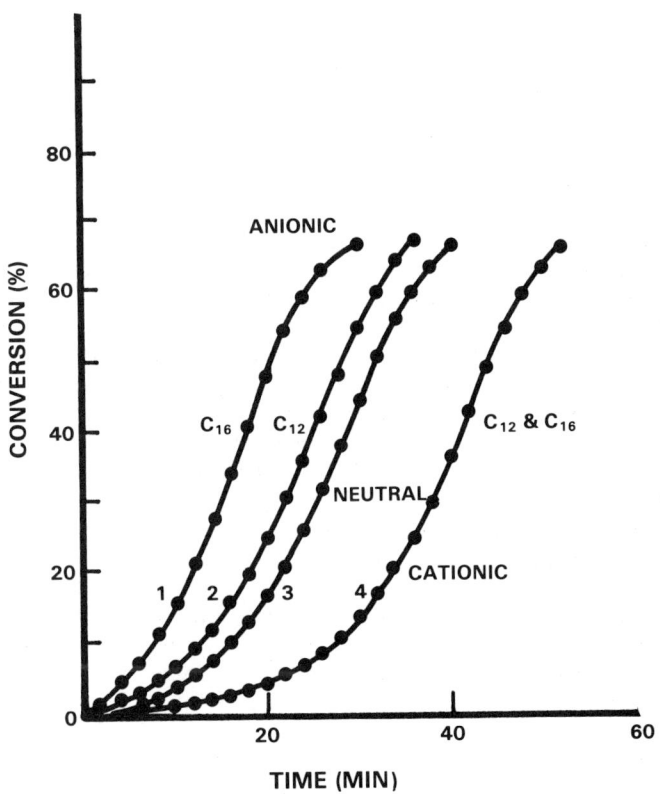

FIG. 3 Effects of soaps on the conversion versus time curve: 1, sodium cetyl sulfate; 2, sodium dodecyl sulfate; 3, Triton X-100; 4, cetyltrimethylammonium bromide or dodecyltrimethylammonium bromide. Vinyl acetate, 0.33 m; potassium peroxydisulfate 0.015 M; ph of solution, 3.80. (From Ref. 13.)

Soap	Sodium cetyl sulfate	None	CTAB
Particles (no./cm^3)	1×10^{14}	2×10^{12}	1×10^{11}

The anionic soap adsorbs onto the primary particle, increasing the charge on the particle, thereby achieving stabilization at a smaller particle size than in the absence of the soap. Coalescence of the particles is thus inhibited, resulting in reduction of the mutual termination rate of the growing radical chain so that the rate of polymerization is enhanced. Since adsorption is a function of the hydrophobic portion of the anionic soap molecule, it is expected that the longer-chain cetyl group is more effective than the lauryl homolog. The effect seen with the cationic soap is that a number of cationic charges are effectively neutralized by the anionic sulfate and groups of the polymer chains. The neutralized particles have a lower zeta potential, thereby promoting coalescence, permitting mutual termination between growing radicals, resulting in larger particles and a reduced rate of polymerization. Beyond this stage (i.e., ~20% conversion) the rate of polymerization becomes that of the soapless system. Further, since the adsorption of the cationic soap is a coulombic effect rather than a hydrophobic effect, the chain length of the alkyl portion of the soap has no effect on the polymerization.

In studies with the less water soluble monomer vinyl chloride (VC1), Peggion et al. [14] came to the same basic mechanism for the formation of particles. They studied the polymerization using several types and levels of anionic soaps, and also looked at the effect of initiator level. Significantly, they found that the number of particles remained the same between 6 and 100% conversion. Further, they found that the particle number increased almost linearly with soap level through the CMC for SLS and sodium dibutylsulfosuccinate.

In 1965, Dunn and Taylor [15], also studying the polymerization of vinyl acetate, further confirmed this mechanism of polymerization. They also proposed that the DLVO theory for colloid stability be applied. In varying the level of SLS used in the polymerization, they observed that only small changes in the rate of polymerization took place but that significant change in particle size occurred (Table 11). Note that the major change in particle size occurs with the addition of very small amounts of SLS at levels much below the CMC.

A further requirement for homogeneous particle nucleation was set forth by Parts et al. [16], in which they proposed that the polymer particles competed very favorably with micelles for radical capture. This arose out of calculations made to assess the ratio of growing particles to total particles, at the point in the polymerization where micelles have disappeared. Without this difference in radical capture efficiency between micelles and nucleated particles, this ratio was 1; by introducing an efficiency of capture factor of

TABLE 11 Effect of SLS on Particle Size[a]

SLS(%)	PS (nm)	SLS (%)	PS (nm)
0	314	0.0040	61
0.00024	165	0.011	55
0.00046	132	0.026	47
0.00080	94	0.049	41
0.0024	83	0.116	31
		0.222	33

[a]2.0% VAc, 0.02% K persulfate; the CMC of SLS was 0.115%.
Source: Ref. 15.

0.01 for micelles, this ratio was reduced to 0.516, close to that obtained experimentally. A consequence of this is that a considerable number of particles have captured more than two radicals during this first stage of polymerization.

Fitch et al. [17] reinvestigated the polymerization of MMA in aqueous media, reaffirming the results of Baxendale et al. [7] that homogeneous kinetics characterized the early stages. However, they found that the later stages demonstrated heterogeneous kinetics. Particle formation, as shown by Tyndall scattering, was independent of the rate of polymerization but occurred at a constant conversion (2 to 4%) of monomer to polymer. They proposed that particle formation is thermodynamically controlled and that a critical degree of polymerization of radicals in solution is required before nucleation can occur.

In an elegant set of experiments, Dunn and Chong [18] measured surface electrical charge densities in order to calculate rates of coagulation by application of DLVO theory. They found that in the absence of soap, the primary particles coagulated at about the theoretical Smoluchowski fastest rate, and as they accumulated charge (initiator sulfate groups) by coagulation and capture of charged oligomeric radicals, they became more stable. With added emulsifier, stability was obtained earlier in the reaction, by adsorption of the emulsifier to the particle, so that nucleation stops at a lower conversion level. They also noted that if nucleation occurred beyond this level, these primary particles would be unstable and would coagulate with the larger particles already present. Therefore, nucleation can continue throughout the reaction even though the total number of particles remains constant. Fitch [19] observed that this can account for the generation of smaller particles upon the addition of more surfactant at some later stage which stabilizes the new nuclei as they are formed.

Fitch and Tsai [20] presented a study in which they developed an approximate quantitative theory for homogeneous nucleation. They determined the effect of soap concentration, monomer concentration, and type of initiator system (i.e., that resulting in an ionic or nonionic end group on the oligomeric radical chain). The effect of soap concentration on particle number is shown in Fig. 4 for MMA using persulfate as the initiator.

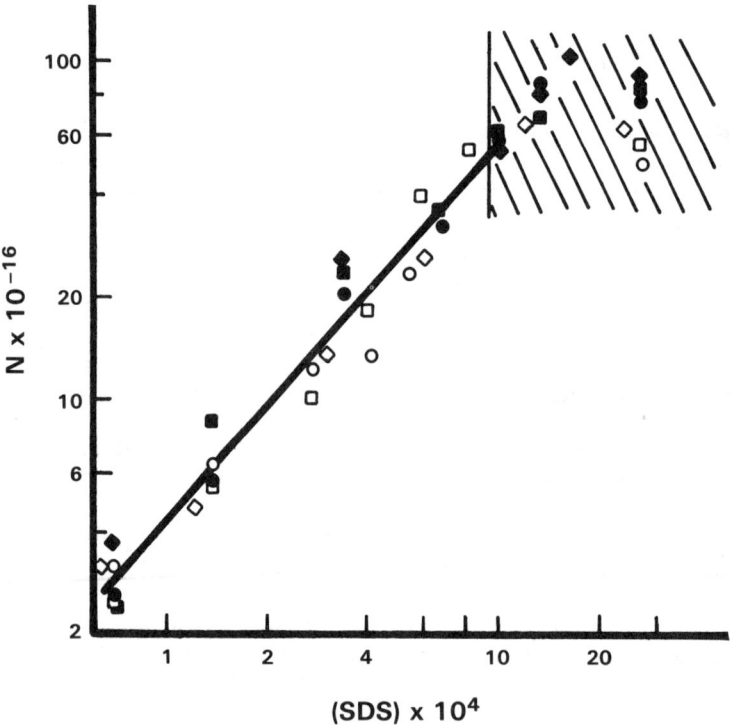

FIG. 4 Dependence of the particle number N on sodium dodecyl sulfate concentration (SDS) in aqueous-phase polymerization of MMA at various monomer concentrations. Monomer concentrations are 0.0038, 0.019, 0.035, 0.0571, 0.761, and 0.0951 M. Initiator system: persulfate–bisulfite iron. Hatched region indicates monodisperse particle size distributions. (From Ref. 20.)

We see here that the particle number is a function of the soap concentration, but that monomer concentration has little effect at low soap levels. At higher soap levels the curve plateaus and there is an indication of an effect of monomer level. A parallel study using hydrogen peroxide as the initiating system is shown in Fig. 5.

This graph is qualitatively similar, but in order to prepare stable emulsions, higher levels of soap were necessary. That is, at SLS = 3.47×10^{-4} mol/L the particle number for the ionic initiator system is 22.9×10^{16}, while that for the nonionic system is 1.4×10^{16}. Further, the slope of the line is much greater in the case of the peroxide initiator, indicating greater dependence on the soap concentration in the nonionic system. This indicates that the ionic chain ends from the persulfate initiator contribute to the stabilization of the

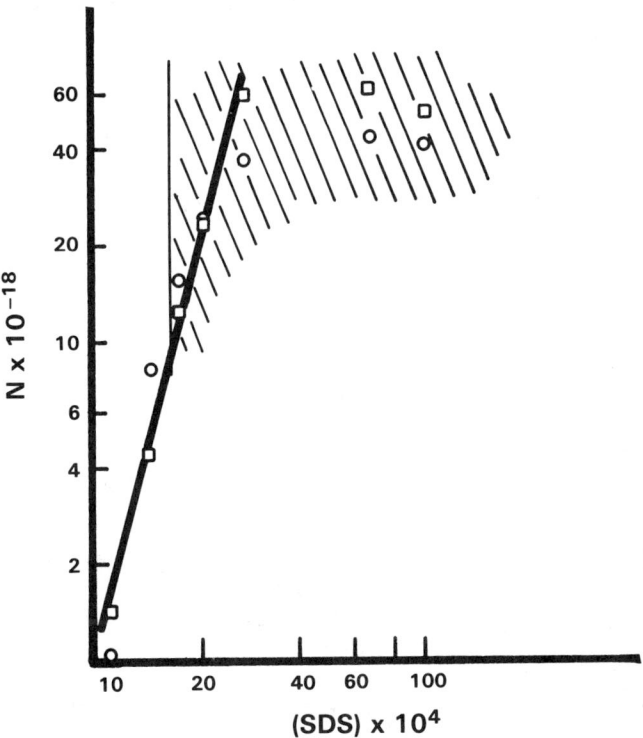

FIG. 5 Dependence of the particle number on sodium dodecyl sulfate concentration in aqueous-phase polymerization of MMA at various monomer concentrations. Monomer concentrations are 0.100 and 0.080 M. Initiator system: hydrogen peroxide/iron(II). Hatched region indicates monodisperse particle size distributions. (From Ref. 20.)

growing chain and the nucleated particle. Fitch and Tsai [20] suggest that the growing oligomeric chain reaches a length at which it precipitates to form the primary particle. If the primary particle is unstable, limited flocculation takes place in which the particle number is reduced. Flocculation continues until some average minimum surface charge potential is established. This point is a function of the charge provided by the initiator residue and the level of soap available for adsorption to the surface of the particle.

In a study of the earliest stages of emulsion polymerization, Fitch et al. [21] demonstrated the effect of soap concentration and monomer type on the increase of Rayleigh scattering. The increase in scatter is a function of the number of particles nucleated and of their size (i.e., if two particles coalesce, the scatter is larger than that of the individual particles). They investigated SLS concentrations from zero to eight times the CMC, at low monomer concentration with photoinitiation. Figure 6 shows the effect of the SLS concentrations on EA polymerization.

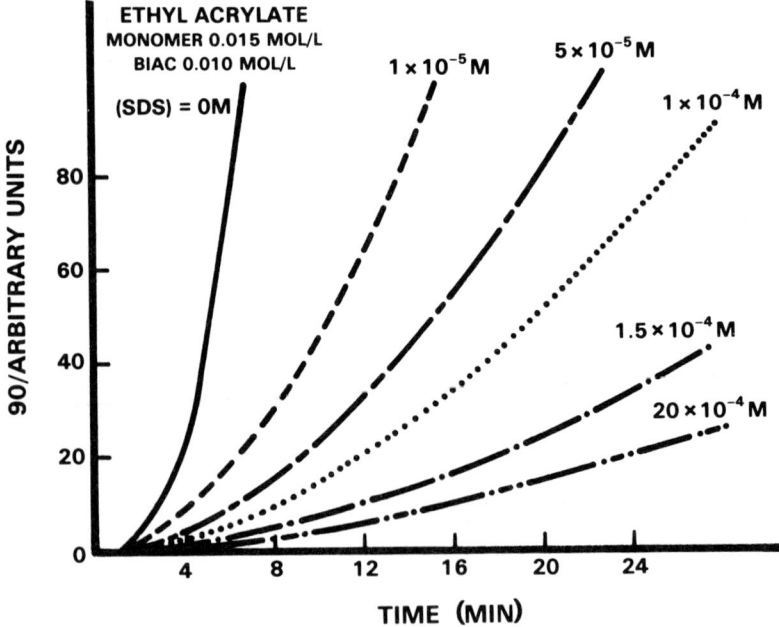

FIG. 6 Light-scattering intensity as a function of time for ethyl acrylate photopolymerization at various SDS concentrations. (From Ref. 21.)

Note that as the soap concentration increases, the scatter decreases. Since the rate of formation of radicals is the same and constant throughout, the difference in the curves depends on the effect of the soap concentration on the stabilization. This is a function both of the oligomeric radical growing to a longer chain length before being precipitated into a particle and the fact that the particles are more stable to coalescence. In a similar vein, Fig. 7 shows the photopolymerization of three acrylic monomers in the absence of soap.

Note here that as the hydrophilicity of the monomer increases, the rate of scattering development decreases. Again this is related to the stability of the oligomeric radical and the particle formed. The more hydrophilic the monomer, the longer the chain length before particle precipitation, and the more stable the particle is to coalescence. Another factor involved here is that once the particles are nucleated, the hydrophobic monomers partition into the particle to a greater degree, resulting in a higher monomer concentration at the site of polymerization. Therefore, it should have a higher rate of polymerization and at equal coagulation rates (i.e., equal soap

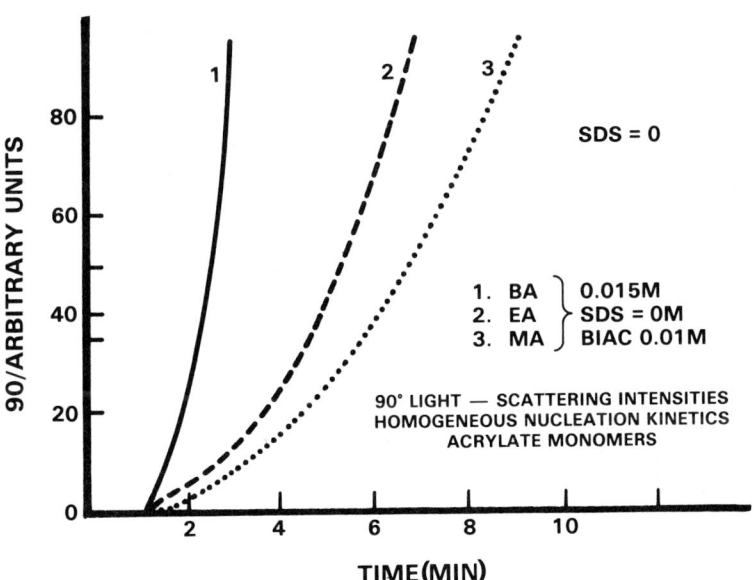

FIG. 7 Light-scattering intensity as a function of time for methyl (MA), ethyl (EA), and butyl (BA) acrylates. (From Ref. 21.)

levels) scatter should increase more rapidly for the more hydrophobic butyl acrylate.

V. COLLOIDAL STABILITY OF LATEX PARTICLES

The colloidal stability of a latex particle is a function of the chemical nature of the groups on the surface. Depending on the application use of the latices, on the process used to synthesize the latex, and on postsynthesis steps, there can be considerable variation. For example, the polymer in the bulk of the latex particle can vary from a polymer of low T_g to high T_g, from one of high hydrophobicity to high hydrophilicity or anywhere in between. The T_g, or glass temperature, is that temperature at which a second-order transition occurs characterized by an abrupt change in the hardness of a polymer. At temperatures above the T_g the polymer is soft; under the T_g it is hard. At temperatures under a polymer's T_g the rate of diffusion of a molecule through the surface is decreased. The surface can reflect the properties of the bulk polymer or can be modified via staged polymerizations to have different properties. The surface environment can be influenced by the catalyst fragment end groups, by grafted soap or other hydrophilic groups (anionic, cationic, or nonionic), or by groups adsorbed onto the surface, all of which, either alone or in concert, can act to modify the stability and properties of the latex particle.

The concept of colloidal stability of these emulsion polymers can take several forms. For example, stability of dispersions can be evaluated in response to the following types of stressing conditions: (1) freeze-thaw cycling; (2) heat, extended shelf life, sterilization; (3) shear or mechanical stability; and (4) electrolyte or solvent addition. Alternatively, in certain applications it may be desirable to reduce the stability of the emulsion in order to obtain aggregates of the particles or to flocculate the particles. In any case the role of surfactant is critical to stabilization and/or destabilization of the latex particle.

A. The Basis of Stability

In the aqueous environment of a dispersion particle, the anionic, cationic, or hydrophilic residues from the initiator are usually found at or near the surface of the particle. Similarly, hydrophilicity or charge imparted to the polymer by monomers such as acrylic acid, methacrylic acid, hydroxyethyl methacrylate, dimethylaminoethyl methacrylate, and so on, is also usually found at or near the surface when conditions permit. Hydrophilic functionalities tend to form a layer around the particles, which then acts to repulse similar particles.

Further, this layer can be extended by the addition of surfactants. This is shown schematically in Fig. 8.

As a result of the surface charge, either positive or negative, the particle has an electrostatic potential. This is used to obtain an expression for the electrostatic repulsive potential energy, V_r [22]:

$$V_r = 3.469 \times 10^{19} \varepsilon \, (kT)^2 a \chi^2 \, \exp(-\chi(h))/v^2 \qquad (24)$$

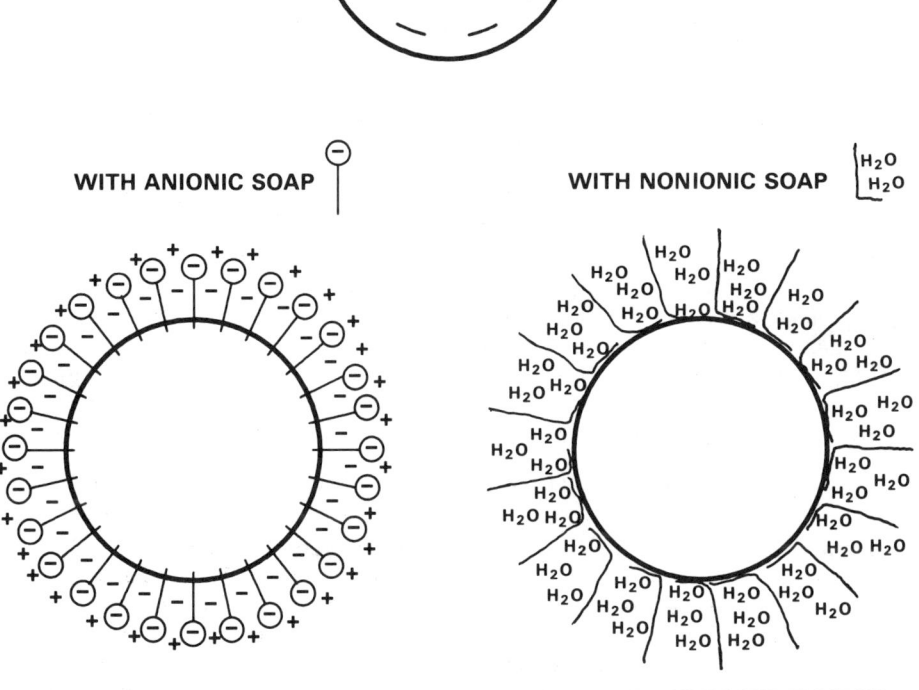

FIG. 8 Schematic representation of surfactant stabilization.

where χ is the reciprocal Debye-Hückel double-layer thickness, ε the dielectric constant of the medium, k Boltzmann's constant, T absolute temperature, v the value of the valence of the counterion (= [exp(veψ_s/2kT) - 1]/[exp(veψ_s/2kT) + 1], where ψ_s is the electrostatic ponential of the particle), and e the electronic charge.

Similarly, using the Hamaker constants (essentially a proportionality constant for the interaction of substance 1 with substance 3 in an intervening medium of substance 2 [23], one can obtain an expression for the attractive part of the interaction energy, V_a [24]:

$$V_a = -\frac{A}{12}\left| \frac{1}{x^2 + 2x} + \frac{1}{x^2 + 2x + 1} + 2\ln\frac{x^2 + 2x}{x^2 + 2x + 1} \right. \quad (25)$$

where A is the composite Hamaker constant for the particles in the medium as given by A = ($\sqrt{A_{11}} - \sqrt{A_{22}}$), with A_{11} = constant for the particles and A_{22} = constant for the medium; and x = h/2a, where a is the particle radius. When x << 1,

$$V_a = -\frac{Aa}{12h}$$

If $|V_r| > |V_a|$ for most particle separation distances h, the potential energy versus distance curve shown in Fig. 9 is obtained.

Note that at short distances there is a deep potential energy minimum, which determines the distance of nearest approach, h_0, known as the primary minimum. At intermediate distances, the repulsive factors contribute significantly, resulting in the primary maximum, or magnitude V_m. At larger distances, the repulsive forces, leading to the secondary minimum, of depth V_{sm}. The secondary minimum is shallow for small particles but can become appreciable for larger particles (>1μm). Inspection of the potential energy curve (Fig. 9) indicates that the stability of the latex is kinetic in origin rather than thermodynamic, in that the lowest free-energy state is in the primary minimum, which is blocked by a large activation energy.

As long as the primary maximum, V_m, remains high, on the order of 10kT, the particles repulse each other and the dispersion remains stable. When V_m becomes small or tends toward zero, the transition to the primary minimum becomes facile, the system becomes unstable, and the particles associate. Accordingly, any change in the double-layer thickness (χ), that is, contraction via an increase in ionic strength of the aqueous phase, increase in the charge density on the latex surface, or increase in hydrophilic layer thickness, will result in a positive change in the stability of the latex particle.

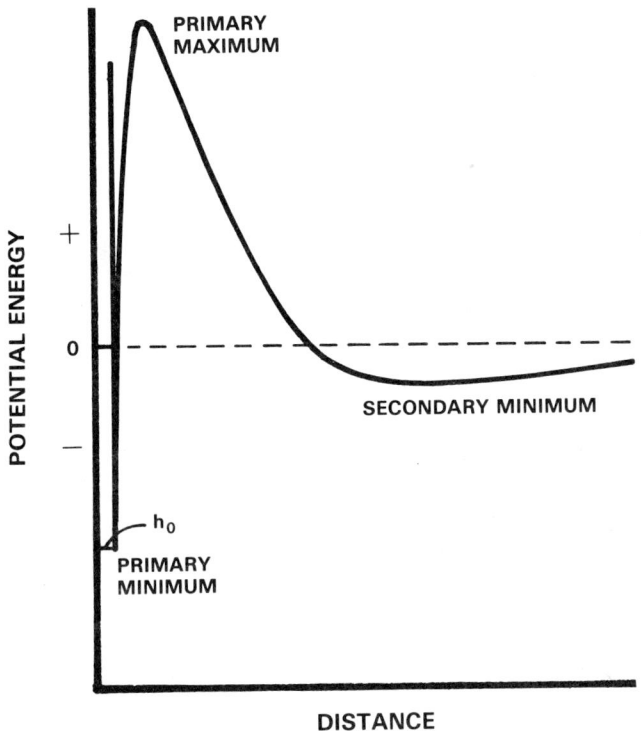

FIG. 9 Schematic illustrating the form of the curve of potential energy versus distance for the interaction between two spheres. Total interaction involving electrostatic repulsion and van der Waals attraction. (From Ref. 24.)

B. The Effect of Added Surfactants

The addition of a surfactant to a latex results in the adsorption of the surfactant to the latex surface. The adsorption takes the form of a Langmuir isotherm, in that the surface becomes saturated, as shown in Fig. 10. In this manner, as the surfactant is adsorbed the charge density on the surface increases, thereby increasing the electrostatic potential and the V_r. Similarly, if a nonionic surfactant is adsorbed, the thickness of the hydrophilic layer is increased by virtue of the ethoxylate chains extending into the aqueous medium.

The adsorption of the surfactant is a function of the interaction of the hydrophobe of the surfactant and the latex surface. As a general rule, the more hydrophobic or nonpolar the surface of the latex, the greater the adsorption of the surfactant. From a knowl-

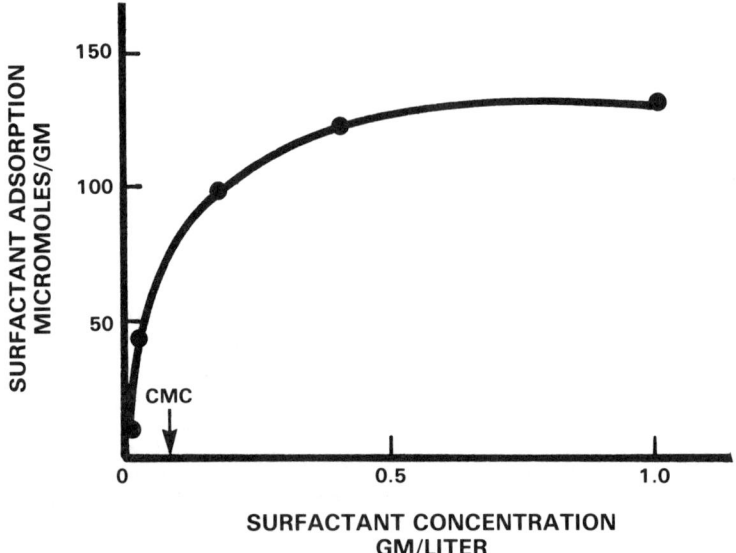

FIG. 10 Adsorption isotherm of Igepal CO-630 surfactant at the 85:15 VA/BA latex-water interface. (From Ref. 25.)

ege of the particle size of the latex, density of the latex particle, and measurement of the adsorption isotherm, the area taken up by a molecule of the soap on the latex surface can be calculated. The interaction of surface hydrophobicity and soap adsorption has been demonstrated by Maurice [26], who measured the area per surfactant molecule on the surface of lattices of different backbones at surface saturation. These data are summarized in Table 12.

In both the styrene and the BA system the area change is a linear function of the MMA level in the polymer backbone. Similarly, Ahmed et al. [27] showed that the surface area per molecule of sodium lauryl sulfate increases from 43 Å for a polystyrene latex to 57 Å for a polymethyl methacrylate latex. Another example of the effect of the polymer backbone on soap adsorption has been given by Vijayendran et al. [25] in vinyl acetate/butyl acrylate copolymer emulsions (Table 13).

In a similar vein, the adsorption of the surfactant is also a function of the surfactant hydrophobe and head group. For example, Ahmed et al. [27] have also measured the adsorption of several soaps to a polystyrene latex surface (Table 14). The difference between the similar sulfosuccinate soaps is attributed to a difference in the manner in which the soaps are bound to the surface; the aerosol OT binds via both hydrophobic groups, the Aerosol MA by only one.

TABLE 12 Effect of Emulsion Polymer Composition on Surfactant Adsorption (Surfactant: Sodium Dodecyl Benzenesulfonate)

Composition	Area/molecule (Å^2)
100 Styrene	81
100 Butyl acrylate	84
100 Methyl methacrylate	156
75 BA/25 MMA	98
52 BA/48 MMA	111
25 BA/75 MMA	144
67 Sty/33 MMA	108
33 Sty/67 MMA	129

Source: Ref. 26.

In a like manner, Kronberg [28] measured the adsorption of a series of nonylphenol ethoxylates of different degrees of nominal ethoxylation. Extrapolation from Kronberg's binding curve (Ref. 28, Fig. 9) indicates the effect of the ethoxyl chain length on area per molecule adsorbed (Table 15).

Interestingly, when one calculates the number of ethoxy groups bound to the latex surface, one gets 27.5, 26, and 25 $\mu m/m^2$ of latex surface, respectively. This suggests that the hydrophilic layer surrounding the polystyrene particle, at saturation, is approximately the

TABLE 13 Adsorption of IGEPAL CO-630[a] onto Vinyl-Acrylate Lattices

Latex Composition	Area/molecule (Å^2)
100 VA	187
85 VA/15 BA	113
70 VA/30 BA	96

[a]Igepal CO-630 is an octylphenol ethoxylate.
Source: Ref. 25.

TABLE 14 Adsorption of Several Anionic Soaps to a Polystyrene Latex

Emulsifier	Area/molecule (Å^2)
Sodium lauryl sulfate	43
Aerosol MA (Sodium dihexyl sulfosuccinate)	45
Aerosol OT (Sodium di-2-ethylhexyl sulfosuccinate)	85

Source: Ref. 27.

same thickness regardless of the length of the ethoxy chain. Ottewill and Walker [29] showed that the nonionic soap dodecyl hexaoxyethylene glycol monoether ($C_{12}E_6$) adsorbs onto a polystyrene latex to give a monolayer in which the alkyl groups are laid down on the surface and the ethylene oxide groups extend into the continuous phase. Ultracentrifuge studies indicate that the water content in the EO phase may be as high as 70%. If one then calculates the thickness of the hydrated EO shell for the nonylphenol ethoxylate case, a 200-nm polystyrene emulsion particle would be surrounded by about 19,800 ethoxy residues, which with the water would move the surface out by approximately 33 Å.

This shell of extensively hydrated ethylene oxide residues serves as a steric barrier and effectively prevents the particles from approaching close enough to enter the primary minimum. This stabilization, achieved via adsorption of surfactant (or even hydrophilic polymers), is termed steric stabilization.

The adsorption of surfactants to the surface of more polar/hydrophilic emulsion polymer surfaces is somewhat different. In some cases the adsorption is as described and follows a Langmuir isotherm; however, in other cases a saturation level is achieved, which is then ex-

TABLE 15 Adsorption of Nonylphenol Ethoxylate to Polystyrene

Ethoxyl chain length	Estimated area/molecule (Å^2)
10	60
20	128
50	332

ceeded at higher soap levels. This type of behavior has been examined by Vijayendran and colleagues [25] on vinyl acetate/butyl acrylate copolymer emulsions. A typical example of this behavior is shown in Fig. 11. Alipal EP-110 and 120 are sulfated nonylphenol ethoxylates of different EO chain length, ~10 and 30, respectively. Note how the amount of the Alipal EP-110 adsorbed to the latex particle increases beyond a surface saturation plateau, while that of the Alipal EP-120 does not. Sodium lauryl sulfate is also adsorbed beyond the saturation level, and has been found to have a surface area of 30 Å (lower than that to polystyrene on a polyvinyl acetate surface [27].

Further studies indicate that as the soap is adsorbed, the viscosity of the latex emulsion increases. Accordingly, the increased adsorption beyond the saturation plateau is attributed to the penetration of the soap into the latex particle. Vijayendran has shown that only relatively low molecular weight anionic soaps such as Alipal EP-110 (MW 708) penetrate the surface, whereas Alipal EP-120 (MW 1640) does not, and that the presence of nonionic soaps at the latex surface interferes with the penetration. Further, the effect is greater the higher the VA level; increasing BA results in less viscosity development, while no effect is seen with polystyrene and with p(MMA-EA). It is readily seen that penetration of the surface with surfactant can have an effect on various properties of the polymer (e.g., T_g and the film coalescence properties) [30]. The surfactant, if

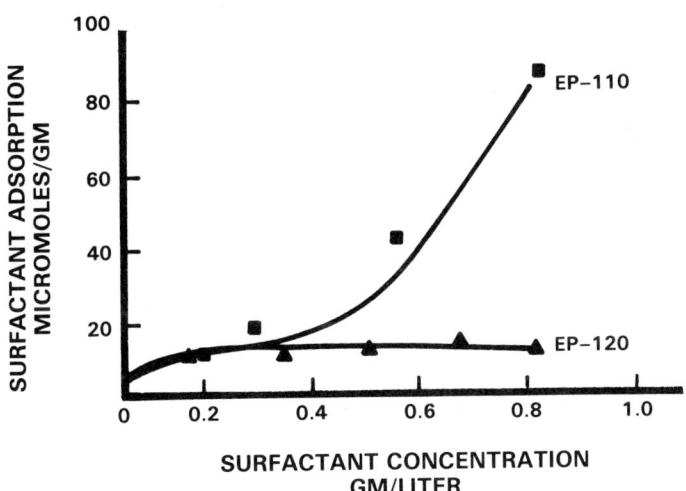

FIG. 11 Adsorption isotherms of Alipal EP-110 and Alipal EP-120 surfactants at the 85:15 VA/BA latex-water interface. (From Ref. 25.)

soluble in the polymer, acts as a plasticizer, thus lowering the T_g of the polymer.

C. Effect of Soap on Dispersion Stability

1. Mechanical Stability

Blackley [31] has reported on the effect of added soaps on the mechanical stability (i.e., high-speed shear) of natural rubber latices. Figure 12 demonstrates the stabilizing effect of fatty acid soaps of

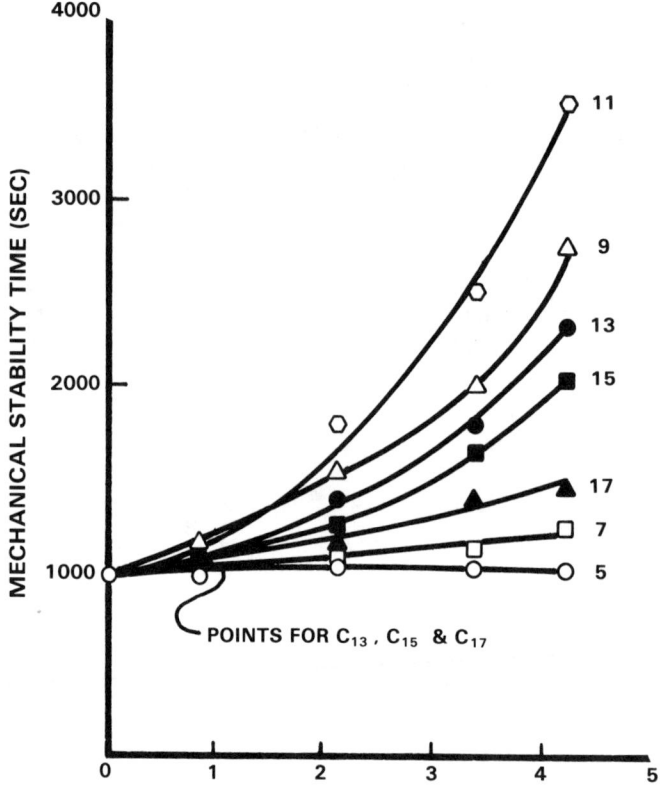

FIG. 12 Effect of added straight-chain potassium fatty acid soaps on mechanical stability of natural rubber latex. Numbers appended to curves are number of carbon atoms in alkyl chain of soap. (From Ref. 31.)

several alkyl chain lengths, at several molar concentrations, on the mechanical stability of dispersions.

It is particularly interesting to note that the effect of chain length is optimum at 11 carbons, which Blackley attributes to the "increasing tendency for the added soap anion to be adsorbed at the rubber-water interface, and a decreasing ability of the added soap anions to make the indigenous soap anions more effective stabilizers." Since small amounts of the soaps are very effective, he suggests that the indigenous soap molecules are clustered on the latex surface instead of being uniformly distributed. Accordingly, the added soap makes the indigenous soaps more effective in stabilization by dispersing the clusters. He further studied the effect of introducing double bonds or hydroxyl groups into the stearate alkyl chain. Generally, the more the structure deviated from the unmodified stearate, the better the stabilization to mechanical action. The sulfated and sulfonated alkyl soaps behaved in a similar manner, the optimum chain length being at 10 carbons rather than 11. Nonionic soaps, ethylene oxide/fatty alcohol condensates, have a somewhat different effect. The effect of small additions is to reduce the stability of the rubber latex; however, further addition causes some improvement. The degree of improvement appears to be quite dependent on the EO chain length of the condensate: The longer the chain, the greater the stabilizing effect at the higher levels. This is shown in Fig. 13.

Note that the effect of chain length is not observed for the smaller EO chains, the effect being similar from 6 to 30 groups, and appears to be more dependent on total number of EO groups. However, above 30 EO groups the effect appears to depend on the chain length, not on the total number of EO groups. The explanation offered is that the initial effect is due to hydration of the EO shell around the particle, while steric stabilization accounts for the effect of the longer chain lengths.

Palgren [32] has studied the mechanical stability of monodisperse polyvinyl chloride lattices as a function of several parameters, particularly the interaction of soap coverage and particle size. Figure 14 shows the mechanical stability obtained. It is seen that as the amount of soap is increased, the lifetime of the emulsion under shear increases. Also, it is seen that the latex stability decreases with increase in particle size, a reflection of the effect of particle size on the secondary minimum in the potential energy versus distance diagram.

2. Freeze-Thaw Stability

The ability of a dispersion to survive several cycles of freeze-thaw treatment is very important for commercialization. Drums of dispersion left on a loading dock through a freeze are frequently found to be coagulated. Implicated in the degree of stability to freezing is

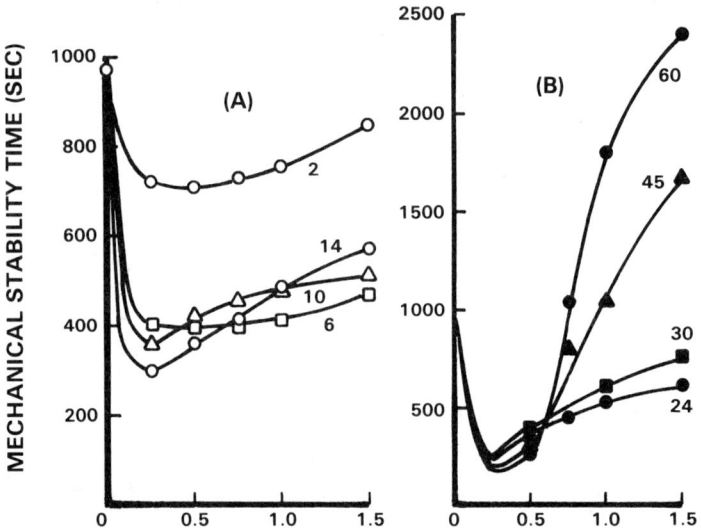

FIG. 13 Effect of added ethylene oxide/fatty alcohol condensates on mechanical stability of natural rubber latex. Levels of condensate are expressed in parts by weight. Numbers appended to curves indicate overall ethylene oxide/fatty alcohol mole ratio in condensate. (From Ref. 31.)

rate of freezing and the ionic strength of the continuous phase. In general [33], reduction of the temperature should result in slower-moving particles and fewer collisions of particles, causing less flocculation and coagulation. Further, at lower temperature the latex will have a harder polymer particle surface and will undergo fewer effective collisions. However, when freezing occurs and ice crystals form, other problems arise. Initially, the ice crystals separate from the unfrozen latex, so that the volume of the continuous phase is reduced, resulting in an increase in the latex particle concentration and an increase in collision frequency. As the continuous-phase volume is reduced further, the latex particles become entrapped between ice crystals, are subjected to high pressures, and aggregated particles become coagulated. Further, as the ice crystals form, the ionic components (salts) in the continuous phase remain and the ionic strength is increased, so that the stability of the latex is further compromised. A similar description for the breaking of a solvent emulsion by freezing was given by Rochow and Mason in 1950 [34].

Surfactants in Emulsion Polymerization

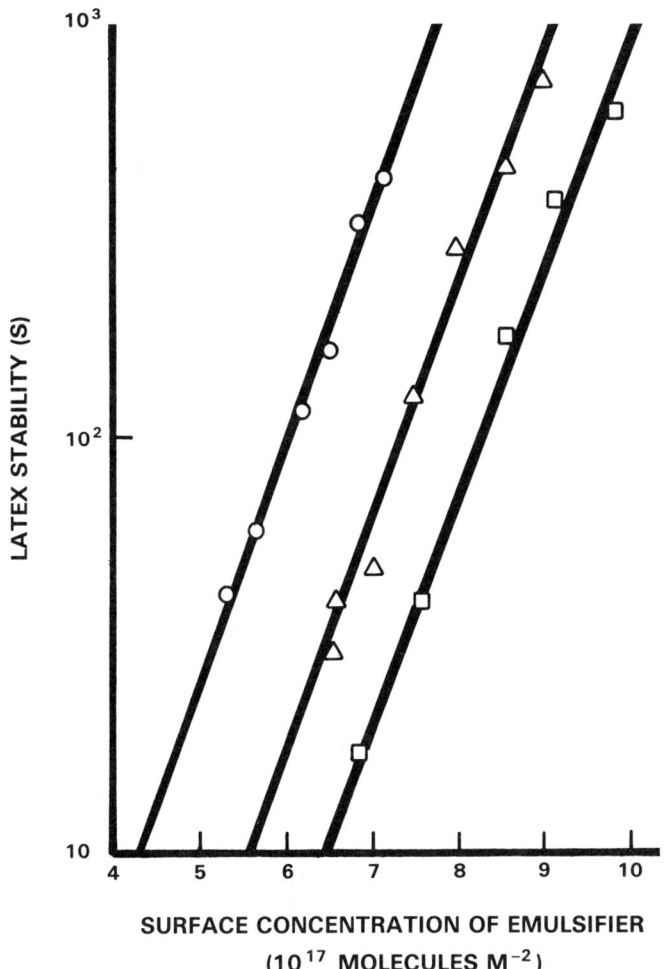

FIG. 14 Mechanical stability of 45% (w/w) PVC lattices as a function of particle size and emulsifier level. Latex particle diameters: from left to right, 123 nm, 202 nm, 283 nm. (From Ref. 32.)

The rate of freezing also has an effect on the result: A latex that has been flash frozen may survive, whereas the same latex slowly frozen will be coagulated. This has been attributed to rapid structure formation, which traps the latex particles within the ice structure so that particle motion is eliminated and collisions cannot occur.

Early work by Digioia and Nelson [35] indicated that the emulsifier for the latex was critical to freeze-thaw stability. They found that anionic-rosin soaps were not very good but that fatty acid soaps were. Nonionics were poor, but mixed with anionics, satisfactory results were obtained. No specific structures were given. It is evident that the role of the soap in the prevention of freeze-thaw coagulation is to prevent collisions that have sufficient energy to cause coagulation. This could be accomplished by increasing the charge stabilization of the latex surface via charged monomers in the backbone [36] or by the addition of ionic soaps that can adsorb to the latex surface. In fact, King and Naidus [37] have shown that the MFT (minimum film firming temperature) of the latex particle and the amount of carboxylate groups copolymerized act jointly to control freeze-thaw stability. That is the minimum level of copolymerized carboxylate groups required to confer stability increases as the MFT decreases or as the polymer becomes softer. Nonionic soaps of long EO chain length have also played a role in improving freeze-thaw stability. Presumably they also adsorb onto the latex surface and provide either steric stabilization or a thicker hydrated shell around the particle in which the water does not freeze to form ice crystals and thereby protect the particle from effective coagulative collisions.

VI. PRACTICAL CONSIDERATIONS FOR SURFACTANT SELECTION

The choice of surfactant is dictated primarily by the application for which the dispersion is intended. Dispersion stability, isolatibility, and formulatability are critical parameters that decide the class or type of surfactant [38]. Dispersions that are prepared primarily as a means of manufacturing a polymer that will be isolated in a later operation are usually made with a surfactant that facilitates easy and quantitative isolation of the polymer. Rubber used in automobile and aircraft tires is one example of this type. Butadiene-styrene copolymer manufacture in terms of rubber properties and ease of manufacture is facilitated by the emulsion polymerization method. Usually, the weak acid types, a carboxylic surfactant such as sodium stearate, sodium laurate, rosin acid soaps, and so on, are employed during the polymerization step. During the polymerization reaction the pH of the system is maintained in a range where the surfactant is somewhat soluble and is surface active. The pH is left in this range until the isolation of the polymer is desired. Then, with the lowering of the pH and at times the introduction of electrolyte, the surfactant is insolubilized and the polymeric dispersion is coagulated. With proper control of the coagulation step, the correct physical form (i.e., lump or grain size) of the polymer is achieved.

Products where the polymer will not be isolated by a coagulation step are usually manufactured using more chemically stable dispersants, such as the strong acid type or the nonionic type, used either alone or in combination with each other. Compared with the anionic and nonionic types, cationic or amphoteric dispersants are used only infrequently for specialty applications.

Table 16 shows examples that illustrate the chemical structure of typical commercial surfactants. These frequently are termed "primary" surfactants or dispersants. There is a class of materials termed "secondary" dispersants which are used primarily in the preparation of butadiene-styrene, butadiene-acrylonitrile, and chloroprene emulsion polymers. The secondary dispersant is usually a salt of naphthalene sulfonate/formaldehyde condensates (Table 16). There are several products marketed under this broad generic classification; however, the products are very specific and selection of one is highly dependent on the specific polymer composition, the ingredients of the polymerization recipe, and the conditions of the reaction. The use concentration ranges from about 0.001 to 0.2% based on polymer. Gum formation in the reactor and its surfaces is greatly reduced or eliminated by the presence of the secondary dispersant during the polymerization reaction. These products are especially effective in minimizing gum in and on transfer pumps, transfer lines, valves, and in the monomer vacuum stripper. However,

TABLE 16 Structures of Representative Surfactants

Alkyl Sulfates

$CH_3(CH_2)_n CH_2O\ SO_3^- Na^+$

Alkyl Benzene Sulfonates

(C_nH_{2n+1})-C$_6$H$_4$-$SO_3^- Na^+$

Phosphates

alkylaryl ethoxylate $-O-\overset{O}{\underset{\underset{Na^+}{O^-}}{P}}-O$ { Na^+ or alkylaryl ethoxylate }

Alkylaryl Ethoxylates/Sulfates

$\left.\begin{array}{c}C_9H_{19}\\ t\text{-}C_8H_{17}\end{array}\right\}$-C$_6H_4$-$O-(CH_2CH_2O)_m-CH_2CH_2-O$ { H or $SO_3^- M^+$ }

Alkyl Ethoxylates/Sulfates

$(C_nH_{2n+1})-O-(CH_2CH_2O)_m-CH_2CH_2-O$ { H or $SO_3^- M^+$ }

Sulfosuccinates

$Na^+O_3S-\overset{\overset{\displaystyle C-O}{\underset{\displaystyle |}{|}}}{\underset{\underset{\displaystyle C-O}{\|}}{CH_2}}\begin{array}{l}\text{{ }}M^+ \text{ or alkyl or alkylaryl ethoxylate}\\ \\ \text{— alkyl or alkylaryl ethoxylate}\end{array}$

secondary dispersants cannot be substituted for primary dispersants; they do not markedly affect the rate of polymerization or the particle size. It is believed that the secondary dispersant adsorbs strongly onto specific surfaces, thereby supplying additional negative charge to the polymer surface.

VII. EXAMPLES OF EMULSION POLYMERIZATION

Polymerization recipes are very specific and usually are closely guarded industrial secrets. Simplified, representative recipes, such as those that follow, may be found in patent examples and in product literature, usually published by surfactant and monomer suppliers. It should be cautioned that as stated on the literature, recipes are usually not sufficiently developed to be considered acceptable for commercial production.

A typical recipe for the preparation of a 43% solids anionic ethyl acrylate polymeric dispersion by a reflux process is as follows [39]:

Ingredient	Parts
Deionized water	1000
Triton X-200(28%)[a]	96
Ethyl acrylate[b]	800

[a]Trademark Rohm & Haas Co. sodium alkylaryl polyether sulfonate.
[b]Inhibited with 15 ppm monomethyl ether of hydroquinone-MEHQ.

Procedure To a closed container charge the above foregoing ingredients less 200 parts of water. Agitate until an emulsion is formed. To a stirred reactor, fitted with a condenser, thermometer, and an addition vessel, add the 200 parts of water and 200 parts of the monomer emulsion. Heat the reaction mix to 82°C, then add 1.6 parts of ammonium persulfate (APS). The temperature will rise to 90°C in about 8 min. Some refluxing will occur. When the reflux subsides, add the remaining monomer emulsion over a period of 1.5 h, while maintaining the reaction temperature between 88 and 94°C. After the last monomer addition, allow the temperature to rise to 97°C to complete the conversion of monomer. Cool the product to room temperature with stirring and strain to remove coagulum.

Ingredient	Parts
Stage I	
Deionized water	1000
Triton X-305[a] (70%)	31.6
Ethyl acrylate (15 ppm MEHQ)	253
Methyl methacrylate (10 ppm MEHQ)	168
Glacial methacrylic acid (100 ppm MEHQ)	4
Stage IA	
Ammonium persulfate	0.5 in 6 parts water
Sodium hydrosulfite	0.6 in 6 parts water
Stage II	
Triton X-305[a] (70%)	35
Ethyl acrylate (15 ppm MEHQ)	283
Methyl methacrylate (10 ppm MEHQ)	188
Glacial methacrylic acid (100 ppm MEHQ)	5
Stage IIA	
Ammonium persulfate	0.6 in 1.5 parts water
Sodium hydrosulfite	0.8 in 6 parts water

[a]Trademark Rohm & Haas Co. octylphenol ethoxylate, OPE-30.

Following is a two-stage recipe for the preparation of a nonionic ethyl acrylate/methyl methacrylate/methacrylic acid terpolymer [39].

Procedure Charge stage I to a stirred reactor fitted with a thermometer. Cool the reactor contents to 15°C before adding the initiator and reducing agent (stage IA). A peak reaction temperature is reached within about 30 min. After about 5 min at the peak temperature, cool the reaction to 15°C. Then add the stage II ingredients in the order listed with the methacrylic acid dissolved in the ethyl acrylate. Add the stage IIA initiator and activator. The temperature should again peak at about 65°C. Allow the kettle contents to stir for about 60 min at peak temperature, then cool to 30°C. Adjust the pH to 9.5 with 28% ammonia and strain the product to remove gum. The solids concent should be about 46%.

A self-cross-linking vinyl acetate/acrylic terpolymer recipe for use in textile finishing illustrates the complexity of emulsion polymerizations. Many commercial processes are much more complex than even this example [40].

Ingredients	Parts
1. Deionized water	145.0
2. Polystep B-23 (Stepan)[a]	14.0
3. Itaconic acid (Eastman Kodak)	5.0
4. Mercaptoacetic acid	0.5
5. tert-Butyl hydroperoxide	0.25
6. Sodium phosphate, dibasic	1.25
7. Vinyl acetate	35.0
8. Butyl acrylate	15.0
9. Vinyl acetate	102.5
10. Butyl acrylate	35.0
11. tert-Butyl hydroperoxide	0.5
12. Methylolacrylamide (60%)	24.2
13. Polystep B-23 (Stepan)[a]	7.5
14. Deionized water	50.0
15. Sequestrene NaFe (Geigy)	0.05
16. Sodium formaldehyde sulfoxalate (2%)	50.0

[a]Sodium lauryl (EO)12 sulfate.

Procedure

1. Purge all of the deionized water to be used with nitrogen. Thereafter maintain a nitrogen blanket over the reactor.
2. Charge (1), (2), (3), (4), (5), and (6) to the reaction flask, and mix.
3. Add a blend of (7) and (8) to the flask and begin heating.
4. Prepare a preemulsion by adding a blend of (9), (10), and (11) to a blend of (12), (13), and (14). This must be done slowly and with sufficient agitation.
5. When the flask temperature reaches 50 to 55°C, add (15) and begin the flow of (16), slowly.
6. When the reaction reaches a 5°C exotherm, begin the flow of the preemulsion at a rate sufficient to maintain an exotherm.
7. After all the preemulsion is added, approximately 2.5 h, heat the reaction to 60 to 65°C to finish.

Results:

Initial solids	41.0%
Initial viscosity (spindle no./rpm)	3760 cP (2/6)
Initial pH	4.2
pH adjusted with ammonia	7.0
Solids at pH 7.0	39.8%
Viscosity at pH 7.0 (spindle no./rpm)	6920 cP (3/6)

VIII. AN EMULSION POLYMERIZATION PLANT

Polymerizations are usually conducted by the batch process in jacketed stainless steel or glass-lined kettles to withstand an internal pressure of at least 65 psig (446 kPa) for high-boiling monomers such as the acrylate or vinyl acetate, and to greater than 4500 psig for ethylene [41]. The reaction train usually consists of an emulsion feed tank, a polymerization reactor, and a drumming tank. Agitators are usually constructed of the same materials as the reactor. The peripheral speed of the agitator usually ranges from 150 to 600 ft/min. A variable-speed drive is suggested for optimum control of agitation. The power requirement for the motor is estimated at 1 hp per 200 gal of kettle capacity. Highly viscous systems may require more power. Baffles are sometimes used to improve mixing; excessive shear must be avoided, however, to prevent the formation of coagulum. Reaction temperature control is obtained by circulating steam or cold water through the reactor jacket. Cooling water is best admitted by means of agitating nozzles which produce a turbulent high-velocity stream.

Feed inlets should be located so that free-fall onto the batch surface of the feed is avoided. The lines should discharge against the reactor walls at a level which is above the level batch so that the material lows into the reaction mix as a film; the tendency for foam and coagulum formation is lessened by this technique. Dip pipes suffer from plugging due to gum formation in and on the pipe, and for this reason are not popular. Nozzles should be provided in the kettle lid for, monomer, water, initiator solution and activator solution additions. Inert gas is best added by bubbling it through a connection on the bottom valves rather than a sparge pipe; gum formation and plugging is again the problem.

Additional control equipment usually found on a reactor is a temperature recorder, manometer, sight glass, and an emergency stack equipped with a rupture disk. An electronic recorder is preferred because of fast response. The temperature probe should be located at the bottom of the reactor so that the temperature of the small volume of reactants at the start of the process (i.e., when particle size is being set) may be monitored accurately. The probe must be kept clean. The emergency stack of spiral weld pipe or galvanized sheet metal or stainless steel is piped to discharge at a safe location; it is filled with a rupture disk. The area of the disk expressed in square inches should be about one-eight of the kettle capacity expressed in gallons. If vacuum is called for in some stage of the process, such as monomer stripping, a disk equipped with a vacuum support is required. Any vent pipe in the system should have a flame arrester. Care should be taken to avoid flame-arrester designs that plug up easily. A simple flame arrester which is constructed by placing three layers of stainless steel or aluminum screening over the pipe outlet is recommended [39].

Monomer or monomer emulsion feeds are stored and fed from a separate stainless steel tank, also equipped with an agitator, manometer, level gauge, cooling coils or jackets, a sparge for inert gas, a temperature recorder, a rupture disk, and a flame arrester, as well as numerous nozzles for charging ingredients. All lines into the tank should be fitted with a filter.

At the discharge end of the kettle there is a drain tank, usually of stainless steel. The tank also serves as a station where adjustment of solids and pH is made, and preservative, thickeners, and other additives are added. A low-speed paddle-type agitator is recommended. A cooling jacket on the tank greatly speeds up the productivity of the system. Once cooled, the finished dispersion is passed through a coarse filter prior to packaging.

REFERENCES

1. A. Rose and E. Rose, eds., *The Condensed Chemical Dictionary*, 6th ed., Reinhold, New York, 1961.
2. M. E. S. McBain, private communication to W. D. Harkins, 1932; private communication W. D. Harkins to W. D. Harkins, 1932.
3. L. I. Fikentscher, Angew. Chem., 51:433 (1938).
4. W. Heller and H. B. Klevens, Copolymer Research Reports to the Office of Rubber Reserve, War Production Board, 124 and 237, 1943; 241, 1944; 563 and 570, 1945.
5. W. D. Harkins, J. Amer. Chem. Soc., 69:1428 (1947).
6. W. V. Smith and R. H. Ewart, J. Chem. Phys., 16:592 (1948).
7. J. H. Baxendale, M. G. Evans, and J. K. Kilham, Trans. Faraday Soc., 42:668 (1946): J. H. Baxendale, S. Bywater, and M. G. Evans, ibid., 42:675 (1946).
8. J. R. Powers, U.S. Patent 2,520,959 (1950).
9. J. W. Vanderhoff, J. Polym. Sci. Polym. Symp., 72:161 (1985).
10. D. P. Durbin, M. S. El-Aasser, G. W. Poehlein, and J. W. Vanderhoff, J. Appl. Polym. Sci., 24:703 (1979).
11. W. J. Priest, J. Phys. Chem., 56:1077 (1952).
12. D. H. Napper and A. G. Parts, J. Polym. Sci., 61:113 (1962).
13. D. H. Napper and A. E. Alexander, J. Polym. Sci., 61:127 (1962).
14. E. Peggion, F. Testa, and G. Talami, Makromol. Chem., 71:173 (1964).

15. A. S. Dunn and P. A. Taylor, Makromol. Chem., 83:207 (1965).
16. A. G. Parts, D. E. Moore, and J. G. Watterson, Makromol. Chem., 89:156 (1965).
17. R. M. Fitch, M. B. Prenosil, and K. J. Sprick, J. Polym. Sci. Part C, 27:95 (1969).
18. A. S. Dunn and L. C. H. Chong, Br. Polym. J., 2:49 (1970).
19. R. M. Fitch, in *Emulsion Polymers and Emulsion Polymerization* (D. R. Bassett and A. E. Hamielec, eds.), ACS Symposium Series 165, American Chemical Society, Washington, D.C., 1981.
20. R. M. Fitch and C. H. Tsai, in *Polymer Colloids* (R. M. Fitch, ed.), Plenum Press, New York, 1971.
21. R. M. Fitch, T. H. Palmgren, T. Aoyagi, and A. Zuikov, Angew. Makromol. Chem., 123/124:261 (1984).
22. H. Reerink and J. Th. G. Overbeek, Discuss. Faraday Soc., 18:74 (1954).
23. R. D. Vold and M. J. Vold *Colloid and Interface Chemistry*, Addison-Wesley, London, 1983.
24. R. H. Ottewill, in *Emulsion Polymers and Emulsion Polymerization* (D. R. Bassett and A. E. Hamielec, eds.), ACS Symposium Series 165, American Chemical Society, Washington, D.C., 1981.
25. B. R. Vijayendran, T. Bone, and C. Gajria, in *Emulsion Polymers and Emulsion Polymerization* (D. R. Bassett and A. E. Hamielec, eds.), ACS Symposium Series 165, American Chemical Society, Washington, D. C., 1981.
26. A. M. Maurice, J. Appl. Polym. Sci., 30:473 (1985).
27. S. M. Ahmed, M. S. El-Aasser, R. J. Micale, G. W. Poehlein, and J. W. Vanderhoff, in *Polymer Colloids II* (R. M. Fitch, ed.), Plenum Press, New York, 1980.
28. B. Kronberg, J. Colloid Interface Sci., 96:55 (1983).
29. R. H. Ottewill and T. Walker, KolloidZ Z. Polym., 227:108 (1968).
30. B. R. Vijayendron, T. Bone, and C. Gajria, Vinyl Acetate Symp. Proc., Lehigh University, Bethlehem, Pa., April 1980.
31. D. C. Blackley, in *Emulsion Polymers and Emulsion Polymerization* (D. R. Bassett and A. E. Hamielec, eds.), ACS Symposium Series 165, American Chemical Society, Washington, D.C., 1981.
32. O. Palmgren, in *Emulsion Polymerization* (I. Piirma and J. L. Gardon, eds.), ACS Symposium Series 24, American Chemical Society, Washington, D.C., 1976.

33. D. C. Blackley, in *Polymer Colloids* (R. Buscall, T. Corner, and J. F. Stageman, eds.), Elsevier, London, 1985.
34. T. G. Rochow and C. W. Mason, Ind. Eng. Chem., *28*:1296 (1936).
35. F. A. Digioia and R. E. Nelson, Ind. Eng. Chem., *45*:745 (1953).
36. W. R. Conn, B. B. Kine, and W. C. Prentiss, U.S. Patent 2,795,564 (1957).
37. A. P. King and H. Naidus, J. Polym. Sci. Part C, *27*:311 (1969).
38. C. Bondy, J. Oil Colour Chem. Assoc., *49*:1045 (1966).
39. *Emulsion Polymerization of Acrylic Monomers*, CM 104A/cf, Rohm and Haas Co., Philadelphia, Pa.
40. *Polystep Surfactants for Emulsion Polymerization*, Form 580-P, Stepan Chemical Co., Northfield, Ill.
41. D. C. Blackley, *Emulsion Polymerization, Theory and Practice*, Wiley, New York, 1975.
42. B. Derjaguin and L. P. Landua, Acta Phys. Chem., *14*:633 (1941).
43. J. W. Verwey and J. Th. Overbeck, *Theory of the Stability of Lyophobic Colloids*, Elsevier, New York, 1948.
44. J. Brandrup and E. H. Immergut, eds., *Polymer Handbook*, 2nd ed., Wiley, New York, 1975.

9
Microemulsions: Formation, Structure, Properties, and Novel Applications

ROGER LEUNG,* MEAN JENG HOU, and DINESH O. SHAH
University of Florida, Gainesville, Florida

I.	Introduction	315
II.	Formation, Structure, and Properties of Microemulsions	319
	A. Spontaneous Emulsification and Thermodynamic Stability of Microemulsions	319
	B. Geometric Aspects and Structure of Microemulsions	326
	C. Solubilization and Phase Equilibria of Microemulsions	333
	D. Experimental Studies and Properties of Microemulsions	342
III.	Microemulsions as Media for Chemical Reactions	345
	A. Polymerization in Microemulsions	346
	B. Photochemical Reactions in Microemulsions	347
	C. Enzymic Reactions in Microemulsions	349
	D. Precipitation in Microemulsions	351
IV.	Novel Applications	353
	References	354

I. INTRODUCTION

It is well known that two immiscible liquids (e.g., oil and water) can form a macroscopically clear, homogeneous mixture upon addition

---
Present affiliation: Aluminum Company of America, Alcoa Center, Pennsylvania.

of a third liquid (dispersing agent) which is miscible with both liquids
[1]. This has conventionally been represented by a ternary-phase
diagram, as shown in Fig. 1. The single-phase region in the diagram
is considered a simple molecular solution. However, when surface-
active molecules (e.g., surfactants or detergents) are used as the
dispersing agent, the single-phase region may consist of microdomains
of dispersed phase and complex association structures of molecules.
Such a single-phase region is often referred to as a "microemulsion"

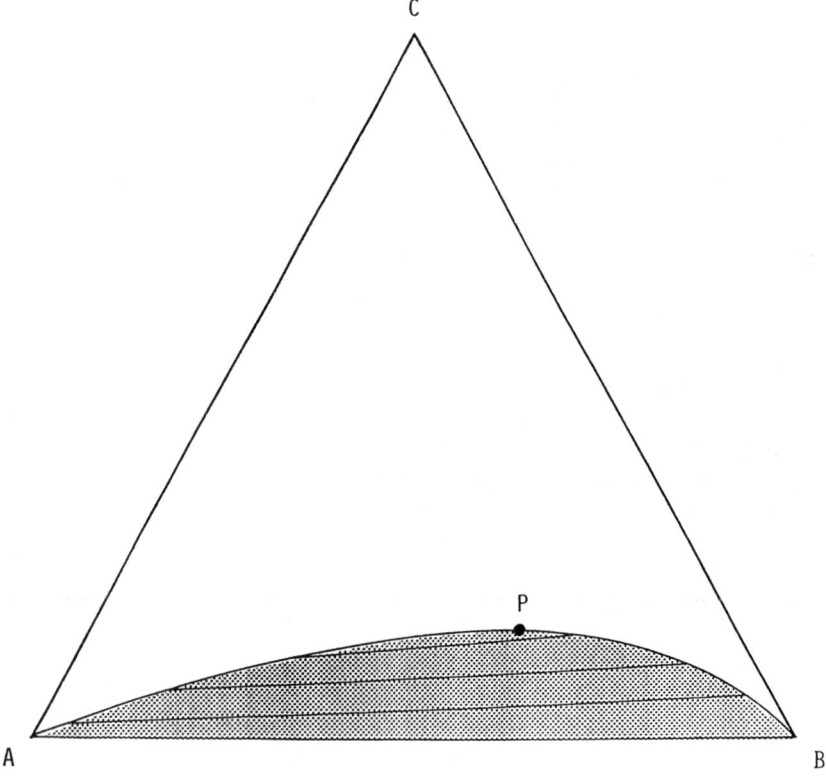

FIG. 1 Schematic ternary-phase diagram representing the formation
of a homogeneous mixture of two immiscible liquids (A and B) upon
addition of a third liquid (C) which is miscible with both liquids.
Shaded area represents a two-phase region; p is the plait point.

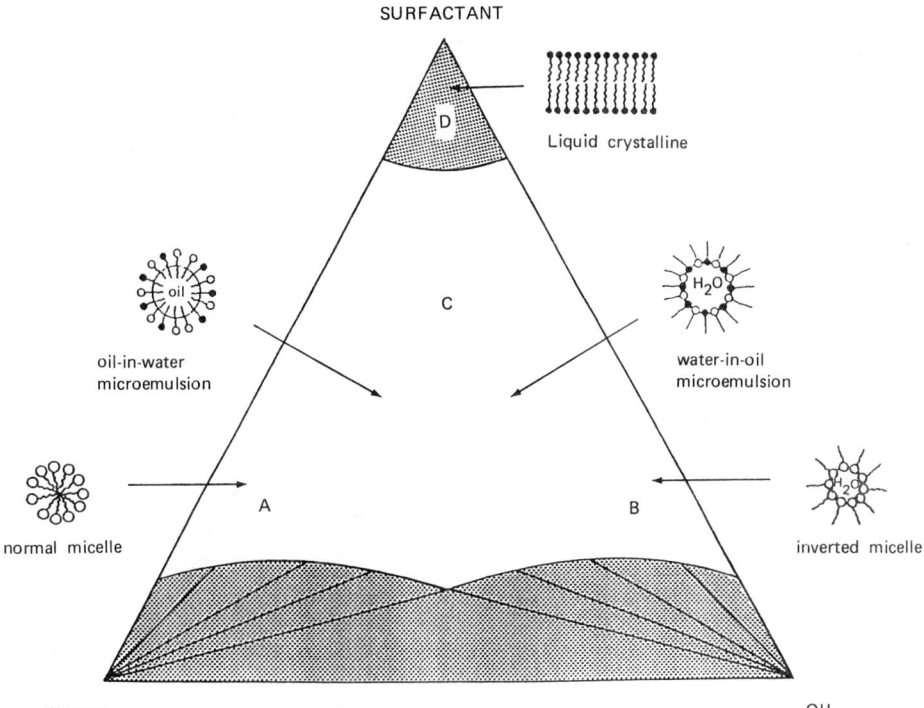

FIG. 2 Schematic ternary-phase diagram of an oil-water-surfactant microemulsion system consisting of various associated microstructures. A, Normal micelles or o/w microemulsions; B, reverse micelles or w/o microemulsions; C, concentrated microemulsion domain; D, liquid-crystal or gel phase. Shaded areas represent multiphase regions.

[2]. Figure 2 represents a schematic ternary-phase diagram for microemulsions composed of microdomains with various possible association structures.

In 1943, Hoar and Schulman [3] first described a microemulsion* as a transparent or translucent system formed spontaneously upon mixing oil and water with a relatively large amount of ionic surfactant together with a cosurfactant [e.g., an alcohol of medium chain length

---

*It should be noted that the term "microemulsion" was not coined until 16 years later, in a paper cited in Ref. 4.

(C_4 to C_7)]. The system contained dispersion of very small oil-in-water (o/w) or water-in-oil (w/o) droplets with radii on the order of 100 to 1000 Å. Figure 3 represents schematically these two basic microemulsion structures. Since Hoar and Schulman's report, considerable interest and attention have been focused on microemulsions. This can be attributed to the fact that microemulsions possess special characteristics of relatively large interfacial area, ultralow interfacial tension, and large solubilization capacity as compared to many other colloidal systems. These special features offer great potential for a wide range of industrial and technological applications (e.g., tertiary oil recovery, detergency, catalysis, drug delivery, etc.).

In general, the formation of microemulsions involves a combination of three to five components: oil, water, surfactant, cosurfactant, and salt. The chemical structure of surfactant, cosurfactant, and oil strongly influences a microemulsion phase diagram[5–7]. In fact, the complexity and diversity in properties, structures, and phase behavior of microemulsions always pose a persistent challenge for many theoretical and experimental researchers. During the past decade, the scientific literature on microemulsions has grown at a fast pace. Several books, symposium proceedings, and review articles have been published [8–22]. An exhaustive coverage of all aspects of microemulsions is virtually impossible in this condensed review chapter. Hence we plan to focus only on some fundamental questions and to describe some recent developments on microemulsions which are of great scientific interest or technological relevance.

At present, there exists no precise, or commonly agreed upon, definition of microemulsions. As a matter of fact, there has been much debate about the terminology of "microemulsions," and as a consequence, many other terms, such as "swollen micelles" or

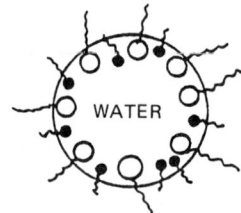

OIL–IN–WATER   WATER–IN–OIL

FIG. 3 Schematic diagram for oil-in-water (o/w) and water-in-oil (w/o) microemulsion structures. The smaller molecules represent cosurfactant.

"solubilized micelles," have been suggested [21]. The debate centers on distinguishing microemulsions from a true micellar solution [23–25]. Historically, microemulsions were originally defined from a phenomenological viewpoint, that is, the observation of a homogeneous, transparent, and low-viscosity system containing a considerable amount of dispersed phase with the presence of suitable surfactant and cosurfactant. At a very low volume fraction of dispersed phase, however, the system actually resembles a true micellar solution. The transition between these two structures generally shows no apparent break in many physical properties of pure surfactant systems [26], but may exhibit a discontinuity for commercial mixed surfactant systems [27]. Based on a temperature-dependence study of photon correlation spectroscopy, Zulauf and Eicke [28] have established a clear transition from Aerosol-OT reverse micelles in iso-octane to w/o microemulsions at a water-to-Aerosol-OT molecular ratio of about 10. But the relationship between normal micelles and o/w microemulsions is not as straightforward. It has been shown that the kinetics of solubilization of oil is much slower for o/w microemulsions than for normal micelles [29]. Hence the key to this long-argued problem lies more with kinetic or dynamic measurements of the system than with static measurements.

Despite the controversy mentioned above, the designation as a microemulsion of a clear isotropic single-phase region in a phase diagram does offer practical convenience in terminology. In our opinion, a microemulsion can be defined phenomenologically [158] as a *"thermodynamically stable, isotropically clear* dispersion of two immiscible liquids, consisting of microdomains of one or both liquids stabilized by an interfacial film of surface-active molecules."

II. FORMATION, STRUCTURE, AND PROPERTIES OF MICROEMULSIONS

A. Spontaneous Emulsification and Thermodynamic Stability of Microemulsions

The two most fundamental questions dealing with a microemulsion are probably those related to the mechanism of formation of a microemulsion and its thermodynamic stability as compared to a conventional emulsion (i.e., a macroemulsion). A macroemulsion, upon standing, has been known to coalesce and eventually to separate into an oil and water phase due to a lack of thermodynamic stability [30]. However, it has been pointed out that some emulsion systems may be thermodynamically unstable but could exhibit long-term stability for practical purposes [10]. This has been referred to as the "kinetic stability" of the system due to a high-energy barrier for coalescence between droplets [10]. The distinction between thermodynamic stability and kinetic stability of a system is probably a matter of concern only from a thermodynamic point of view.

1. Ultralow Interfacial Tension

One of the early, important contributions of Schulman and co-workers was to realize that the stabilization of microemulsions required a low solubility of surfactant (or surfactant mixture) in both oil and water phases [31], resulting in the adsorption of the surfactant at the water-oil interface to lower the interfacial tension. This can be described by the well-known Gibbs adsorption isotherm for multiple-component systems [32]:

$$d\gamma = -\sum_i \Gamma_i \, d\mu_i = -\sum_i \Gamma_i RT \, d \ln a_i \tag{1}$$

where γ is the interfacial tension, Γ_i is the surface excess of component i (amount of component i adsorbed per unit area), μ_i the chemical potential of component i, and a_i the activity of the solute i. Equation (1) basically dictates that the increase of surfactant activity a_i in the solution would result in a decrease of interfacial tension if the surface excess of the surfactant is positive. Moreover, the addition of a second positively adsorbed surfactant to the system would always cause a further decrease in interfacial tension. Hence it has been proposed that the role of cosurfactant, together with the surfactant, is to lower the interfacial tension down to a very small, even transient negative value, at which the interface would expand to form fine dispersed droplets, and subsequently adsorb more surfactants and cosurfactants until their bulk concentration is depleted enough to make interfacial tension positive again. This process, known as "spontaneous emulsification," forms the microemulsion.

The concepts of transient negative interfacial tension and its relation to spontaneous emulsification have been proposed and experimentally examined for some time [33-36]. Their value is to emphasize the importance of ultralow interfacial tension for the formation and thermodynamic stability of microemulsions. In fact, the mechanism of microemulsion formation has been analyzed by Ostrovsky and Good [37] based on a dynamic equilibrium process in which the rate of self-emulsification is equal to the rate of coalescence of microemulsion droplets. The analysis established a boundary of interfacial tension between a thermodynamically stable microemulsion and an unstable macroemulsion. For tensions lower than 10^{-2} dyn/cm, stable microemulsions can be obtained. In other thermodynamic models [38-45], even lower interfacial tensions, on the order of 10^{-4} to 10^{-5} dyn/cm, have been employed to satisfy the stable condition of microemulsions.

Microemulsions

It is known that some surfactants (e.g., many double chain surfactants and nonionic surfactants) can form microemulsions without the addition of a cosurfactant [46,47]. Although this has been attributed to different ability of surfactants in lowering the interfacial tension [19], it seems that factors in addition to the ultralow interfacial tension may have to be considered for a complete explanation. In fact, the interfacial bending instability resulting from the thermal fluctuations of interfaces and the dispersion entropy of droplets in the solution may also contribute significantly to the formation of microemulsions when the interfacial tension is low.

2. Interfacial Curvature, Fluidity, and Entropy

The formation of small microemulsion droplets requires a bending of the interface. It has been shown by Murphy [48] that the bending of an interface requires work to be done against both interfacial tension and bending stress of the interface. Although always present, the bending stress is important only for very low interfacial tension or highly curved interfaces. This can be described schematically by Figure 4. At an equilibrium condition with very low interfacial tension, an interface would assume an optimal configuration and curvature, known as the spontaneous curvature $1/R_o$, at which the bending energy of the interface is minimized. Bending the interface farther away from this spontaneous curvature will cause an increase in bending energy, which can be represented by a constant K, known as the curvature elasticity (or bending elasticity) of the interface. The constant K with the unit of energy actually dictates the ease of interfacial deformation. A large K value corresponds to a "rigid" interface for which large

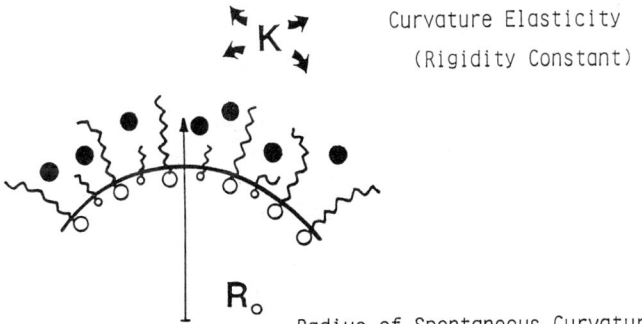

FIG. 4 Schematic diagram for spontaneous curvature and curvature elasticity of an interfacial film. The filled circles represent oil molecules penetrating the interfacial film.

energy is required to bend the interface. A small K value represents a "fluid" interface for which little energy is necessary for bending. Hence K is also called the "rigidity constant" of an interface. When K is close to $k_b T$, where k_b is the Boltzmann constant, the interface is subject to a bending instability resulting from thermal fluctuations.

Safran and Turkevich [49] have expressed the interfacial free energy F_I of microemulsion droplets in terms of both interfacial tension and bending energy for an uncharged interface:

$$F_I = n \left[4\pi\gamma R^2 + 16\pi K \left(1 - \frac{R}{R_0}\right)^2 \right] \qquad (2)$$

where n is the number density of droplets, γ the interfacial tension, R the droplet radius, and R_0 the radius of spontaneous curvature (or the natural radius). Apparently, Eq. (2) contains only the energetic term, and the entropic term of interface will be discussed later. Equation (2) is applicable to ionic w/o or nonionic microemulsions where the electrostatic energy can be neglected. K has been found to be in the order of 10^{-14} erg for microemulsions [50]; hence the bending energy term is important only when γ is close to zero. Accordingly, Murphy [48] has concluded that a planar interface having a low but positive interfacial tension could nevertheless be unstable with respect to thermal fluctuations if the reduction in interfacial free energy due to bending exceeds the increase in free energy due to expansion of the interface. Therefore, he suggested that the bending instability at low interfacial tension might be responsible for spontaneous emulsification.

The preceding discussion focused on the effect of thermal fluctuations on a "fluid" interface (small K). Although the role of membrane "fluidity" for the formation of microemulsions was noted earlier [34], its significance is better elucidated by recent theories and experimental results [50–52]. It has been shown [50,51] that when K is larger than $k_b T$, oil, water and surfactant may form a birefringent lamellar phase, and that only when K is small, are isotropic disordered microemulsions obtained. A lamellar birefringent phase is often observed in the vicinity of a microemulsion phase in a phase diagram [53]. The addition of cosurfactant is often found to increase the fluidity of the interface, leading to a structural transition from birefringent lamellar phase to isotropic microemulsions [50,52]. In practice, the fluidity of an interface can be increased by choosing a surfactant and cosurfactant with widely different sizes of the hydrocarbon moiety [34], or by setting a temperature so that there is a balance between the hydrophilic and lipophilic properties of the surfactant [54].

The thermal fluctuations of a fluid interface lead to an increase in the entropy of interfacial film. The entropy of such a fluctuating interface has been approximated by the mixing entropy of oil and water [10]. The decrease in free energy of the system due to this dispersion entropy may exceed the increase of free energy caused by newly created interfacial area due to emulsification, thus resulting in spontaneous formation and stabilization of a microemulsion. This has been quantitatively accounted for on the basis of phenomenological thermodynamics by many researchers [38–45]. Because excellent reviews on various thermodynamic models of microemulsions have been published [10,19], we describe only briefly here some important concepts and results.

Ruckenstein and Chi [38] have expressed the Gibbs free-energy change of microemulsion formation by three terms:

$$\Delta G_M(R) = \Delta G_1 + \Delta G_2 + \Delta G_3 \tag{3}$$

where ΔG_1 is the interfacial free energy, including a positive term due to the creation of uncharged interface and a negative term due to the formation of an electric double layer; ΔG_2 is the free energy of interdroplet interactions, composed of a negative term due to van der Waals attraction and a positive term due to repulsive double-layer interaction; and ΔG_3 is the entropy term accounting for dispersion of microemulsion droplets in the continuous medium. From Eq. (3), the condition for spontaneous formation of microemulsions with the most stable droplet size (R^*) at a given volume fraction of dispersed phase may be obtained by:

$$\left. \frac{\partial \Delta G_M}{\partial R} \right|_{R = R^*} = 0 \tag{4}$$

$$\left. \frac{\partial^2 \Delta G_M}{\partial R^2} \right|_{R = R^*} > 0 \tag{5}$$

Equations (4) and (5) indicate that a negative, minimum $\Delta G_M(R^*)$ is required to obtain a stable microemulsion, as shown in curve A of Fig. 5. Curve B in Fig. 5 represents a kinetically stable macroemulsion provided that the height of energy maximum is significant, and curve C corresponds to an unstable emulsion. Figure 6 shows the influence of interfacial tension on the formation of microemulsions. When interfacial tension is less than 2×10^{-2} dyn/cm, a stable microemulsion can be formed. Figure 7 shows the individual contribution of the three terms in Eq. (3) to the stability of microemulsions. The dispersion entropy predominantly contributes to the thermodynamic stability of microemulsions. Rosano and Lyons [55], using a titration method, have shown that the formation of microemulsions is indeed

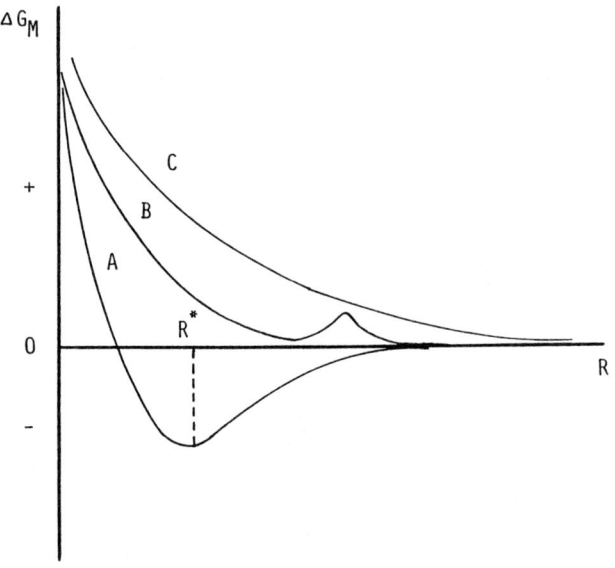

FIG. 5 Schematic illustration of the Gibbs free energy of microemulsions formation ΔG_M as a function of droplet radii R. A, stable microemulsion with droplet radius R* at the minimum ΔG_M; B, kinetically stable emulsion; C, unstable emulsion.

entropically driven. Ruckenstein's model further predicts a phase inversion from one type of microemulsion to another (i.e., w/o to o/w), as well as a phase separation [38–43].

As a criterion for the formation of a thermodynamically stable dispersion system with low interfacial tension, an inequality has been proposed [37]:

$$- \frac{d\ln \gamma}{d\ln R} \geq 2 \tag{6}$$

where γ is the interfacial tension and R is the droplet radius. Although the form of this inequality may differ depending on different thermodynamic treatments [37], many analyses do agree on a similar trend—that for microemulsions the average equilibrium radius of droplets increases with a decrease in interfacial tension [45,56] but the reverse is true for macroemulsions [37].

To recapitulate the discussion so far, we conclude that the spontaneous formation of microemulsions with a decrease of total free energy of the system can be expected only if the interfacial tension

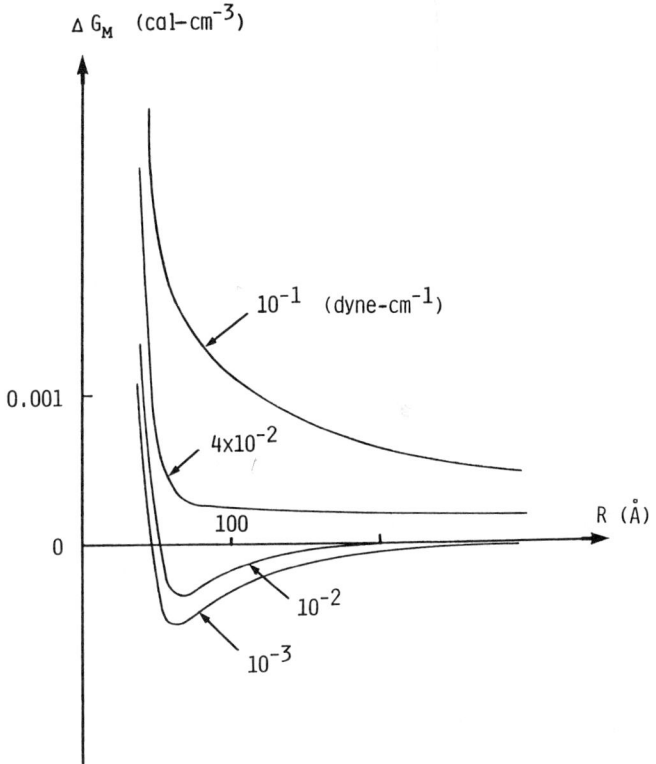

FIG. 6 Influence of interfacial tension on the formation of microemulsions. A small but positive value of interfacial tension can result in a stable microemulsion. [Courtesy of Elsevier, Adv. Colloid Interface Sci., 20:167 (1984).]

is so low that the free energy of the newly created interface can be overcompensated by the dispersion entropy of droplets in the solution. The bending instability resulting from the thermal fluctuations of interfaces with low tension and high fluidity could be responsible for spontaneous emulsification. Two necessary conditions for the formation of microemulsions are as follows:

1. *Large adsorption of surfactant or surfactant mixture at the water-oil interface.* This can be achieved by choosing a surfactant mixture with proper hydrophilic-lipophilic balance (HLB) value for the system. One can also employ various methods to adjust the HLB of a given surfactant mixture, such as adding a cosurfactant, changing salinity or temperature.

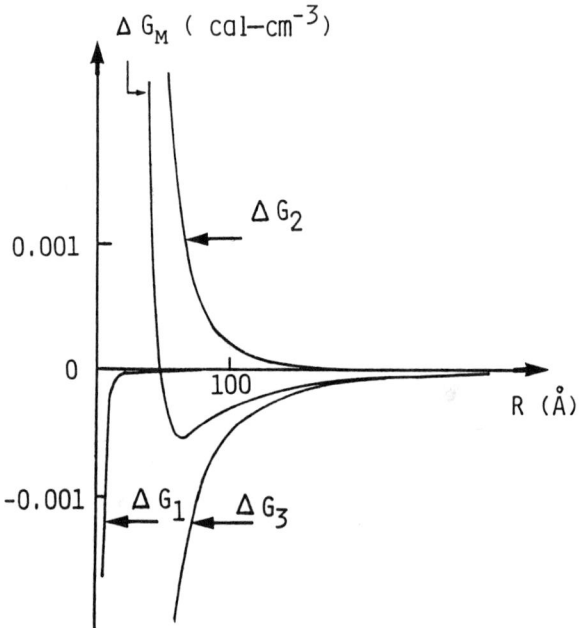

FIG. 7 Contribution of ΔG_1, ΔG_2, and ΔG_3 to the free energy of microemulsion formation. [Courtesy of Elsevier, Adv. Colloid Interface Sci., 20:167 (1984).]

2. *High fluidity of the interface.* The interfacial fluidity can be enhanced by using a proper cosurfactant or an optimum temperature.

The role of cosurfactant in microemulsion formation is to (1) decrease the interfacial tension, (2) increase the fluidity of interfaces, and (3) adjust the HLB value and spontaneous curvature of the interface, leading to the spontaneous formation of microemulsions.

B. Geometric Aspects and Structure of Microemulsions

Two types of most commonly encountered microemulsions are o/w and w/o globular droplets, as shown in Fig. 3. Some theories, such as mixed (or duplex) film theory [33,34,57,58], R theory [59], and the concept of hydrophilic-lipophilic balance of surfactant [60,61], have been proposed in an attempt to delineate the factors that determine the formation of a specific structure (i.e., w/o or o/w) for a given water-oil-surfactant system. Recently, a geometric model concerning the surfactant packing at the interface has also been proposed [56, 62]. All these theories define certain parameters that can dictate the curvature of a given interfacial film and hence predict the corresponding structure. Since reviews of these theories are available in the literature [10,17,63], only the geometric model will be discussed.

1. Geometric Packing Considerations in Amphiphile Aggregation

Basically, the geometric model emphasizes the importance of geometric constraints in packing of amphiphiles at the interface for determining the structure and shape of amphiphilic aggregates. Following the concept of duplex film, which was first proposed by Bancroft [64] and Clowes [65] and later applied to microemulsions by Bowcott and Schulman [33], the model essentially considers the interfacial film as duplex in nature; that is, the polar heads and hydrocarbon tails of amphiphiles are acting as separate uniform liquid interfaces, with water hydration in the head layer and oil penetration in the tail layer. The key element of describing the geometric packing of surfactants at the interface is a packing ratio defined as the ratio of cross-sectional area of hydrocarbon chain to that of polar head of a surfactant molecule at the interface, $v/a_o \ell_c$, where v is the volume of hydrocarbon chain of the surfactant, a_o the optimal cross-sectional area per polar head in a planar interface, and ℓ_c is approximately 80 to 90% of the fully extended length of the surfactant chain [62].

The direction and degree of interfacial curvature are basically a result of this packing ratio and are influenced further by the differential tendency of water to swell the head area and oil to swell the tail area. It is intuitively clear that a greater cross-sectional area of tail than that of head ($v/a_o \ell_c > 1$) will favor the formation of w/o droplets, while a smaller cross area of tail than that of head ($v/a_o \ell_c < 1$) would favor the o/w droplets. A planar interface requires $v/a_o \ell_c = 1$, which leads to the formation of lamellar structure. Figure 8 schematically depicts the foregoing description.

Assuming that the optimal head area a_o will not change with interfacial curvature, Mitchell and Ninham [62] suggested a necessary geometric condition for the existence of o/w droplets,

$$\frac{1}{3} < \frac{v}{a_o \ell_c} < 1 \tag{7}$$

Equation (7) predicts the formation of (1) normal micelles for $v/a_o \ell_c < 1/3$, (2) o/w droplets for $1/3 < v/a_o \ell_c < 1$, and (3) w/o (inverted) droplets for $v/a_o \ell_c > 1$. It should be pointed out that the increase in packing ratio $v/a_o \ell_c$ corresponds to an increase in o/w droplet size but to a decrease in w/o droplet size. The boundary at $v/a_o \ell_c = 1$ indicates a structural transition from o/w to w/o droplets. The structure and molecular mechansim of this phase inversion domain remain poorly understood and will be discussed next. Similar geometric

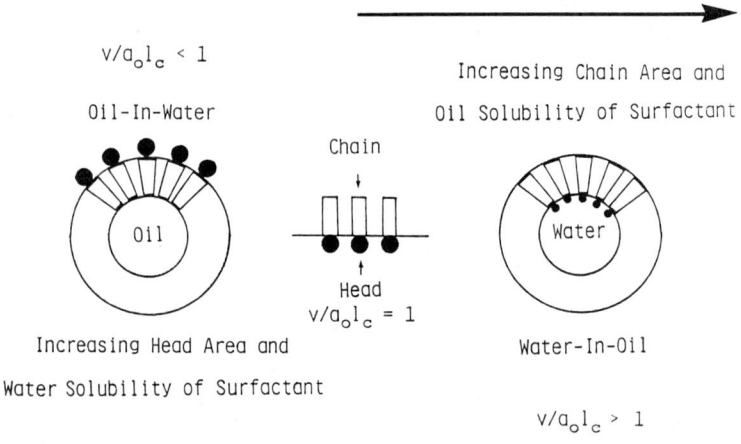

FIG. 8 Schematic diagram representing the interfacial curvature of o/w and w/o microemulsions and phase inversion based on the geometric model.

criteria have also been proposed to describe the structure of biological lipid aggregates [66]. These results seem to suggest that the geometric packing of amphiphiles plays an important role in determining the structure and shape of aggregates.

One of the advantages of this geometric model is that the packing ratio can quantitatively account for the HLB of a surfactant. A low HLB value, in the range 4 to 7, favoring w/o emulsions, corresponds to $v/a_o \ell_c > 1$, whereas a high HLB value, in the range 9 to 20, favoring o/w emulsions, corresponds to $v/a_o \ell_c < 1$. Further, taking the geometric packing term into account in a thermodynamic model can serve as a simple approach to establish a unified thermodynamic framework for amphiphilic aggregation [62]. In addition, the concept of geometric model can easily account for the influence of salt, cosurfactant, and oil on interfacial curvature. For a simple water-oil-surfactant system, a surfactant with bulky head group and relatively small tail area, such as some of the single-chain surfactants, tends to form o/w droplets. To obtain w/o droplets in this case, one has to employ cosurfactant (e.g., medium-chain alcohols), high salinity, or oil with smaller molecular volume or higher aromaticity in the system. The incorporation of cosurfactant in the interface is expected to increase the mean hydrocarbon volume per surfactant molecule without affecting appreciably either a_o or ℓ_c [62]. The addition of salt is expected to decrease the head area a_o due to the suppression of the electric double layer. Oil with smaller molecular volume or higher aromaticity can enhance the penetration of oil into the surfactant layer, thus increasing the surfactant hydrocarbon volume [62,67–69]. All these effects tend to increase the packing ratio, hence favoring the formation of w/o droplets. On the other hand, a surfactant with a laterally bulky hydrocarbon part and a relatively small head group, such as some of the double-chain surfactants, favors the formation of w/o droplets. This can explain why Aerosol-OT (sodium bis-2-ethyl hexyl sulfosuccinate) forms w/o microemulsions spontaneously without the addition of a cosurfactant.

The effect of temperature on the packing ratio is difficult to predict, due to a lack of understanding of all forces in the system. However, experimental data for biological lipid and nonionic surfactant systems seem to suggest an increase in $v/a_o \ell_c$ with increasing temperature [62]. This can be explained partially by the decrease in water hydration of the head group (decreasing a_o) at elevated temperature [70]. One thus expects a growth of normal micelles formed by nonionic surfactants with increasing temperature due to increasing $v/a_o \ell_c$ until the cloud point [71], beyond which phase separation occurs. On the other hand, flocculation of micelles due to attractive interaction between micelles at elevated temperature has also been observed [72–74]. At even higher temperatures, known as the phase inversion temperature (PIT), phase inversion from normal to inverted micelles occurs [47,75,76]. At this PIT, one expects $v/a_o \ell_c = 1$ and thus a zero curvature.

2. Nonglobular Domain and Microemulsion Structure in Phase Inversion Region

Apart from the consideration of geometric packing presented above, two additional geometric constraints have to be observed for the existence of globular structure in a ternary-phase diagram [77]. First, there exists an upper limit of 0.64 as the maximum volume fraction of dispersed droplets in the solution according to a simple random close-packing model of hard spheres [78]. Second, there must also exist a lower limit of the polar head area a_o below which electrostatic repulsion between polar heads increases. Such a limit imposes a lower bound on the size of w/o droplets because decreasing size requires a decreasing polar head area a_o and/or an increasing hydrocarbon volume v according to the geometric model. But this constraint is not relevant to o/w droplets because a_o increases as the droplet size diminishes.

Taking these two constraints into account, Biais et al. [77] have identified some domains that cannot have any globule in the ternary phase diagram shown in Fig. 9a. Figure 9b shows the experimentally determined region where no globules are observed [79]. Instead, a lamellar structure has been found in region 2. In region 1, the

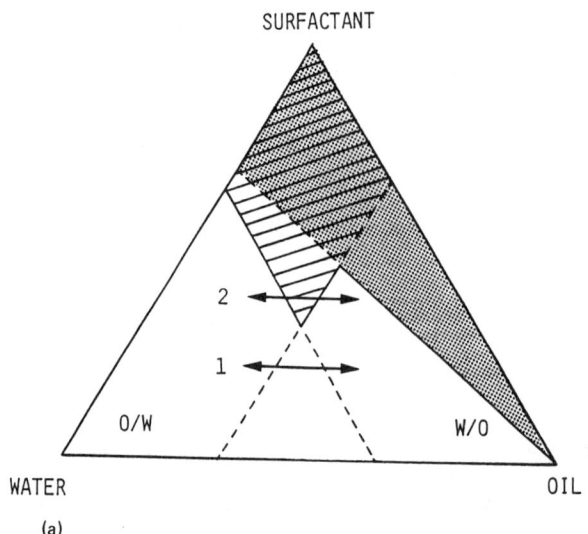

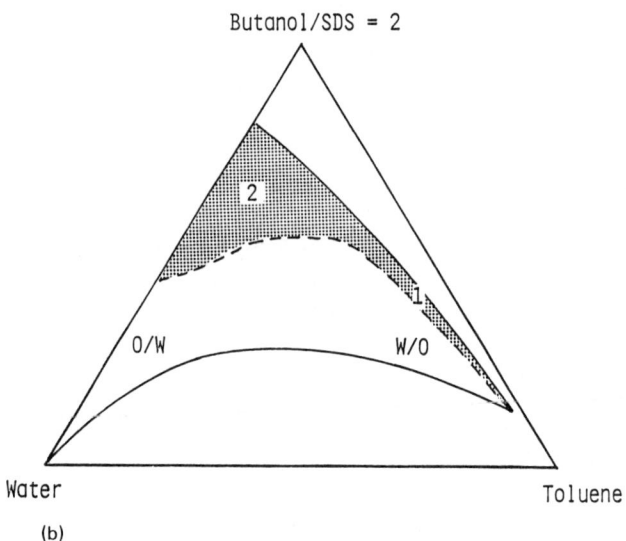

FIG. 9 (a) Nonglobular domains due to close-packing constraint (hatched area) and limitation of minimum head area (shaded area). Two possible mechanisms of phase inversion are shown. (b) Experimentally determined nonglobular domain (shaded area) in water-toluene-butanol-SDS microemulsion system. [Courtesy of Academic Press, J. Colloid Interface Sci., 80:136 (1981), and 78:275 (1980).]

solution probably consists of small hydrated soap aggregates solvated by alcohol molecules and dispersed in the oil medium.

It has also been proposed, according to Fig. 9a, that at least two different mechanisms of phase inversion are possible [77]. Path 1 indicates a continuous transition from w/o to o/w, with an intermediate region in which o/w and w/o droplets may coexist [76,80,81], or a bicontinuous structure has been suggested [82]. A discontinuous transition is also possible along path 2 through a structure that cannot be spheres. Usually, a birefringent lamellar structure has been observed in this case [79].

According to the prediction of geometric model, zero curvature is expected at phase inversion, thus justifying the existence of lamellar structure along path 2. However, the rationale for the continuous transition along path 1 is not so obvious. In fact, the mechanism and structure of this continuous phase inversion remain poorly understood. Talmon and Prager [83] have proposed a statistical mechanical model of bicontinuous structure to account for this continuous phase inversion without a priori concern about the geometric features of aggregates. It was based on a Voronoi tesselation followed by random segregation of the oil and water domains. Subsequently, de Gennes and Taupin [51] and Jouffroy et al. [84] proposed a modification by taking a cubic lattice instead of the Voronoi tesselation. It was shown [51] that a persistence length ξ_k can be defined as the characteristic length of the water-oil interface. ξ_k increases exponentially with increasing curvature elasticity K of the interface. An isotropic microemulsion phase can exist when K and consequently ξ_k are small; otherwise, periodic ordered structures such as lamellae are expected. This result elucidates the importance of fluidity and thermal fluctuations of interfaces for the formation of microemulsions. It further delineates the correlation between the isotropic random microemulsion phase and the periodic ordered phase of lyotropic nematics. In addition, the physical meaning of the elementary size of a system without well-defined geometry such as the bicontinuous structure has been clarified. The bicontinuous structure [82] is envisioned as containing continuous interpenetrating domains of both oil and water, with neither one surrounding the other. Although the equilibrium mean curvature of the interface is zero, complying with the prediction of the geometric model, the interface is constantly subject to thermal fluctuations, resulting in continuous sinusoidal bending with no specific preference toward either water or oil phases. It should be mentioned that in addition to the spherical and bicontinuous structures discussed above, other microemulsion structures, such as cylinders and lamellae, have been proposed [85].

Phase-inversion phenomena bear important technological relevance. In general, one can obtain phase inversion by changing a large number of variables in a systematic manner. Of greatest importance among these changes are increasing the volume fraction of dispersed phase, varying the salinity of the system, and adjusting the temperature. When salinity is varied, a middle-phase microemulsion with equal solubilization of brine and oil can be obtained at phase-inversion salinity—the so-called "optimal salinity." This has important implications for tertiary oil recovery because maximum solubilization and ultralow interfacial tension can be obtained at this optimal salinity. More details of this will be discussed later. Shinoda and Kunieda [47] have also established that maximum solubilization can be obtained for nonionic surfactant systems at phase-inversion temperature (PIT). Maximum or optimal detergency is often obtained at the PIT vicinity (i.e., the cloud point) [86,87].

3. Design Characteristics of Microemulsions

Based on the preceding discussion, it can be concluded that the geometric feature (or HLB) of a surfactant plays an important role in determining the formation and structure of microemulsions. To design a microemulsion, the use of surfactant or surfactant mixture is required to lower the interfacial tension according to Eq. (1). But the addition of alcohol is not a theoretical requirement, although alcohol is often used to fluidize the interfacial film (decrease K). Actually, one can also obtain a fluid interfacial film by using a double- or branched-chain surfactant at a temperature above the thermotropic phase transition temperature [88]. However, when a cosurfactant is not used in microemulsions, it is a necessary but not a sufficient condition that surfactant hydrocarbon volume v, effective chain length ℓ_c, and head group a_0 should satisfy the relation $v/a_0\ell_c = 1$, as the elementary design characteristic for a simple three-component (oil, water, and surfactant) microemulsion system [46]. Several other variables, such as the chemical nature of cosurfactant and oil, salinity, and temperature, can alter the packing ratio. Thus many parameters are available for manipulation in the design and formulation of microemulsions.

4. Shape Fluctuations and Structural Dynamics of Microemulsions

Thus far our discussion has focused mainly on the equilibrium structure and character of microemulsions. It should be mentioned that, on the one hand, the thermal fluctuations of interfaces result in a thermodynamic stability of microemulsions, but on the other hand, a highly dynamic character of microemulsions also results due to thermally induced size and shape fluctuations (polydispersity) of spherical microemulsions [89,239]. In fact, a microemulsion should be viewed as a dynamic structure [89]. They are thermodynamically stable, but there is a constant coalescence, breakdown, and deformation of microemulsion droplets. The picture of microemulsions as persistent entities

Microemulsions 333

having definite geometric shape is not accurate. A detailed discussion of the study of the structural dynamics of microemulsions is beyond the scope of this chapter, but some relevant results will be mentioned briefly here.

The structural dynamics of microemulsions has been investigated by a variety of techniques and methods, such as nuclear magnetic resonance (NMR), electron spin resonance (ESR), chemical relaxation techniques, chemical reaction or fluorescence quenching kinetics in microemulsions [90], and quasi-elastic light scattering [91,92]. The results of NMR and ESR studies confirm that there exists a constant and fast exchange (characteristic time on the order of 10^{-8} to 10^{-9} s) of microemulsion components (e.g., surfactant and cosurfactant) between the interfacial film and the continuous phase [90]. This corroborates the view that the interfacial film of microemulsions is highly fluid. Further, the content of microemulsion droplets, especially w/o droplets, is found to be rapidly exchanged between the droplets through collisions and formation of "transient dimers." This is evidenced by studying the kinetics of chemical reactions and fluorescence quenching in microemulsions [90,93–96]. The formation of dimers has been attributed to "sticky collisions" between droplets, resulting from attractive interdroplet interactions as suggested by neutron and light-scattering studies [97–100]. Such an exchange process and formation of dimers have important relevance to the chemical reactions occurring in microemulsions. This will be discussed in Section III. We conclude that the study of dynamic aspects of microemulsions has advanced the fundamental understanding on the stability, fluidity of interfaces, interaction forces and collision rate between microemulsion droplets.

C. Solubilization and Phase Equilibria of Microemulsions

1. Solubilization and Structure of Microemulsions

Solubilization is one of the most salient features of microemulsion sustems from which most applications stem. Many early studies of solubilization reported in a classic book by Laing et al. [101] are based on simple soap (micellar) solutions (i.e., the ability of surfactants to increase the solubility of hydrophobic compounds in water). Since Marsden and McBain [102] published one of the very first phase diagrams illustrating solubilization phenomena in a solution, the field of solubilization has expanded considerably.

Figure 10 presents a series of schematic ternary or pseudoternary (in which two components are grouped at the same vertex) microemulsion phase diagrams. These diagrams show the changes of general features of microemulsions when varying the alcohol chain length, and varying the surfactant from single chain to double chain or from ionic

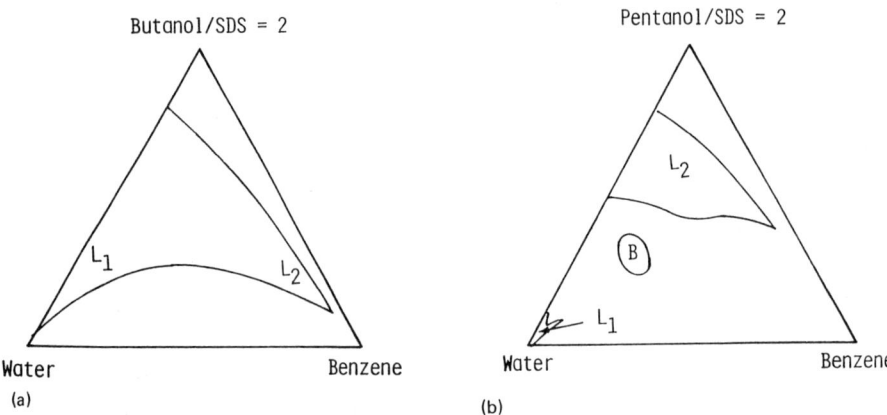

FIG. 10 Schematic ternary (or pseudoternary) phase diagrams of various microemulsion systems. L_1, Normal micelle or o/w microemulsion region; L_2, reverse micelle or w/o microemulsion region. M, middle-phase microemulsion; B, anisotropic phases. Part (f) is shown at phase-inversion temperature. Note that detailed liquid-crystalline regions are not shown. [(a) and (b) are based on Ref. 5; (c) and (d) on Ref. 129.]

to nonionic surfactant. Each clear microemulsion phase region represents a solubilization area corresponding to a specific structure. Two mechanisms of phase inversion from o/w (L_1) to w/o (L_2) microemulsions described earlier can be seen in Fig. 10. The continuous phase inversion is often observed when a short-chain cosurfactant is used (K is very small), resulting in a large connecting homogeneous solubilization area (Fig. 10a and c). A discontinuous phase inversion is seen in most cases, with L_1 and L_2 regions separated by some intermediate liquid-crystal regions.

The factors that determine solubilization have not been completely delineated. However, based on current theories and understanding of solubilization [42,43,49,83,84,101,103—108], we can identify some important parameters on a qualitative basis. It has been shown that three solubilization sites are possible in a surfactant aggregate [109—112]. Taking a normal micelle as an example, hydrocarbons and other nonpolar compounds are thought to be incorporated in the micelle interior (swollen micelles, Fig. 11a). Some solubilizate molecules may distribute themselves among the surfactant molecules at the interface (Fig. 11b). Polar solubilizate molecules may adsorb at the micellar surface (Fig. 11c). Different solubilization sites may influence the

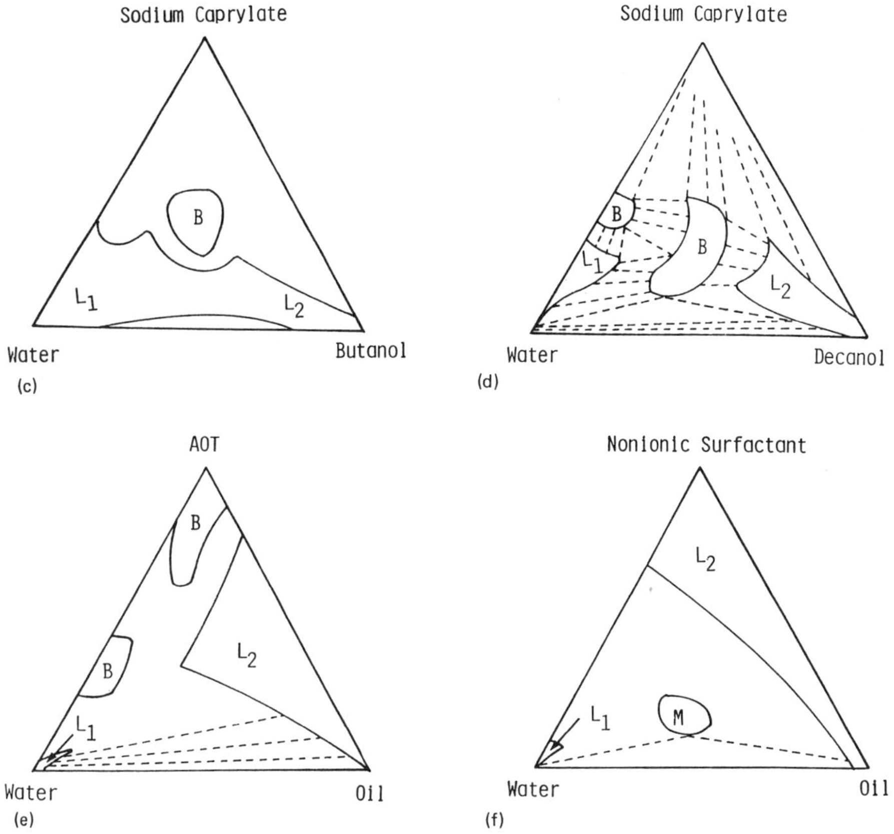

Microemulsions

FIG. 10 Continued

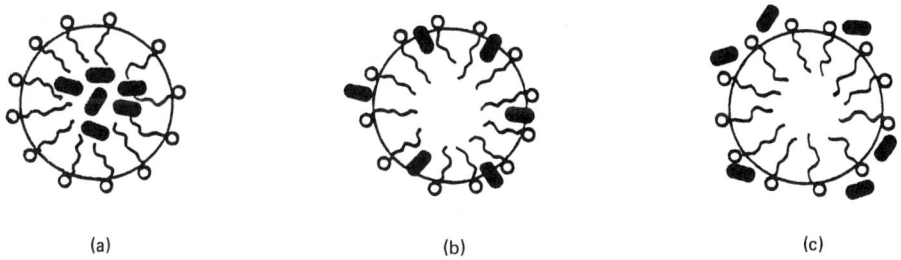

FIG. 11 Schematic view of three possible solubilization sites in surfactant aggregates: (a) micelle interior; (b) palisade layer; (c) micellar surface.

kinetics of chemical reactions in microemulsions, which will be discussed in Section III. Here we limit our discussion of solubilization only to the first case, which is relevant to the formation of swollen micelles or microemulsions.

From a simple geometric calculation, it can be shown that the total solubilization volume V in a microemulsion is equal to

$$V = \frac{AR}{3} \tag{8}$$

where A is the total interfacial area and R is the radius of droplets. A is related to the total emulsifier concentration in a system. At constant total emulsifier concentration, the solubilization is directly related to the droplet radius and hence the curvature of interface [49,84,104–107]. Therefore, solubilization depends on the structure of microemulsions. Equation (8) predicts that solubilization is large when R approaches infinity (zero curvature). This explains the maximum solubilization observed at the phase inversion region, as discussed earlier. Further, one often observes a smaller solubilization area in o/w (L_1) than in w/o (L_2) microemulsions, as shown in Fig. 10. This can be attributed partly to the highly curved interface (large a_o) associated with o/w droplets resulting from strong electric repulsion between polar heads and from strong water hydration of polar heads (for nonionic surfactants). Increasing salinity or temperature (for nonionic surfactants) can decrease a_o and consequently decrease the curvature, thus increasing the solubilization. It is also generally observed that w/o microemulsions form more readily than o/w microemulsions [47,113].

The analysis above focuses only on the influence of interfacial curvature (or bending energy) on solubilization. The interaction between microemulsion droplets can also influence the stability, structure, and hence the solubilization of microemulsions [51,108,114]. The long-range electrostatic repulsive force in aqueous micellar solutions at higher surfactant concentrations can lead to a structural transition from isotropic micelles to an anisotropic ordered structure such as hexagonal or lamellar phase [115]. On the other hand, the attractive force between droplets can cause coagulation or coalescence between droplets, and consequently a phase separation of microemulsions [51,108,114]. Coagulation of o/w droplets can usually be obtained by increasing the salinity of the system [116,117], while increasing the fluidity of interfaces leads to coalescence of w/o droplets [118,119]. In any of the foregoing events, a corresponding change in solubilization is usually observed.

Apart from the viewpoint that solubilization depends on structure and properties of microemulsions, solubilization itself also induces changes in shape, size, and structure of microemulsions [120–123]. Hence solubilization, structure, and properties of microemulsions are all interrelated.

2. Phase Equilibria of Microemulsions

When the limit of solubilization of a microemulsion is reached, phase separation occurs and the microemulsion phase can coexist in equilibrium with other phases. The phase equilibria of microemulsions are conventionally described by a phase diagram with tie lines as shown schematically in Fig. 10d.

According to the Gibbs phase rule, the degrees of freedom of a given system at constant temperature and pressure are equal to

$$F_{T,P} = C - P \tag{9}$$

where C is the number of components and P is the number of phases in the system. Thus a general four-component microemulsion system—namely, an oil-water-surfactant-cosurfactant system—can be constituted by one, two, three, or four phases in equilibrium. Consequently, the approach of studying microemulsions becomes a matter of choice depending on the problem of concern. The most popular approach is to study the one-phase microemulsion region. However, the study of two- and three-phase equilibria is important for understanding the stability and interaction forces in microemulsions. This will become more clear as our discussion proceeds. Such a phase-equilibrium approach is also useful for determining the composition of the phase boundary.

At this time, there is very little known about four-phase equilibria of microemulsions [124,125]; hence only two- and three-phase equilibria will be discussed. At least three types of two-phase equilibria of microemulsions have been elucidated:

1. *Microemulsions in equilibrium with excess internal phase* (i.e., w/o microemulsions with water or o/w microemulsions with oil). This type of phase equilibria is driven by the bending stress (or curvature) of interfacial film [49,84,104–107], and phase separation occurs due to the resistance of interfacial film to bending for further growth of microemulsion droplets.
2. *Two isotropic microemulsion phases* (containing a high and low density of droplets, respectively) *coexists*. This phase separation is driven by attractive interdroplet interactions [51,108,114]. Critical-like behavior and sometimes a critical point may be observed in this case [1,118,119,126–128].
3. *Coexistence of both w/o and o/w microemulsions* [106]. This phase equilibrium is driven by the balance of hydrophilic and lipophilic property of a surfactant (i.e., equal solubility of surfactant in both oil and water). Experimentally, one often observes this type of phase equilibria only at very low surfactant concentration [129]. At sufficiently high surfactant concentrations, birefringent mesophases are often present between w/o (L_2) and o/w (L_1) micro-

emulsions (discontinuous phase inversion), hence no direct phase equilibrium between w/o and o/w is observed.

It has further been shown that the frist two types of phase equilibria together can give rise to three-phase equilibria of microemulsions (i.e., microemulsions in equilibrium with both excess oil and water) when both bending stress and attractive force act in parallel on the system [49,51,84,130]. Some theoretical treatments of the above-mentioned phase equilibria of microemulsions can be found in the literature [49,51,84,104–108,114,130]. We conclude that the study of phase equilibria leads to a better understanding of the stability of microemulsions and serves as a simple measure to assess the driving force for phase separation of microemulsions. This is important for the design and formulation of microemulsions.

3. Phase Behavior of Winsor-Type Microemulsion Systems

The most studied phase equilibria of microemulsions are probably the Winsor-type microemulsions [131–133] using a salinity scan, as shown in Fig. 12a. One can prepare such a system by mixing an equal volume of brine and oil with a proper surfactant and cosurfactant. By increasing the salinity, one observes a progressive change in phase diagram and behavior, as described by Fig. 12a and b.

In a low-salinity region, the Winsor type I system represents a lower-phase o/w microemulsions in equilibrium with excess oil. In a high-salinity region, the Winsor type II system consists of an upper-phase w/o microemulsion in equilibrium with excess brine. It is clear that both Winsor I and Winsor II phase equilibria are driven by the bending stress of interfacial film.

In an intermediate-salinity region, the Winsor type III system is composed of a middle-phase microemulsion in equilibrium with both excess oil and brine. The optimal salinity is defined as the salinity at which equal volumes of brine and oil are solubilized in the middle-phase microemulsion. The structure of this middle-phase microemulsion has not been determined conclusively. Based on the data of ultracentrifugation, Hwan et al. [81] proposed that the middle phase is a o/w microemulsion near the boundary close to the low-salinity ragion, and a w/o microemulsion near the boundary close to the high-salinity region. Thus, a middle-phase microemulsion at the optimal salinity would represent a continuous phase inversion from o/w to w/o structure. A bicontinuous structure [82] has been proposed for the middle-phase microemulsion at optimal salinity and has been widely examined both experimentally and theoretically [80,108,134–138,164].

It is the attractive force between microemulsion droplets that leads to a transition of both Winsor I and Winsor II to Winsor III microemulsions [108]. The transition from Winsor I to Winsor III microemulsions has been attributed to the coacervation of normal micelle

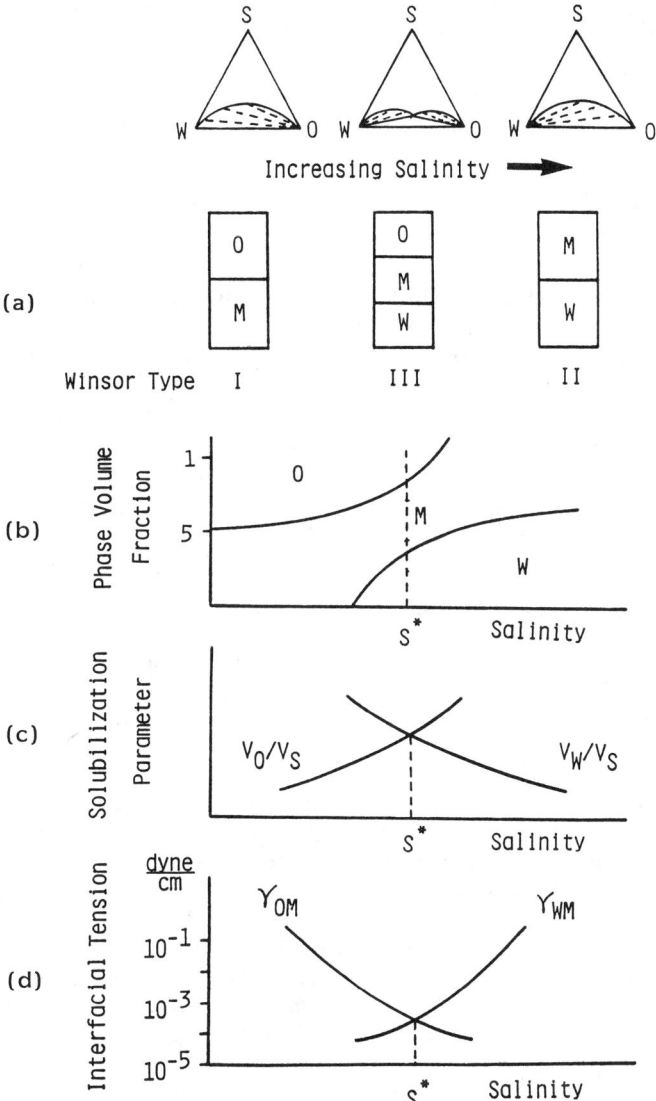

FIG. 12 Schematic presentation of a typical Winsor-type microemulsion showing the progression of phase diagrams, phase volumes, and interfacial tensions by salinity scan. M, W, and O represent microemulsion, excess water, and excess oil phases, respectively. γ_{OM} represents the interfacial tension between microemulsion and excess oil phases, and γ_{WM} is the interfacial tension between the microemulsion and excess water phases.

[117], while the transition from Winsor II to Winsor III microemulsions is associated with the percolation phenomena of w/o droplets [139–144]. Both these transitions have also been associated with critical phenomena [126]. Thus the phase equilibria of Winsor III systems are governed by both attractive force between droplets and interfacial bending stress.

Apart from the conventional salinity scan, the transition of a Winsor-type system from o/w to w/o structure can also be produced by changing any of the following variables in a systematic way [117, 130, 138]: (1) increasing alkyl chain length or molecular weight of surfactant, (2) increasing surfactant concentration, (3) increasing aromaticity of oil, (4) decreasing the chain length of oil, (5) increasing alcohol chain length (more oil soluble) or concentration, (6) increasing temperature of nonionic surfactant system or decreasing temperature of ionic surfactant system, and (7) decreasing the number of hydrophilic group (e.g., ehtylene oxide) of nonionic surfactants. All these changes may be accounted for by a corresponding change of packing ratio $v/a_0 \ell_c$ according to the geometric model. Some important properties of middle-phase microemulsions and their relation to tertiary oil recovery remain to be discussed later.

4. Pseudophase Hypothesis and Dilution Method

All the phase equilibria discussed so far refer to the equilibria between "macroscopic" phases. However, it is a well-accepted concept today that a bulk homogeneous microemulsion phase consists of three microscopic domains, namely a dispersed domain separated from a surrounding continuous domain by a domain of interfacial film. The three-compartment model [33] of microemulsions treats each microdomain as a "microscopic" phase in equilibrium with the other two. Components of microemulsions, such as surfactant and cosurfactant molecules, will partition in all three domains under an equilibirum condition. Since it is assumed that equilibria between the microdomains obey the same thermodynamic laws as the equilibria between macroscopic phases [77], each domain has been referred to as a thermodynamic "pseudophase." This is the essence of the pseudophase model [77,145] on which many thermodynamic frameworks [10,146,147] for micellar and microemulsion systems have been proposed.

Based on the pseudophase model and some simple geometric considerations, Biais et al. [77] have justified the existence of dilution lines and the use of a dilution method for w/o microemulsions in a pseudoternary-phase diagram. A dilution line in a pseudoternary-phase diagram represents a locus along which the volume of continuous phase of a microemulsion can be increased without significantly altering the size, shape, and composition of the droplets. The existence of dilution lines is important for the structural study of elementary microemulsion droplets by scattering techniques or centrifugation.

Since the data obtained from these experiments are themselves a function of droplet concentration, a dilution of droplets and extrapolation to zero droplet concentration are often employed in experiments to exclude the concentration dependence. Further, a dilution procedure is used to obtain information about interactions between droplets (10, 148]. By diluting a w/o microemulsions, one can also determine the composition of each pseudophase. Most important, the distribution of alcohol between continuous and interfacial domains can be determined, which by no means can be obtained from other measurements.

One of the great difficulties in diluting a microemulsion is to ensure the constancy of structure and composition of the droplets during dilution. In the course of dilution, water-to-surfactant ratio in the dispersed phase has to be kept unchanged to ensure a constant droplet size. A dilution procedure first proposed by Bowcott and Schulman [33] and modified by Graciaa [149] can be implemented as follows: first, oil is added to a transparent microemulsion until turbidity occurs; then the transparency is reinstated by adding alcohol together with a certain amount of water. By repeating this titration many times and plotting the volume of added alcohol versus that of added oil, one can obtain a titration curve as shown in Fig. 13. Only at a correct alcohol/water ratio corresponding to that in the continuous phase can a linear dilution line be obtained (curve b in

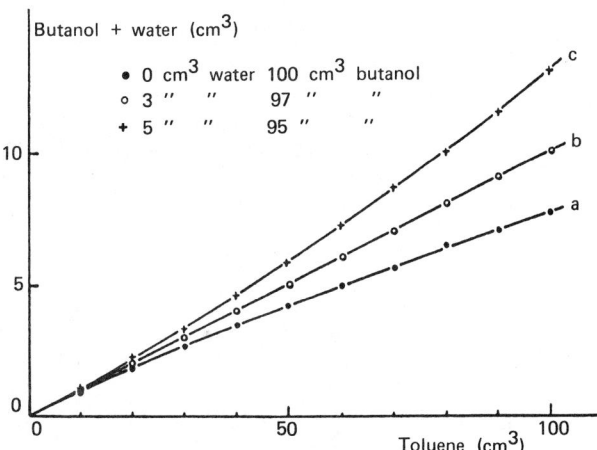

FIG. 13 Dilution curves of a water-SDS-butanol-toluene microemulsion. The curvature in curves a and c indicates a change of continuous-phase composition, and hence the droplet composition of microemulsions varies during dilution. Only the linear curve b corresponds to a dilution line. [Courtesy of C. R. Acad. Sci. Paris, Ser. B, 282:547 (1976).]

Fig. 13). The dilution line can be described by the equation

$$V_a = kV_s + rV_o \tag{10}$$

where V_s, V_a, and V_o are the volume of surfactant, added alcohol, and added oil, respectively. Assuming that the dispersed phase contains only water and that the surfactant molecules partition only at the interface, r gives the volumetric ratio of alcohol to oil in the continuous phase and k provides the volumetric ratio of alcohol to surfactant at the interface. The composition of all pseudophases can thus be deduced [10,77,149].

The validity of this dilution procedure has been examined experimentally using neutron scattering [150]. It was concluded that the dilution method applies only to a microemulsion with well-defined droplet structure exhibiting weak interdroplet interactions. As a result, the use of dilution method is limited to systems with small volume fraction of droplets [151]. At higher volume fraction of dispersed phase, such as a middle-phase microemulsion which cannot be described by a droplet structure, the dilution method fails. Many o/w microemulsion systems cannot be diluted unless enough salt is added to screen the electric repulsive force between droplets [10].

It may also be pointed out that a dilution line in a pseudoternary-phase diagram always corresponds to a demixing line where a phase separation from one-phase to two-phase occurs [10]. No dilution line can be observed in a one-phase region. The dilution line should also be a straight line due to the constancy of composition and structure of droplets during dilution. Further, dilution can only be applied to the demixtion (or phase separation) of microemulsions resulting from the interfacial bending stress, not from the interactions between droplets.

D. Experimental Studies and Properties of Microemulsions

1. Experimental Techniques for Characterization of Microemulsions

Microemulsions have been studied using a great variety of techniques. The shape, size, structure, and many other physicochemical properties of microemulsions have been determined for various systems [152–160]. A widely used method for structural study of microemulsions is probably the scattering method, including static and dynamic light scattering [148,161], small-angle neutron scattering, and X-ray scattering [162–166]. These scattering techniques not only provide detailed structural information about microemulsions, but also measure the interactions between droplets which can influence the structure, properties, and behavior of microemulsions [167]. Some other structural studies of microemulsions include sedimentation and ultracentrifugation [81,168], electron microscopy [117,169−171], positron annihilation [172,173], static and dynamic fluorescence methods [95, 120,174], and NMR [175,176]. The techniques probing the dynamics of microemulsions are NMR [67,157], ESR [50,80], ultrasonic adsorption [177,178], electric birefringence [179,180], and so on. The measurements of various properties of microemulsions include conductivity and dielectric measurements [143,181,182], viscometry [126,134,168], interfacial tension and ellipticity measurements [183,184], density and heat capacity measurements [153,154], and vapor pressure measurements [160]. It is not intended here to describe elaborately these techniques or measurements and the information thereby obtained, because reviews on some of these techniques are available in the literature [9,10,17,19,20,90]. Many of these techniques are complementary. A result obtained from one technique often requires a comparison with other techniques to avoid possible artifacts associated with each technique. The remaining discussion will be devoted to describing some important properties of microemulsions which are of technological relevance, and some applications of microemulsions.

2. Middle-Phase Microemulsions and Ultralow Interfacial Tension

The middle-phase microemulsion has been widely studied due to its relevance to tertiary oil recovery processes [185]. After the primary and secondary oil recovery, a large amount of oil remains trapped as oil ganglia in the porous rocks of reservoir due to capillary forces [186]. A surfactant solution is then injected into the reservoir to mobilize the oil ganglia by lowering the interfacial tension between the oil and water phases. In tertiary oil recovery, a lowering of oil-water interfacial tension from about 20 to 30 dyn/cm to at least 10^{-2} to 10^{-3} dyn/cm is required under practical reservoir conditions [117]. The formation of an in situ middle-phase microemulsion with sufficient solubilization of oil and brine in the reservoir by the injected surfactant solution can fulfill this requirement.

It has been shown that as salt concentration approaches the optimal salinity, the solubilization parameter of microemulsions (defined as the ratio of volumes of solubilized phase to that of the surfactant, V_o/V_s and V_w/V_s) increases in both lower- and upper-phase microemulsions, as shown in Figure 12c. At the same time, interfacial tension between the microemulsion phase and the excess phase decreases as shown in Figure 12d. Apparently, interfacial tension is related to the solubilization parameter of microemulsions. A higher solubilization parameter corresponds to a lower interfacial tension. At the optimal salinity, equal solubilization of brine and oil in microemulsions as well as equal oil and water phases are observed. These are the most important properties of middle-phase microemulsions as related to tertiary oil recovery. Other properties of middle-phase microemulsions, such as conductivity and viscosity, can be found in the literature [139].

Some empirical rules have been proposed to predict the optimal salinity for a given oil and surfactant system [187–190], but the precise mechanism responsible for the ultralow interfacial tension is not well established. The study of low interfacial tension systems can be divided into two regimes [10,117]: (1) a two-phase system with low surfactant concentrations (0.1 to 2% by weight), basically a micellar system, and (2) a three-phase (Winsor type) system with high surfactant concentrations (2 to 10%), containing a middle-phase microemulsion. In both cases the low interfacial tension has been attributed to the presence of a thin adsorbed surfactant and/or cosurfactant layer (Langmuir film) with high surface pressure at the interface [191,192]. This can be described by the Gibbs adsorption isotherm of Eq. (1). It has also been proposed that a surfactant-rich phase at the interfacial region containing liquid-crystalline structures may be responsible for the low interfacial tension observed in some systems [193]. However, in the high-surfactant-concentration regime near the optimal salinity (S*), extremely low interfacial tensions of γ_{WM} below S* and γ_{OM} above S* (see Fig. 12d) have been attributed to a thick diffuse interfacial region associated with critical phenomena [10,119,126], and the ultralow interfacial tension has been described satisfactorily by the critical scaling laws [119].

Several theoretical models [51,56,114,137,194,195] have been proposed to predict the low interfacial tension between two bulk phases in which micelles or microemulsion droplets are present and a surfactant monolayer is adsorbed at the interface between the two bulk phases. For most two-phase systems, the result seems to confirm that low interfacial tension can be accounted for by the presence of a surfactant layer at the interface. The value of interfacial tension is influenced mainly by the curvature (or size) of micelle or microemulsion droplets. However, for the critical diffuse interface of a middle-phase microemulsion near the optimal salinity, the dispersion entropy and interactions of droplets may become dominant in determining the interfacial tension. Theoretical prediction of interfacial tension becomes less satisfactory in this case.

Although many microscopic properties, such as the interfacial curvature, dispersion entropy, and interactions of microemulsion droplets, can influence the interfacial tension as predicted by many theoretical models, the presence of microemulsion droplets in two-phase systems is not required for maintaining the low interfacial tension once the equilibrium between two bulk phases has been reached. It has been shown [119] that the interfacial tension of a two-phase system say a Winsor I microemulsion system, remains unchanged after diluting continuously the o/w microemulsion phase by brine (but surfactant concentration has to remain above the critical micelle concentration in the aqueous phase). This conclusion is also valid for a Winsor II system [119]. These striking results seem to confirm further the role of a surfactant layer at the interface in obtaining a low interfacial tension. It is not clear at this time, however, whether the presence of middle-phase microemulsion structure is important for maintaining the ultralow interfacial tension of a Winsor III system because dilution method cannot be applied.

III. MICROEMULSIONS AS MEDIA FOR CHEMICAL REACTIONS

Microemulsions have recently aroused much interest as novel media for chemical reactions. The reasons behind such popularity are easily seen from the unique features of microemulsions. To mention a few, well-defined microphase with large interfacial area provides an excellent device for alteration of reaction rates, paths, and stereochemistry [196,197]. On the other hand, monodispersed microemulsion droplets with radii as small as 100 to 1000 Å offer promising means for better controlling the product morphology [2]. Moreover, microemulsions, being optically isotropic, are suitable for photochemical study [95,198].

Reaction kinetics in micellar and microemulsion systems have been studies extensively by many scientists and are well reviewed in the literature [196,197,199-201]. Basically, in the kinetic treatment of reactivities in microemulsions, several factors have to be considered. The first one is the time scale involved in the dynamics of surfactant aggregates. The mean residence time of a surfactant molecule in a microemulsion is on the order of microseconds [28,202]. An additional factor is the interdroplet exchange of solubilized water pools. This event occurs, at least for Aerosol-OT w/o microemulsions, on the time scale of milliseconds [28,93,94,147]. Thus reactions occurring in microemulsions can be classified in terms of three time scales: ultrafast, fast, and slow reactions [203]. Ultrafast photophysical reactions occur faster than the fluctuations of aggregates. On this time scale, microemulsions can be considered to be frozen. Fast chemical reactions (e.g., electron and photon transfer) can go to completion faster then the exchange of water pools among neighboring droplets. Finally, some chemical reactions are relatively slow and hence their rates are governed by statistical distribution and interdroplet exchange of water pools and reactants. Kinetically, reactions occurring at rates faster than the fusion rate of microemulsion droplets have been treated in terms of reactant entry or exit [204,205]. On the other hand, reactions occurring at rates slower than the fusion rate of droplets can be treated in terms of regular rate equations.

The location of reactants and the sites of reaction are important factors to consider in the interpretation of kinetic data. If the reactants partition considerably in both bulk (continuous) phase and aggregates, the overall reaction rate R_{total} is equal to the sum of the reaction rates occurring in the bulk continuous phase (R_b) and in the aggregate pseudophase (R_m): $R_{total} = R_b + R_m$. Such cases can be seen in many studies involving alteration of rates, paths, and stereochemistry. Usually, the role of surfactant aggregates is to concentrate the reactants in the microenvironment constituted by the aggregates. Such studies have been throughly reviewed and well documented in the literature [196,197]. In the case where reactants are soluble only in dispersed phase, there is no ambiguity regarding the reaction site, and hence the kinetic data can be interpreted more accurately. A detailed description can be found in Robinson's excellent article on the problems of concentration and reactivity in microemulsions [206].

Instead of elaborately describing the kinetic principles involved in reactions in microemulsions, we now turn our attention to the discussion of several recent studies which, in our opinion, are relevant to technological and industrial applications.

A. Polymerization in Microemulsions

Emulsion polymerization is a widely used method for the preparation of polymer latex. However, it has some disadvantages. For example, the latex particles produced are usually larger than 1000 Å and the particle size distribution is not uniform. Also, because of the turbidity of emulsions, polymerization is not easy to induce photochemically. In these regards, microemulsions are, without doubt, superior systems for polymerization. Recently, research conducted by our laboratory and by other research groups has independently demonstrated the capability of producing ultrafine polymer latex particles (100 to 500 Å) with uniform size distribution by polymerization in microemulsions [207–213]. The basic principle of polymerization in microemulsions is to incorporate polymerizable monomers into a microemulsion system, then polymerize the system by decomposing the added initiators thermally or photochemically. To achieve such microemulsification, monomers can be introduced into either the continuous or dispersed phase of microemulsions.

The polymerization of oil-soluble monomers in the continuous phase of microemulsions has been studied [214–218]. In all these studies, phase separation has been observed during the course of polymerization. The kinetics of polymerization in continuous phase is also found to be similar to that of solution polymerization [214,215]. Hence recent research in this area has been focused on polymerization in the dispersed phase of microemulsions. For example, Leong et al. [207–209] and Candau et al. [210,211] have reported the formation of stable transparent latex with low viscosity by polymerizing acrylamide solubilized in w/o microemulsions. The latex particles formed were characterized to be uniformly distributed in size with diameter smaller than 500 Å (distribution variance $\mu = \pm 0.01$). Johnson and Gulari [212] found that the polystyrene produced contained fractions of two different sizes in a study of the effect of oil-soluble versus water-soluble initiators in the polymerization of styrene in o/w microemulsions. They thus concluded that polymerization occurred in both microemulsion droplets and micelles, with microemulsion droplets being the primary site for polymerization. These authors also observed that the size of polymer products seemed to correlate with the size of microemulsion droplets in the case of an oil-soluble initiator, whereas no correlation was observed in the case of using water-soluble initiators. Although the added volume fraction of monomers was limited to less then 10% in the above-mentioned studies, Jayakrishnan and Shah [213] have been able to solubilize styrene up to 30% by combining an anionic and a nonionic surfactant (Aerosol MA-80 and Pluronic L-31). The polystyrene latex particles produced are close to 100 Å with a fairly uniform particle size distribution [213].

In a recent study of the effect of microemulsion structure on the polymerization of microemulsified methyl methacrylate (MMA), Hou and Shah [219] found that the kinetics of polymerization is strikingly different in o/w and w/o microemulsions. An anomalous decrease in the polymerization rate was observed in the structural transition from w/o to o/w microemulsions [219]. Unusual kinetics was also observed by Candau and Leong [211] in a study of dependence of polymerization rate on monomer (acrylamide), surfactant (AOT), and initiator concentrations. In all these cases, the kinetics cannot be described by the classical theory of Smith and Ewart [220]. These papers thus reveal a need for a systematic study of the effect of microemulsion structure (i.e., size, surface area, and the permeability of surfactant layer for radicals) on polymerization kinetics in microemulsions. Such studies should contribute to a better understanding of the mechanism of polymerization in microemulsions.

B. Photochemical Reactions in Microemulsions

The transparency of microemulsions has provided photochemists with an excellent medium for photochemical reactions. A recent focal point of photochemistry of microemulsions concerns itself with conversion of light energy into other forms of energy (artificial photosynthesis). Many model systems have been suggested which may lead to useful practical systems in the future [95]. A much favored system is the photoproduction of ions or charge separation, which can lead to subsequent electrical power or to the cleavage of water. Micellar and microemulsion systems have been found to be excellent vehicles for efficient production of photoionization and photoinduced charge separation [95,203].

In artifical photosynthesis, sensitizers S and electron relays R are used to mimic two natural photosystems to capture visible light and to transform it to chemical energy. The thermodynamics of converting photochemical energy to chemical energy is shown in Fig. 14. It is important to recognize that the excited state of the sensitizers, S^+, is a better electron donor and, at the same time, a better electron acceptor than the ground state, S. Therefore, absorption of light (hν) can drive a redox reaction and result in the storage of energy, ΔG, in S^+ and R^-. In Fig. 14, S* functions as an electron donor. The role of microemulsions is to perform the function of thylakoid membrane in natural photosynthesis. Judicious organization of S and R by microemulsions should bring about favorable energy deposition and transmission and, most important, prevent back electron

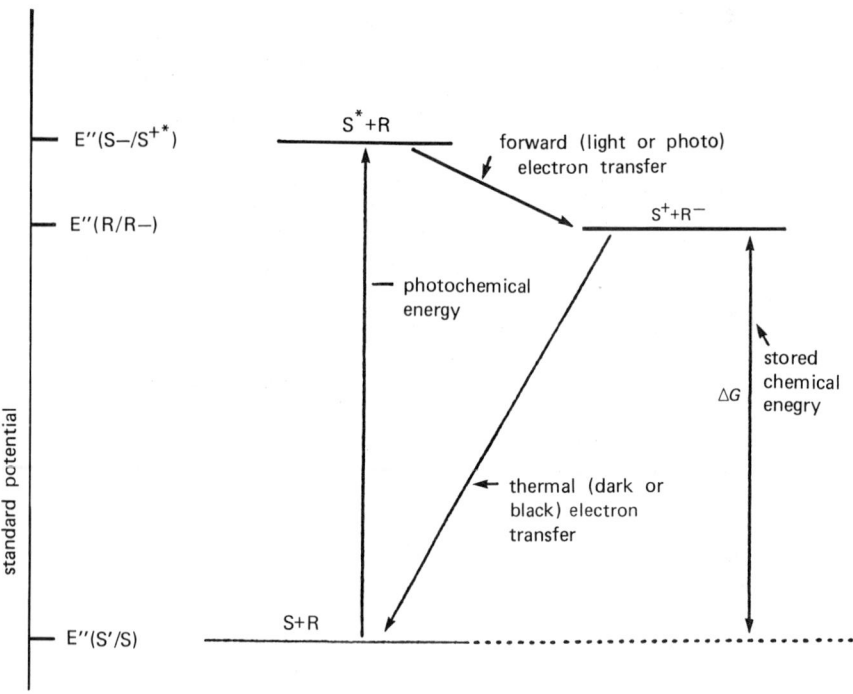

FIG. 14 Thermodynamics of converting photochemical energy to chemical energy in artificial photosynthesis. [Courtesy of Plenum Press, *Reverse Micelles*, P. L. Luisi and B. E. Straub, eds., 1984, p. 305.]

transfer between R^- and S^+ as presented in Fig. 14. A more detailed description of the mechanism involved in artificial photosynthesis can be found in an excellent article by Fendler [203].

Photosensitized electron transfer and the inhibition of undesirable back reactions have been realized in several systems [221–224]. For example, photosensitized tris(2,2´-bipyridine)ruthenium cation, (Ru^{2+}), localized in water pools of dodecylammonium proprionate (DAP) microemulsions, transferred an electron to 1,1´-dihexadecal-4,4´-bipyridinium chloride, (HV^{2+}), intercalated in the interface [221]. When EDTA was used as an electron donor (localized in the water pools), photosensitized reduction of 4-di-methyl aminoazabenzene, was mediated by benzylnicotinamide, BNA^+ [221].

More recently, hydrogenase trapped inside the w/o microemulsion was shown to generate hydrogen by Ru^{2+} photosensitized electron transfer from the thiophenal donor (distributed in the organic solvent) via the methylviogen ralay [224].

C. Enzymic Reactions in Microemulsions

Research on the enzymic reactions in microemulsions is still at its beginning stage. Only a few scattered reports can be found in the literature [225–229]. Furthermore, they are complicated by such problems as impurities in surfactants, characterization of water pools, and location of the guest biopolymers [230,231]. However, some conclusions can be drawn from this limited number of studies. Basically, the substrates can be solubilized into either a continuous of a dispersed phase of microemulsions. The reaction is then mediated by enzymes trapped in the water pools of microemulsions. The typical enzymes used are cytochrome c, chymotrypsin, rhodopsin, ribonuclease, peroxidase, phospholipase, lysozyme, lactate dehydrogenase, and pyruvate kinase.

Luisi et al. [231] have studied the reactivity of ridonuclease in AOT microemulsions. They found that the reactivity of the enzymes in microemulsions is highly pH dependent, as expected in aqueous solution. However, the pH-activity profile obtained in microemulsions is very different from that in aqueous media, as shown in Fig. 15. The reactivity of enzyme is also found to depend on the water content of microemulsions. It is now well established that the maximal reactivity of enzymes in AOT w/o microemulsions is not found at the largest possible water concentration, as one would expect, but rather at a water-to-AOT molar ratio (W_0) well below 20 [228,231]. It may be premature at this time to offer an explanation for this observation.

In general, the enzymic activity is retained in w/o microemulsions, although in some cases it is enhanced. Therefore, enzyme-catalyzed reactions can be fruitfully investigated in w/o microemulsions. This certainly is a better model for in vivo situation than that provided by aqueous solutions.

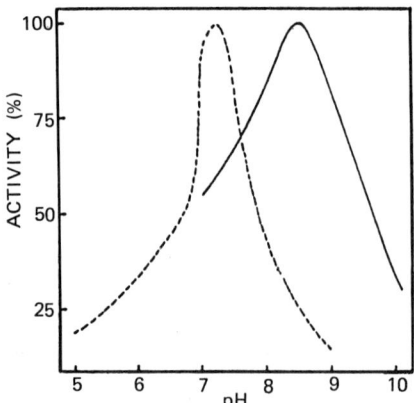

--- in aqueous solution

—— in AOT–isooctane reverse micelles at $W_0 = 7.4$

FIG. 15 pH-activity profiles for RNase activity in aqueous solution and in AOT-isooctane reverse micelles at $W_0 = 7.4$. [Courtesy of J. Solid Phase Biochem., 5:269 (1980).]

Finally, as pointed out by Luisi et al. [231], reverse micelles or microemulsions are excellent "microreactors" for enzymic reactions. This is based on the observation that enzymes (e.g., α-chymotrypsin, lysozyme, ribonuclease, etc.) hosted in microemulsions or reverse micelles can accept and catalyze not only water-soluble substrates but also water insoluble (or sparingly water soluble) ones such as steroids [225] or linoleic acid in the case of lipooxygenase [232], or hexanol or steroids in the case of horse liver alcohol dehydrogenase [232]. This surprising behavior probably can be explained on the basis of the peculiar properties of the solvent in the water pool. This solvent, although basically aqueous, has a lower dielectric constant [196] and has other features that make it compatible with hydrophobic compounds. This is very important for biotechnological applications, in that it provides an excellent means for the catalytic transformation of water-insoluble compounds by enzymes entrapped in microemulsions.

D. Precipitation in Microemulsions

Preparation of monodispersed, ultrafine particles has been one of the most pursued goals in many industries. However, it is difficult to obtain small and nearly monodispersed particles by classical methods, particularly for a particle size smaller than 100 Å. This difficulty stems from the fact that a lot of metallic particles and inorganic particles sinter easily during the reaction of their precursor species. In this respect, microemulsions offer the best reaction media that one can possibly find. Utilizing the small size and uniform distribution of microemulsion droplets, several researchers have successfully produced monodispersed, ultrafine particles in microemulsions [176, 233–236].

The basic idea of such a novel technology is illustrated in Fig. 16. In step 1, the precursor reagent A is solubilized into the water pool of a w/o microemulsion. The other reacting reagent, B, can be solubilized separately into another w/o microemulsion or be introduced

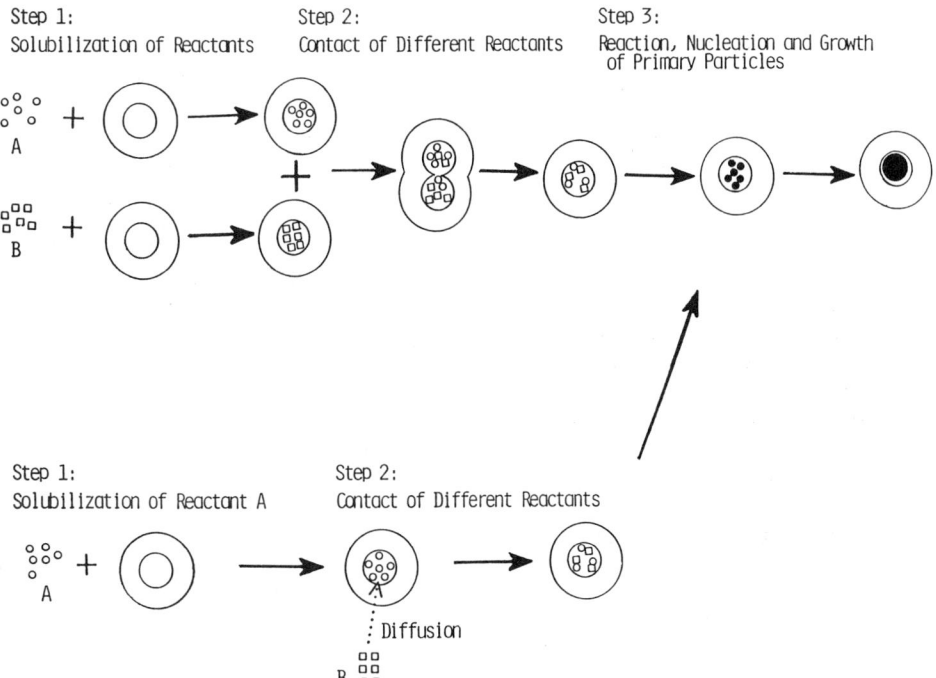

FIG. 16 Schematic illustration of various stages in the growth of ultrafine particles in microemulsions.

directly to the same w/o microemulsion. For the former case, different reactants can be brought into contact in step 2 through the energetic collisions between microemulsion droplets containing different reactants. Such contact can also be achieved in the latter case by the diffusion of reactant B to the microemulsion droplets containing A. In any case, such a contact should lead to a reaction and subsequent nucleation of desired particles inside the microemulsion droplets, as shown in step 3. In the following steps, the particles can grow via coagulation, with concomitant rearrangement of the surfactant molecules and water pools. However, the coagulation between particles residing in different microemulsion droplets will be very much limited by the steric barrier provided by the surfactant layer of microemulsion droplets. Therefore, the growth rate and particle size can be controlled by manipulation of microemulsion droplets as well as reactant concentration.

The growth of colloidal particles in microemulsions has been studied by many scientists. As early as 1981, Kon-no et al. [233] reported a precipitation of iron chlorides in AOT microemulsions by ammonia solution, leading to very monodispersed magnetic dispersions. The diameter of the particles was characterized to be around 1000 Å. Boutonnet et al. [234] have also reported the preparation of colloidal Pt, Pd, Rh, and Ir particles by reducing their precursor species with hydrogen or hydrazine in microemulsions. The size of these particles was in the range 50 to 500 Å. In a recent study of growth of AgCl in AOT microemulsions by Hou and Shah [235], AgCl particles prepared are found to be as small as 50 Å in diameter, with uniform size distribution. Through a systematic study on the effects of varying microemulsion components, they also found that the growth rate of AgCl increased when the chain length of hydrocarbon increased, the amount of the alcohol increased, or the alcohol chain length decreased. Furthermore, the addition of a long-chain nonionic surfactant as cosurfactant decreased the coagulation rate of AgCl. The results have been explained by the interaction between microemulsion droplets. More fluid interface causes greater attractive interaction, resulting in more inelastic collisions between microemulsion droplets. In another study of the growth of iron boride particles in CTAB-hexanol-water systems, Lufimpadio et al. [176] reported that particle size increased with the local concentration of reactants as well as the amount of n-hexanol used, which was found to increase the lability of the interface as suggested by EPR spectra. We also notice that an attempt has recently been made by Dvolaitzky et al. [236] to study particle growth in oil-external AOT and nonionic microemulsions. By combining x-rays and light-scattering techniques, these authors have reported two distinct processes during the growth of particles: a crystal formation with small diameter (60 Å) and a flocculation of these crystals. It is worthwhile to point out that the flocculation process can be controlled, as shown by Hou and Shah [235], through manipulation of microemulsion systems and reactant concentration. Thus the production of ultrafine particles with desired size can be achieved.

IV. NOVEL APPLICATIONS

Apart from the use of microemulsions as the media for chemical reactions, microemulsions offer a great variety of technological, industrial, and biomedical applications. Some advantages of the microemulsion technology are its spontaneous formation (easy to prepare), thermodynamic stability (long shelf life), isotropically clear appearance (easy to monitor spectroscopically), low viscosity (easy to transport and mix), molecularly ordered interfaces (easy to control the diffusivity as membrane), large interfacial area (accelerates surface reactions), low interfacial tension (flexible, high penetrating power), and large mutual solubilization of water and oil (thus possessing both hydrophilic and lipophilic characteristics). It is on these special characteristics that many applications of microemulsions are based. Some of potential engineering applications of microemulsions are: (1) enhanced oil recovery, (2) lubrication and metal cutting fluid, (3) detergency, (4) improved combustion efficiency of fuels, (5) novel heat transfer fluid, and (6) corrosion inhibition. Some potential biomedical applications of microemulsions include: (1) agricultural spray, (2) improved radiation detection fluid, (3) cosmetic and health care products, (4) drug-delivery systems, and (5) blood substitutes and organ preservation fluid. Surveys of some of these applications can be found in the literature [2,17,19,203,237,238].

In conclusion, it is clear that there has been a rapid development and better understanding of microemulsions and their applications since their introduction decades ago. Today, microemulsions still offer worthwhile scientific challenges for researchers. Many novel applications of microemulsions will probably emerge in the coming years.

ACKNOWLEDGMENTS

The authors are grateful for financial support from the National Science Foundation (Grant NSF-CPE 8005851), the American Chemical Society—Petroleum Research Fund (Grant PRF-14718-AC5), and the ALCOA Foundation (Aluminum Company of America, Alcoa Center, Pennsylvania).

REFERENCES

1. C. U. Herrmann, U. Wurz, and M. Kahlweit, Ber. Bunsenges. Phys. Chem., 82:560 (1978).
2. B. H. Robinson, Nature, 320:309 (1986).
3. T. P. Hoar and J. H. Schulman, Nature, 152:102 (1943).
4. J. H. Schulman, W. Stoekenius, and L. M. Prince, J. Phys. Chem., 63:1677 (1959).
5. M. Clausse, J. Peyrelasse, C. Boned, J. Heil, L. Nicolas-Morgantini, and A. Zradba, in Surfactants in Solution, Vol.3, (K. L. Mittal and B. Lindman, eds.), Plenum Press, New York, 1984, p. 1583.
6. M. Kahlweit, E. Lessner, and R. Strey, J. Phys. Chem., 87:5032 (1983).
7. J. Lang, R. Ruett, M. Dinh-Cao, and R. Zana, J. Colloid Interface Sci., 101:184 (1984).
8. S. Friberg and P. Bothorel, Microemulsions: Structure and Dynamics, CRC Press, Boca Raton, Fla., 1987.
9. V. Degiorgio and M. Corti, Physics of Amphiphiles: Micelles, Vesicles and Microemulsions, International School of Physics "Enrico Fermi," Course 90, Elsevier, New York, 1985.
10. A. M. Bellocq, J. Biais, P. Bothorel, B. Clin, G. Fourche, P. Lalanne, B. Lemaire, B. Lemanceau, and D. Roux, Adv. Colloid Interface Sci., 20:167 (1984).
11. D. O. Shah, Macro- and Microemulsions: Theory and Applications, ACS Symposium Series 272, American Chemical Society, Washington, D. C., 1985.
12. K. L. Mittal and B. Lindman, Surfactants in Solution, Vol. 3, Plenum Press, New York, 1984.
13. K. L. Mittal and E. J. Fendler, Solution Behavior of Surfactants, Plenum, New York, 1982.
14. K. L. Mittal, Solution Chemistry of Surfactants, Vol. 2, Plenum Press, New York, 1979.
15. K. L. Mittal, Micellization, Solubilization and Microemulsions, Plenum Press, New York, 1977.
16. I. D. Robb, Microemulsions, Plenum Press, New York, 1982.
17. L. M. Prince, Microemulsions, Academic Press, New York, 1977.

18. M. Kerker, *Colloid and Interface Science*, Vol. 2, Academic Press, New York, 1976.

19. J. T. G. Overbeek, P. L. de Bruyn, and F. Verhoeckx, in *Surfactants*, (Th. F. Tadros, ed.), Academic Press, New York, 1984, p. 111.

20. Th. F. Tadros, in *Surfactants in Solution*, Vol. 3 (K. L. Mittal and B. Lindman, eds.), Plenum Press, New York, 1984, p. 1501.

21. K. Shinoda and S. Friberg, Adv. Colloid Interface Sci., *4*:281 (1975).

22. L. Rosano, J. Soc. Cosmet. Chem., *25*:609 (1974).

23. L. M. Prince, J. Colloid Interface Sci., *52*:182 (1975).

24. M. Rosoff and A. Giniger, in *Colloid and Interface Science*, Vol. 5 (M. Kerker, ed.), Academic Press, New York, 1976, p. 475.

25. M. P. Pileni and S. Chevalier, J. Colloid Interface Sci., *92*:326 (1983).

26. H. F. Eicke, in *Microemulsions*, (I. D. Rodd, ed.), Plenum Press, New York, 1982, p. 17.

27. S. I. Chou, Ph. D. thesis, University of Florida, 1980, p. 39.

28. M. Zulauf and H. F. Eicke, J. Phys. Chem., *83*:480 (1979).

29. D. B. Siano, J. Colloid Interface Sci., *93*:1 (1983).

30. P. Becher, ed., *Encyclopedia of Emulsion Technology*, Vol. 1, Marcel Dekker, New York, 1983.

31. J. H. Schulman and J. B. Montagne, Ann. N. Y. Acad. Sci., *92*:366 (1961).

32. A. W. Adamson, *Physical Chemistry of Surfaces*, 4th ed., Wiley, New York, 1982, p. 70.

33. J. E. Bowcott and J. H. Schulman, J. Electrochem., *59*:283 (1955).

34. J. H. Schulman, W. Stoekenius, and L. M. Prince, J. Phys. Chem., *63*:1677 (1959).

35. J. T. Davis and D. A. Haydon, Ind. Int. Congr. Surf. Activity, *1*:417 (1957).

36. W. Gerbacia and H. L. Rosano, J. Colloid Interface Sci., *44*: 242 (1973).

37. M. V. Ostrovsky and R. J. Good, J. Colloid Interface Sci., *102*:206 (1984).

38. E. Ruckenstein and J. C. Chi, J. Chem. Soc. Faraday Trans. 2, 71:1690 (1975).
39. E. Ruckenstein, in *Micellization, Solubilization, Microemulsions*, Vol. 2 (K. L. Mittal, ed.), Plenum Press, New York, 1977, p. 755.
40. E. Ruckenstein, J. Colloid Interface Sci., 66:369 (1978).
41. E. Ruckenstein, Chem. Phys. Lett., 57:517 (1978).
42. E. Ruckenstein and R. Krishnan, J. Colloid Interface Sci., 71:321 (1979).
43. E. Ruckenstein and R. Krishnan, J. Colloid Interface Sci., 75:476 (1980); 76:188 (1980); 76:201 (1980).
44. J. Th. G. Overbeek, Faraday Discuss. Chem. Soc., 65:7 (1978).
45. E. Ruckenstein, in *Surfactants in Solution*, Vol. 3 (K. L. Mittal and B. Lindman, eds.), Plenum Press, New York, 1984, p. 1551.
46. L. R. Angel, D. F. Evans, and B. W. Ninham, J. Phys. Chem., 87:538 (1983).
47. K. Shinoda and H. Kunieda, in *Microemulsions* (L. M. Prince, ed.), Academic Press, New York, 1977, p. 57.
48. C. L. Murphy, Ph. D. thesis, University of Minnesota, 1966.
49. S. A. Safran and L. A. Turkevich, Phys. Rev. Lett., 50:1930 (1983).
50. J. M. di Meglio, M. Dvolaitzky, and C. Taupin, J. Phys. Chem., 89:871 (1985).
51. P. G. de Gennes and C. Taupin, J. Phys. Chem., 86:2294 (1982).
52. J. M. de Miglio, M. Dvolaitzky, L. Leger, and C. Taupin, Phys. Rev. Lett., 54:1686 (1985).
53. J. M. de Miglio, M. Dvolaitzky, R. Ober, and C. Taupin, J. Phys. Paris Lett., 44:L-229 (1983).
54. K. Shinoda, Prog. Colloid Poly. Sci., 68:1 (1983).
55. H. L. Rosano and G. B. Lyons, J. Phys. Chem., 89:363 (1985).
56. M. L. Robbins, in *Micellization, Solubilization, Microemulsions*, Vol. 2 (K. L. Mittal, ed.), Plenum Press, New York, 1977, p. 713.
57. L. M. Prince, J. Colloid Interface Sci., 23:165 (1967).
58. L. M. Prince, J. Soc. Cosmet. Chem., 21:193 (1970).

59. P. A. Winsor, Chem. Rev., 68:1 (1968).
60. L. M. Prince, in *The Chemistry and Manufacture of Cosmetics*, Vol. 3 (M. G. de Navarre, ed.), Continental Press, Orlando, Fla., 1975, p. 25.
61. P. Becher and W. C. Griffin, in *Detergents and Emulsifiers*, Allured, Ridgewood, N. J., 1974.
62. D. J. Mitchell and B. W. Ninham, J. Chem. Soc. Faraday Trans. 2, 77:601 (1981).
63. P. Becher, in *Surfactants in Solution*, Vol. 3 (K. L. Mittal and B. Lindman, eds.), Plenum Press, New York, 1984, p. 1925.
64. W. D. Bancroft, J. Phys. Chem., 17:501 (1913).
65. G. H. A. Clowes, J. Phys. Chem., 20:407 (1916).
66. J. N. Israelachvili, S. Marcelja, and R. G. Horn, Q. Rev. Biophys., 13:121 (1980).
67. L. J. Magid and C. A. Martin, in *Reverse Micelles* (P. L. Luisi and B. E. Straub, eds.), Plenum Press, New York, 1984, p. 181.
68. S. E. Friberg, H. Christenson, G. Bertrand, and P. W. Larsen, in *Reverse Micelles* (P. L. Luisi and B. E. Straub, eds.), Plenum Press, New York, 1984, p. 105.
69. A. Maitra, G. Vasta, and H. F. Eicke, J. Colloid Interface Sci., 93:383 (1983).
70. C. A. Miller and P. Neogi, *Interfacial Phenomena: Equilibrium and Dynamic Effects*, Marcel Dekker, New York, 1985, p. 151.
71. K. Shinoda, *Principles of Solubilization and Solubility*, Marcel Dekker, New York, 1978, p. 185.
72. J. C. Ravey, J. Colloid Interface Sci., 94:289 (1983).
73. M. Corti, C. Minero, and V. Degiorgio, J. Phys. Chem., 88:309 (1984).
74. J. M. di Meglio, L. Paz, M. Dvolaitzky, and C. Taupin, J. Phys. Chem., 88:6036 (1984).
75. C. V. Hermann, G. Klar, and M. Kahlweit, in *Microemulsions*, (I. D. Robb, ed.), Plenum Press, New York, 1982, p. 1.
76. S. Friberg, I. Lapcznska, and G. Giliberg, J. Colloid Interface Sci., 56:19 (1976).
77. J. Biais, P. Bothorel, B. Clin, and P. Lalanne, J. Colloid Interface Sci., 80:136 (1981).
78. J. L. Finney, Nature, 266:309 (1977).

79. A. M. Bellocq and G. Fourche, J. Colloid Interface Sci., *78*: 275 (1980).
80. C. Ramachandran, S. Vijayan, and D. O. Shah, J. Phys. Chem., *84*:1561 (1980).
81. R. N. Hwan, C. A. Miller, and T. Fort, Jr., J. Colloid Interface Sci., *68*:221 (1979).
82. L. E. Scriven, Nature, *263*:123 (1976); and in *Micellization, Solution, Microemulsion*, Vol. 2 (K. L. Mittal, ed.), Plenum Press, New York, 1977, p. 877.
83. Y. Talmon and S. Prager, J. Chem. Phys., *69*:2984 (1978).
84. J. Jouffroy, P. Levinson, and P. G. de Gennes, J. Phys. Paris, *43*:1241 (1982).
85. S. A. Safran, L. A. Turkevich, and P. Pincus, J. Phys. Paris Lett., *45*:L-69 (1984).
86. H. L. Benson, paper presented at the annual meeting of the American Oil Chemists' Society, May 2–6, 1982, Toronto, Ontario, Canada. (reprint available from Shell Chemical Company, Technical Bulletin SC:731-83, Houston, Tex.)
87. H. L. Benson, K. R. Cox, T. K. Brunck, and J. E. Zweig, paper presented at the 1984 World Surfactant Congress, Munich, Germany. (reprint available from Shell Chemical Company, Technical Bulletin SC:741-84, Houston, Tex.)
88. D. Chapman, Q. Rev. Biophys., *8*:185 (1975).
89. B. Lindman, N. Kamenka, B. Brun, and P. G. Nilsson, in *Microemulsions* (I. D. Robb, ed.), Plenum Press, New York, 1982, p. 115.
90. R. Zana and J. Lang, in *Solution Behavior of Surfactants*, Vol. 2 (K. L. Mittal and E. J. Fendler, eds.), Plenum Press, New York, 1982, p. 1195.
91. P. N. Pusey and R. J. A. Tough, Adv. Colloid Interface Sci., *16*:143 (1982).
92. A. M. Cazabat, D. Chatenay, D. Langevin, J. Meunier, and L. Leger, in *Surfactants in Solution*, Vol. 3 (K. L. Mittal and B. Lindman, eds.), Plenum Press, New York, 1984, p. 1729.
93. P. D. I. Fletcher, B. H. Robinson, F. Bermejo-Berrera, and D. G. Oakenfull, in *Microemulsions*, (I. P. Robb, ed.), Plenum Press, New York, 1982, p. 221.
94. P. D. I. Fletcher, A. M. Howe, N. M. Perrins, B. H. Robinson, C. Toprakcioglu, and J. C. Dore, in *Surfactants in Solution*,

Vol. 3 (K. L. Mittal and B. Lindman, eds.), Plenum Press, New York, p. 1745.

95. S. S. Atik and J. K. Thomas, Chem. Phys. Lett., 79:351 (1981).

96. S. S. Atik and J. K. Thomas, J. Phys. Chem., 85:3921 (1981).

97. D. Roux, A. M. Bellocq, and P. Bothorel, Prog. Colloid Polym. Sci., 69:1 (1984).

98. S. Brunettl, D. Roux, A. M. Bellocq, G. Fourche, and P. Brothorel, J. Phys. Chem., 87:1028 (1983).

99. R. Ober and C. Taupin, J. Phys. Chem., 84:2418 (1980).

100. T. Dichristina, D. Roux, A. M. Bellocq, and P. Bothorel, J. Phys. Chem., 89:1433 (1985).

101. M. E. Laing, M. E. L. McBain, and E. Hutchinson, *Solubilization and Related Phenomena*, Academic Press, New York, 1955.

102. S. S. Marsden and J. W. McBain, J. Phys. Chem., 52:110 (1948).

103. K. Shinoda, *Principles of Solution and Solubility*, Marcel Dekker, New York, 1978, Chap. 10.

104. S. Mukherjee, C. A. Miller, and T. Fort, J. Colloid Interface Sci., 91:223 (1983).

105. J. F. Jeng and C. A. Miller, in *Surfactants in Solution*, Vol. 3 (K. L. Mittal and B. Lindman, eds.), Plenum Press, New York, 1984, p. 1829.

106. C. A. Miller and P. Neogi, AIChEJ., 26:212 (1980).

107. A. W. Adamson, J. Colloid Interface Sci., 29:261 (1969).

108. C. Huh, J. Colloid Interface Sci., 97:201 (1984).

109. K. Shinoda, T. Nakagawa, B. Tamamushi, and T. Isemura, *Collodial Surfactants*, Academic Press, New York, 1963.

110. M. J. Shick, *Nonionic Surfactants*, Marcel Dekker, New York, 1967.

111. P. H. Elworthy, A. T. Florence, and C. B. McFarlane, *Solubilization by Surface Active Agents and It's Application in Chemistry and the Biological Sciences*, Chapman & Hall, London, 1968.

112. A. E. Alexander and P. Johnson, in *Colloid Science*, Vol. 2, Clarendon Press, Oxford, 1949.

113. P. Stenius, in *Reverse Micelles* (P. L. Luisi and B. E. Straub, eds.), Plenum Press, New York, 1984, p. 1.
114. C. A. Miller, R. N. Hwan, W. J. Benton, and T. Fort, J. Colloid Interface Sci., *61*:554 (1977).
115. G. S. Hartley, Nature, *163*:787 (1949).
116. S. I. Chou and D. O. Shah, J. Colloid Interface Sci., *80*:311 (1981).
117. D. O. Shah, Proc. European Symp. Enhanced Oil Recovery, Bournemouth, England, September 21–23, 1981, Elsevier, Lausanne, 1981.
118. A. M. Cazabat, D. Langevin, J. Meunier, O. Abillon, and D. Chatenay, in *Macro- and Microemulsions* (D. O. Shah, ed.), ACS Symposium Series 272, American Chemical Society, Washington, D. C., 1985, p. 75.
119. D. Chatenary, O. Abillon, J. Meunier, D. Langevin, and A. M. Cazabat, in *Macro- and Microemulsions* (D. O. Shah, ed.), ACS Symposium Series 272, American Chemical Society, Washington, D. C., 1985, p. 119.
120. P. Lianos, J. Lang, C. Struzielle, and R. Zana, J. Phys. Chem., *86*:1019 (1982).
121. P. Lianos, J. Lang, and R. Zana, J. Phys. Chem., *86*:4809 (1982).
122. P. Lianos, J. Lang, J. Sturn, and R. Zana, J. Phys. Chem., *88*:819 (1984).
123. R. Zana, J. Lang, and P. Lianos, in *Surfactants in Solution*, Vol. 3 (K. L. Mittal and B. Lindman, eds.), Plenum Press, New York, 1984, p. 1627.
124. K. E. Bennett, H. T. Davis, and L. E. Scriven, J. Phys. Chem., *86*:3917 (1982).
125. D. Roux and A. M. Bellocq, in *Macro- and Microemulsions*, (D. O. Shah, ed.), ACS Symposium Series 272, American Chemical Society, Washinston, D. C., 1985, p. 105.
126. A. M. Cazabat, D. Langevin, J. Meunier, and A. Pouchelon, Adv. Colloid Interface Sci., *16*:175 (1982).
127. P. Honorat, D. Roux, and A. M. Bellocq, J. Phys. Paris Lett., *45*:L-961 (1984).
128. H. Kunieda and S. E. Friberg, Bull. Chem. Soc. Jpn., *54*:1010 (1981).

129. P. Ekwall, in *Advances in Liquid Crystals*, Vol. 1 (G. H. Brown, ed.), Academic Press, New York, 1975, p. 1.

130. C. Huh, Soc. Pet. Eng. J., *829*, (1983).

131. P. A. Winsor, *Solvent Properties of Amphiphilic Compounds*, Butterworth, London, 1954.

132. R. L. Reed and R. N. Healy, in *Improved Oil Recovery by Surfactant and Polymer Flooding* (D. O. Shah and R. S. Schechter, eds.), Academic Press, New York, 1977, p. 383.

133. G. J. Jirasaki, H. R. Van Domselaar, and R. S. Nelson, Soc. Pet. Eng. J., *23*:486 (1983).

134. C. Blom and J. Mellema, J. Dispersion Sci. Technol., 5:193 (1984).

135. P. Guering and B. Lindman, Langmuir, *1*:404 (1985).

136. C. Boned, J. Peyrelasse, A. Graciaa, and J. Lachaise, paper presented at 5th International Symposium on Surfactants in Solution, Bordeaux, France, July 9—13, 1984.

137. C. Huh, J. Colloid Interface Sci., *71*:408 (1979).

138. K. S. Chan and D. O. Shah, paper (SPE 7869) presented at the 1979 SPE of AIME International Symposium on Oilfield and Geothermal Chemistry, Jan. 22—24, Houston, Tex.

139. K. E. Bennett, J. C. Hatfield, H. T. Davis, C. W. Macosko, and L. E. Scriven, in *Microemulsions* (I. D. Robb, ed.), Plenum Press, New York, 1982, p. 65.

140. A. M. Cazabat, D. Langevin, J. Meunier, and A. Pouchelon, J. Phys. Paris Lett., *43*:L-89 (1982).

141. A. M. Cazabat, D. Chatenay, D. Langevin, and A. Pouchelon, J. Phys. Paris Lett., *41*:L-441 (1980).

142. M. Laques, R. Ober, and C. Taupin, J. Phys. Paris Lett., *39*:L-487 (1978).

143. M. Laques and C. Sauterey, J. Phys. Chem., *84*:3503 (1980).

144. A. M. Cazabat, D. Chatenay, P. Guering, D. Langevin, J. Meunier, O. Sorba, J. Lang, R. Zana, and M. Paillette, in *Surfactants in Solution*, Vol. 3 (K. L. Mittal and B. Lindman, eds.), Plenum Press, New York, 1984, p. 1737.

145. L. Damaszewki and R. A. Mackay, J. Colloid Interface Sci., 97, 166 (1984); and J. Biasis, L. Odberg, and P. Stenius, J. Colloid Interface Sci., *86*:350 (1982).

146. B. Lindman and H. Wennerstrom, Top. Curr. Chem., *87*:3 (1980).

147. H. F. Eicke, Top. Curr. Chem., 87:86 (1980).

148. A. M. Cazabat, D. Langevin, and A. Pouchelon, J. Colloid Interface Sci., 73:1 (1980).

149. A. Graciaa, Ph. D. thesis, University of de Pau, France, 1978.

150. M. Laques, R. Ober, and C. Taupin, J. Phys. Paris Lett., 39:L-487 (1974).

151. A. M. Cazabat, J. Phys. Paris Lett., 44:L-593 (1983).

152. A. M. Bellocq, J. Biais, B. Clin, P. Lalanne, and B. Lemanceau, J. Colloid Interface Sci., 70:524 (1979).

153. A. H. Roux, G. Roux-Desgranges, J-P. E. Grolier, and A. Viallard, J. Colloid Interface Sci., 84:250 (1981).

154. G. Roux-Desgranges, A. H. Roux, J-P. E. Grolier, and A. Viallard, J. Colloid Interface Sci., 84:536 (1981).

155. M. Sanchez-Rubio, L. M. Santos-Vidals, D. S. Rushforth, and J. E. Puig, J. Phys. Chem., 89:411 (1985).

156. C. Boned, M. Clausse, B. Lagourette, J. Peyrelasse, V. E. R. McClean, and R. J. Sheppard, J. Phys. Chem., 84:1520 (1980).

157. B. Lindman, P. Stilbs, and M. E. Moseley, J. Colloid Interface Sci., 83:569 (1981).

158. R. Leung, M. J. Hou, C. Manohar, D. O. Shah, and P. W. Chun, in *Macro- and Microemulsions: Theory and Applications* (D. O. Shah, ed.), ACS Symposium Series 272, American Chemical Society, Washington, D. C., 1985, p. 325.

159. F. D. Blum, S. Pickup, B. Ninham, S. J. Chen, and D. F. Evans, J. Phys. Chem., 89:711 (1985).

160. E. Sjoblom, B. Jonsson, A. Jonsson, P. Stenius, P. Saris, and L. Odverg, J. Phys. Chem., 90:119 (1986).

161. M. Corti, in *Physics of Amphiphiles: Micelles, Vesicles and Microemulsions* (V. Degiorgio and M. Corti, eds.), Elsevier, New York, 1985, p. 123.

162. J. C. Ravey and M. Buzier, in *Surfactants in Solution*, Vol. 3 (K. L. Mittal and B. Lindman, eds.), Plenum Press, New York, 1984, p. 1759.

163. E. W. Kaler, K. E. Bennett, H. T. Davis, and L. E. Scriven, J. Chem. Phys., 79:5673 (1983).

164. E. W. Kaler, H. T. Davis, and L. E. Scriven, J. Chem. Phys., 79:5685 (1983).

165. M. Kotlarchyk, J. S. Huang, and S. H. Chen, J. Phys. Chem., 89:4382 (1985).
166. J. B. Hayter, in *Physics of Amphiphiles: Micelles, Vesicles and Microemulsions* (V. Degiorgio and M. Corti, eds.), Elsevier, New York, 1985, p. 59.
167. A. M. Cazabat and D. Langevin, J. Chem. Phys., 74:3148 (1981).
168. M. Dvolaitzky, M. Guyot, M. Laques, J. P. LePesant, R. Ober, C. Sauterey, and C. Taupin, J. Chem. Phys., 69:3279 (1978).
169. E. Sjoblom and S. Friberg, J. Colloid Interface Sci., 83:569 (1981).
170. A. Verkleij, Biochim. Biophys. Acta, 779:43 (1984).
171. Y. Talmon, J. Colloid Interface Sci., 93:366 (1983).
172. S. Millan, R. Reynoso, J. Serrano, R. Lopez, and L. A. Pucuqauchi, in *Surfactants in Solution*, Vol. 3 (K. L. Mittal and B. Lindman, eds.), Plenum Press, New York, 1984, p. 1675.
173. K. Serrano, R. Reynoso, R. Lopez, O. Olea, B. Djermounl, and L. A. Fucuqauchi, J. Phys. Chem., 87:707 (1983).
174. E. Keh and B. Valeur, J. Colloid Interface Sci., 79:465 (1981).
175. A. Maitra, J. Phys. Chem., 88:5122 (1984).
176. N. Lufimpadio, J. B. Nagy, and E. G. Derouane, in *Surfactants in Solution*, Vol. 3 (K. L. Mittal and B. Lindman, eds.), Plenum Press, New York, 1984, p. 1483.
177. J. Lang, A. Djavanbakht, and R. Zana, in *Microemulsions* (I. D. Robb, ed.), Plenum Press, New York, 1982, p. 233.
178. J. Lang, A. Djavanbakht, and R. Zana, J. Phys. Chem., 84:1541 (1980).
179. W. Schorr and H. Hoffmann, J. Phys. Chem., 85:3160 (1981).
180. R. Guering and A. M. Cazabat, J. Phys. Paris Lett., 44:L-601 (1983).
181. J. V. Nieuwkoop and G. Snoei, J. Colloid Interface Sci., 103:417 (1985).
182. J. Peyrelasse and C. Boned, J. Phys. Chem., 89:370 (1985).
183. A. Pouchelon, J. Meunier, D. Langevin, and A. M. Cazabat, J. Phys. Paris Lett., 41:L-239 (1980).

184. D. Beaglehole, M. T. Clarkson, and A. Upton, J. Colloid Interface Sci., *101*:330 (1984).

185. D. O. Shah and R. S. Schechter, *Improved Oil Recovery by Surfactant and Polymer Flooding*, Academic Press, New York, 1977.

186. V. K. Bansal and D. O. Shah, in *Microemulsions* (L. M. Prince, ed.), Academic Press, New York, 1977, p. 149.

187. J. L. Salager, M. Bourrel, R. S. Schechter, and W. H. Wade, Soc. Pet. Eng. J., *19*:271 (1979).

188. R. L. Cash, J. L. Cayias, G. Fournier, D. J. MaCallister, T. Scharer, R. S. Schechter, and W. H. Wade, J. Colloid Interface Sci., *59*:39 (1977).

189. J. L. Cayias, R. S. Schechter, and W. H. Wade, Soc. Pet. Eng. J., *16*:351 (1976).

190. M. C. Puerto and R. L. Reed, Soc. Pet. Eng. J., *23*:669 (1983).

191. K. S. Chan and D. O. Shah, J. Dispersion Sci. Technol., *1*:55 (1980).

192. E. I. Franses, M. S. Bidner, and L. E. Scriven, in *Micellization, Solubilization and Microemulsions*, Vol. 2 (K. L. Mittal, ed.), Plenum Press, New York, 1977, p. 855.

193. E. I. Franses, J. E. Puig, Y. Talmon, W. G. Miller, L. E. Scriven, and H. T. Davis, J. Phys. Chem., *84*:1547 (1980).

194. Y. Talmon and S. Prager, J. Chem. Phys., *76*:1535 (1982).

195. P. D. Fleming and J. E. Vinatieri, J. Colloid Interface Sci., *81*:319 (1981).

196. J. H. Fendler, *Membrane Mimetic Chemistry*, Wiley, New York, 1982.

197. J. H. Fendler and E. J. Fendler, *Catalysis in Micellar and Macromolecular Systems*, Academic Press, New York, 1975.

198. M. Wong and J. K. Thomas, in *Micellization, Solubilization and Microemulsions*, Vol. 2 (K. L. Mittal, ed.), Plenum Press, New York, 1977, p. 647.

199. S. L. Holt, *Inorganic Reactions in Organized Media*, ACS Symposium Series 177, American Chemical Society, Washington, D. C., 1982.

200. I. V. Berezin, K. Martinek, and A. K. Yatsimirskii, Russ. Chem. Rev., *42*:787 (1973).

201. E. Cordes, *Reaction Kinetics in Micelles*, Plenum Press, New York, 1973.
202. M. Almgren, F. Grieser, and J. K. Thomas, J. Am. Chem. Soc., *102*:3188 (1980).
203. J. H. Fendler, in *Reverse Micelles* (P. L. Luisi and B. E. Straub, eds.), Plenum Press, New York, 1984, p. 305.
204. P. P. Infetta and M. J. Gratsel, J. Chem. Phys., *70*:179 (1979).
205. A. Yetha, M. Aikawa, and N. J. Turro, Chem. Phys., *72*:4358 (1980); *74*:1098 (1981); *74*:5627 (1981).
206. B. H. Robinson, in *Reverse Micelles* (P. L. Luisi and B. E. Straub, eds.), Plenum Press, New York, 1984, p. 73.
207. Y. S. Leong, Ph. D. thesis, Strasburg University, France, 1983.
208. Y. S. Leong, G. Riess, and F. Candau, J. Chem. Phys., *78*:279 (1981).
209. Y. S. Leong and F. Candau, in *Surfactants in Solution*, Vol. 3 (K. L. Mittal and B. Lindman, eds.), Plenum Press, New York, 1984, p. 1897.
210. F. Candau, Y. S. Leong, G. Pouyet, and S. Camdau, J. Colloid Interface Sci., *101*:167 (1984).
211. F. Candau and Y. S. Leong, J. Polym. Sci. Polym. Chem. Ed., *23*:193 (1985).
212. P. L. Johnson and Es. Gulari, J. Polym. Sci. Polym. Chem. Ed., *22*:3967 (1984).
213. A. Jayakrishnan and D. O. Shah, J. Polym. Sci. Polym. Lett. Ed., *22* (1984).
214. J. O. Stoffer and T. Bone, J. Dispersion Sci. Technol., *1*:37 (1980).
215. J. O. Stoffer and T. Bone, J. Polym. Chem. Ed., *18*:264 (1980).
216. L. M. Gan, C. H. Chew, and S. E. Friberg, J. Polym. Sci. Polym. Chem. Ed., *19*:1585 (1983).
217. L. M. Gan, C. H. Chew, and S. E. Friberg, J. Polym. Sci. Polym. Chem. Ed., *21*:513 (1983).
218. L. M. Gan, C. H. Chew, and S. E. Friberg, J. Macromol. Sci. Chem., *19*(5):783 (1983).

219. M. J. Hou and D. O. Shah, paper presented at the Symposium on Microemulsions, AIChE Annual Meeting, Chicago, Nov. 1985.
220. W. V. Smith and P. H. Ewart, J. Chem. Phys., *J-2*:592 (1948).
221. I. Willner, C. Laane, J. W. Otvos, and M. Calvin, in *Inorganic Reactions in Organized Media* (S. L. Holt, ed.), ACS Symposium Series 177, American Chemical Society, Washington, D. C., 1982, p. 71.
222. I. Willner, W. E. Ford, J. W. Otvos, and M. Calvin, Nature, *280*:823 (1979).
223. J. Kiwi and M. Gratzel, J. Phys. Chem., *84*:1503 (1980).
224. R. Hihorst, C. Laane, and C. Veeger, Proc. Natl. Acad. Sci. USA, *79*:3927 (1980).
225. P. L. Luisi, F. Henninger, M. Joppich, A. Dossena, and G. Casonati, Biochem. Biophys. Res. Commun., *74*:1384 (1977).
226. P. L. Luisi and R. Wolf, in *Solution Behavior of Surfactants*, Vol. 2 (K. L. Mittal and E. J. Fendler, eds.), Plenum Press, New York, 1982, p. 887.
227. F. M. Menger and K. Yamada, J. Am. Chem. Soc., *101*:6731 (1979).
228. K. Martinek, A. V. Levashov, N. L. Klyachko, V. I. Pantin, and I. V. Berezin, Biochim. Biophys. Acta, *657*:277 (1981).
229. C. Balny and P. Douzou, Biochimie, *61*:445 (1979).
230. P. D. I. Fletcher, N. M. Perrins, B. H. Robinson, and C. Toprakcioglu, in *Reverse Micelle* (P. L. Luisi and B. E. Straub, eds.), Plenum Press, New York, 1984, p. 69.
231. P. L. Luisi, P. Meier, V. E. Imre, and A. Pande, in *Reverse Micelle* (P. L. Luisi and B. E. Straub, eds.), Plenum Press, New York, 1984, p. 323.
232. P. Meier and P. L. Luisi, J. Solid Phase Biochem., *5*:269 (1980).
233. K. Kon-no, M. Gobe, K. Kandoni, and A. Kitahara, paper presented at the International Conference on Surface and Colloid Science, Jerusalem, 1981.
234. M. Boutonnet, J. Kizling, P. Stenius, and G. Maire, Colloids Surf., *5*:209 (1982).
235. M. J. Hou and D. O. Shah, paper presented at the 60th National Symposium of Colliod and Surface Science, Atlanta, Ga., June 15–18, 1986.

236. M. Dvolaitzky, R. Anthare, X. Auvray, R. Ober, C. Petipas, C. Taupin, and C. Williams, C. R. Acad. Sci. Paris and J. Dispersion Sci. Technol., in press.

237. D. Langevin, in *Reverse Micelles* (P. L. Luisi and B. E. Straub, eds.), Plenum Press, New York, 1984, p. 287.

238. K. L. Mittal and P. Mukerjee, in *Micellization, Solubilization and Microemulsions*, Vol. 1 (K. L. Mittal, ed.), Plenum Press, New York, 1977, p. 1.

239. S. A. Safran, J. Chem. Phys., 78:2073 (1983).

10
Importance of Surfactants and Surface Phenomena on Separating Dilute Oil-Water Emulsions and Dispersions

DONALD R. WOODS and EVAN DIAMADOPOULOS   *McMaster University, Hamilton, Ontario, Canada*

I.	Introduction	370
	A. Why a Dispersion Should Break or Separate	370
	B. What May Keep a Dispersion Stable	383
	C. Role of Surfactants	421
	D. Role of Inorganics	426
	E. Role of Polymers	434
	F. Solubility	436
II.	Characterizing a Dispersion	440
	A. Surface-Interfacial Tension	440
	B. Predicting the Drop Size	440
	C. Predicting the Continuous Phase	461
III.	Overview for Separating	463
IV.	Physical Separation	464
	A. Fundamentals and Equipment Considerations	464
	B. Variations in Design	465
	C. Pretreatment	466
	D. Design Procedures and Results	466
V.	Causing Instability: Chemical Displacement	468
	A. Type of Surfactant	468
	B. Type of Solvent	470
	C. Other Types of Displacing Additives	471
VI.	Causing Instability: Chemical Destabilization-Coagulation and Flocculation	471
	A. Types of Coagulants or Chemical Destabilizers	473

    B. Determining the Chemistry 473
    C. Equipment Selection, Sizing, and Results 477

VII. Removal by Air Attachment Without and with Destabilization 481

    A. DAF Without Upstream Addition of Destabilizers 489
    B. DAF With Upstream Addition of Chemical Destabilizers 494
    C. Equipment Selection and Sizing 497

VIII. Removal by Filtration 498

    A. Fundamental Principles and Characteristics 498
    B. Characteristics of Deep-Bed Filtration 508
    C. Performance Without the Addition of Destabilizing Chemicals 508
    D. Performance with the Addition of Destabilizing Chemicals 521
    E. Regeneration of the Filter 524
    F. General Selection/Sizing Procedure 525

IX. Summary 529

References 529

I. INTRODUCTION

A dispersion of oil drops in water should inherently separate into two phases. But often they do not; often they form a stable dispersion. A dispersion of water drops in oil should likewise inherently separate. Sometimes they do; sometimes they are stable. A study of the microscale action that occurs when separation occurs reveals that (1) the drops migrate or contact each other or contact some "collecting" device and (2) the drops may coalesce or they may stick together or stick to the collecting device. Coalescence is illustrated in Fig. 1. In this review we discuss the background fundamentals, the characteristics of dispersions and emulsions, and the design principles—with examples—for some devices to separate dilute oil/water dispersions.

The important fundamentals include why a dispersion should be unstable and the factors that could keep a dispersion stable. Both equilibrium and rate phenomena can be important. From this, eleven different mechanisms can be identified. Surfactants, soluble and insoluble inorganics, and polymers all play key roles. Finally, the mutual solubilities of the two phases should be accounted for. Consider each of these fundamental ideas in turn.

A. Why a Dispersion Should Break or Separate

A dispersion creates much surface area. Figure 2 shows the amount of area created per unit volume for different concentrations of dispersions of different diameters. Since thermodynamics predicts that any

Dilute Oil-Water Emulsions and Dispersions 371

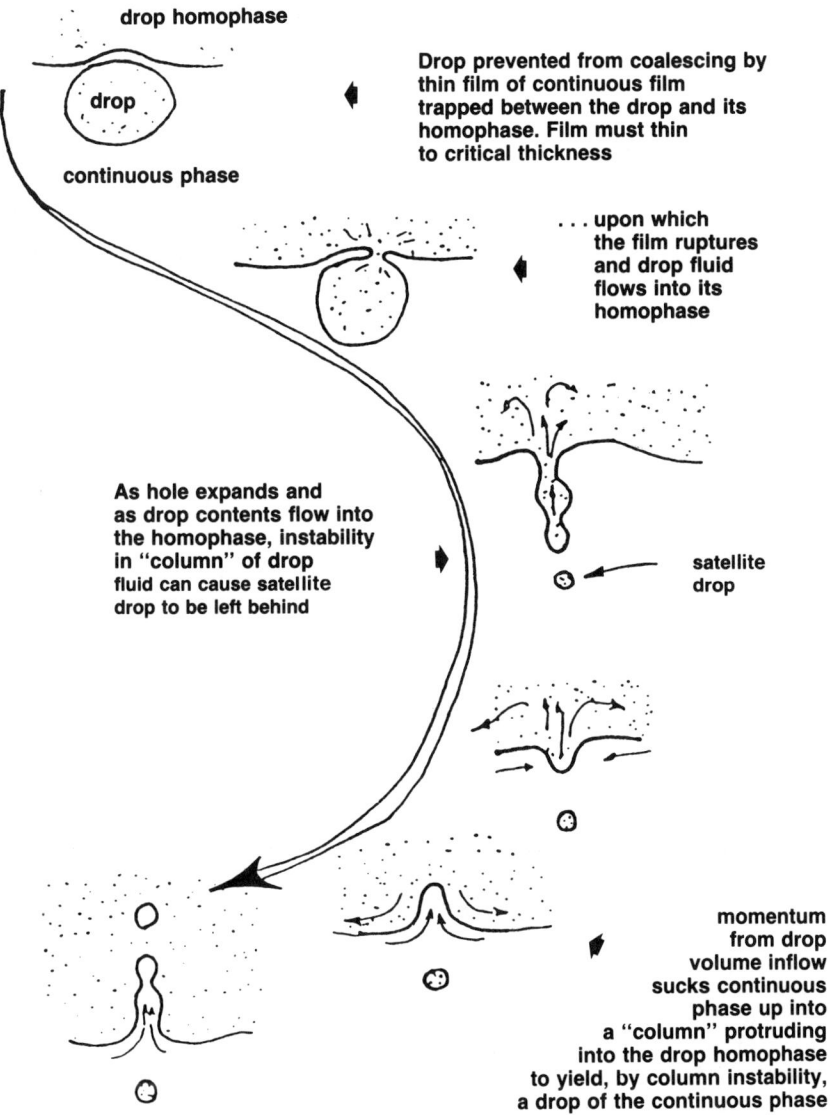

FIG. 1 Coalescence in action.

system will move such as to minimize the amount of surface area, the small drops should coalesce to form larger drops and eventually leave just the interface between the oil layer and the water layer.

What forces pull two surfaces together? On the microscale, as two surfaces, say two oil drops, approach one another, van der Waals

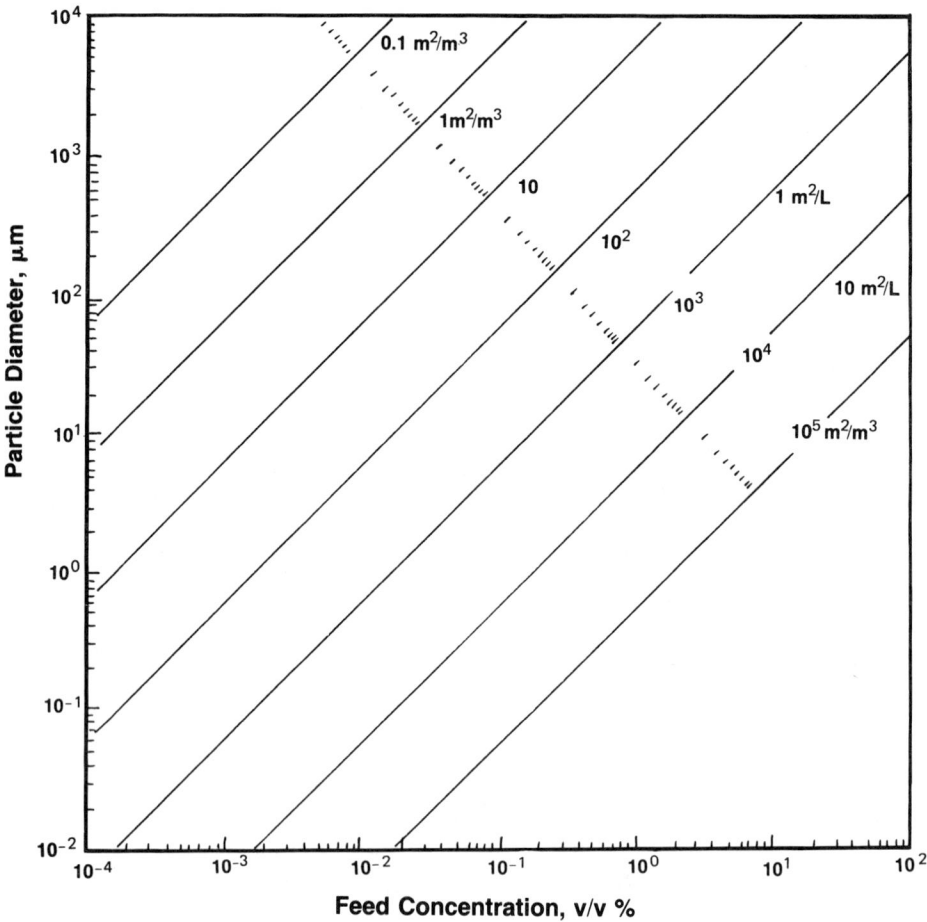

FIG. 2 Estimate of the surface area per unit volume for dispersions.

forces of attraction draw the surfaces together. Such forces act over distances of separation of about 100 nm; the closer the surfaces are to each other, the stronger is the force. Usually, we express this attraction as an energy of interaction U. The force F is the derivative of the energy with respect to the separation distance.

Several theories have been developed to estimate this attractive interaction energy (Lifshitz, 1955; Ninham and Parsegian, 1970; Mahanty and Ninham, 1976; Parsegian, 1975; Hamaker, 1936, 1937). For this chapter the Hamaker approach will suffice. For two spheres of radius R approaching each other, the interaction energy is

Dilute Oil-Water Emulsions and Dispersions

$$U = -\frac{A}{12}\left(\frac{1}{R^{+2}-1} + \frac{1}{R^{+2}} + 2\ln\frac{R^{+2}-1}{R^{+2}}\right) \quad [=]\,J \quad (1)$$

where A is the Hamaker "constant" and where

R^+ = dimensionless distance

$$= \frac{2R+h}{2R}$$

$$= 1 + \frac{1}{2}h^+$$

h^+ = dimensionless distance of separation

$$= \frac{h}{R}$$

For close separations this is approximately

$$U = -\frac{A}{12}\frac{1}{h^+} \quad [=]\,J \quad (2)$$

Analogous equations are available for spheres of different radius.

For a drop approaching an infinitely flat surface, the relationship is

$$U = -\frac{A}{12}\left(\frac{4}{h^+} + \frac{4}{h^++4} + 2\ln\frac{h^+}{h^++4}\right) \quad [=]\,J \quad (3)$$

Values of A depend on the materials making up the drops and the material in the intervening space between the drops. The rest of the equation depends on the geometry; thus Eq. (1) differs from Eq. (3) because of the difference in the geometry. Table 1 summarizes values for different material combinations and as calculated from Hamaker's approach. Thus for c_{12} alkane drops approaching each other in water, the Hamaker constant is about 5 βJ = 5 x 10^{-21} J. In general, tabular data report vacuum, air, or water in between the approaching surfaces. Methods are available to estimate the effect of different media between the two approaching surfaces (see Gregory, 1969; Hough and White, 1980; Verwey and Overbeek, 1948; Bargeman and van Voorst Vader, 1972. As an approximation, for material 1 approaching material 1 with material 3 in the gap, the appropriate Hamaker parameter would be

TABLE 1 Some Values for Hamaker's Parameter, 10^{-21} J = βJ

Air	4.1 x 10^{-5}	Carbon disulfide	
Alkane, C		C/vac/C	65
A/vac/A		n-Decane	
		D/vac/D	4.7, 5.6
A/air/A		D/H_2O/D	0.9
C5/air/C5	37.5	Diethyl ether	
C6/air/C6	40.7	D/vac/D	42
C7/air/C7	43.2		
C8/air/C8	45.0	Dioxan	
C9/air/C9	46.6	D/vac/D	59
C10/air/C10	48.2	Dodecane	
C11/air/C11	48.8	D/air/D	9.49
C12/air/C12	50.4		
C13/air/C13	50.5	Carbon Tetrachloride	
C14/air/C14	51.0	C/vac/C	63
C15/air/C15	51.6	Chloroform	
C16/air/C16	52.3	C/vac/C	60
A/H_2O/A		Cyclohexane	
C5/H_2O/C5	3.36	C/vac/C	59
C6/H_2O/C6	3.60	Eisosane	
C7/H_2O/C7	3.86	E/air/E	20.7
C8/H_2O/C8	4.10	Ethyl acetate	
		E/vac/E	42
C9/H_2O/C9	4.35	Ethylene glycol	
C10/H_2O/C10	4.62	E/vac/E	62
C11/H_2O/C11	4.71	Glycerol	
C12/H_2O/C12	5.02	G/vac/G	71
C13/H_2O/C13	5.04	Graphite	
C14/H_2O/C14	5.14	G/H_2O/G	37
C15/H_2O/C15	5.26	Heptane	
		H/air/H	10.5
C16/H_2O/C16	5.40	Hexane	
Benzene		H/air/H	59.7
B/vac/B	57	Methyl ethyl ketone	
Carbon		M/vac/M	47
C/vac/C	990	Nitrobenzene	
C/H_2O/C	600	N/vac/N	57

TABLE 1 (Continued)

Octane			Stearic acid	
O/air/O	60.3		S/H$_2$O/S	0.79–3.68
Paraffin			Toluene	
P/H$_2$O/P	15		T/vac/T	61
Paraffin oil			Water	
P/H$_2$O/P	43.8		H$_2$O/vac/H$_2$O	43.6
Silica			H$_2$O/air/H$_2$O	37
Si/vac/Si	248			
Si/air/Si	41		H$_2$O/HC/H$_2$O	5.6
Si/H$_2$O/Si				

$$A_{11(3)} = [(A_{11})^{1/2} - (A_{33})^{1/2}]^2 \qquad (4)$$

Methods are also available to estimate the effect of different materials on the surface: for example, a surfactant "layer" on the surface of an oil drop (see Vold, 1961; Vincent, 1973; Osmond et al., 1973; Mahanty and Ninham, 1976).

Sometimes we prefer to consider a dimensionless interaction energy U^+ by dividing the energy by a reference energy that is related to the Brownian motion within a fluid system. The term kT is such a reference state, where k is Boltzmann's constant and T is the absolute temperature. For room temperature kT = 4.11 βJ. The dimensionless energy is thus

$$U^+ = \frac{U}{kT} \qquad (5)$$

Example 1 Calculate the dispersion interaction between two spheres of diameter 100 nm separated by distances of 0 to 10 nm if the Hamaker parameter is 100 βJ and the temperature is 25°C.

Solution: The result is actually independent of the diameter of the particles as long as the separation distance, h, is small relative to the particle diameter. As an approximation, assume (with $h^+ = h/R$) that

$$U_\infty = -\frac{A}{12} \frac{1}{h^+}$$

$$U^+ = -\frac{A}{12}\frac{1}{h^+} \times \frac{1}{kT}$$

$$= -\frac{100}{12h^+}\beta J \times \frac{1}{4.11\ \beta J}$$

$$U^+ = -\frac{2.034}{h^+}$$

For sphere-sphere interaction, the result is dimensionless and no area term is needed. The results are tabulated as follows:

h (nm)	h⁺	U⁺
0.5	0.01	−203
2.5	0.05	−40.5
5	0.10	−20.3
25	0.5	−4.05
50	1	−2.03

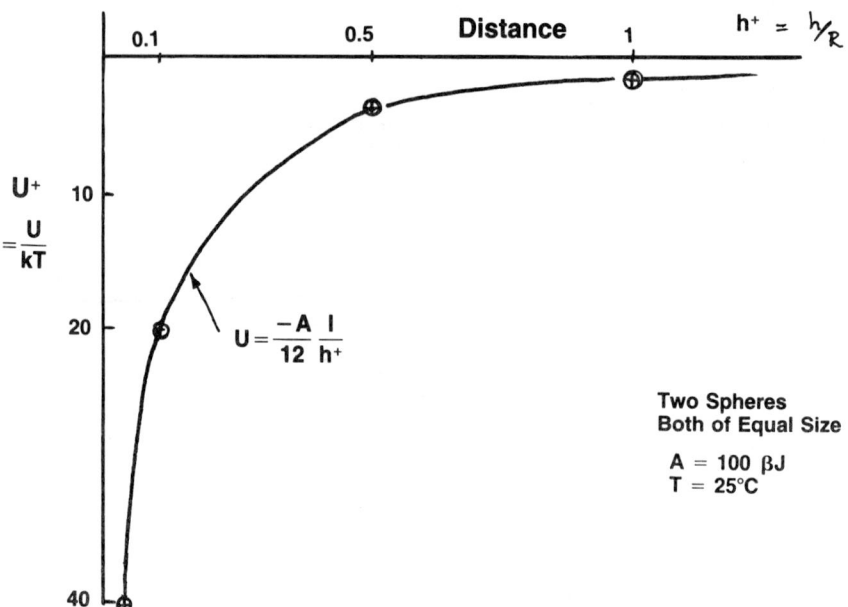

FIG. 3 Attraction for sphere-sphere interaction.

The results are plotted in Fig. 3. We note, however, that our assumption that $h^+ \ll 1$ is not valid for the larger separation distances. Hence we recalculate the results using the more rigorous equation.

Example 2 Recalculate Example 1 with the more rigorous equation

$$U = -\frac{A}{12}\left[\frac{1}{R^{+2}-1} + \frac{1}{R^{+2}} + 2\ln\frac{R^{+2}-1}{R^{+2}}\right]$$

where

$$R^+ = 1 + \frac{h^+}{2}$$

Solution: We will calculate U^+ by dividing U by kT or $A/12kT = 100 \ \beta J / 4.11 \ \beta J \times 12 = 2.03$.

h^+	R^+	$R^{+2}-1$	R^{+-2}	2 ln	U^+
0.01	1.005	100	0.99	−9.23	−186.3
0.05	1.025	19.61	0.99	−6.05	−29.45
0.10	1.05	10	0.91	−4.80	−9.87
0.5	1.25	1.79	0.64	−2.05	−0.77
1.0	1.5	0.80	0.44	−1.18	−0.12

A comparison of these results with those calculated in Example 1 shows a significant difference. The data are plotted in Fig. 4.

Example 3 What is the force of attraction between two 100-μm-diameter drops of chlorobenzene in water if there is a 1-nm-thick layer of adsorbed surfactant poly(oxyethylene hexadecyl ether)? The distance of separation is 50 nm. As an approximation, neglect the surfactant layer.

Solution: To illustrate the effects, we solve the problem first by neglecting the adsorbed surfactant. We will assume that Hamaker's approach is accurate enough. For two spheres, from Eq. (2), the equation is

$$U = -\frac{A}{12}\frac{1}{h^+}$$

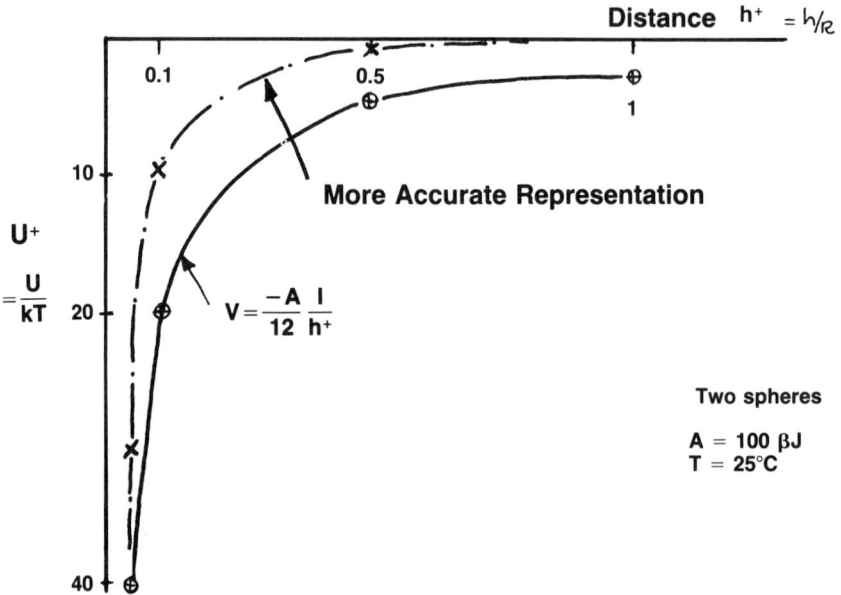

FIG. 4 Comparison for sphere-sphere.

where

$$h^+ = \frac{h}{R} = \frac{50 \times 10^{-9} m}{50 \times 10^{-6} m} = 10^{-3}$$

As an approximation, A_{113} is

$$A_{11(3)} = \left((A_{11})^{1/2} - (A_{33})^{1/2}\right)^2$$

From Table 1, where all data refer to vacuum in the gap,

$A_{11} = 60 \ \beta J$ (chlorobenzene)

$A_{33} = 38 \ \beta J$ (water)

$A_{11(3)} = [(60 \times 10^{-21})^{1/2} - (38 \times 10^{-21})^{1/2}]^2$

$$= [(2.45 - 1.95) \times 10^{-10}]^2 \; \beta J$$

$$= 2.5 \times 10^{-21} J$$

$$U = -\frac{2.5 \times 10^{-21}}{12 \times 10^{-3}} J$$

$$= -0.21 \times 10^{-18} J \text{ for the system}$$

The total area for the system is

$$2\pi D^2 = 2\pi(100 \times 10^{-6})^2 m^2$$

$$= 2\pi \times 10^{-8} m^2$$

Hence

$$U = \frac{-0.21}{2 \times \pi \times 10^{-8}} \times 10^{-8} J$$

$$= -3.34 \times \frac{10^{-2} \times 10^{-18}}{2 \times \pi \times 10^{-8}}$$

$$= -3.34 \times 10^{-12}$$

$$= -3.34 \times \frac{pJ}{m^2}$$

The force is the first derivative of the energy with respect to the separation distance. Thus

$$F = \frac{\partial U}{\partial h} = \frac{\partial}{\partial h}\left(\frac{-A}{12}\frac{R}{h}\right)$$

$$= \frac{+A}{12}\frac{R}{h^2}$$

$$= \frac{A}{12}\left(\frac{R}{h}\right)\frac{1}{h}$$

$$= \frac{U}{h}$$

$$= \frac{3.34 \text{pJ}}{\text{m}^2} \times \frac{1}{50 \times 10^{-9} \text{m}}$$

$$= 6.68 \times \frac{10^{-19} \text{N}}{\text{m}^2}$$

The area in this system is $2\pi \times 10^{-8}$ m^2. Hence the force is

$$6.68 \times \frac{10^{-19} \text{N}}{\text{m}^2} \times 2\pi \times 10^{-8} \text{m}^2 = 42 \times 10^{-27} \text{N}$$

There are some special cases when the van der Waals forces of "attraction" that two surfaces experience as they approach each other actually yields a repulsion (or a positive interaction energy). For this to occur the Hamaker's parameter for the intervening material would have a numerical value in between the Hamaker value for the two approaching surfaces. This is shown in Fig. 5. An example of *negative* overall Hamaker's parameters is crystalline quartz approaching an air surface with water as the intervening material: $A_{123} = -18.3$ βJ, or a C_{10} alkane approaching an air surface with water as the intervening material: $A_{123} = -0.034$ βJ (Hough and White, 1980).

Another possibility is that a dispersion may retain its large drops, yet separate into a band of drops if the gravity settling is fast enough. Figure 6 shows the settling (or rising) velocity of oil droplets in water at 25°C. In general, drops behave like solid spheres because of the buildup of surfactants at the oil/water interface, and the movement is characterized by Stokes' law. If, however, the surfactant buildup is negligible, then the surfaces can move

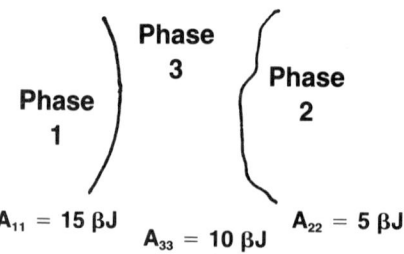

FIG. 5 Example conditions for a negative Hamaker parameter.

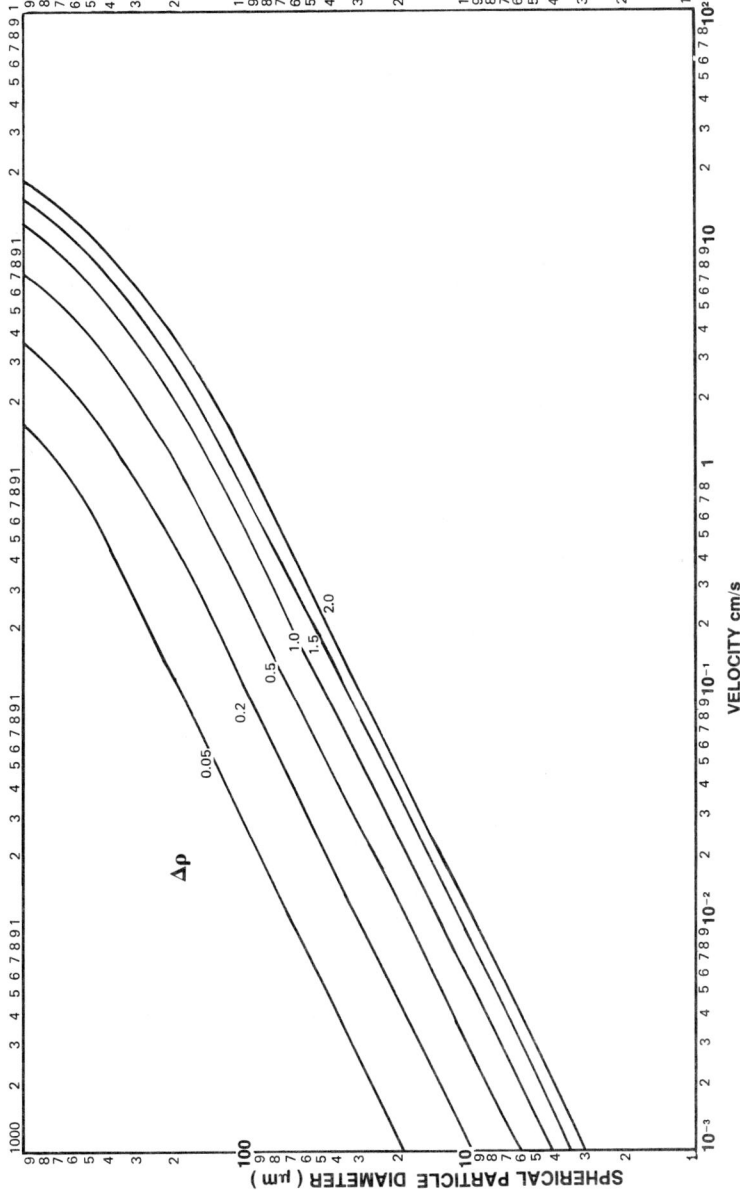

FIG. 6 Terminal velocity of noncirculating spheres in water at 25°C.

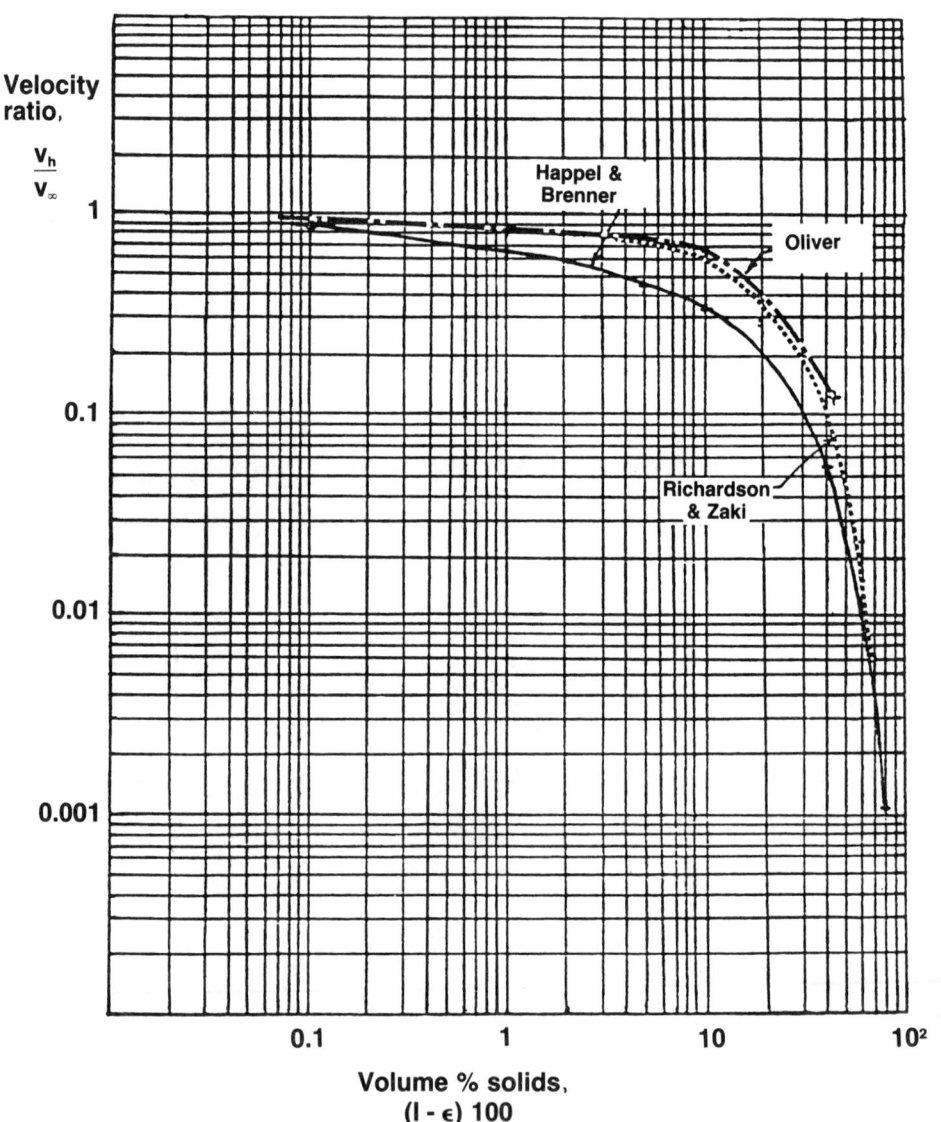

FIG. 7 Corrections because of hindered settling.

so that the drops move faster, as predicted by Hadamard-Rybczynski. If the concentration of drops is large, the drops no longer behave as individual drops in an infinite fluid; the migration velocity decreases as given by correlations such as given in Fig. 7. Thus, for a fast

Dilute Oil-Water Emulsions and Dispersions

migration velocity, the drops should be large. Sometimes, the clusters of drops that make up a "large" drop break up because the van der Waals forces holding them together are overcome by the fluid forces. To prevent this and to strengthen the cluster, polymers can be added. The idea is that the polymer adsorbs across the drops like a bridge. The addition of polymer to strengthen the floc is called flocculation.

In summary, dispersions of drops should separate to minimize the total energy in the system and because when surfaces approach each other, van der Waals usually causes them to be attracted together. We can estimate these values.

B. What May Keep a Dispersion Stable

Three considerations are the effects of equilibrium, rate, and fluid dynamics. (1) Under *equilibrium* conditions, repulsive forces between the two surfaces may counteract the attractive forces discussed in Section II.A; (2) under *rate* considerations, insufficient time is allowed for the intervening fluid to thin down to distances where attractive forces become significant; and (3) the fluid dynamics may create *shear* fields that break the surfaces or tear the drops apart just after we have gotten them together.

1. Equilibrium Considerations

Two major sources of repulsion are surface charge effects (or the electrochemical double layer) and adsorbed polymers (or the steric stabilization).

   a. Electrochemical Double Layer Usually all surfaces have a local charge. It is only when we encounter systems where we have a large surface area-to-volume ratio that these charge effects become important. When we create a surface out of an *overall neutral* system, usually in the surface region, ions adsorb and distribute themselves so that there is a charge in the surface layer and an equal and opposite countercharge in the surrounding region. Thus overall neutrality is maintained, but locally there are positive and negative regions. The ions that adsorb on the surface and "cause" the charged surface in the first place we call the potential determining ions (PDI). The ions that make up the countercharge we usually call the indifferent electrolyte—indifferent because it does not cause the charge initially. However, we cannot isolate and separate these two types of ions because they affect each other. The usual causes for a charge are as follows:

1. A sparingly ionic crystal is in equilibrium with a liquid (e.g., silver iodide, silver chloride, metal oxides, and hydroxides).

Whenever a precipitate forms in the presence of an excess of one ion, the surface charge will be that of the excess ion. The PDI would be that excess ion. For silver iodide, the PDI is either silver or iodide ion.
2. Ions or ionized surfactants adsorb to the surface to produce a net charge. For silica in water, the silanol group on the surface can change from a positive to a negative charge depending on the pH. Thus hydrogen or OH ions are the PDI.
3. For solids, the charge could be caused be internal defects, lattice substitutions, or ionizable end groups on polymer latex surfaces.

Thus our task is to try to identify which ions are the PDI and which serve as the indifferent electrolyte, and to model the charge distribution in the countercharge region. In general, the PDI are H^+, OH^-, and surfactants. Usually, the indifferent electrolytes are salts or inorganic ions. More ideas about how to identify these and their implications are given later in this section.

Consider the modeling of the electrochemical double layer or more specifically, the modeling of the countercharge region. In its simplest form, we visualize it as being made up of two parts: a Stern layer or "fixed" countercharge right next to the surface and a diffuse region sometimes called the Gouy-Chapman region. The sum of the charge accounted for in these two regions must equal the charge at the surface. The models used for these are: The charge in the Stern layer, σ_s, is

$$\sigma_s = e\Omega \sum_{i=1}^{I} \frac{z_i}{1 + 1/x_i^I \exp[(z_i e \phi_s - \phi_{pw})/kT]} \quad (6)$$

where the charge in the diffuse layer is given by

$$\sigma_{DL} = \left(2\varepsilon_r \varepsilon_0 kTn \bigg|_{\phi^+ = 0}\right)^{1/2} 2 \sinh \frac{z\phi_\delta^+}{2} \quad (7)$$

Although we model the system based on a charge balance, we often like to visualize the behavior by considering how the inner or Galvani potential varies in the region of the double layer. Figure 8 shows the ideal aystem with an oil phase contacting a water phase. The Galvani potential inside each bulk phase represents the work done to build up the ionic and chemical constituents of that phase by taking them from an initial, zero-energy state of infinity and a vacuum. At the boundary between the two phases we have an in-

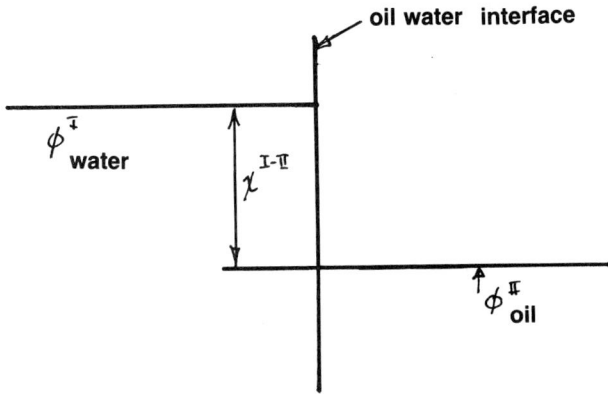

FIG. 8 Two-phase system with no charge at the surface.

terfacial potential of χ. The system shown in Figure 8 *does not* have a charge at the interface. The absolute values of the χ and Galvani potentials are difficult to measure; fortunately, it is the relative values of the Galvani potential that are important to us. Now, when the interface becomes charged, the countercharge resides in either or both of the bounding phases. We can show the charge by realizing that the potential is related to the charge by the Poisson equation

$$\nabla^2 \phi = - \frac{\sigma^I}{\varepsilon_r \varepsilon_0} \tag{8}$$

where ε_0 is the permittivity in free space (=8.85 × 10^{-12} C^2/J·m). This equation is interesting because the mathematical form of the equation shows that the curvature of the potential-distance plot suggests the sign of the charge, and the slope of the line at the interface represents the magnitude of the total charge in the region. The latter implication is based on the integration of Eq. (8) subject to the boundary conditions that at the surface we know the value of the potential and at infinity the slope of the potential decay should be zero. Hence

$$\sigma^I = \varepsilon_r \varepsilon_0 \left. \frac{\partial \phi}{\partial y} \right|_{y=0} \tag{9}$$

This says that the *magnitude* of the charge in the bulk region is related to the slope of the potential at the charged surface. This helps us to visualize and draw graphs of the potential variation in the surface region.

Figures 9 to 11 show some possible conditions that might result near a surface. In Fig. 9 a negatively charged species is in the surface, and a positive countercharge region exists in both the water and oil phases. Figure 10 shows a positively charged adsorbed species at the surface. Figure 11 shows the situation where there is no charge at the surface but there are negative ions that prefer the oil phase and an equal and opposite countercharge in the water phase. At first glance for the situation in Fig. 11 we might think that the slopes of the potential curves at the surface must be equal to show the overall neutrality as suggested by Eq. (9). However, it is the following condition that needs to be satisfied:

$$\epsilon_r^I \left.\frac{\partial \Phi^I}{\partial y}\right|_{y=0} = \epsilon_r^I \left.\frac{\partial \Phi^{II}}{\partial y}\right|_{y=0} \tag{10}$$

For an example oil-water system, with the relative permittivity of water being eight times that of the oil and the ionic concentration in water being 81 times that in the oil, Verwey and Niessen (1939) show the potential curves as in Fig. 11. Thus the charged layer appears to be only in the oil phase.

Now that we have explored how to visualize the variation in potential, let us recast Eqs. (6) and (7) into expressions for how the potential varies in the countercharge region. In the Stern layer

$$\Phi_0 - \Phi_\delta = \frac{\sigma_\delta \delta}{\epsilon_r \epsilon_0} \tag{11}$$

where Φ_0 is the inner or Galvani potential at the surface at $y = 0$, Φ_δ the inner or Galvani potential at the outside of the Stern layer

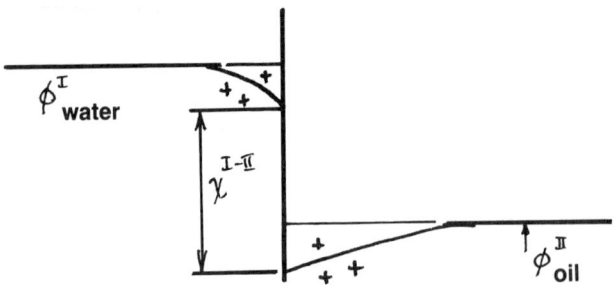

FIG. 9 Two-phase system with negative charge at the surface.

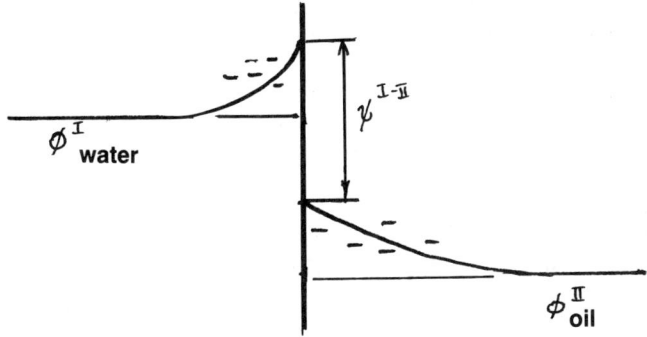

FIG. 10 Two-phase system with positive charge at the surface.

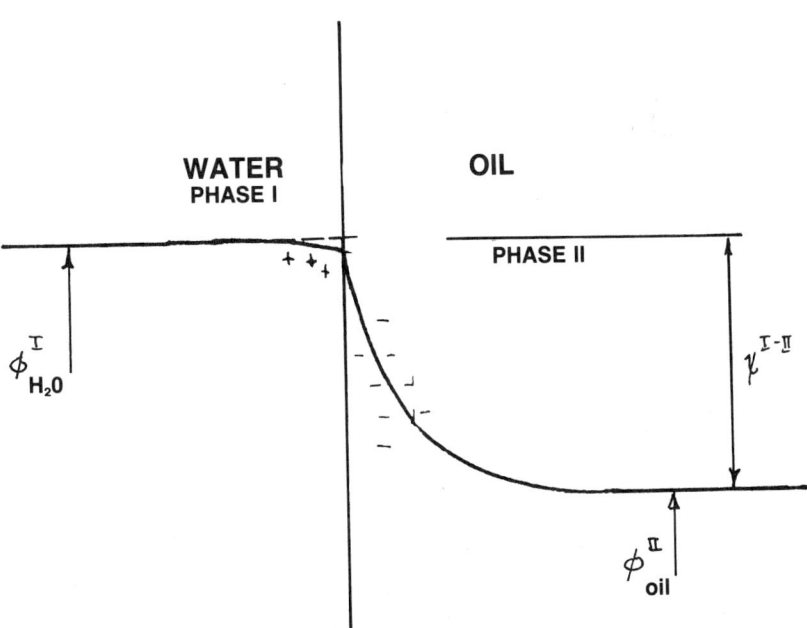

FIG. 11 Two-phase system with no charge at the surface but with ionic distribution between the two phases. [Sample distribution of potential ϕ_{DL} between two phases $\varepsilon_r^I = 9\varepsilon_r^{II}$ and $n_i|_{\phi=0}^I = 81\, n_i|_{\phi=0}^{II}$. From Verwey and Niessen (1939) with permission.]

at $y = \delta$, δ the thickness of the Stern layer, σ_δ the charge in the Stern layer, ε_r the relative permittivity, and ε_0 the permittivity of free space.

This is illustrated in Fig. 12 and assumes that the potential drops linearly across the Stern layer and that we can estimate the thickness of the layer and the relative permittivity in the layer. Usually, the layer is one molecule thick, so 0.5 nm is a reasonable thickness; usually, the relative permittivity in the layer is about 5 to 10. In the diffuse layer, the result is a little more difficult to see in terms of potentials. The result is illustrated in Fig. 13. The formulation is

$$y^+ = \ln \frac{\tanh z\Phi_\delta^+/4}{\tanh z\Phi^+/4} \tag{12}$$

where

$$\Phi_\delta^+ = e\Phi_\delta/kT \tag{13}$$

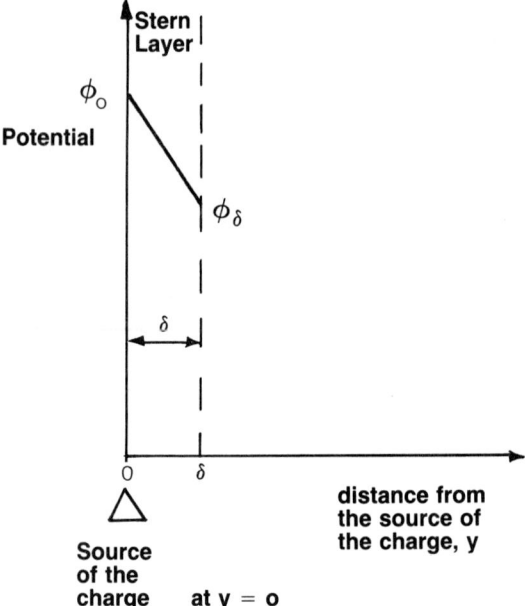

FIG. 12 Model of the Stern layer.

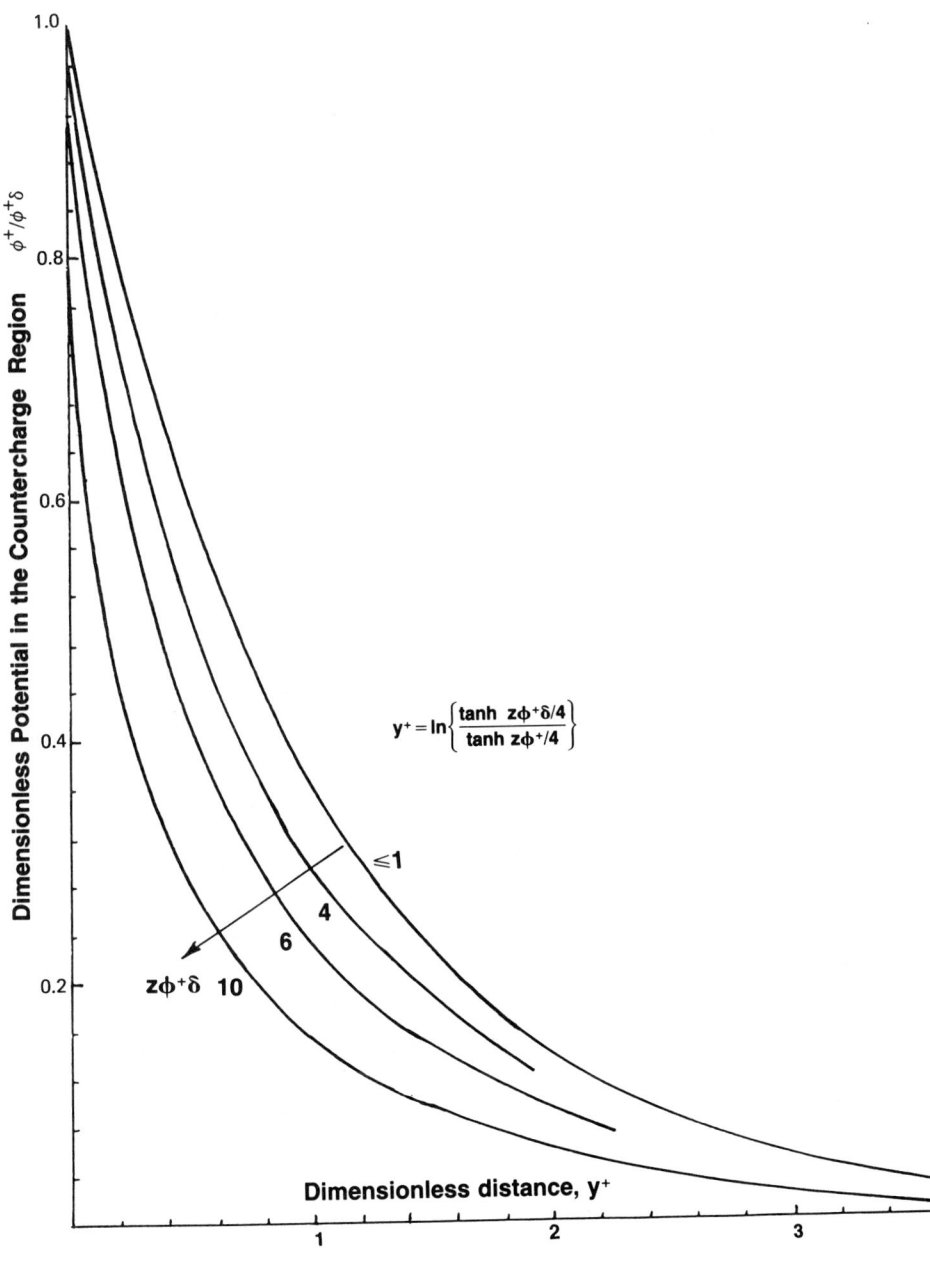

FIG. 13 Distribution of the potential in the diffuse layer.

is the dimensionless potential at the boundary between the diffuse and Stern layer, Φ^+ the dimensionless potential in the diffuse layer at a distance y^+, z the valence, and y^+ the dimensionless distance from the Stern layer. Since y is a measure of the distance from the surface and since the Stern layer's thickness is δ, the dimensional distance is $(y - \delta)$. To make it dimensionless, we use the "thickness" of the diffuse double layer as a reference length, K^{-1}. Thus the dimensionless distance is $K(y - \delta)$. To account for the valence effects we include a dimensionless valence in the definition of y^+.

$$y^+ = \frac{K(y - \delta)}{z^+} \tag{14}$$

z^+ is the dimensionless valence

$$z^+ = \left[\frac{(z_+) + (z_-)}{z|z|} \right]^{1/2} \tag{15}$$

The following relationships are sometimes useful:

$$\sinh x = \frac{e^x - e^{-x}}{2} \qquad \cosh x = \frac{e^x + e^{-x}}{2} \qquad \tanh x = \frac{e^x - e^{-x}}{e^x + e^{-x}}$$

$$\sinh^{-1} x = \ln\left(x + \sqrt{x^2 + 1}\right) \qquad \tanh^{-1} x = \frac{1}{2} \ln \frac{1 + x}{1 - x}$$

For small potentials (i.e., $4 > z\Phi^+_\delta > 1$ and $z\phi_\delta^+ \ll 0.5$, respectively) the following approximations can be made:

$$y^+ = \ln \frac{\tanh z\Phi_\delta^+/4}{z\Phi_\delta^+/4} \tag{16}$$

and the Debye-Hückel approximation,

$$y^+ = \ln \frac{\Phi_\delta^+}{\Phi^+} \tag{17}$$

or

$$\frac{K(y - \delta)}{z^+} = \ln \frac{\Phi_\delta}{\Phi} \tag{18}$$

Dilute Oil-Water Emulsions and Dispersions

So far we have considered only the charge effects around a single interface, whereas what we want to know is what happens when two surfaces approach each other. Nevertheless, some understanding of some of the interesting effects of the double layer can be appreciated by looking at the single surface first. Key things to note are as follows:

1. The thickness of the double layer, as measured by the Debye-Hückel parameter K, is a function of the concentration and valence of the counterions:

$$K^2 = \frac{e^2}{\varepsilon_r \varepsilon_0 kT} \sum_i z_i^2 n_i \bigg|_{\phi^+ = \phi^I} \quad [=] \frac{1}{\text{length}^2} \quad (19)$$

where e is the unit charge on an electron, ε_r the relative permittivity of fluid in the continuous phase (for water $\varepsilon_r = 78$), ε_0 the permittivity of free space, k is Boltzmann's constant, z_i is the valence of species i, and n_i the number of ions of species i per unit volume. For water at 298K and for a single counterion species,

$$K = 2.28 \, z_i \sqrt{c_i} \quad [=] \frac{1}{\text{nm}} \quad (20)$$

where c_i is the concentration of indifferent electrolyte (mol/L).

Example 4 Estimate the thickness of the double layer at 298K in water containing 10^{-4} mol of $BaCl_2$ per liter.

Solution: We cannot start the calculation unless we know whether the surface is charged positively or negatively because we need to know which ion is the counterion. If the surface is negatively charged (e.g., oil drops at pH 8), then Ba^{2+} is the counterion. Hence

$$K = 2.28 \times 2 \sqrt{10^{-4}}$$

$$= 4.56 \times 10^{-2} \quad [=] \frac{1}{\text{nm}}$$

Hence the "thickness" is

$$\frac{1}{4.56 \times 10^{-2}} \text{ or } 22 \text{ nm}$$

The physical significance of the thickness is that at a distance of 22 nm from the Stern layer, the potential has dropped to 1/e or 0.37 times its value at the Stern layer.

Example 5 Reestimate the thickness in Example 4 if the concentration had been 10^{-4} mol of NaCl per liter.

Solution: The thickness for a monovalent counterion would be

$$K = 2.28 \times 1 \times \sqrt{10^{-4}}$$

$$= 43 \text{ nm}$$

How the concentration and valence of the counterion affects the changes in potential in the diffuse double layer are illustrated in Fig. 14. Thus the use of high-valence, high-concentration counterions will "compress" the double layer and hence the effects of charged surfaces will not extend far beyond the surface. For oil phases, where the ionic concentration may be small, the double layer may be extended far into the oil phase from the surface.

2. The potential at the surface is difficult to probe; indeed, even the potential at the outer surface of the Stern layer is hard to

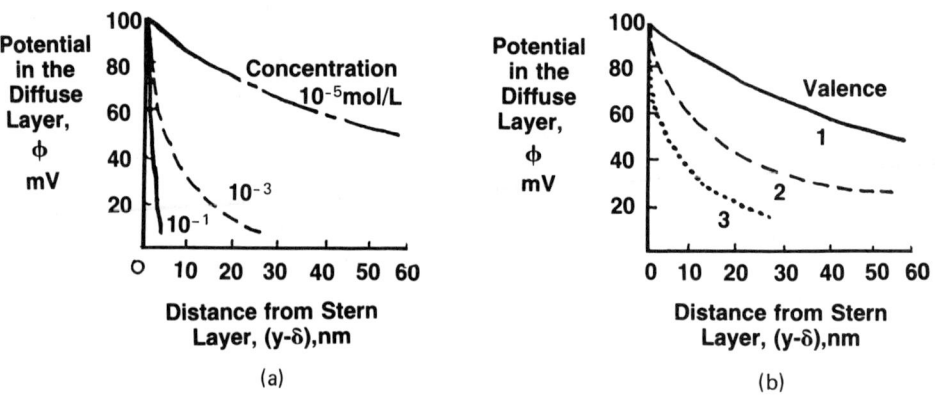

FIG. 14 Effect of concentration and valence on the potential in the diffuse double layer. (a) Effect of concentration of indifferent electrolyte. (b) Effect of valence of the counter-ion.

evaluate. One way that we have of evaluating the charge potential is to put a drop in a dc electrical field and see if the drop moves. Thus, if the drop had a positive surface charge, it would move toward the negative electrode. The higher the charge, the faster the movement. But what potential or charge does such behavior actually show? The answer is: the potential at the fluid dynamic slip plane just where the drop starts to move relative to the continuous phase. This slip plane is usually beyond the thickness of the Stern layer and is "somewhere" in the diffuse layer. This is illustrated in Fig. 15. The potential at the slip plane is called the zeta (ζ) potential and is usually in the range 0 to 100 mV. When the diffuse layer becomes more compressed, the potential at the slip plane will change (although the change is relatively minor). Some data to illustrate this point are given in Fig. 16. This shows that for the titanium dioxide/water system at about the same pH (with H^+ being the PDI) the slip plane shifts from 2.5 nm beyond the Stern layer to 1.2 nm beyond the Stern layer as the concentration of indifferent electrolyte changes from 0.0004 M to 0.01 M. The potential at the slip plane, the zeta potential, changes from about 115 mV to 65 mV for this change in indifferent electrolyte, even though the potential at the Stern layer has remained at 150 to 170 mV.

Figure 17 shows the results of zeta-potential measurements for various oil water systems. Before we consider the details of Fig. 17, first, the actual calculation of the potential is a challenge because the conversion between the drops movement or mobility (in the dc field) and the potential depends on the diameter and shape of the drop, the conductance of the liquid in the drop, and the thickness of the diffuse layer relative to the diameter.

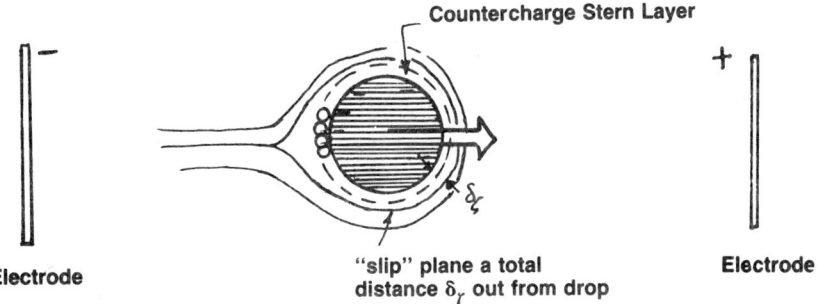

FIG. 15 Illustration of the slip plane.

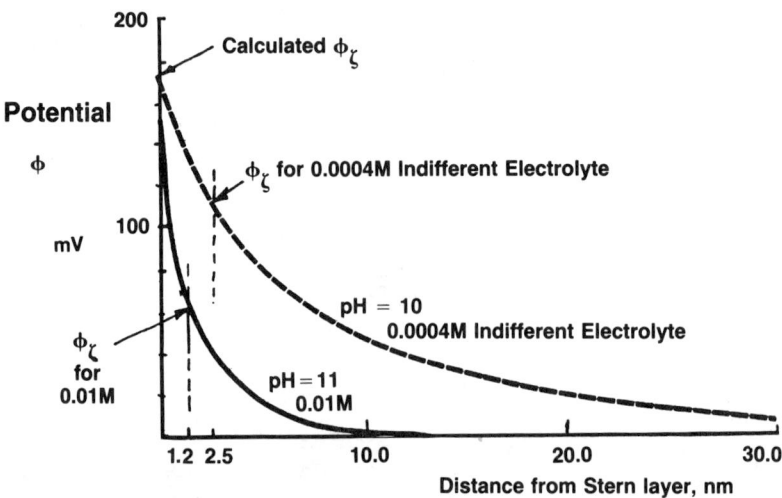

FIG. 16 Estimations of the slip plane location and potential for TiO_2/water with different concentrations of indifferent electrolyte.

For a sphere of a dielectric fluid, the zeta potential is

$$\phi_\zeta = \frac{v}{E} \frac{\mu}{\varepsilon_r \varepsilon_0} \left[1 - \frac{\sigma_{DL}^2}{kKr\mu} \right] f \qquad (21)$$

where

v/E = mobility
μ = shear viscosity of the continuous phase
σ_{DL} = charge in the double layer
k = electrical conductivity of the continuous phase
$1/K$ = thickness of the diffuse double layer
e/kT = 38.96 V at 298°C
k = Boltzmann constant (1.38×10^{-23} J/°C)
v = velocity
E = applied voltage per unit distance
z = valence of the counterions
e = unit charge on an electron
r = drop or particle radius
f = correction factor to account for shape, small diameter

For many applications, we are interested in the relative zeta potentials, especially the location where the zeta potential is zero [the

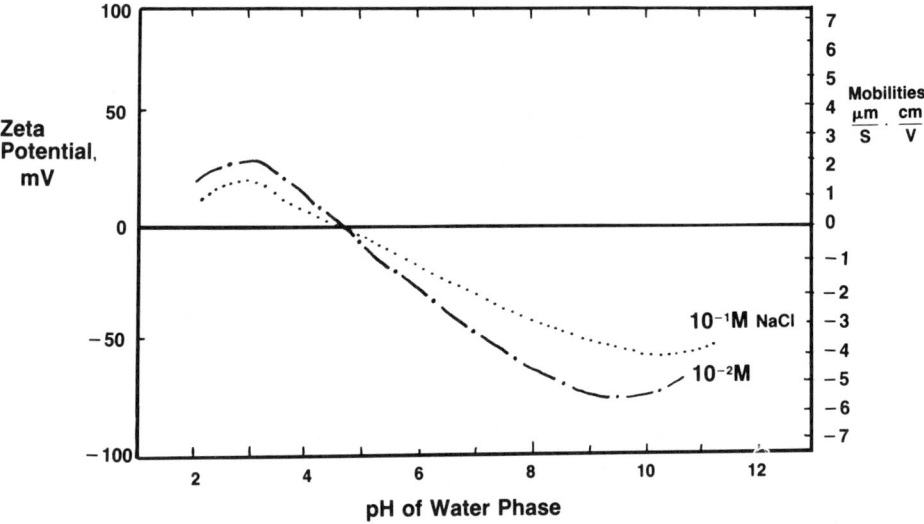

FIG. 17a API separator waste float oil/distilled water.

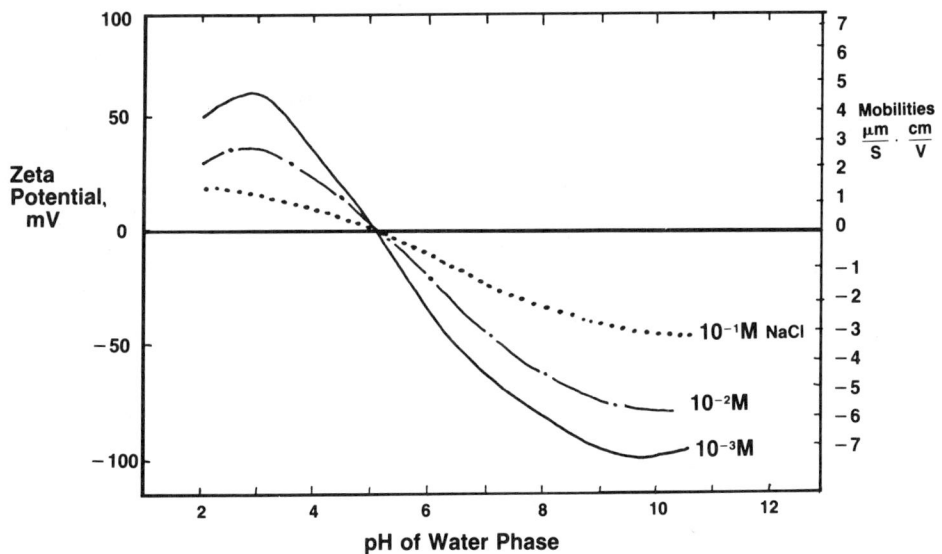

FIG. 17b Heavy southern California crude oil/distilled water.

point of zero charge (ZPC)]. Hence we can leave the data as mobilities (μm/s·V/cm). This way we do not need to determine which corrections are needed. The right-hand ordinate in Fig. 17 shows the mobility; the left-hand ordinate estimates the zeta potential assuming that f = 1.0 and that the second term in the brackets is negligible. Next, the potential-determining ions are those that, when changed, cause the major changes in the zeta potential. In this example, we suspected H^+ or OH^- to be the PDI and so plot that on the abscissa. Indeed, as the pH increases, the zeta potential changes from positive to negative. The ZPC is about pH 4 to 5. Third, the effect of changing the thickness of the double layer by increasing the concentration of salt shows up in the mobility data. However, all the mobility data for all concentrations of indifferent electrolyte must be coincident at the ZPC.

Sometimes natural systems include surface species with different charge behavior. Casassa and co-workers (1985), working with coal in water, illustrate this well in Fig. 18. The coal consists of at least three major components: anthracite, ferrous compounds, and gypsum. The electrophoretic mobility data for these three pure components are shown in Fig. 18a. For a mixture bound together in coal the resulting data are given in Fig. 18b. Thus, for oil-water systems that have clay particles at the surface we would expect behavior similar to that shown in Fig. 18c. A collection of data for various oil-water systems is displayed in Fig. 19.

Now, how does all the modeling of single charged surfaces relate to what happens when two surfaces approach each other? When the surfaces approach, the diffuse double layers overlap and in essence the potential variation with distance changes to a new location. The energy needed to bring in the ionic species to create the change represents the repulsive energy. In essence, we are resolving the previous equation but with the new boundary condition that $\partial \Phi / \partial y = 0$ at a noninfinite distance from either surface. If both surfaces had the same potential, that location would be midway between the two surfaces. The calculations are complex. Tables of data have been published for sphere-sphere interaction by Loeb et al. (1961) and for the flat plate in flat plate interaction by Devereux and deBryn (1963). Table 2 gives some approximate models for the double-layer sphere-sphere interaction. Values of the parameter Y are given in Table 3.

The DLVO Model As a starting approximation for the interaction between two surfaces, researchers Derjaguin, Landau, Verwey, and Overbeek (DLVO) added the van der Waals attraction (discussed in Section I.A) to the electrochemical double-layer repulsion to yield a working model. An example plot is given in Fig. 20 showing the in-

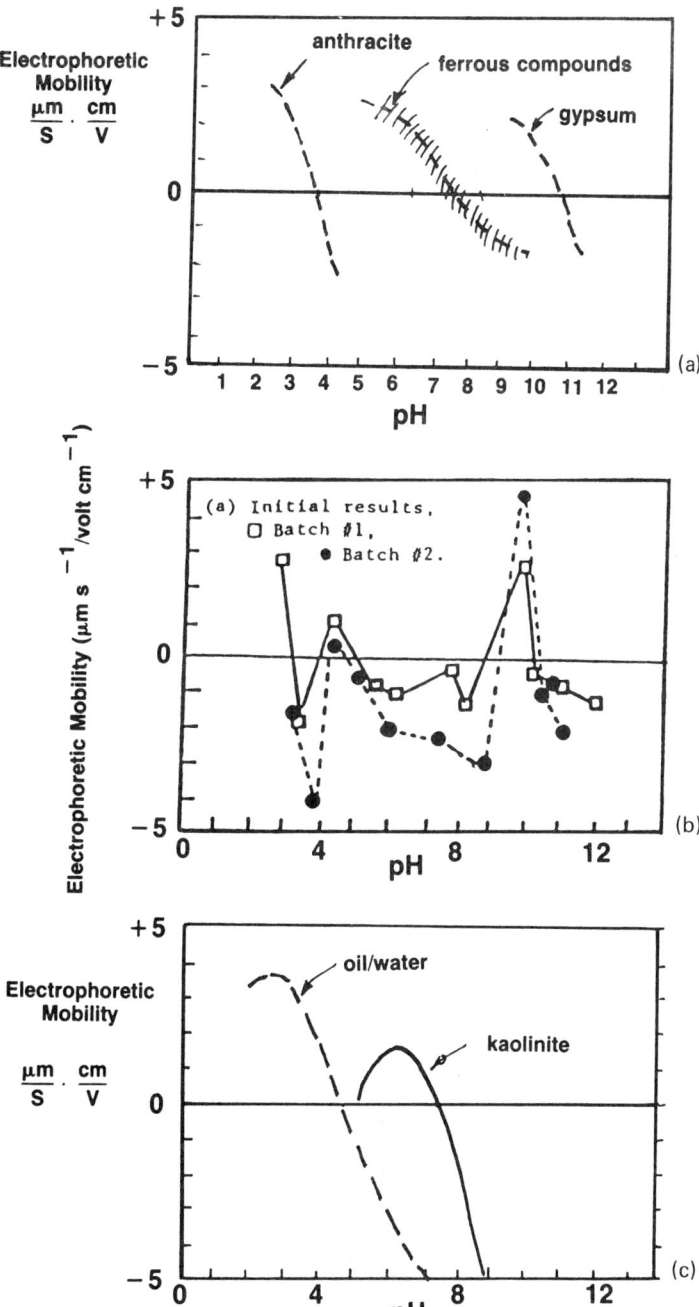

FIG. 18 Effect of other species on electrophoretic mobility. Top graph: (a) Estimates of mobilities for "pure" components in water. (b) Behavior of coal-water system [Casassa et el. (1985) with permission]. (c) Estimates of mobilities for "pure" oil and clay in water.

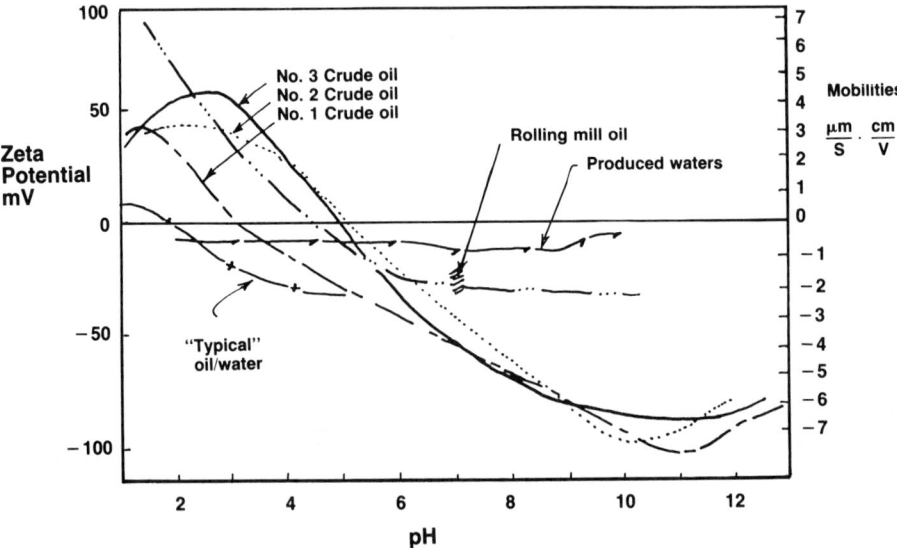

FIG. 19 Summary of different oil-water systems.

teraction of oil drops of radius $2\mu m$, with Hamaker's parameter of $A = 10$ βJ and with a potential at the Stern layer of -30 mV. The variable is the electrolyte valence and concentration as reflected by the value of K.

Figure 21 shows a variety of results that we might see for the energy interaction curves. Figure 21a means that if drops move toward each other, the attractive force draws the drops together to coalesce or destabilize the suspension. The equilibrium state is one of two surfaces held closely together with negligible separation between them. Figure 21b illustrates the condition where an energy barrier at a distance h_1 is such that as surfaces approach each other, they are prevented from going closer than h_1. The equilibrium state is a stable dispersion. For fluid-fluid systems as long as $h_1 > 50$ nm (an estimate of the distance at which coalescence occurs) a stable dispersion results. Figure 21c illustrates the so-called secondary minimum. As surfaces approach each other, they become held apart at a separation distance of h_2. The equilibrium state is a "loose" floc or dispersion. The surfaces are attracted together to form a cluster with free space between them. For liquid-liquid systems, if the distance h_2 is greater than 50 nm, no coalescence occurs and the drops cling together in clusters. The conditions for Fig. 21a and c are illustrated in Fig. 22. Figure 21d shows a potential secondary minimum. However, it is not deep enough to withstand the agitation of the continuous phase, and hence any flocs that do form

TABLE 2 Constant Potential: Two Spheres of Radius r_1

	Total interactive energy, U_R
$Kr_1 \geqslant 5$, binary symmetrical electrolyte (small potentials $z\phi_0 < 25$ mV)	$U_R = \dfrac{\varepsilon r_1 \phi_0^2}{2}\left[\dfrac{2(h+r_1)}{h+2r_1}\right]\ln\left[1 + \dfrac{r_1}{h+r_1}\exp(-Kh)\right]$ and $h \ll 2r_1$ $= \dfrac{\varepsilon r_1 \phi_0^2}{2}\ln\left[1 + \exp(-Kh)\right]$ $= \dfrac{64\pi n k T r_1}{K^2}Y^2 \exp(-Kh)$
$Kr_1 \leqslant 3$	$U_R = \dfrac{\varepsilon r_1 \phi_0^2}{2}\beta\left(\dfrac{2r_1}{h+2r_1}\right)\exp(-Kh)$ where β = electric field distortion factor, $0.6 \leqslant \beta \leqslant 1.0$
$Kr_1 \ll 1$	$= \dfrac{\varepsilon r_1^2 \phi_0^2}{h+2r_1}\exp(-Kh)$
$3 < Kr_1 < 10$	$= \dfrac{8}{e^2}\dfrac{\varepsilon(kT)^2}{z^2}r_1 Y^2\left[\exp(-Kh)\right]$
low potentials	$= \dfrac{\varepsilon r_1 \phi_0^2}{2}\exp(-Kh)$

TABLE 3 Values of the Potential Parameter

Surface potential		Potential parameter Y_0 for	
ϕ_0 (mV)	ϕ^+ at 25°C	1-1 elect. at 25°C	2-2 at 25°C
260	10.1	0.9874	0.9999
240	9.4	0.9814	0.9998
220	8.6	0.9727	0.9996
200	7.8	0.9600	0.9992
180	7.0	0.9415	0.9982
160	6.2	0.9149	0.9961
140	5.4	0.8765	0.9915
120	4.7	0.8230	0.9815
100	3.9	0.7500	0.9602
80	3.1	0.6528	0.9149
60	2.3	0.5249	0.8239
40	1.6	0.3711	0.6522
20	0.8	0.1968	0.3711

will be torn apart again because the attractive force is insufficient to withstand the shear stress on the floc. At equilibrium, this system may be *partially* coagulated. Figure 21e shows repulsion dominating. At equilibrium, a stable dispersion results.

We can use this model to understand how *we* can control the stability of a dispersion.

1. A focus on the effect of the valence of the counterion on the height of the repulsive "hill" on the energy interaction curve led Schultz-Hardy to extract from the DLVO theory that the concentration of indifferent electrolyte required to eliminate (and thus cause destabilization) varied as $(1/z)^6$. Table 4 illustrates the significance of this effect.

2. We can use the model to predict the critical coagulation (destabilization) concentration (CCC) or alternatively—having measured the CCC experimentally—to back-calculate the surface potentials. For water as the continuous phase

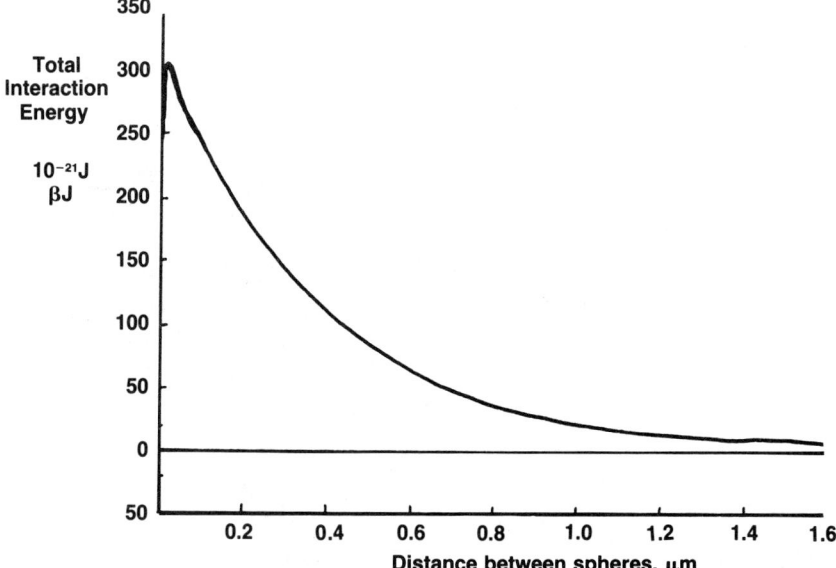

FIG. 20 Interaction energy for two spheres. Sphere-sphere interaction: sphere diameter = $2\mu m$; temperature = 25°C; Hamaker constant = $10\beta J$; surface potential = 30 mV; divalent counterions = 0.13 × 10^{21}.

$$\text{CCC} = \frac{8 \times 10^{-36} Y_0^4}{A^2 z^6} \quad [=] \text{ mmol/L} \tag{22}$$

For very small potentials, this is

$$\text{CCC} = 8 \times 10^{-44} \frac{\Phi_0^{+4}}{A^2 z^2} \tag{23}$$

where Φ_0^+ is the dimensionless potential at the original charged surface = $e(\Phi_0/kT)$.

3. We can either coagulate an unwanted suspension or conversely, create a stable suspension, emulsion, or dispersion as follows:

a. The attractive force (the Hamaker constant) is relatively beyond our control. For any two-surface system this is relatively temperature independent and independent of the species in the intervening phase. This will be continually trying to *destabilize* the system.

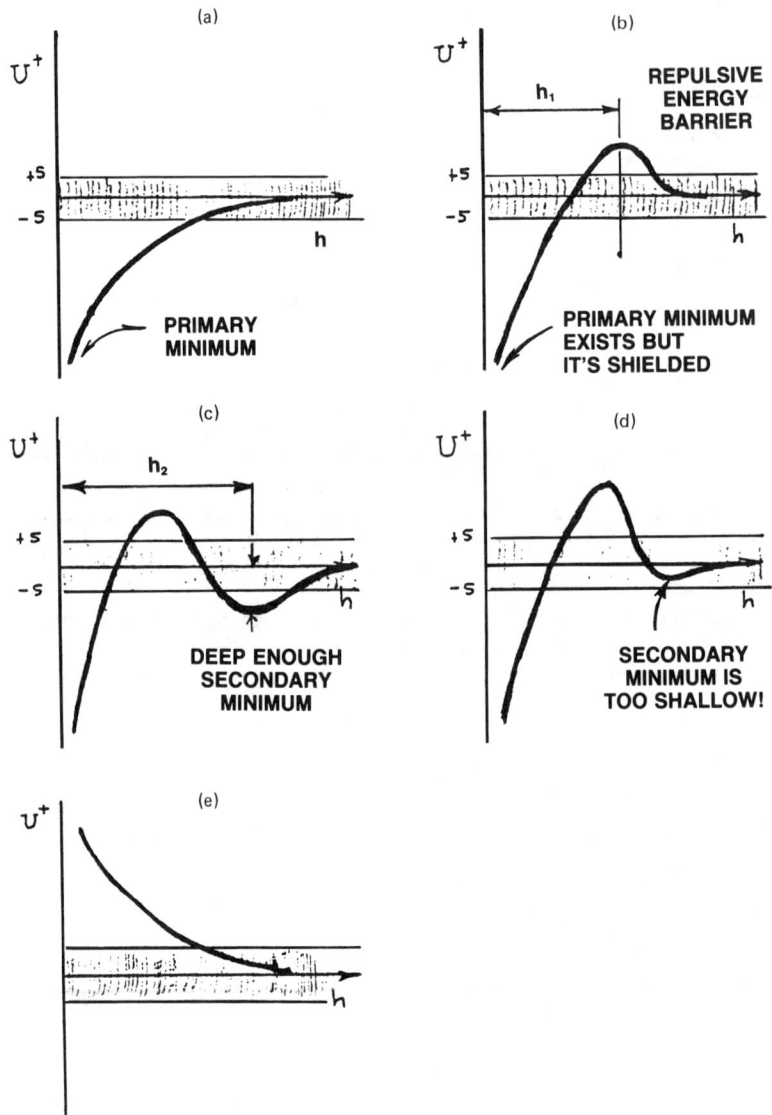

FIG. 21 Conditions for van der Waals and electrostatics.

b. The charge effects can be handled by three alternatives: two methods of altering the seat of charge and one method for the diffuse layer. Control of this counterbalances the attractive force.

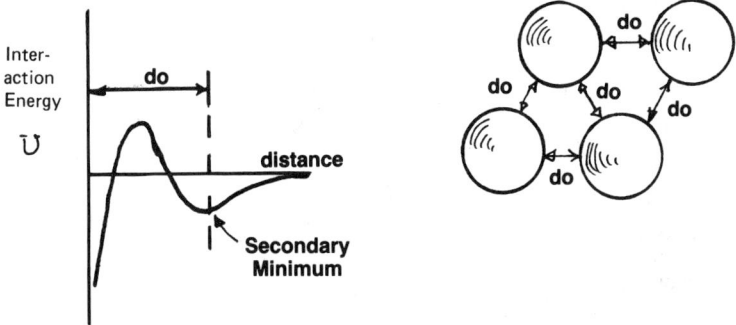

FIG. 22 Stable floc caused by a secondary minimum.

c. Alter the charge by changing the PDI in the system. (Use zeta meter measurements to determine the PDI and the ZPC. Perhaps an adsorbed surfactant is the PDI. Then alter this concentration.) For example, with H^+ or OH^- as PDI and clay as the solid surface, then
    pH < ZPC surface positive
    pH = ZPC surface no charge
    pH > ZPC surface negative
Will it work? Yes, usually provided that we can determine the PDI and provided that within the process we can control the PDI. This is illustrated in Fig. 23.
d. Forget the PDI and determine the sign of charge on the surface and adsorb something of opposite sign to neutralize the surface charge. For example, for negatively charged polystyrene, add cationic surfactant (which will adsorb into the surface). This is

TABLE 4 Effect of the Counterion Valence

Valence of counterion or gegenion	Experimentally observed amount of indifferent electrolyte needed to destabilize (millimol/L)	Predicted from DLVO theory for $\phi_0 = 100$ mV; $A = 200$ βJ
1	25–150	50
2	0.5–2	2
3	0.01–0.1	0.2

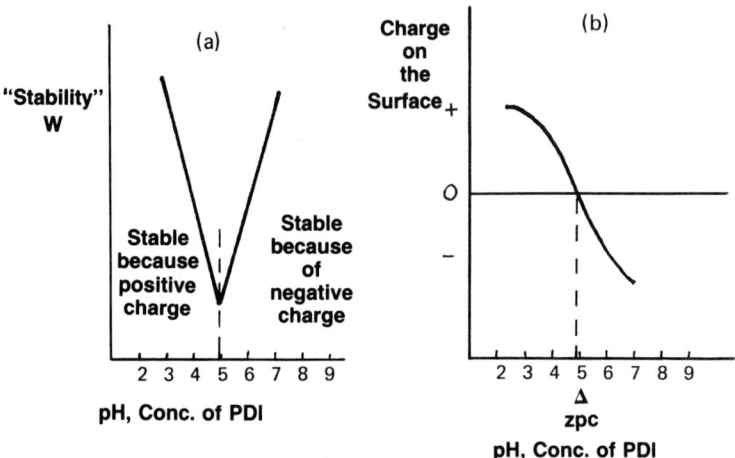

FIG. 23 Controlling stability by altering the PDI. (a) Illustration if H^+ or OH^- is the PDI: stability. (b) Charge.

illustrated in Fig. 24. This works well provided that we do not add too much. If too much is added, the particles are repeptized or stable because of the charge from the adsorbed species (e.g., adding alum to clay).
e. Leave the surface charge alone. Add electrolyte to compress the surface charge effects. We need to know the sign of the charge and use high-valence counterions. This is illustrated in Fig. 25.
f. Displace the surface species by altering the solubility or by adding surfactants that displace them. For example, if the species is an initial surfactant that stabilizes an oil-in-water dispersion, the addition or a second surfactant that tends to produce water in oil dispersions may break the dispersion (Lissant, 1983, p. 88; Davies and Rideal, 1963, p. 385; Pushkarev et al., 1983, p. 168; Osipow, 1962, p. 336). Thus, if an oil-water dispersion exists, one might try adding a surfactant with a small hydrophil-lipophile balance (HLB < 6). Examples of the types of surfactants and their HLB values are given in Table 5. Alternatively, species could be added to change the solubility of the stabilizing surfactant, or if the stability is caused by a film or membrane, by dissolving the membrane (Gopal, 1968, p. 68; Berkman and Egloff, 1941; Overbeek, 1968, p. 341).

b. *Adsorbed Polymers.* If large bulky molecules adsorb into the surface, some of the species may reside as a "train" or "anchor" in the surface and part may extend like fingers out away of the sur-

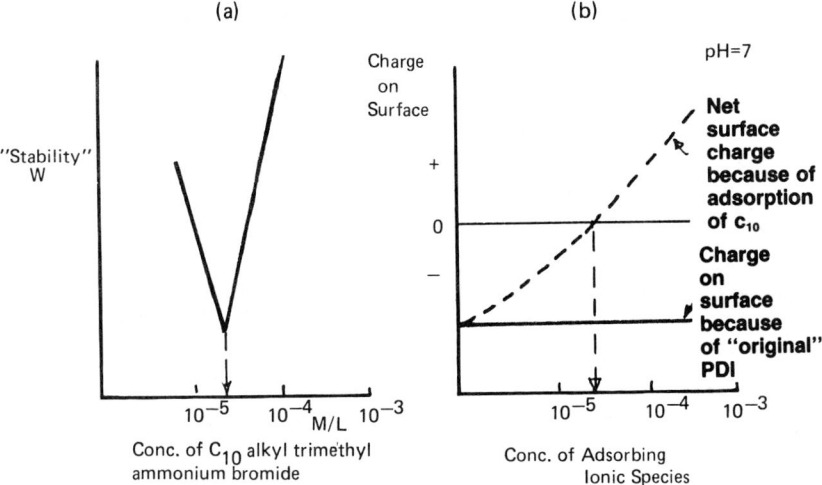

FIG. 24 Controlling stability by adsorption. (a) Stability. (b) Charge.

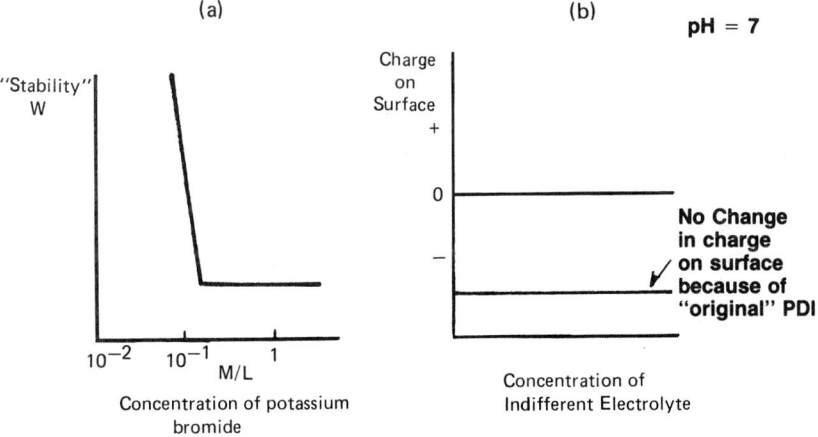

FIG. 25 Controlling stability by adding indifferent electrolyte. (a) Stability. (b) Charge.

TABLE 5 Some Surfactants and Their HLB Values

Surfactant	HLB	Application
Sorbitan tetrastearate	0.5	
Oleic acid	1	
Cetyl alcohol	1	
Span 85 (sorbitan trioleate)	1.8	
Sorbitan tristearate	2.1	
Propylene glycol monostearate	1.8–3.4	
Arlacel C (sorbitan sesquioleate)	3.7	Water-in-oil
Glycerol monostearate	3.8	Emulsifiers
Octyl phenol-1 ethanoxide	4.0	
Propylene glycol monolaurate	4.5	
Span 80 (sorbitan monooleate)	4.3–5.7	
Sorbitan monostearate	5.9	
Sorbitan monopalmitate	6.7	
n-Butanol	7.0	Wetting agents
n-Propanol	7.4	
Ethanol	7.9	
Methanol	8.3	Oil-in-water emulsifier
Span 20 (sorbitan monolaurate)	8.6–10	
Octyl phenol-4 ethanoxide	9.6	
Tween 81 (sorbitan monolaurate-6 ethanoxide)	10–15	
Alkyl aryl sulfonate	11.7	
Triethanolamine oleate	12	Detergents
Octyl phenol-10 ethanoxide	14	
Tween 80 (sorbitan monolaurate-20 ethanoxide)	15	Solubilizers
Sodium oleate	18	
Potassium oleate	20	
Sodium dodecyl sulfate	40	

face. This is illustrated in Fig. 26. When two surfaces approach each other, these polymer fingers or tails could

1. Physically impede the approach of the two surfaces (called volume restriction-repulsion).
2. Intermingle, and in intermingling change the energy fields that each polymer experiences. This creates "free energy of mixing" or "osmotic pressure" repulsion.
3. Possess charges and interact as described in section I.B.1.a but at different locations around the polymer tail out in the continuous phase rather than back at the surface.

Models have been made and are being developed so that we can predict the repulsive behavior from such adsorbed polymers (see Napper, 1970). Berg (1981) has summarized the characteristics of dispersions stabilized by adsorbed polymers.

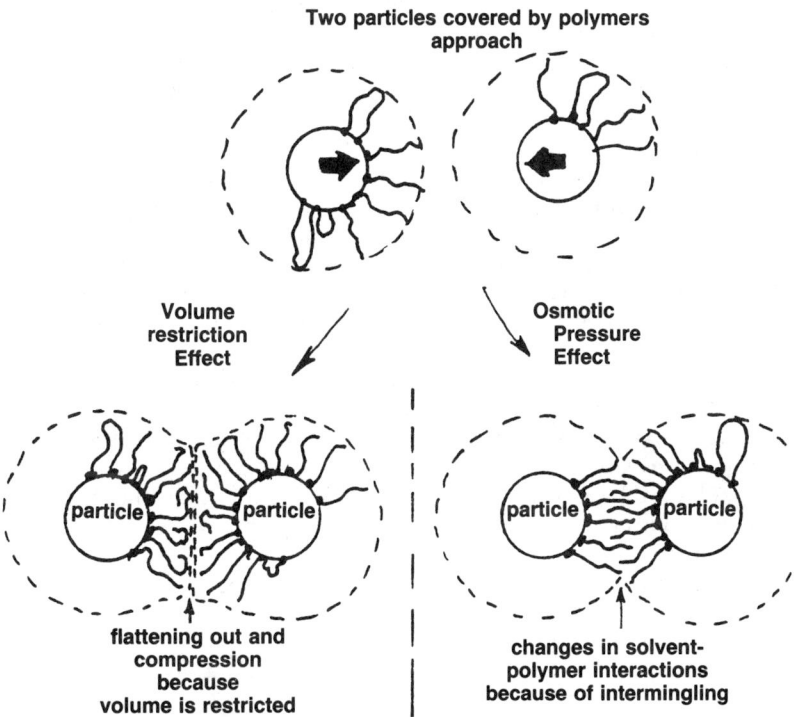

FIG. 26 Effect of adsorbed polymers.

1. The stability of such dispersions is often highly sensitive to changes in temperature, the transition from high stability to catastrophic flocculation occurring often over a range of 1 to 2°C. The temperature at which this occurs is termed the critical flocculation temperature (CFT). *Aqueous* dispersions are most often (but not always) flocculated by *raising* the temperature, whereas nonaqueous dispersions are usually (but not always) flocculated by *decreasing* temperature. Some cases have been found that do not have an accessible CFT and one (polyacrylonitrile latex particles stabilized by polystyrene in methyl acetate), at least, has been found to exhibit both an upper and a lower CFT. This temperature sensitivity stands in sharp contrast to the relative temperature *insensitivity* of electrostatically stabilized dispersions discussed in Section I.B.1.a.
2. The temperature-induced instability mentioned above is almost always readily reversible (spontaneous) upon reheating or recooling the flocculated dispersion. This also contrasts with the commonly observed irreversibility of flocculated dispersions stabilized initially by electrostatic forces.
3. Catastrophic instability can be caused at a sharply defined point by dissolving other materials in the dispersion medium. If the material added to the solvent is liquid, one calls the volume required to induce flocculation the critical flocculation volume (CFV), and if the additive solute is solid, the corresponding term is the critical flocculation concentration (CFC). The additives that have such an effect are always ones which reduce the solvency of the dispersion medium for the stabilizing parts of the absorbate. Once again, the resulting flocculation is usually spontaneously reversible when the original solvency conditions are restored.
4. Changes in pressure can also induce sudden changes in stability.
5. Combinations of changes in the above-mentioned variables can be contrived to induce or reverse flocculation of a sterically stabilized dispersion, the sharply defined conditions for which define a "critical flocculation point" (CFPT).
6. Provided that there is good anchoring and total coverage, stability conditions appear to depend only on the nature of the stabilizing part of the polymer, essentially independent of the specific anchor group(s).
7. For large enough molecular weights of the stabilizing part (say, 10^4), stability conditions are essentially independent of molar mass.

   c. *Summary*. From an equilibrium viewpoint, the options for controlling the stability are summarized in Table 6.

TABLE 6 Mechanisms for Ideal Dispersions

	Stabilizing	Destabilizing
van der Waals; always present, relatively temperature independent, difficult to adjust	*Mechanism 1:* rare situations of negative Hamaker parameters	*Mechanism 1:* usually positive Hamaker parameters
Alter PDI, usually easy but can be expensive depending on cost of the excursion from usual conditions; relatively temperature independent (e.g., pH)	*Mechanism 2:* adjust away from ZPC	*Mechanism 2:* adjust to ZPC; overadjustment can restabilize
Alter indifferent electrolyte; valence or/ and concentration of *counterions*; temperature independent (e.g., NaCl)	*Mechanism 3:* difficult to do because we would want to *reduce* the concentration and reduce the valence	*Mechanism 3:* add high-valency *counterions* or high concentrations of counterions; Schultz-Hardy rule should apply; overdosing gives no unwanted events other than cost; for sterically stabilized dispersions, the Schultz-Hardy rule does not seem to apply
Adsorb ions organic or inorganic species if electrostatically stabilized (e.g., alum, sodium dodecyl sulfate)	*Mechanism 4:* add excess to yield charge because of the adsorbed ions	*Mechanism 4:* add until obtain neutral charge; often evaluate in a "jar test;" overdosing may restabilize
If sterically stabilized	*Mechanism 5:* cause polymer tails to extend	*Mechanism 5:* cause polymer tails to collapse

TABLE 6 (Continued)

	Stabilizing	Destabilizing
Change temperature; if electrostatically stabilized this will have little effect	*Mechanism 6:* for sterically stabilized we can reversibly alter the stability by raising or lowering the temperature depending on the dispersed phase	*Mechanism 6*
Add polymer	*Mechanism 7:* if polymer adsorbs (and does not bridge), then sterically stabilizes	*Mechanism 7:* if polymer bridges (and not adsorbs on individual drops), then it might destabilize; usually it is used to strengthen an already destabilized dispersion
	Mechanism 8: polymer added to bridge to strength floc but wrong conditions chosen and mechnism 7 occurs instead	*Mechanism 8:* polymer bridges and causes destabilized floc to be strong enough to withstand local turbulence; used in combination with other mechanisms
Add material that precipitates or that catches the droplets		*Mechanism 9:* drops are captured and either coalesce and build up on the material or move along with the material
Add chemicals that dissolve or displace the stabilzing species		*Mechanism 10:* displace the stabilizing agent *Mechanism 11:* dissolve the stabilizing agent

2. Rate Considerations

Equilibrium conditions identify the target or ultimate goal state; rate considerations examine how rapidly the system approaches equilibrium. Thus we can predict that under equilibrium conditions the dispersion will be unstable but still end up with a stable dispersion if the rate of coalescence, or the rate of adsorption or neutralizing species, is not fast enough.

a. Coalescence Phenomena. The *time required for coalescence* to occur depends primarily on the rate at which the thin film between the drops can thin to about 30 to 50 nm thickness (at which thickness the separating film ruptures and the drops join together). This is illustrated in Fig. 1. This thinning process is complicated because of the very large effect trace quantities of surfactants and electrolyte can have on the thinning process. Surfactants can slow it down, and increased electrolyte can speed it up. Studies on single drops suggest that the coalescence time, τ, is correlated as follows:

For relatively clean surfaces:

$$\tau = \frac{3\mu_c D_p^2}{2\gamma} \left(\frac{1}{h_2} - \frac{1}{h_1} \right) \quad (24)$$

For dirty systems:

$$\tau = \frac{\phi 4}{16} \frac{\mu_c \Delta\rho g}{\gamma^2} \frac{D_p^5}{32} \left(\frac{1}{h_2^2} - \frac{1}{h_1^2} \right) \quad (25)$$

where

τ = time, s
μ_c = continuous phase viscosity, mPa·s
γ = surface tension, mN·m^{-1}
D_p = diameter of drops
h_1 = initial thickness of film when drops approach each other; use 2000 nm
h_2 = final thickness of film when rupture occurs; for most oil-water systems, use 30 nm
ϕ = mobility of moving surface ($1 < \phi < 4$ depending on the surfactant concentrations)

Davies et al. (1971), have correlated a wide variety of coalescing conditions for a drop at a flat interface as follows:

$$\frac{\tau \gamma}{D_p \mu_c} = k \left(\frac{\Delta \rho \ g}{\gamma} D_p^2 \right)^{0.25} \qquad (26)$$

where k is a constant dependent on the system (0.3×10^5 to 1.3×10^5), or

$$\tau = k \frac{\Delta \rho^{0.25} g^{0.25}}{\gamma^{1.25}} \mu_c D_p^{1.5} \qquad (27)$$

$$= \gamma \mu_c D_p^{1.5} \qquad (28)$$

Thus Davies et al. (1971) seem to correlate the time in terms of some combination of Eqs. (24) and (25) (except that the exponent on the drop diameter seems to be low; we would expect a number between 2 and 5).

Barnea and Mizrahi (1975) extended Eq. (28) to study the coalescence behavior in a bed of drops. The result of this analysis was that the coalescence time depends on the diameter raised to the j power, where $j > 3$. A value of $j = 3$ means that the drops do not bounce together in the band; rather, they just sit side by side. They suggested that $3 < j < 9$. Thus

$$\tau \propto D_p^j \qquad (29)$$

All these data were developed based on relatively large drops, 200 µm in diameter and larger. Thus models suggest that to decrease the coalescence time we would want large surface tensions and small-diameter drops. Thus, if we visualize drop-drop coalescence, the time for each successive coalescence would be longer as the size of the drops grew.

Example 6 Contrast the predicted values obtained for coalescence from Eqs. (24), (25), and (27) for single drops or beds of drops of 10 µm of benzene in water at 25°C.

Solution: For the clean benzene-water system the surface tension is 30.5 mN/m, the density difference is 0.124 kg/liter, and the water viscosity is 1.39 mPa·s. For relatively clean surfaces and single drops at a planar interface, Eq. (24) yields

$$\tau = \frac{3}{2} \frac{1.39 \times 10^{-3} \ Pa \cdot s}{30.5 \times 10^{-3} \ N} \ m \ (10 \times 10^{-6} m)^2 \left(\frac{1}{30 \times 10^{-9} m} - \frac{1}{2000 \times 10^{-9} m} \right)$$

$$= 6.8 \times 10^{-12} \left[\frac{328 \times 10^5}{m} \right]$$

$$= 2.2 \times 10^{-4} s$$

Equation (25) yields

$$\tau = \frac{\phi 4}{16} \frac{1.39 \times 10^{-3} Pa \cdot s}{(30.5 \times 10^{-3})^2} \times 0.124 \times 10^3 \frac{kg}{m^3} \times 9.8 \frac{m}{s^2}$$

$$\times \left[\frac{(10 \times 10^{-6})^5}{32} \right] \times \left[\left(\frac{1}{30 \times 10^{-9} m} \right)^2 - \left(\frac{1}{2000 \times 10^{-9} m} \right)^2 \right]$$

$$= \phi\, 1.4 \times 10^{-21} \times 1.11 \times 10^{15}$$

$$= 1.5 \text{ to } 6 \times 10^{-6} s$$

Equation (27) yields

$$\tau = 1 \times 10^5 \frac{(0.124 \times 10^3)^{0.25} (9.8)^{0.25}}{(30.5 \times 10^{-3})^{1.25}} \times 1.39 \times 10^{-3} \, Pa \cdot s$$

$$(10 \times 10^{-6})^{1.5}$$

$$= 1 \times 10^5 \times \frac{3.33 \times 1.76}{1.27 \times 10^{-2}} \times 4.39 \times 10^{-11}$$

$$= 2 \times 10^{-3} s$$

Thus, the three predictions are 0.2 ms, 1 to 6 μs, and 2 ms.

However, these estimations should be used with care. For the batch behavior of many drops, the drops form a central band, with the thickness of the band decreasing with time as illustrated in Fig. 27. Coalescence occurs at the active interface (and to a limited extent within the band) and settling occurs at the other interface.

Golob and Modic (1977) correlated the velocity of active interface as follows: for a fixed oil/water ratio of 0.66,

$$v_i = 0.098 \times (72.3)^{0.85} \frac{\mu_c^{0.5}}{\mu_d^{1.5}} \gamma^{0.15} \left(\frac{\Delta\rho}{\rho_c}\right)^{0.41} \tag{30}$$

This work suggests that the rate of coalescence (or disappearance of the band boundary) is independent of initial drop diameter and is relatively independent of all but the densities and viscosities. However, Golob and Modic did not measure the drop diameter; they did vary the power input to the drop dispersion unit from 0.8 to 2kW/m^3 and found a negligible difference in the coalescence rate. In concentrated dispersions, in reactor systems and in solvent extraction work where oil/water separations have been researched, the drop size is usually 500 μm to 1 to 2 cm.

Example 7 Reestimate the single-drop coalescence time with Eqs. (24), (25), and (27) for the butanol/saline water system used by Barnea and Mizrahi (1975) for a 600-μm drop. Contrast that with the required time for a 30-cm band of drops to coalesce fully. Assume 22°C.

Solution: The pertinent properties are surface tension, 5.3 mN/m; density difference, 0.40 kg/liter; and a continuous phase viscosity of 6.40 mPa·s. From Eq. (24),

$$\tau = \frac{3}{2} \times \frac{6.4 \times 10^{-3}}{5.3 \times 10^{-3}} (600 \times 10^{-6})^2 (328 \times 10^5)$$

$$= 21.4 \text{ s}$$

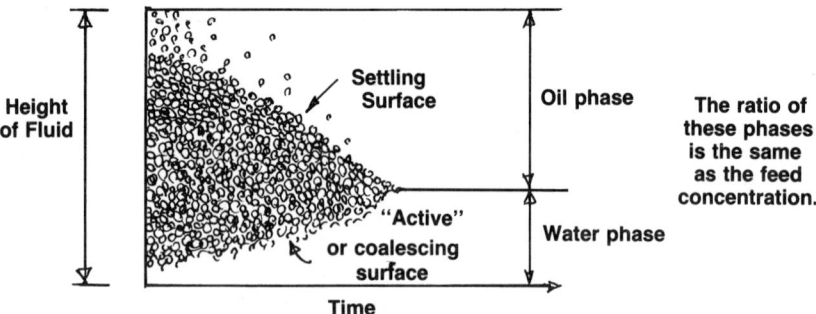

FIG. 27 Time history of a coalescing dispersion of water in oil.

From Eq. (25),

$$\tau = \frac{\Phi 4}{16} \times \frac{6.4 \times 10^{-3}}{(5.3 \times 10^{-3})^2} \times 0.4 \times 10^{-3} \times 9.8 \frac{(600 \times 10^{-6})^5}{32}$$

$$\times 1.11 \times 10^{15}$$

$$= 600 \, \Phi \, s$$

$$= 600 \text{ to } 2400 \text{ s}$$

From Eq. (27),

$$\tau = 1 \times 10^5 \frac{(0.40 \times 10^3)^{0.25}(9.8)^{0.25} 6.4 \times 10^{-3} \times (600 \times 10^{-6})^{1.5}}{(5.3 \times 10^{-3})^{1.25}}$$

$$= 52 \text{ s}$$

These predictions are for single drops at a planar interface. For the 30-cm band of drops to coalesce completely, the experimental results range from 117 to 130 s at 22°C. Thus

Equation	Single drops, τ (s)	Experimental data for 30-cm band (s)
(24)	21	
(25)	600–2400	117–130
(27)	52	

Thus, when the correlations are used for the size of drops consistent with the size used to develop the correlations, the agreement is fair. For many industrial situations, the drop size is very small and we suspect that the correlations give unreasonably small coalescence times for drops of less than 100 μm. Furthermore, these models are idealized and although surfactants were present, no particles or polymerized membranes existed at the surfaces.

b. *Chemical Adsorption Phenomena.* As described in Section I.B.1.a, when dispersions are stabilized by the electrochemical double layer, one approach is to adsorb ions onto the surface to "neutralize" the charge. The idea is that the chemicals are added to the dispersion and mixed rapidly to distribute the chemicals throughout the system and onto the oil-water surfaces. Amirtharajah and Mills (1982) studied these effects in the context of the addition of aluminum sulfate or alum. Alum is an interesting chemical to study because—for different conditions—it may hydrolyze and the ions may migrate into the countercharge region, alum may hydrolyze and polymerize, or it may precipitate. Table 7 summarizes their estimate of the times for these occurrences. Thus for most applications these rates are relatively fast.

c. *Coagulation or Attachment Phenomena.* To promote coagulation, we expect some movement in the continuous phase to waft the drops together. For drops of less than 1 μm, the mechanism usually is fluid dynamical or, more specifically, velocity gradients.

Brownian Movement Although we could visualize the fundamental mechanism to be the same as in coalescence (namely, the fluid dynamical thinning of the thin film of liquid trapped between the two approaching drop surfaces), we tend to model it as a diffusion coefficient. That is, the rate of disappearance of the number of particles per unit volume, n, is given by

$$\frac{\partial n}{\partial t} = D_{ij} \frac{\partial^2 n}{\partial r^2} \qquad (31)$$

at

$$r = R_{ij} \quad n = 0 \quad t > 0$$

$$r = \infty \quad n = n_0$$

where D_{ij} is the diffusivity of drop i relative to drop j and R_{ij} is the radius of drop i plus radius of drop j.

TABLE 7 Estimates for Various Reactions of Alum

Hydrolysis	<0.1 ms
Adsorption	1 ms
Polymerization dimers (micellization)	1 s
Complex polymers	Al_{13}, min
Precipate forms	1 to 7 s

The result is that for equal-sized drops,

$$-\frac{\partial n}{\partial t} = \frac{1}{2} \eta (8\pi D_i R_{ii}) n_0^2 \quad \text{for } t^+ = \frac{tD_{ij}}{R_{ii}^2} \gg 1 \tag{32}$$

where $D_i = \frac{1}{2}D_{ii}$ is the diffusion of drop i relative to fixed coordinates and η is the efficiency of sticking on collision.

Einstein suggested that

$$D_i = \frac{kT}{6\pi\mu(R_{ii}/2)} \tag{33}$$

Since we would like to put this phenomenon on a time scale, we define $t_{1/2}$ as the time it takes for the number of drops to half (i.e., all the drops have collided with and joined into pairs). This is sometimes referred to as the "time of coagulation." This is, for the Einstein value of the "diffusivity" of drop, given as

$$t_{1/2} = \frac{3}{4} \frac{\mu}{\eta kT n_0} \tag{34}$$

This result is interesting because it suggests that the rate of coagulation depends only on the number of drops. This result assumes that every collision results in attachment.

Example 8 An oily stream contains 1000 ppm of oil with drops 0.1 μm in diameter. What is the time of coagulation in water at 25°C?

Solution: To use Eq. (34), $kT = 4.11$ βJ and we need the initial number of particles. For an initial volume fraction Φ of $1000/10^6$ and 0.1-μm-diameter drops, the number is, for 1 m³,

$$n_0 = \frac{1000}{10^6} \times \frac{1}{\pi/6(0.1 \times 10^{-6})^3}$$

$$= 1.9 \times 10^{18} \text{ drops/m}^3$$

Assume that $\eta = 1.0$; then

$$t_{1/2} = \frac{3}{4} \times 1 \times 10^{-3} \frac{N \cdot s}{m^2} \times \frac{1}{1.0} \times \frac{1}{4.11 \times 10^{-21} N \cdot m} \times \frac{1}{1.9 \times 10^{18}} m^3$$

$$= 0.1 \text{ s}$$

The mechanism based on Brownian movement is often called perikinetic coagulation.

Velocity Gradients We can waft drops together by imposing laminar velocity gradients on the continuous phase. We refer to this as orthokinetic coagulation. If the velocity gradient is $G = (\partial v_r/\partial z)$, the rate of disappearance is given as

$$-\frac{dn}{dt} = \frac{4}{3}\eta R_{ij}^3 G n_0^2 \qquad (35)$$

For equal-sized drops, $R_{ij} = R_{ii} = 2R_i = D_{pi}$ and with a definition of the initial volume fraction of drops, Φ, the results are

$$-\frac{dn}{dt} = \frac{4}{3}\eta \Phi G n_0 \qquad (36)$$

$$t_{1/2} = \frac{0.54}{\eta \Phi G} \qquad (37)$$

where

$$\Phi = \frac{\pi}{6} D_{p_0}^3 n_0 \qquad (38)$$

We can estimate when the rate of orthokinetic coagulation becomes more significant than perikenitic coagulation as follows:

$$\frac{\text{ortho}}{\text{perikinetic}} = \frac{\mu G D_{p_0}^3}{2kT} \qquad (39)$$

Example 9 If $G = 60 \text{ s}^{-1}$ and the drop diameter is 5 μm, what is the relative importance of ortho to perikinetic coagulation for oil drops in water at 25°C? Is the fluid dynamics important?

Solution: For water at 25°C,

$$\mu = 1 \times 10^{-3} \frac{\text{N} \cdot \text{s}}{\text{m}^2}$$

$$kT = 4.11 \times 10^{-21} \text{ N} \cdot \text{m}$$

Hence, from Eq. (38),

$$\frac{\text{ortho}}{\text{perikinitic}} = 1 \times 10^{-3} \frac{N \cdot s}{m^2} \times \frac{60}{2} \times (5 \times 10^{-6} m)^3 \times 2 \times 4.11$$

$$\times 10^{-21} N \cdot m$$

$$= 912$$

Hence the fluid dynamics is important under these conditions.

The fluid dynamics would be equal to the impact of the Brownian movement if the drop size was, from Eq. (38),

$$D_p^3 = \frac{\text{ortho}}{\text{perikinetic}} \frac{2kT}{\mu G}$$

For ortho/perik = 1,

$$D_p = 0.5 \ \mu m$$

The value of G depends on the practical devices used for introducing G and upon the operating conditions. Some alternative devices are shown in Fig. 28. Corresponding ranges of values for G are given in Table 8. To use Eqs. (36) or (37) requires that we estimate the efficiency, η. Table 9 gives values of η based on different experimental systems for the optimum addition of the chemical coagulants. Figure 29 illustrates how η varies with the chemical dosage for an oil/water dispersion treated with alum.

3. Fluid Dynamics Considerations

Sometimes we can undo all the good we have done if we subject the coalesced, coagulated, or separated dispersion to too much shear. Although details have not been quantified for all types of interaction, here are some examples to illustrate this effect. For Brownian motion alone, Ottewill suggests that the fluid dynamic energy is about 5 to 10 kT. Thus the equilibrium phenomena predicted in such figures as Fig. 21 are inapplicable in the region $U^+ = \pm 5$ to 10 because the fluid dynamics creates a stable dispersion here.

For the design of decanters, we need to design the exit lines carefully so as not to reentrain drops that have already been separated. For deep-bed filters, Brutsch and Mallatt (1976) suggest that the liquid flow rate through the filter be decreased by about half for low-temperatures operations because at the lower temperature the increase in liquid viscosity increases the shear and tears the particles off the filter. Some other quantitative predictions are given in Section II.

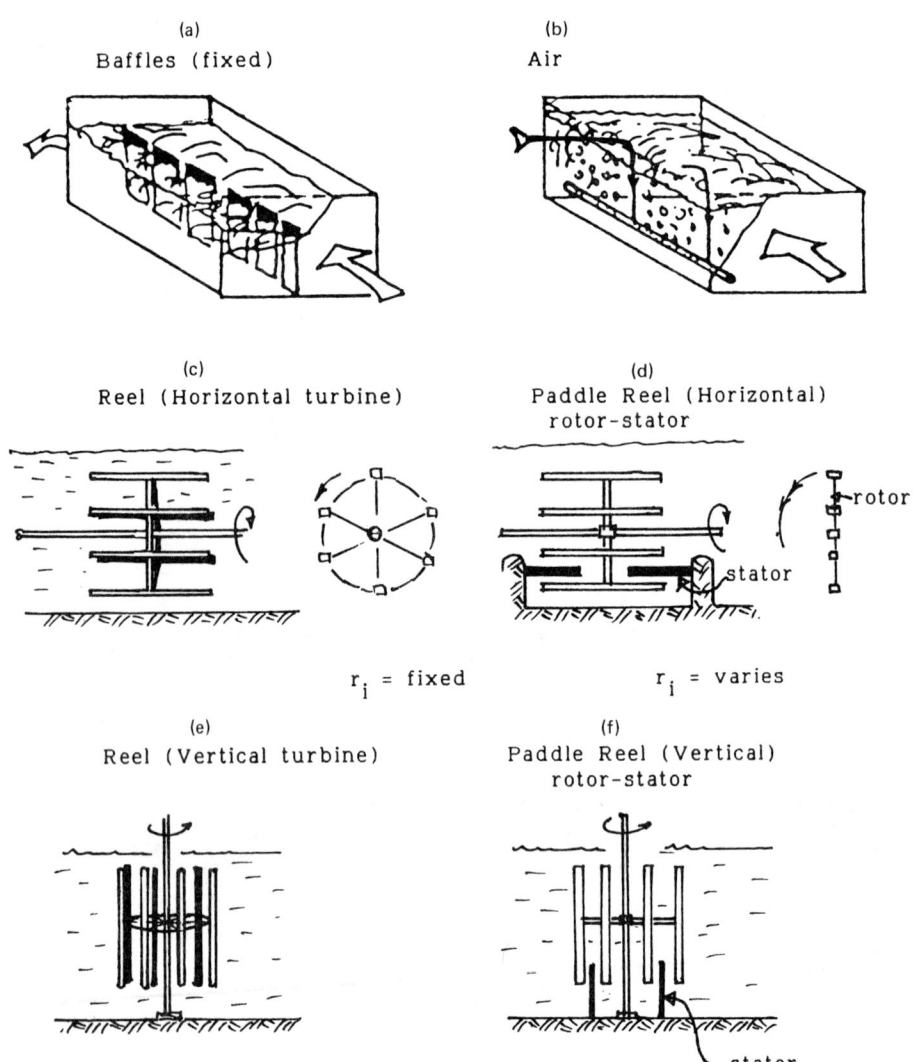

FIG. 28 (a) and (b), Some configurations to provide velocity gradients G. (c) through (h), Some mechanical designs to provide velocity gradients G.

Dilute Oil-Water Emulsions and Dispersions

(g)
Oscillating

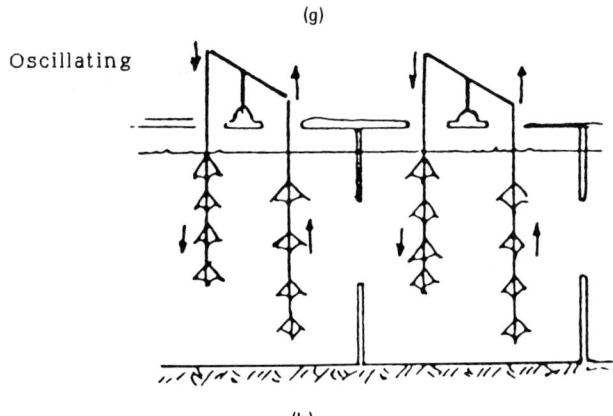

(h)
Vertical Propeller/turbine

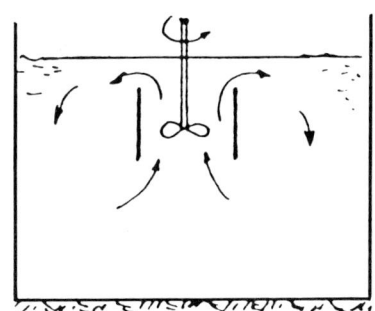

FIG. 28 (Continued)

C. Role of Surfactants

Surfactants are usually present in all industrial oil/water systems, and these tend to stabilize the dispersions. On the other hand, with the addition of an astute choice of surfactant we might break the dispersion.

1. Surfactants Already Present

Only in the confines of a research laboratory have we been able to prepare oil/water systems that are free from surfactants. Usual "pure" tap water, for example, contains about 0.01 mg/L of anionic surfactant as typified by sodium dodecyl sulfate. Industrial water from refinery contained 7 mg/liter equivalent of sodium dodecyl sulfate. Naturally, some oil/water systems contain more "natural" sur-

TABLE 8 Methods of Introducing G

	Method of Estimating[a]	Estimates of G (s^{-1})
Baffles		
Aeration	$G_{max} = \dfrac{p_2}{\mu} \dfrac{1}{V} F_a \ln \sqrt{\dfrac{p_2}{p_1}}$	100–1000
Reel		
Paddle	$G = \sqrt{\dfrac{P_0}{V\mu}} \quad G = 0.16 N^{1.6}$	
Propeller/turbine		
Oscillating		20–70
Natural lake		0.5–2
Natural river		5–8
Natural estuary		5–15

[a] p_2, pressure inside aeration inlet pipe; F_a, volumetric flow rate of air through aeration pipe; V, volume of liquid in basin to be aerated; p_1, liquid pressure just outside the aeration pipe; P_0, power supplied to turn the reels, paddles, or turbine; μ, liquid viscosity; N, rpm of a rotating reel or paddle.

factants than do others. For example, Ali (1978) has shown that for the conditions used in the hot water flotation process used to separate oil from tar sand, sulfoxides, $C_{14}H_{23}S, C_{14}H_{25}S, C_{15}H_{25}S$, form. These are surface active.

Asphaltenes are surfactants that occur naturally in refinery processing. Siffert et al. (1984) have shown that not all asphaltenes are the same. The most surface-active asphaltenes are those with intermediate aromaticity, low hydrogen bonding ability, and high acid base balance. These are characteristics of an oxidized asphaltene. They also reported that a membrane forms at the oil/water interface if asphaltenes are present. The most surface-active asphaltenes tend to form two-dimensional or lamella structures that stabilize emulsions. Nonoxidized asphaltenes tend to form globular micelles. Thus the naturally occurring surfactants are important. Hence most oil/water dispersions produced from unrefined oils tend to abound with naturally occurring surfactants that often thwart attempts to destabilize the suspensions. For example, Luthy et al. (1977), working with model systems of crude oil/distilled water, altered conditions so that there

TABLE 9 Some Perikinetic and Orthokinetic Coagulation Data[a]

Particulate	Coagulant	ϕ	Perikinetic	Orthokinetic	G (s^{-1})	η	k_A	Gt	k_B (s)
Polystyrene latex	NaCl			✓	11	0.488			
	NaCl			✓	45	0.344			
	NaCl			✓	1–80	0.364			
	NaCl		✓		—	0.315			
	Polyethylenimine			✓	11	0.217			
	Polyethylenimine			✓	45	0.063			
Silica	Al(III)			✓	10	0.011			
	Al(III)		✓		—	0.010			
Oil	Ca(NO$_3$)$_2$		✓		—	0.355			
Clay	Alum	200 mg/L		✓	60	0.52	14×10^{-5}	10^5	
	Alum						1 to 15×10^5		1 to 42 $\times 10^{-8}$

[a]k_A is the agglomeration coefficient $4\eta\phi/\pi$, where ϕ is the initial volume fraction dispersed phase. Sources: Weber (1972), p. 98) and Omelia (1978, p. 258).

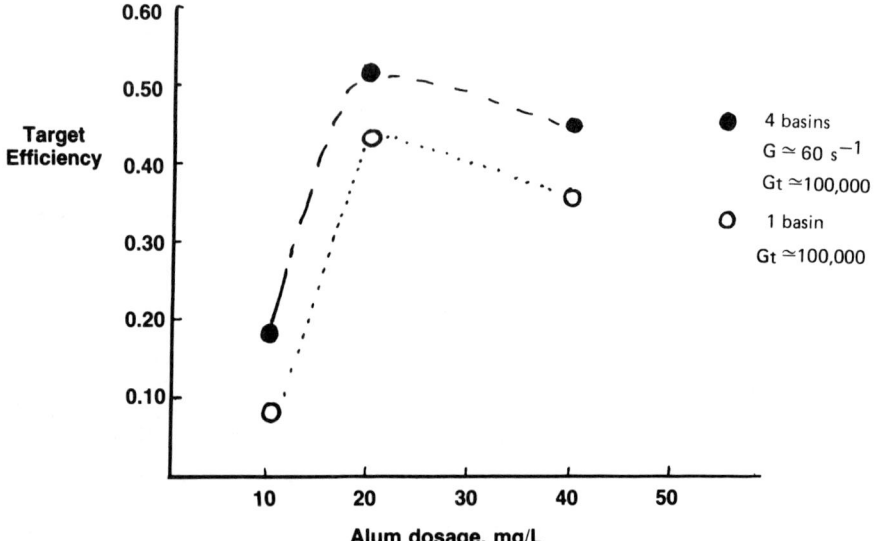

FIG. 29 Removal of 200 mg/L of oil to be removed from water by aluminum addition.

was no charge on the surface; the oil drops should coagulate and coalesce. Yet no coalescence occurred. Similar findings are reported by Churchill and Kaufman (1973).

Even for refined oils that are relatively free of asphaltenes, most will have some surface-active materials present; some systems are more sensitive to their presence than others. Table 10 gives some examples of such systems.

2. The Effect of Adding Surfactants

One of the mechanisms used to control stability is the addition of ions (mechanism 4 and 10). Ionic surfactants are popular choices because relatively small additions can make drastic changes. In principle, this approach works for many solid-liquid systems. However, for oil/water systems the effectiveness of this as a method of treatment has been mixed.

Luthy et al. (1977) added different types of cationic surfactants to ideal dispersions of light Arabian crude oil in distilled water. For dosages of less than 2 g/liter, pH 8, 10^{-2}M NaCl, they could get charge reversal for all surfactants used. These included dodecyl trimethyl ammonium chloride, hexadecyl trimethyl ammonium chloride, and various dialkyl quaternary ammonium chlorides. However, although they caused charge reversal, coalescence and separation did not occur.

TABLE 10 Some Complications for Decanter Sizing

Well-behaved systems ("insensitive systems")	Ill-behaved systems ("sensitive systems")
Do not drift much with time.	May or usually drift with time.
Rate of dispersion band separation is relatively independent of upstream mixing/pumping conditions. (For mixer-settler combinations, the mixing time can vary from 10 to 200 s and the power input can be 2 to 3 times greater than minimum value.)	Rate of dispersion-band separation is dependent on feed drop size distribution or the energy input into upstream mixer-pump.
Can treat decanter design independent of upstream mixer.	Design combination of mixer/pump and decanter.
Scaleup straightforward.	Scaleup extremely difficult.
	Build in flexibility to adjust system.
Example: -di(tridecyl)amine/ NaCl solution	Examples: most systems containing: Kerosene $HCl-H_2O$-iamyl alcohol at low pH Kerosene/acidic $CuSO_4$ solution + 75 ppm Alamine Shellsol-T + 5% LIX-64N/ aqueous $CuSO_4$ solution.

Churchill and Kaufman (1973) added 1 to 120 mg/liter of Tretolite J-80, a quarternary amine, to synthetic crude oil/water mixtures but found that oil removal was only about 20 to 30%. No pH or electrophoretic mobility data were reported. Thus both of these groups of researchers found that the addition of surfactants did not help to break the emulsions. These were dilute, synthetic emulsions of crude oil in water.

On the other hand, Majid and Farnard (1980) worked with concentrated oil/water dispersions (50 to 65% oil) from fire flood wells.

The focus was batch "dewatering." Over 95% dewatering occurred for the conditions shown in Table 11. On the other hand, other surfactants were ineffective.

D. Role of Inorganics

Inorganics are present in the water phase of most industrial oil/water systems. In general, they tend to buffer pH changes and cause instability. Often, the addition of inorganics will break the dispersion.

1. Inorganics Already Present

The most pertinent ions that might be present naturally in the water are calcium, magnesium, and iron. These can have a significant impact for several reasons. Because they are ionic and di- or trivalent, they act to compress a double layer for a negatively charged surface (as in mechanism 3) and thus promote destabilization for higher concentrations. In addition, the evidence is mounting from a variety of sources that calcium ion tends to promote the adsorption of other species. Hence it is important to monitor the inorganics already present. Insoluble inorganics could also be present. Clay is often present. Indeed, Majid and Farnand (1980) found that bitumen tends to adsorb strongly on clay fines found in most oil/sand formations. Gewers (1968) suggests that small amounts of clay, present in heavy oils, form membranes at the oil-water interface so as to stabilize the emulsions. The clay or solids will not only affect the coalescence behavior. The electrochemical properties of the clay or solids in the interface might dominate the surface charge.

2. The Effect of Adding Inorganics

Again, by mechanism 4, we should be able to control the stability by adsorbing an appropriate soluble inorganic oin. Alternatively, the inorganic ions could serve as mechanism 3—compression of the double layer. The most popular choices of chemicals are alum, ferric salts,

TABLE 11 Surfactants that Dewater Emulsions from Fireflooded Wells

Tetrolite E5J 303	0.2 wt % + 10 wt % toluene	
Triton X100	1 wt %	
Igepal Co-630	0.5 wt %	
Nonionic 1 + Lb 13	0.3–0.5 wt %	With added dewatering if toluene added

and calcium oxide. However, especially for alum and ferric salts, the water chemistry of these inorganics is complex. The ions are present in completely different forms, depending on the concentration and on the pH. Because of the importance of alum for destabilization, the focus in this section is on alum.

Alum is the commercial name for aluminum sulfate. Aluminum ions exist in aqueous solutions as hydrates coordinating six molecules of water per metal ion. The aluminum ions tend to react with the OH^- part of the water, while the H^+ ions, released as a result of this reaction, lower the pH. This hydrolysis effect accounts for the acidic conditions produced when aluminum is added. The products of hydrolysis may be monomeric and polymeric.

The distribution of the soluble aluminum species in water can be calculated under two circumstances: first, when the soluble aluminum species are in equilibrium with a solid phase, such as amorphous of crystalline aluminum hydroxide, and second, when no solid phase exists but the total amount of aluminum in water is known. In the first case, the governing equations are the equilibrium equations of the formation of the soluble species, the solubility product equation of the solid phase and the water equilibrium. The second case requires the equilibrium equations of the species, the water equilibrium and the mass balance with respect to aluminum. These procedures require the knowledge (or assumption) of the exact species present in solution along with the values of the equilibrium constants. These, however, are not always known.

a. Aqueous Aluminum Monomers. The monomers include the species $Al(H_2O)_6^{3+}$, $Al(OH)(H_2O)_5^{2+}$, $Al(OH)_2(H_2O)_4^+$, $Al(OH)_3(aq)$, and $Al(OH)_4(H_2O)_2^-$. Not every author, however, agrees that all these monomers are present. The presence of $Al(OH)_2(H_2O)_4^+$ and $Al(OH)_3(aq)$ has been questioned by Hayden and Rubin (1976) and Pulfer and Kramer (a982). Wide disagreement also exists in the estimated values of the equilibrium constants of the hydrolysis reactions. Differences of at least one order of magnitude have been reported by Pulfer and Kramer (1982). These differences are the result of different experimental techniques, experimental conditions (such as aluminum concentration, ionic strength, temperature, conditions of mixing, and base addition), as well as assumptions regarding the existence of certain species. A summary of values of equilibrium constants and Gibbs free energy of the mononuclear aluminum hydrolysis species is presented in Table 12. For simplicity, the water molecules associated with the aluminum ion have been omitted. The equilibrium constants β_n in Table 12 correspond to the reactions

$$Al^{3+} + nOH^- = Al(OH)_n^{3-n}$$

TABLE 12 Equilibrium Constants and Free Energy of Formation of Mononuclear Aluminum Hydrolysis Species

Complex	log β_n	Free energy of formation, $\Delta G°_{298}$ (kcal/mol)
Al^{3+}	—	-116 ± 1
$AlOH^{2+}$	8.99 ± 0.04	-165.84
$Al(OH)_2^+$	19.3 ± 0.1	-217.5
$Al(OH)_3(aq)$	26.8 ± 0.1	-265.3
$Al(OH)_4^-$	32.7 ± 0.1	-311.0 ± 0.1

Source: Parks (1972).

where n = 1, 2, 3, or 4. From these equations we see that the amount of each aluminum species formed depends on the pH.

Insoluble aluminum hydroxide that may be in equilibrium with the species in solution may be crystalline or amorphous. Although crystalline forms are thermodynamically more stable, kinetic limitations determine that under the usual conditions of coagulation/precipitation, amorphous aluminum hydroxide precipitates. The precipitate would gradually become crystalline with time. The solubility products of the amorphous $Al(OH)_3$ and gibbsite, the equilibrium form of crystalline aluminum hydroxide at standard conditions, are given in Table 13.

If one considers that the aluminum species present in water and in equilibrium with amorphous aluminum hydroxide are Al^{3+}, $AlOH^{2+}$, $Al(OH)_2^+$, and $Al(OH)_4^-$ (omitting the water molecules for simplicity), then the system is fully defined by the three equilibrium equations

$$\beta_1 = \frac{[Al(OH)^{2+}]}{[Al^{3+}][OH^-]} = 10^{8.99} \tag{40}$$

$$\beta_2 = \frac{[Al(OH)_2^+]}{[Al^{3+}][OH^-]^2} = 10^{19.3} \tag{41}$$

$$\beta_3 = \frac{[Al(OH)_4^-]}{[Al^{3+}][OH^-]^4} = 10^{32.7} \tag{42}$$

TABLE 13 Solubility Products of Aluminum Hydroxide

Solid	$\log k_{so} = \log[(Al)\cdot(OH^-)^3]$	Reference
Gibbsite	−34.0	Parks (1972)
	−33.5	Baes and Mesmer (1976)
	−33.89	May et al. (1979)
am. $Al(OH)_3$	−31.2	Parks (1972)
	−32.69	DeHek et al. (1978)
	−31.81	Bersillon (1983)

the solubility product equation

$$k_{so} = [Al^{3+}][OH^-]^3 = 10^{-31.2} \quad (43)$$

and the equilibrium of water

$$k_w = [H^+][OH^-] = 10^{-14} \quad (44)$$

By taking the logarithms of the foregoing equations and after simple mathematical manipulations, we can get an explicit expression for the amount of each aluminum species as a function of pH. These equations are

$$\log[Al^{3+}] = 3.6 - 3(pH) \quad (45)$$

$$\log[Al(OH)^{2+}] = 2.9 - 2(pH) \quad (46)$$

$$\log[Al(OH)_2^+] = 2.1 - (pH) \quad (47)$$

$$\log[Al(OH)_4^-] = 11.4 + 3(pH) \quad (48)$$

These equations are shown on a pH-concentration diagram (Fig. 30), assuming that the activity of each species coincides with the concentration. Figure 30 shows speciation and solubility. If the total aluminum concentration (soluble and dissolved aluminum) is 10^{-2} mol/L (point A on Fig. 30), the solubility has been exceeded and precipita-

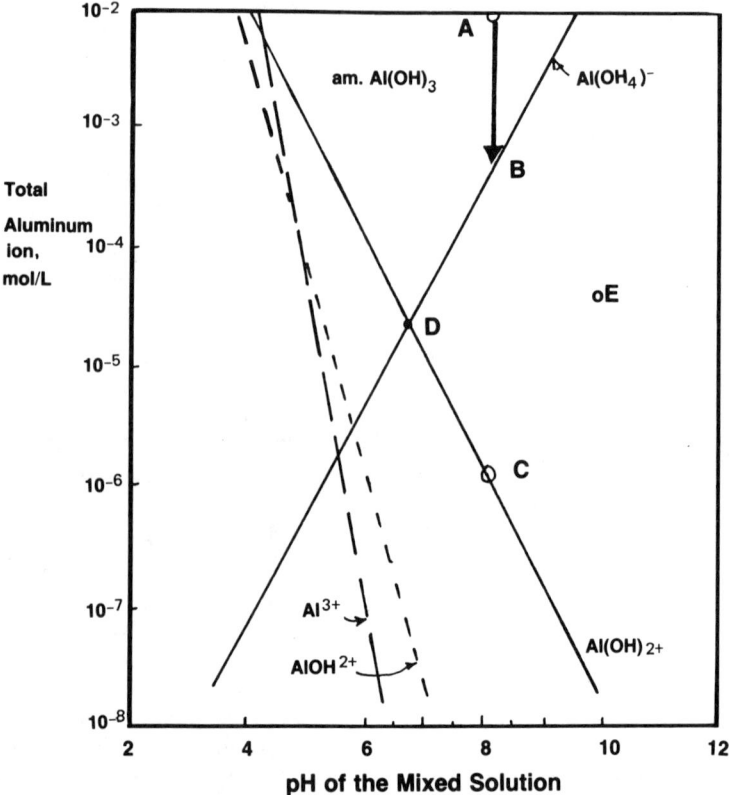

FIG. 30 Aluminum speciation in equilibrium with amorphous Al(OH)$_3$.

tion occurs. If the precipitated solid amorphous Al(OH)$_3$ and sufficient time is given for the system to equilibrate, the soluble aluminum concentration is $10^{-3.4}$ mol/L or 10.75 mg/L (point B). Point B is actually the concentration of the species Al(OH)$_4^-$. The concentration of the other aluminum species will be given by the interception of the line AB with the equilibrium lines of each species. Hence the concentration of the species Al(OH)$_2^+$ is $10^{-5.9}$ mol/L or 0.03 mg/L (point C), while the concentration of the other species is even lower. Therefore, at pH 8 the species Al(OH)$_4^-$ is the predominant species, while the soluble aluminum concentration (being the amount of all soluble species per liter of solution) can be adequately represented by the concentration of the predominant species. This approximation of the soluble aluminum concentration with the concentration of the predominant species is an excellent one, provided that we are not close to the point of interception of two equilibrium lines (point D). Close to this point the concentration of both species should be considered, although the difference is only 0.3 unit on the log scale. If the total aluminum concentration in solution is less than that required for precipitation (point E), the system is no longer described by the same set of equations, since there is no solid phase. In this case the solubility product equation (41) is replaced by the mass balance with respect to aluminum.

$$[Al_T] = [Al^{3+}] + [AlOH^{2+}] + [Al(OH)_2^+] + [Al(OH)_4^-] \qquad (49)$$

and the calculation of the species concentration is also possible.

b. *Aqueous Aluminum Polymers.* Although some researchers were able to explain their data based on a monomeric pathway only (May et al., 1979; Bersillon, 1983; Sullivan and Singley, 1968; Marion et al., 1976), most believe that polynuclear species play a very important role in aluminum hydrolysis as well as its coagulation efficiency. In general, there are two types of problems related to the study of aluminum polymeric species: specifying the stoichiometry and estimating the stability constants.

The simplest aluminum polymer is the dimer $Al_2(OH)_2(H_2O)_8^{4+}$. The structure of the dimer can be described by two aluminum octaedra sharing an edge by hydroxide bridging (each octaedron represents an aluminum hexa-coordinated atom with six ligands). A graphical representation of the dimer is shown in Fig. 31. Many other polynuclear species have been reported in the literature by Pulfer and Kramer (1982). Two theories describe the development of aluminum polymers. One pathway suggests that the hexa-coordi-

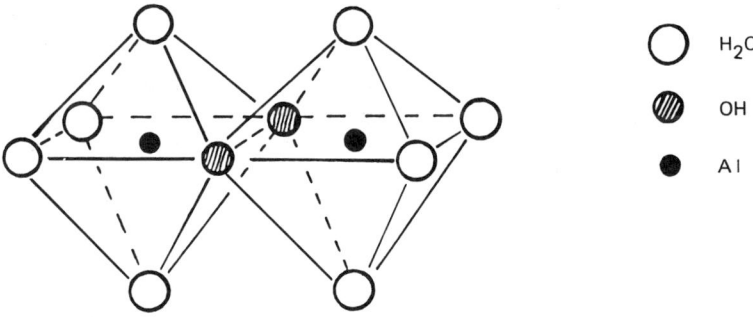

FIG. 31 Structure of the dimer $Al(OH)_2(H_2O)_B^{4+}$.

nated aluminum ions tend to polymerize in a six-membered ring unit (Brosset et al., 1954; Hsu and Bates, 1964; Hsu, 1977). Each ring consists of six octaedra, similar to those of the dimer structure. A single-ring structure has the formula $Al_6(OH)_{15}^{3+}$. This six-membered ring unit is also the unit of the gibbsite crystalline structure. Multiring structures, such as $Al_{10}(OH)_{22}(H_2O)_{16}^{8+}$, may also develop, leading to the formation of sheetlike polymers and eventually, if precipitation occurs, to the formation of crystalline $Al(OH)_3$.

The second pathway suggests the existence of a polynuclear structure with eight aluminum ions. This may be of the formula $Al_8(OH)_{24}^{6}$ (Hayden and Rubin, 1976) or $Al_8(OH)_{20}^{4+}$ (Matijevic et al., 1961).

A special type of polynuclear species has also been reported by Johanson (1960), Akitt et al. (1972), Bottero et al. (1980), and Aveston (1965). Formulas that have been suggested are $Al_{13}O_4(OH)_{24}(H_2O)_{12}^{7+}$, $Al_{13}O_4(OH)_{28}(H_2O)_8^{3+}$, and $Al_{13}(OH)_{32}^{7+}$. This species consists of 12 hexa-coordinated and are tetra-coordinated aluminum atoms. Structurally, it looks like a tetrahedron with the planes formed by the hexameric rings (with sharing Al atoms), while the tetra-coordinated aluminum atom is located at the center of the tetrahedron.

The existence of the various species was based on indirect evidence, while only the dimer and the Al_{13} polymer have been identified by means of analytical techniques, such as x-ray crystallography and ^{27}Al nuclear magnetic resonance. Differences in the experimental conditions, kinetic considerations, and data handling may also contribute to the uncertainty of the speciation. This uncertainty is also reflected in the reported values of the equilibrium constants. For example, values of the equilibrium constant of the $Al_6(OH)_{15}^{3+}$ species varies by nine orders of magnitude (Pulfer and Kramer, 1982). A list of equilibrium constants of formation of some aluminum polynuclear species is presented in Table 14. The equilibrium constants correspond to the reactions

$$mAl^{3+} + nH_2O = Al_m(OH)_n^{3m-n} + nH^+ \qquad (50)$$

Equilibrium calculations in the presence of polynuclear species are similar to those in the presence of mononuclear species. Mathematical manipulations, however, are more difficult, because the stoichiometry of the polynuclear species results in nonlinear sets of equations.

c. *Kinetics of Aluminum Chemistry.* The kinetics of the aluminum reactions are important not only for the formation of certain species, but kinetic phenomena may also affect the efficiency of aluminum as a coagulant. A time scale of various chemical interactions is presented in Table 7. The hydrolysis of aluminum toward mononuclear species is complete in microseconds, while simple aluminum polynuclear

TABLE 14 Equilibrium Constants of Polynuclear Aluminum Species

Complex	log β_{mn}	Reference
$Al_2(OH)_2^{4+}$	−7.1	Hayden and Rubin (1976)
	−7.07	Aveston (1965)
	−6.95	Bottero et al. (1980)
$Al_6(OH)_{15}^{3+}$	−47.0	Brosset et al. (1954)
	−47.0	Bottero et al. (1980)
$Al_8(OH)_{20}^{4+}$	−68.65	Pulfer and Kramer (1982)
$Al_{13}(OH)_{32}^{7+}$	−104.5	Bottero et al. (1980)
		Aveston (1965)

species are formed within 1 s. The precipitation of amorphous Al(OH)$_3$ starts in approximately 1 s and may slowly continue for months, until equilibrium values are reached. For practical applications, however, the aluminum hydroxide precipitation is complete within 1 to 7 s (Letterman et al., 1973). By comparison, the crystallization of aluminum hydroxide may occur within days to years, while the aluminum/fulvic acid interaction is considered complete in less than 3 min (Bersillon, 1983).

 d. *Coagulation with Aluminum.* Aluminum coagulants, as well as other metal coagulants, can destabilize suspended or colloidal particles through three mechanisms:

Mechanism 3: Double-layer compression: Al^{3+} ions may compress the electric double layer of the colloidal particles. As the result of this, the distance between colloids is reduced and attractive forces may bring the particles together. However, due to the fact that the aluminum ion undergoes hydrolysis, this mechanism is expected to occur at pH less than 4 (Matijevic et al., 1961). The fact that this pH range is very untypical of practical applications and that the most common environmental colloids, such as clays, undergo charge reversal at this low pH implies that this mechanism is not very important in the case of hydrolyzing coagulants.

Mechanism 4: Adsorption of the aluminum hydrolysis species on the surface of the colloids: Polynuclear hydroxocomplexes of aluminum, formed as kinetic intermediates during the formation of aluminum hydroxide precipitate, have been shown to adsorb on the surface of the negatively charged colloids (Matijevic et al., 1966). As a result of that, the charge of the colloidal particles may be reduced, neutralized, or even reversed, depending on the amount of metal species added to the solution. Colloid destabilization by adsorption is characterized by two important observations: first, the coagulant dosage depends on the colloid concentration, and second, overdosing produces restabilization.

Mechanism 9: Sweep-zone precipitation: When a large excess of coagulant is added to the water, a voluminous precipitate of the metal hydroxide is formed which enmeshes the colloidal particles. This bulky floc, together with the enmeshed colloids, is removed by sedimentation.

The two important mechanisms of colloids destabilization by aluminum, adsorption and sweep-zone precipitation, apply in different domains on an Al dosage-pH diagram. This is presented in Fig. 32. The boundaries of the restabilization zone depend on the type and concentration of the colloid, as well as the type and concentration of background ions. In practice, most typical applications employ an alum dosage of at least 10 mg/L in the pH range 6 to 9.

The operating region in Fig. 32 (in terms of pH and applied dosage), along with the kinetic considerations of Table 7, specify the optimum mixing conditions. Since the formation of polynuclear aluminum species is accomplished in 1 s, operation in the adsorption-destabilization zone requires the rapid dispersion of the hydroxopolymers in short periods of time. In operation in the sweep zone, where the kinetically slower aluminum hydroxide is important, the effect of mixing is not very significant. These considerations have been experimentally validated by Matijevic et al. (1966).

E. Role of Polymers

Polymers may be present in the feed or they can be added.

1. Polymers Already Present

Natural occuring polymers could include humic or fulvic acids present in the waters used in the process; another source of polymers could be upstream additions made for upstream processing. For example, the oil/water stream from a dissolved air flotation (DAF) unit may contain synthetic polymers that were added for the DAF separation. Thus in treating this stream we should be aware that some polymers

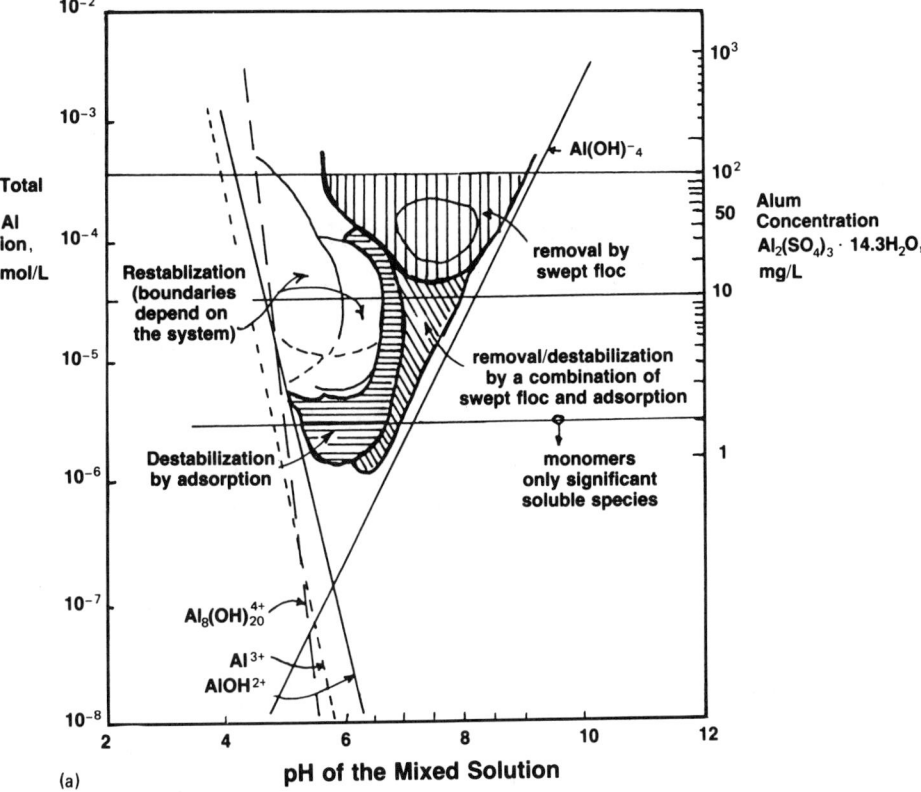

FIG. 32a Operation diagram for aluminum coagulation. [From Amirtharajah and Mills (1982) by permission. Copyright © 1982, American Water Works Association.]

are present in the stream. These could be adsorbed onto the surface and stabilize the dispersion by mechanism 7.

2. The Effect of Adding Polymers

In general, the astute addition of polymers can destabilize oil-water emulsions and are often called "emulsion breakers." The choice and effectiveness depend on the type of separation process being considered. The main factors considered include the type of polymer (cationic, anionic, or nonionic), the molar mass, the location and method of addition, and the synergistic uses with inorganic or surfactant additions. Table 15 summarizes some of the characteristics of polymers and their effectiveness in destabilizing emulsions.

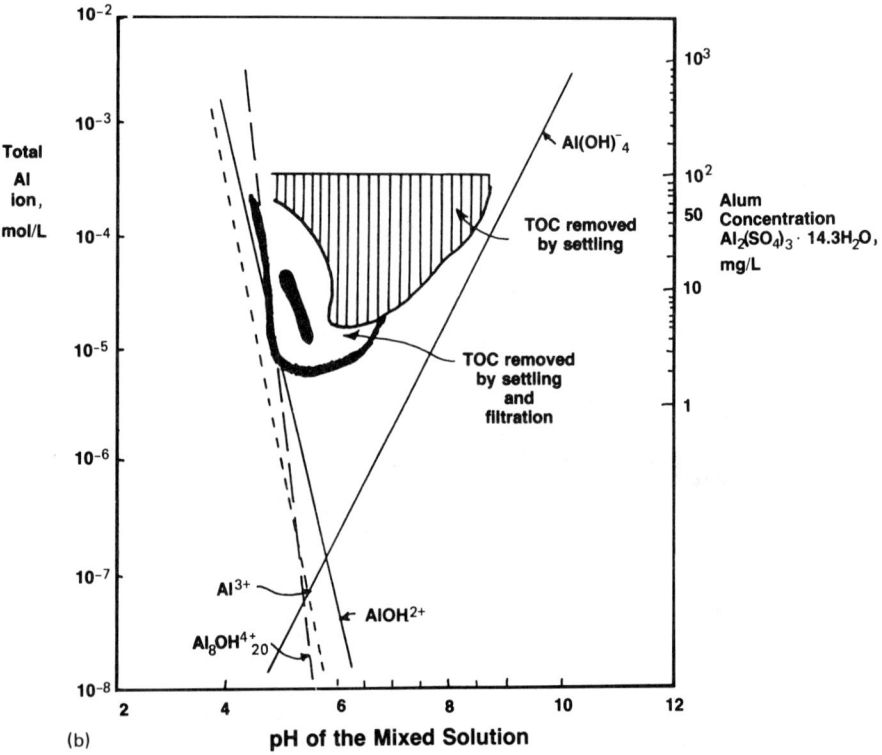

FIG. 32b Another operation diagram for aluminum coagulation. [From Dempsey et al. (1984) by permission. Copyright © 1984, American Water Works Association.]

F. Solubility

Most oils are to some extent soluble in water, and vice versa. Hence, when selecting operating conditions and when analyzing the composition of the streams, it is important to distinguish between these two forms of oil. Figure 33 shows the solubility of organics in water as a function of temperature; Fig. 34 shows the solubility of water in organics. The solubility of other species is also shown. For crude oil in equilibrium with water, limited data are available, although Humenick and Davis (1978) suggest that a value of 6 to 9 mg/L is reasonable.

If the surfactants are present at concentrations above their CMC values, or a surfactant and a cosurfactant are present, then up to five times more nonimmiscible oil could be present than we would expect from the usual estimates of solubility. For more, see Shinoda (1978). Although the solubility affects the interpretation of concen-

TABLE 15 Coagulants Often Used in Environmental Engineering

Inorganic

    Hydrolyzing: $Al_2(SO_4)_3$ $AlCl_3$ $NaAlO_4$ $FeCl_3$
                              (alum)

    Non-hydrolizing: $Ca(OH)_2$
                               lime

Organic

    Cationic:

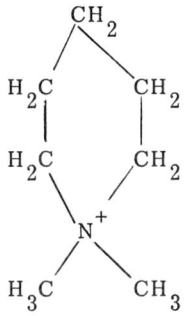

polydiallyl dimethylammonium

    Nonionic:

$$-CH_2-CH- \atop \underset{NH_2}{\underset{|}{\underset{C=O}{|}}}$$
          polyacrylamide $-CH_2-CH_2-O-$
                                                                  polyethylene oxide

    Anionic:

$-CH_2-CH-$, $C=O$, O^- $\left[-CH_2-CH-\atop{|\atop{C=O\atop{|\atop O^-}}}\right]_x$ $\left[-CH_2-CH-\atop{|\atop{C=O\atop{|\atop NH_2}}}\right]_z$

polyacrylic acid   hydrolyzed polyacrylamide

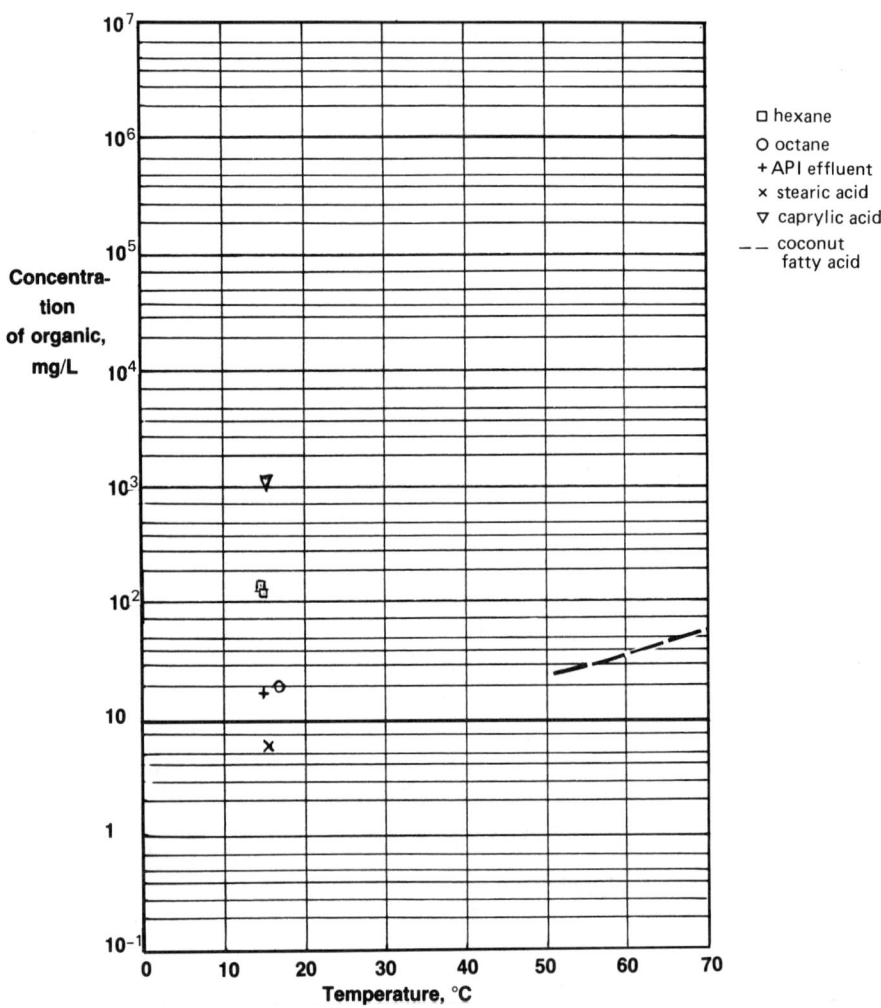

FIG. 33 Solubility of organics in water.

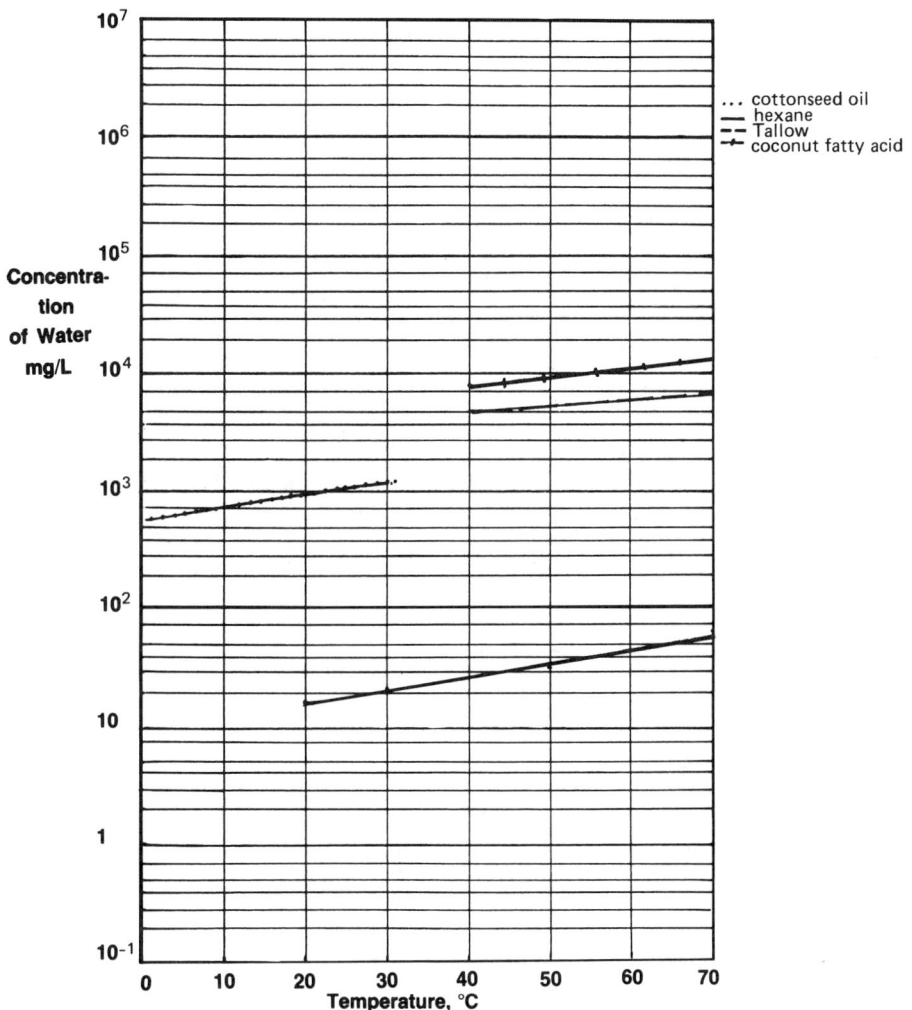

FIG. 34 Solubility of water in organics.

tration data, it also affects the properties of the dispersion important for stability calculations. In particular, it affects the density difference between the two phases.

II. CHARACTERIZING A DISPERSION

Although we should know values for all the parameters that affect the fundamentals outlined in Section I, here we focus on surface tension, drop diameter, and electrical effects.

A. Surface-Interfacial Tension

The size of the drops, the effect of fluid dynamics, and coalescence phenomena depend on the surface tension at the oil/water interface. Surface tension is defined as interfacial force per unit surface boundary (or the interfacial energy per unit surface area) that resists the creation of more area. Methods of measuring it are reviewed by Padday (1973), Adamson (1967), Baird et al. (1981), Cayias et al. (1975), and Fraser et al. (1971). Data are given in Table 16. Sometimes we have the data for air/liquid systems (from sources such as Jasper, 1972) and we need to estimate the oil/water values.

Several methods are available; the method of Girifalco and Good (1957) and Good et al. (1958) is

$$\gamma_{II-III} = \gamma_{III} + \gamma_{II} - 2\Phi \sqrt{\gamma_{III}\gamma_{II}} \tag{51}$$

where Φ is a factor to make the equality hold. Girifalco and Good (1957) found that parameter Φ was a unique property of components II and III, and Good and co-workers have put forward theories for calculating it. General order-of-magnitude values of Φ are given in Table 17.

Certain oil-water surface tensions are a function of pH. This is especially true of complex hydrocarbon systems such as crude oil-water where some of the species are surface active. Figure 35 shows data for some different systems. Strassner (1968) shows how the oil-water surface is "rigid" under certain conditions. Thus a key parameter in characterizing oil-water dispersions is the surface tension. For all systems, the air-liquid tension decreases with an increase in temperature; for liquid-liquid surface tensions the value usually decreases with temperature increase. For some systems it is a function of pH.

B. Predicting the Drop Size

The surface tension holds the drop together; viscous shear forces try to tear it apart. Thus we can estimate the size of drop if we can de-

TABLE 16 Surface Tensions for Liquid-Liquid Systems

Compound	Temp. (°C)	Surface tension (mN/m)	Temp. dep. (mN/m·°C)	$\Delta\rho$ oil-water	μ_{water} (mPa·s)	μ_{water}	References[a]
Water/compound systems							
Acetylene tetrabromide	20	38.82		1.96			
Amyl alcohol (pentanol-1)	30	4.9		0.176	4.881	0.95	
i-Amyl butyrate ($C_9H_{18}O_2$)		23		-0.1339			
Amyl chloride		15.44		-0.12404			
Amyl alcohol	20	5.2		-0.1707	4.025	1.112	Golob and Modic (1977)
Amyl acetate	20	9.2		-0.1120	0.809	1.036	Golob and Modic (1977)
Aniline	20	5.77		0.217			
	20	5.8		0.0237	4.334	1.064	Golob and Modic (1977)
Anisole ($CH_3OC_6H_5$)	20	25.82		0.009	0.9574	0.91	
Benzaldehyde	20	15.5		0.0464			
	20	35		-0.1181	0.635	1.006	Golob and Modic (1977)
Benzene	25	30.5	-0.058	-0.124	0.59	1.39	de Chazal and Ryan (1971)
	21.5	34.1		-0.1225	0.598	1.478	

TABLE 16 (Continued)

Compound	Temp. (°C)	Surface tension (mN/m)	Temp. dep. (mN/m·°C)	$\Delta \rho$ oil-water	μ_{oil} (mPa·s)	μ_{water} (mPa·s)	References[a]
Benzyl alcohol (C_7H_7OH)	22.5	4.75		+0.038	4.08	0.91	
Bromobenzene (C_6H_5Br)	23	31.2		+0.493	1.075	0.934	
Bromoform	20	40.85		+1.8814			
α-Bromonaphthalene	20	42.07		+0.4756			
o-Bromotoluene	20	41.15		+0.43357			
	20	4.5		0.1239	0.714	1.002	Golob and Modic (1977)
n-Butanol	20	1.9		−0.134	3.3	1.28	
	20	1.6		0.1479	3.185	1.414	Golob and Modic (1977)
Butyl acetate	25	12.9		0.116			
Butyl acrylate		30−31					
i-Butyl alcohol	18	2.1		−1.41			
Sec-Butyl alcohol	28	0.6		−0.104	2.78	1.56	
Butyl benzoate		41.5					Sherony (1969)
i-Butyl chloride	20	24.43		−0.1207	0.428	0.924	ICT; Golob and Modic (1977)

tert-Butyl chloride	20	24.43			
	20	23.75	−0.1207		
Butyl methacrylate					
Butyronitrile		10.38	−0.00225		
Caprylic acid (octanoic acid)	20	8.217			
Carbon disulfide	20	48.36	+0.2624	0.378	1.00	ICT
Carbon tetrabromide	20	38.8				Fowkes (1965)
Carbon tetrachloride	20	45	+0.5874			ICT
	20	45.5	+0.5941	1.012	0.999	Golob and Modic (1977)
Chloroacetone	25	41.9	+0.597			
	20	7.11	+0.1552			ICT
Chlorobenzene	25	35.4	+0.1038	0.7625	0.8904	Vijayan (1978)
	20	37.4				
	30	34				
Chloroform	20	32.8	+0.4829			ICT
α-Chloronaphthalene	20	40.7	+0.1718			
Cocoanut fatty acid	30	8.22	0.111			
Corn oil	28	23.4				
Cottonseed oil		23.8−24.2	−0.0877			Null and Johnson (1958)
m-Cresol	25	4.0	+0.039	15.2	0.8904	Vijayan (1978)
Cyclohexane	25	41.1	−0.219			

TABLE 16 (Continued)

Compound	Temp. (°C)	Surface tension (mN/m)	Temp. dep. (mN/m·°C)	Δρ oil-water	μoil (mPa·s)	μwater (mPa·s)	References[a]
Cyclohexanol	20	3.9		−0.051	32.8		Sherony (1969); Fowkes (1965)
	21	3.9		−0.0456	17.25	1.024	de Chazal and Ryan (1971)
	25	3.92		−0.035	68	0.91	Vijayan (1978)
	20	4.1		−0.0468	2.475	1.262	Golob and Modic (1977)
Cyclohexanone	20	3.8		−0.046	2.28	1.03	Vijayan (1978)
p-Cymene ($C_{10}H_{14}$)	13.5	39.41		−0.1407			
	20	34.61					
n-Decane	20	50.2		0.2719	0.886	1.016	Golob and Modic (1977)
Di-bromoethane	25.8	31.7		+1.479	0.888	0.975	Vijayan (1978)
Di-n-butyl amine	20	10.3					Fowkes (1965)
1,1-Dichloroacetone	20	14.43					ICT
β,β-Dichloroethyl sulfide	20	28.36					
Diethyl carbonate	20	12.86		−0.02392			Golob and Modic (1977)
	20	13.8		−0.0241	0.794	1.043	
Diethyl ether	20	10					Kaye and Laby (1966)

Dilute Oil-Water Emulsions and Dispersions

Compound	Temp					Reference
Diethyl phthalate	20	10.7	0.2678	0.259	1.118	Golob and Modic (1977)
Diisoamyl ($C_{10}H_{22}$)	20	16.27	−0.27443			ICT
	20	46.80				ICT
Diisoamylamine ($C_{10}H_{22}N$)	20	13.5	−0.22187			ICT; Fowkes (1965)
Diisobutylamine ($C_8H_{19}N$)	20	10.28	−0.24917			ICT
		16.4				
Diisobutylcarbinol	20	17.3	0.194			
Diisopropyl ketone				0.616	0.909	
Dipropyl amine ($C_6H_{15}N$)	20	1.66	−0.17224			ICT
"Escaid 100"/aqueous 1 M sodium sulfate	20	29.1	−0.21 ≈			
100/acidified $CuSO_4$		22.3	−0.21 ≈			
Ether		10.9	−0.261			
Ethyl acetate	20	2.9	−0.0926	0.498	0.216	Golob and Modic (1977)
	25	5.12	−0.104	0.89	0.50	Vijayan (1980)
	25	6.5				Grilc et al. (1984)
Ethyl acetoacetate	25	3.5				

TABLE 16 (Continued)

Compound	Temp. (°C)	Surface tension (mN/m)	Temp. dep. (mN/m·°C)	$\Delta \rho$ oil-water	μ_{oil} (mPa·s)	μ_{water} (mPa·s)	References[a]
Ethyl acrylate		21–22					Vijayendran (1980)
Ethyl benzene	17.5	31.35					ICT
	26	35					
Ethyl bromide	20	31.20		+0.4459			ICT
Ethyl caproate ($C_8H_{16}O_2$)	20	19.8		−0.1268			ICT
Ethyl chloroacetate	25	14.4		+0.153	1.19	0.8904	Vijayan (1980)
Ethyl cinnamate ($C_{11}H_{10}O_2$)	25	21.5		+0.051	7.95	0.8904	Vijayan (1980)
	19.5	21.36					ICT
Ethyl ether	20	10.7		−0.2694	0.235	1.021	ICT; Keith and Hixon (1955)
2-Ethyl hexane (1–3 diol)	20	4.0		−0.0522	79.23	1.046	Keith and Hixon (1955)
Ethyl hydrocinnamate ($C_{11}H_{14}O_2$)	21.5	20.19					ICT
Ethyl iodide	16	40					ICT
Ethyl isovalerate ($C_7H_{14}O_2$)	20	18.39		−0.1313			ICT

Dilute Oil-Water Emulsions and Dispersions

Compound	T						Reference
Ethyl mercaptan (C_2H_6S)	20	26.12					ICT
Ethyl nonylate ($C_{11}H_{22}O_2$)	20	23.88					ICT
Ethyl oleate ($C_{20}H_{38}O_2$)	20	21.34					ICT
Ethyl propyl ketone ($C_6H_{12}O$)	20	13.58		−0.1812			
Ethylene bromide	24	29.3		+1.169	1.603	0.909	Vijayan (1978)
Ethylene dibromide	20	36.54		+1.1782			Vijayan (1978)
Furfural	26.5	1.5		+0.134	1.34	0.96	Vijayan (1978)
Glycol diacetate	24.2	2.3					
	20	51.7		−0.2907	0.451	0.984	Golob and Modic (1977)
n-Heptane	20	51					Keith and Laby (1966)
	25	47		−0.317	0.40	0.910	Vijayan (1978)
/H_2O + 0.05 M decanoic acid	25	32.4		−0.311			Vijayan (1978)
0.5 M	25	22.5		−0.295			Vijayan (1978)
1 M	25	18.6		−0.273			Vijayan (1978)
Heptylic acid ($C_7H_{14}O_2$)	20	7.0	−0.037				ICT
	24.5	6.9		−0.078	4.27	0.95	Vijayan (1978)
n-Hexane	20	51.0		−0.3426	0.313	1.012	Golob and Modic (1977)

TABLE 16 (Continued)

Compound	Temp. (°C)	Surface tension (mN/m)	Temp. dep. (mN/m·°C)	$\Delta\rho$ oil-water	μ_{oil} (mPa·s)	μ_{water} (mPa·s)	References[a]
n-Hexane	20	51.1	−0.026	−0.336			ICT
	20	49.5					
	30	50		−0.336	0.294	0.797	Vijayan (1978)
	21.5	49.5		−0.341	0.294	0.887	de Chazal and Ryan (1971)
Iodobenzene (C_6H_5I)	16.8	45.67					ICT
	20	41.84		+0.82718			
Kerosene	28	40		−0.184	1.422	0.836	Vijayan (1978)
	18	40		−0.191	1.47	1.08	Vijayan (1978)
		28.2		−0.195			Grilc et al. (1984)
	28	40.4					
	20	47					Sherony (1969)
Mercury	20	375					ICT
Mesityl oxide	30	7		−0.135	0.650	0.7975	Vijayan (1978)
Mesitylene (C_9H_{12})	20	38.7		−0.13577			ICT
Methyl acrylate		13−14					Vijayendran (1980)
Methyl butyl ketone ($C_6H_{12}O$)	20	9.73		−0.1796			ICT

Dilute Oil–Water Emulsions and Dispersions

Compound						Reference
Methyl ethyl ketone	26	9.8				
	23	0.3	−0.123	0.60	1.45	Vijayan (1978)
Methyl isobutyl ketone	20	9.8	−0.179	0.60	0.93	Vijayan (1978)
	21	9.7	−0.1968	0.478	0.894	de Chazal and Ryan (1971)
	20	10.5	−0.201	0.59	1.00	Vijayan (1978)
	20	10.2	−0.1874	0.615	1.060	Keith and Hixon (1955)
Methyl tert-butyl ketone	20	10.81	−0.1864			ICT
Methyl hexyl carbinol	20	9.42	−0.1717			ICT
Methyl hexyl ketone	20	14.09	−0.1775			ICT
Methyl propyl ketone	20	6.28	−0.1772			ICT
Methylene chloride (CH_2Cl_2)	20	28.31	+0.3268			ICT
Methylene iodide	20	48.5	+2.3189			
Nitrobenzene	25.2	22	−0.573	1.771	0.892	Vijayan (1978)
	20	25.66				ICT
Nitromethane (CH_3NO_2)	20	9.66				ICT
	20	9.66	+0.1104			Vijayan (1978)
	27.5	9.61	+0.099	0.56	0.875	Vijayan (1978)
o-Nitrotoluene ($C_7H_7NO_2$)	20	27.19	+0.1627			ICT
	25	26.6	+0.163	2.46	0.8904	Vijayan (1978)
m-Nitrotoluene	20	27.68	+0.1576			ICT
Nonyl alcohol	20	4.9	−0.191	16.2	1.00	Vijayan (1978)

TABLE 16 (Continued)

Compound	Temp. (°C)	Surface tension (mN/m)	Temp. dep. (mN/m·°C)	$\Delta\rho$ oil-water	μ_{oil} (mPa·s)	μ_{water} (mPa·s)	References[a]
Nujol	25	50.8					Golob and Modic (1977)
i-Octane	20	50.8		−0.3043	0.522	0.998	ICT; Kaye and Laby (1966)
n-Octane	20	50.81	−0.048	−0.295			Fowkes (1965)
Octanoic acid (M = 144.2)	20	8.5					
n-Octyl alcohol ($C_8H_{17}OH$)	20	8.5	+0.039	−0.168			
Oil							
Coker slop.							
No. 1		10.34		−0.11	2.74		Chiev et al. (1977)
No. 2		13.02		−0.138	8.16		Chiev et al. (1977)
heating		16.31		−0.10	4.5		Chiev et al. (1977)
API separator skimming (pH dependent)		10.69		−0.138	4.4		Chiev et al. (1977)
Oleic acid ($C_{18}H_{34}O_2$)	20	15.59					ICT
Paraffin oil	25	43.7					Grilc et al. (1984)

Dilute Oil-Water Emulsions and Dispersions 451

Pentachloroethane	18	42			
	25	42.4	+0.676	2.03	0.95	Vijayan (1978)
i-Pentane	20	49.64	−0.3784			ICT
Perchlorethylene		32.3	+0.62			
Pentetole ($C_8H_{10}O$)	20	29.4	−0.03346			ICT
Phenetole	24.4	29.1	−0.0373	0.956	0.942	de Chazal and Ryan (1971)
Phenyl ether	25	40.8	0.079			Vijayan (1978)
Phenyl isothiocyanate (C_7H_5NS)	20	39.04	0.13515			ICT
Polydimethylsiloxane	20	41.3				Fowkes (1965)
Propyl acrylate		26–27				Vijayendran (1980)
i-Propyl ether		18.3	−0.2724	0.324	0.930	Keith and Hixon (1955)
Ricinoleic acid ($C_{18}H_{34}O_3$)	20	21.34	∼0.124			ICT
Shell Sol K	25	40				Grilc et al. (1984)
Styrene	19	35.48				ISEC (1970)
	20	49.3	0.2033	1.865	1.005	Golob and Modic (1977)
		40–43				Vijayendran (1980)

TABLE 16 (Continued)

Compound	Temp. (°C)	Surface tension (mN/m)	Temp. dep. (mN/m·°C)	$\Delta\rho$ oil-water	μ_{oil} (mPa·s)	μ_{water} (mPa·s)	References[a]
Tetrabromoethane	25	36.2		+1.957	9.46	0.89	Vijayan (1980)
Tetrachloroethane	25	31.3		+0.584	1.502	0.89	Vijayan (1980)
Tetrachloroethylene (C_2Cl_4)	20	47.48		+0.6235			Vijayan (1980)
	25	37.0		0.618	0.835	0.876	Vijayan (1980)
Toluene	20	36.1					Fowkes (1965)
	20	36.0		−0.1341	0.627	1.003	Golob and Modic (1977)
	23	34.6		−0.134			Vijayan (1978)
	25	28.0		−0.130	0.558	0.910	Vijayan (1978)
1,2,3-Tribromo propane ($C_3H_5Br_3$)	20	38.50		+1.4162			ICT
Tributyl phosphate	25	9.9		0.003			Vijayan (1978)
Trichloroethylene							
Trimethyl ethylene (C_5H_{10})	20	17.26					ICT
Trimethyl pentane	20	46.9		−0.305	0.517	1.00	Vijayan (1978)
Undecylinic acid ($C_{11}H_{20}O_2$)	25	10.14		−0.0885			ICT

Compound	Temp				Reference
i-Valeric acid [$(CH_3)_2$ $CHCH_2CO_2H$]	20	2.73	−0.0541		ICT
i-Valeronitrile (C_5H_9N)	20	14.4	−0.20328		ICT
Vinyl acetate	20	18–19			Vijayendran (1980)
Vinyl chloride	50	32			Nilsson et al. (1985)
White oil	20	55.1	−0.127	184	Vijayan (1978)
o-Xylene (C_8H_{10})	20	36.06	−0.1190		ICT
p-Xylene	20	37.77	−0.1318		ICT
	20	37.5	−0.1266	0.64 1.004	Keith and Hixon (1955)
Ethylene glycol/compound systems					
Benzene	27	7.4			
Hexane	26.7	14.2			
Methyl isobutyl ketone		3.0	−0.258	0.65 15.18	
Nitrobenzene	26.7	4.9			

[a]ICT, International Critical Tables.

TABLE 17 Estimating the Surface Tension for Liquid-Liquid Systems

Liquid-liquid pair	Φ_{exp} for use in Eq. (51)	
Fluorocarbon-hydrocarbon near critical mixing temperature	1	
Liquid metal/liquid metal	1	
Hydrogen-bonded organic liquid/water	1	
Nonpolar, saturated organics/water	0.55	
Aromatic hydrocarbons/water	0.7	
Dipolar organics/water	0.7 to 1	Increasing with increasing dipole moment of the organic, μd

Source: Good and Elbing (1971).

fine the fluid dynamics in the continuous phase. The drop size in an agitated vessel is described elsewhere and by Woods (1984) and Coulaloglou and Tavalarides (1976). Here we consider flow in a pipe and through a pump.

1. Pipe Flow

This behavior has been studied by Karabelas (1978), Kolmogoroff (1949), Hinze (1955), Shinnar (1961), and Sleicher (1962). Most of the workers correlate their data in terms of the maximum size drop that can exist in a given set of conditions. Since the drop size being considered is small enough, most drops will be spherical, so the shape factors do not have to be considered. The results naturally depend on the configuration and on what mechanism is important: viscous shear, turbulent shear, or turbulent pressure variation. Karabelas's correlation is

$$\frac{D_{p95}}{D} = \left(\frac{1}{We}\right)^{0.6} \qquad (52)$$

where

Dilute Oil-Water Emulsions and Dispersions

$$We = \frac{D<v>^2 \rho_c}{\gamma} \tag{53}$$

and where D is the inside pipe diameter, $<v>$ the average fluid velocity, γ the interfacial tension, ρ_c the density of the continuous phase, and D_{p95} the drop diameter at the 95th percentile.

Example 10 The exit from the reactor has a $<D_p>_{32}$ of 200 μm. If the viscosity of sulfuric acid is 1.3 mPa·s and a density of 1.85 Mg·m^{-3} and the mixture flows at an average velocity of 5 m·s^{-1} in a pipe of inside diameter 8 cm, does the drop size remain the same? Assume that the drop size distribution can be represented by the upper-limit log-normal distribution.

Solution: If we assume that Karabelas's correlation is the best, then for sulfuric acid as the continuous phase,

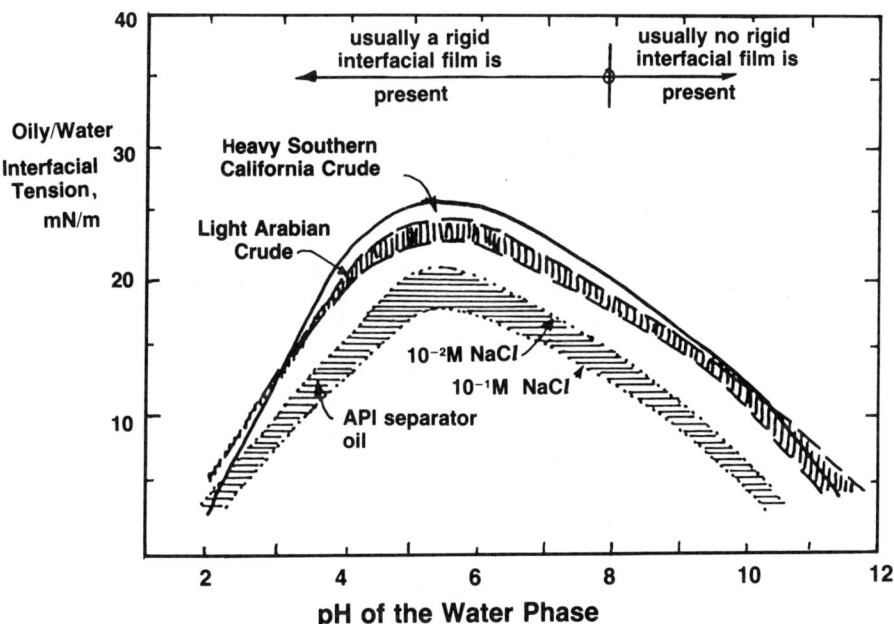

FIG. 35 Oil-water surface tension as a function of pH.

$$We = \frac{\rho_c D <v>^2}{\gamma}$$

$$= 1.85 \times \frac{10^3 \text{kg}}{\text{m}^3} \times 0.08 \text{ m} \times \frac{25 \text{ m}^2}{\text{s}^2} \times \frac{1}{\gamma}$$

The surface tension was estimated to be 54.1 mN·m^{-1}. Hence

$$We = 6.84 \times 10^4$$

Hence

$$\frac{D_{p95}}{D} = 4\left(\frac{1}{We}\right)^{0.6}$$

$$D_{p95} = 4\left(\frac{1}{796}\right) 0.08 \text{ m}$$

$$= 402 \text{ }\mu\text{m}$$

For an upper-limit log-normal distribution,

$$D_{p95} = \frac{(D_{pmax}/D_{p50}^+) \exp(1.163/\delta)}{1 + (1/D_{p50}^+) \exp(1.163/\delta)}$$

If we assume that $\delta = 0.90$ and $D_{p50}^+ = 1.33$, we can calculate D_{pmax}.

$$\frac{D_{pmax}}{1.33} = \frac{(402 \text{ }\mu\text{m})[1 + (1/1.33) \exp(1.163/0.90)]}{\exp(1.163/0.90)}$$

$$D_{pmax} = 550 \text{ }\mu\text{m}$$

$$<D>_{32} = \frac{D_{pmax}}{1 + D_{p50}^+ \exp(1/4\delta^2)}$$

$$= \frac{550 \text{ }\mu\text{m}}{1 + 1.33 \exp[1/4(0.81)]}$$

$$= 195 \text{ }\mu\text{m}$$

Dilute Oil-Water Emulsions and Dispersions

Hence, according to this analysis and these assumptions, the drop size has been changed because of the turbulent conditions in the pipeline. The former distribution had $<D>_{32} = 200$ μm; now it is $<D>_{32} = 195$ μm.

Comment: The velocity of 5 m/s used in this example is about five times higher than we would expect in practice.

2. Dispersions Flowing Through a Valve

Burrill (1967) studied pipe flow where oil and water mixed at a tee junction and then the mixture flowed through a valve. He correlated the drop size distribution in terms of the pressure drop across the valve. The total fluid velocity was constant at 2.4 m/s; the valve opening was changed. His results were relatively independent of the oil concentration for the range of oil/water ratios 0.13 to 0.21. For the range of pressure drops across the valve of 2.3 to 11.7 kPa, the geometric mass average diameter varied as follows:

$$_G<D_p>_{43} = 524.9 - 12.7 \, (\Delta p, kPa) \quad [=] \text{ micrometers} \qquad (54)$$

The geometric standard deviation was 2.00. Some results are shown in Fig. 36 for $\Delta p = 9$ kPa. The system was carbon tetrachloride in water ($\gamma = 38.5$ mN/m; $\mu_c = 0.894$ mPa·s; $\Delta \rho = 0.583$ Mg/m^3).

3. Dispersions Flowing Past an In-line Mixer

Manchanda (1966) studied the flow of a slow-moving liquid mixture of coconut fatty acid/water ($\gamma = 8.22$ mN/m; $\Delta \rho = 0.111$ Mg/m^3) past a small, in-line propeller mixer turning at 1000 rpm. The resulting distribution is shown in Fig. 37. The geometric standard deviation is 1.62.

4. Pumping Dispersions

Shakleton et al. (1960) report the data for pumping oil-water dispersions through different types of pumps. First, the drop size distributions still tend to be log normal but with very large geometric standard deviations (on the order of 3.00). Centrifugal pumps produce the smallest-diameter dispersions with a geometric mass average diameter of about 180 to 220 μm. On the other hand, vane pumps gave geometric mass average diameters of 850 to 15,000 μm, depending on the exit pressure. A typical dispersion is shown in Fig. 38 together with the data from Figs. 36 and 37 corrected for physical properties so that the data are about what would be expected if the same oil/water mixtures had been used for valves, in-line mixers, and pumps.

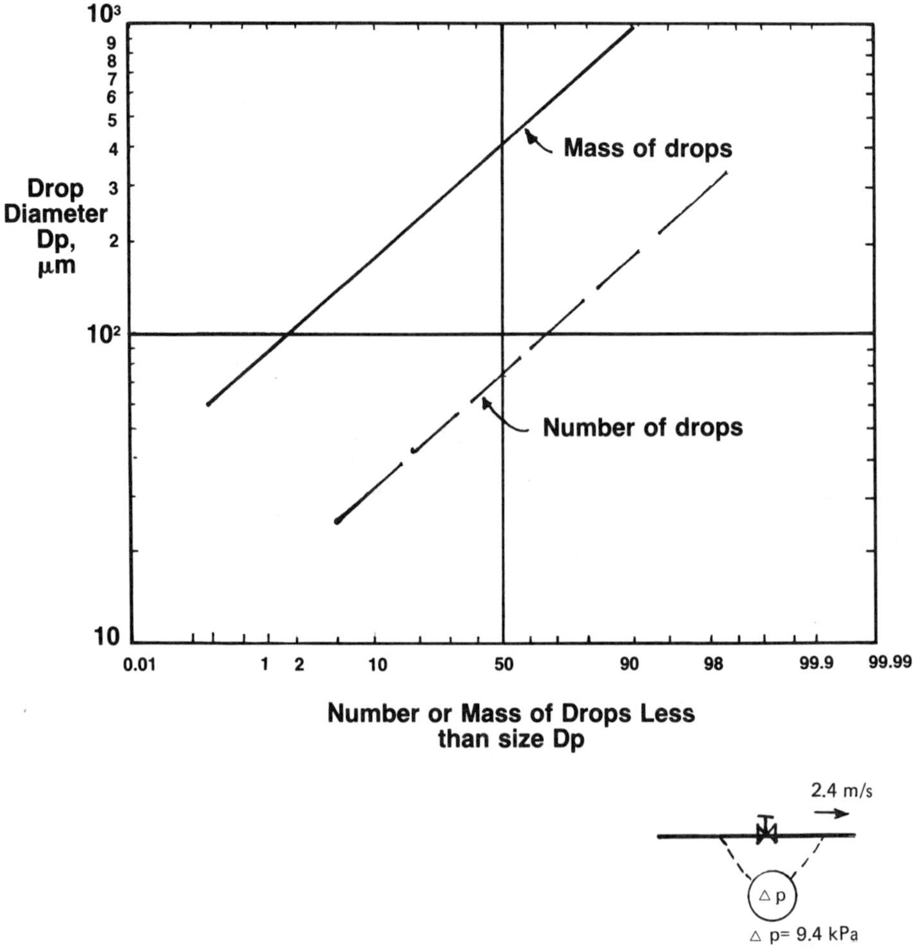

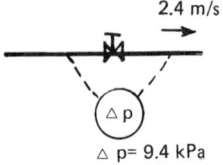

FIG. 36 Drop-size distribution for dispersions flowing through a valve. Diagram below graph: for carbon tetrachloride in water, O/W 0.13 to 0.21 [from Burrill (1967)].

5. Estimations

For design purposes we can estimate the diameter of the drops in the dispersion with the information in Sections II.B.1 to II.B.4 as guidelines. In general, the size will depend on the interfacial tension, the turbulence the dispersion experiences, and the density of the continuous phase. Equation (52) can be used as a guideline. For typical refinery oils with interfacial tensions of about 30 mN/m, Fig. 38 gives a comparison of the effect of turbulence caused by

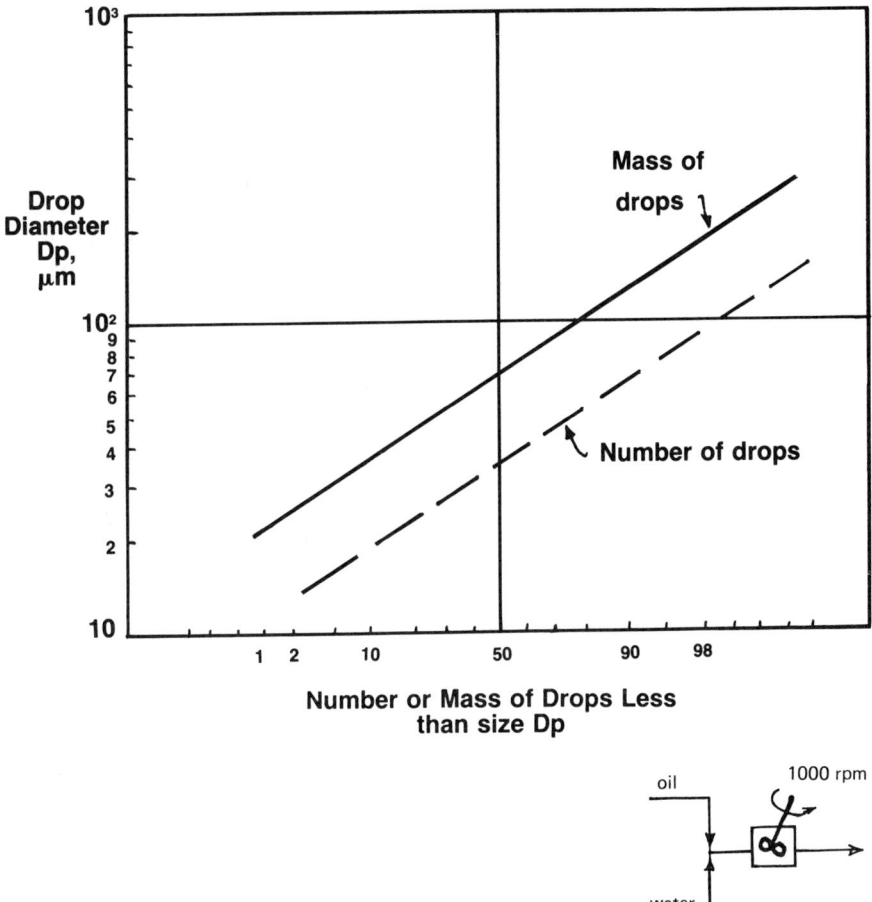

FIG. 37 Drop-size distribution for dispersions flowing through an in-line mixer. Diagram below graph: coconut fatty acid in water.

pumps, valves, and in-line mixers. These results are measured immediately downstream of the device. If the drops are unstable, coalescence will occur as the dispersion moves downstream, so that the data in Fig. 38 represent the "worst-case" situation.

Example 11 A refinery waste is pumped by a centrifugal pump 100 m downstream to a settling basin. Does this mean that if the basin is designed to remove drops 200 μm in diameter, only 50% of the drops will be removed?

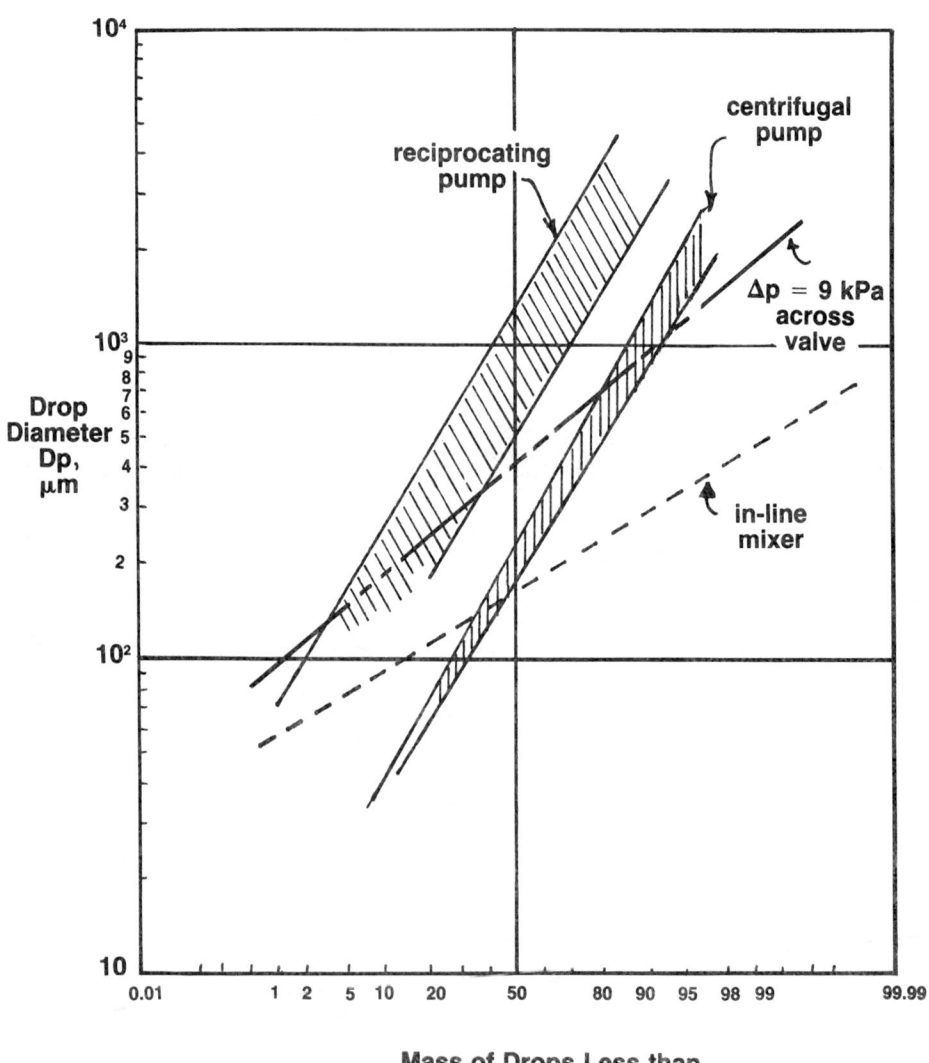

FIG. 38 Drop-size distributions for dispersions (oil-water with = 30 mN/m).

Solution: We usually use the recommendations of the API (1963) or Happel (1958a) and use 150 to 250 μm. This usually means that about 90% "of the mass" is removed. That is, we expect a lot of coalescence to occur in the pipeline. Next, we need to be very careful

Dilute Oil-Water Emulsions and Dispersions 461

in interpreting distributions. Distributions can be reported on a number basis (meaning the number of drops removed) or on a volume or mass basis (meaning the total volume of oil removed). These are very different; the amount of difference depends on the slope of the distribution or on the magnitude of the geometric standard deviation of the distribution. All the data reported here have been converted to a consistent basis of mass or volume. Thus to answer what percentage of drops are removed, Fig. 38 could not be used directly. It would have to be replotted with each separate distribution corrected using the Hatch-Choate relationships (see Woods, 1984).

Example 12 The liquids here are water dispersed in coconut fatty acid. What is likely to happen to the drop size distribution for such a dispersion flowing through a piping network, and could it be predicted quantitatively?

Solution: The interfacial tension for this system is about 8 mN/m. This value applies whether the water is dispersed in the oil, or vice versa. From Eq. (52) we expect that the drop size will decrease because of the decrease in both interfacial tension and density of the continuous phase. We might try using Eq. (52) to predict this decrease quantitatively and apply it to diameter at the 95th percentile. Which reference diameter we choose at this percentile depends on what characterizes the "piping network." Is it straight run piping? Then we might focus on the API- and Happel-recommended design values and adjust those. Does it have a pump close to the discharge? Then we might use the centrifugal pump line, including its geometric standard deviation of about 3.

Example 13 If a separator is designed to remove all crude oil in water drops that are greater than 150 μm, and the feed to the separator comes directly from a centrifugal pump, about how much oil would be removed?

Solution: From Fig. 39, the intersection of the diameter 150 μm with the representative distribution for a centrifugal pump is about 40 to 45% by volume (or mass). Thus about 60 to 55% *of the mass* would be removed, assuming that no coalescence occurred.

C. Predicting the Continuous Phase

Another concern is to decide which phase is likely to be the continuous phase. Figure 39 illustrates a dark fluid as the continuous phase and light fluid as the dispersed phase. As the oil concentration increases, the system undergoes phase inversion to yield light fluid as the continuous phase. Which phase is the continuous phase is a complex function of how the two phases are mixed, the wetting properties of the container walls, the energy input, the hydrophile-lipophile bal-

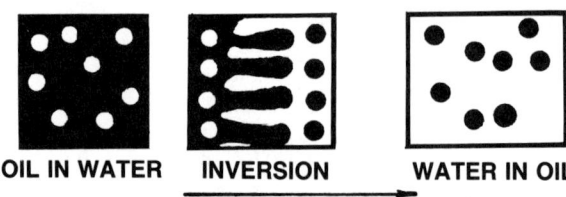

FIG. 39 Inversion phenomena.

ance (HLB) value of any surfactant present, and the volumetric concentration. Selker and Sleicher (1965) give us a simplified guidance based on the properties and phase ratios:

$$f = \frac{F_L}{F_H}\left(\frac{\rho_L}{\mu_L}\frac{\mu_H}{\rho_H}\right)^{0.3} = \frac{\varepsilon_L}{1-\varepsilon_L}\left(\frac{\rho_L}{\mu_L}\frac{\mu_H}{\rho_H}\right)^{0.3} \quad (55)$$

where L is the light phase $\rho_L < \rho_H$, H the heavy phases, F the volumetric flow rates of each, and ε_L the volumetric fraction of light phase. Table 18 helps us interpret the results and predict which is the continuous phase.

TABLE 18 Identifying the Continuous Phase

f Value	Excluding surfactant and wetting phenomena the continuous phase is:	
	Light	Heavy
<0.3		Always
0.3–0.5		Probably
0.5–2.0	Either	
2.0–3.3	Probably	
>3.3	Always	

III. OVERVIEW FOR SEPARATING

The steps for most separations are to understand the fundamentals as to why the dispersion is likely to be stable, to evaluate the strategies that could be used to create instability, to select the chemistry, and then to select and size the device. The key is to understand the fundamentals and select the chemistry. In real systems, we often encounter surprises and apparent contradictions. Indeed, not all the uncertainties have been resolved. We still rely heavily on pilot-scale tests.

The purpose of this and the following sections is to summarize what we know so far and to illustrate the general principles. Figure 40 illustrates the general principles. We first consider whether the dispersion is likely to be stable or unstable and to hypothesize as to the reason. If it is stable, we usually treat it first to cause it to be unstable. Of the various options described in Section I.B, coagulation is usually used, although either dissolved air flotation or deep-bed filtration is possible without chemical pretreatment. Once an unstable dispersion has been created, various physical separation devices can be applied, depending primarily on the concentration of dispersed phase and the drop size. Figure 41 gives the general regions of applicability.

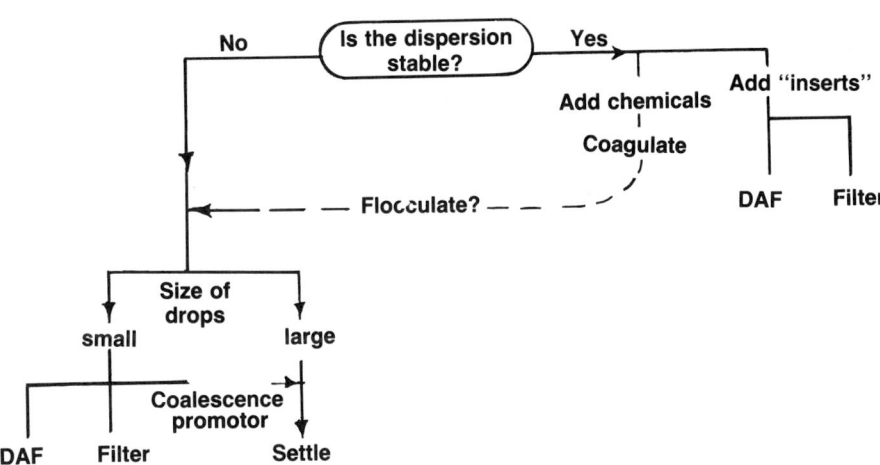

FIG. 40 Some options available.

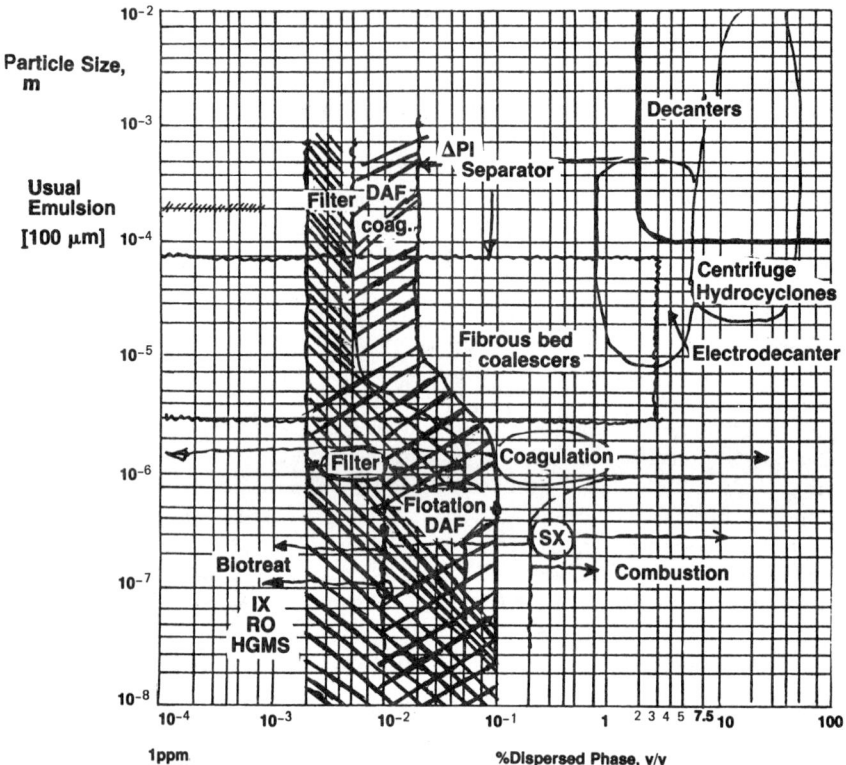

FIG. 41 General hardware applications for liquid-liquid separations.

IV. PHYSICAL SEPARATION

For larger drops, the body forces tend to dominate the surface forces and the drops will migrate to form a band of drops and may indeed coalesce to form a separate layer. This migration of the drops can occur under gravity or in a centrifugal field. However, the latter is rarely used for dilute dispersions.

A. Fundamentals and Equipment Considerations

The fundamental principle is that the drops migrate because of gravity, so that they travel a distance x in time t. We design a vessel of depth x that provides the drops with the required residence time t. For dilute dispersions, the separator is usually a long, horizontal, rectangular basin. Such devices are often referred to as API separators because the general design procedures were developed by the American Petroleum Institute.

Dilute Oil-Water Emulsions and Dispersions

For single drops of diameter D_p we can estimate the migration velocity in quiet water from Stokes' law, assuming that the density difference between the phases is constant and is the driving force, that the continuous phase fluid viscosity is constant and known, that the drops are spherical and such that Stokes' law applies, and that the drop surface is sufficiently contaminated that surface movement causing internal circulation does not occur. If the influent liquids have been equilibrated and negligible temperature gradients occur in the basin, the most challenging assumption is that the continuous phase is "quiet." Because the drops migrate vertically, horizontal, nonturbulent, uniform flow of the continuous phase would be equivalent to having a "quiet" continuous phase. Under these ideal conditions, for a basin of depth h, the required time for the drops to migrate the full depth would be

$$t = \frac{h}{v_\infty} \tag{56}$$

If for the continuous phase, the ideal horizontal, uniform, nonturbulent flow were achieved, then for a volumetric flow rate of F_1, through a tank of width B and length L, the elapsed time of the fluid (and the drops) would be

$$t = \frac{LBh}{F_1} \tag{57}$$

Equating these two times leads to the design concept of an overflow velocity, v_{OF}:

$$v_{OF} = \frac{F_1}{LB} = v_\infty \tag{58}$$

Thus the basins would be designed to provide an overflow velocity consistent with the expected size of droplets entering the basin.

The design procedures suggest how to design the inlet so as to try to achieve the ideal continuous phase conditions, how to correct for turbulence and nonideal flow patterns, and how to design the exit configurations. Other complications are how to collect and remove the heavy materials that might sink to the bottom of the basin—without stirring up the basin—and how to skim off the top layer of oil—without causing circulation in the basin.

B. Variations in Design

A popular approach to cope with the above-mentioned nonidealities is to insert parallel plates in the direction of flow. These minimize short circuiting and turbulence in the continuous phase and reduce the distance the drops have to migrate to about 2 cm (instead of the full depth of the basin). For more on the use of parallel-plate inserts, see Brunsmann et al. (1962), Probstein et al. (1981), Hooper and Jacobs (1979), and Miranda (1977).

C. Pretreatment

The main upstream pretreatment used is to try to stabilize or smooth out the flow rates and compositions of the feed to the basin. Often, an API separator is used as a pretreater for other separation methods used to remove the smaller-diameter droplets.

D. Design Procedures and Results

The design of the API separator can be based on the overflow velocity or a total-detention-time approach.

1. Overflow Velocity Approach

Two different approaches are suggested by the American Petroleum Institute (1963) and by Hart (1947). Although the basic approach is an overflow velocity approach, this design parameter is not used explicitly. The correction factors used in the API approach are obtained from work at Wisconsin in 1951 (research cited in API, 1963) which suggested that the short circuiting could be accounted for by a factor of 1.20; the turbulence factors were those suggested by Camp (1946). These depended on the ratio of the settling to horizontal velocities. Hart's approach was to select a standard width (6 m because of the size of mechanical raking equipment) and a standard depth of 2.4 m. Then he studied scale models to obtain expressions for laminar flow of the fluid with "pure" water properties used in the Reynolds number calculation and with different oils (as characterized by their °API gravity) as the contaminant. His idea of allowing the drops to circulate is probably not correct because the presence of surfactants and contamination is likely to force the drops to behave as rigid spheres.

2. The Detention-Time Approach

Perhaps the most popular approach for more concentrated dispersions is the detention time approach. In this method we somehow have access to a reasonable value of the "total, ideal plug-flow residence time" that the dispersion should spend in the separator. Such numbers are not commonly used in wastewater separations, but this deficiency is not something to worry about, because the usefulness of this single number is questionable. Order-of-magnitude numbers calculated for a designed unit are:

1. For 190 L/s in a 2.4 m X 6 m standard cross-section unit designed by the API (1963) method, the t = 417,400/188 L/s = 37 min.
2. For 102 L/s in a 2.4 m X 6 m standard cross-section unit designed by the Hart (1947) method, the t = 55,100/102 = 9 min.
3. For the example given in Eckenfelder (1966, p. 50) designed by the API (1963) method, the t = 1.8 X 30 X 6.7 X 10^3/60 X 188 L/s = 33 min.

General relationships between the inlet and exit oil concentrations from an API type of gravity settler are given in Fig. 42.

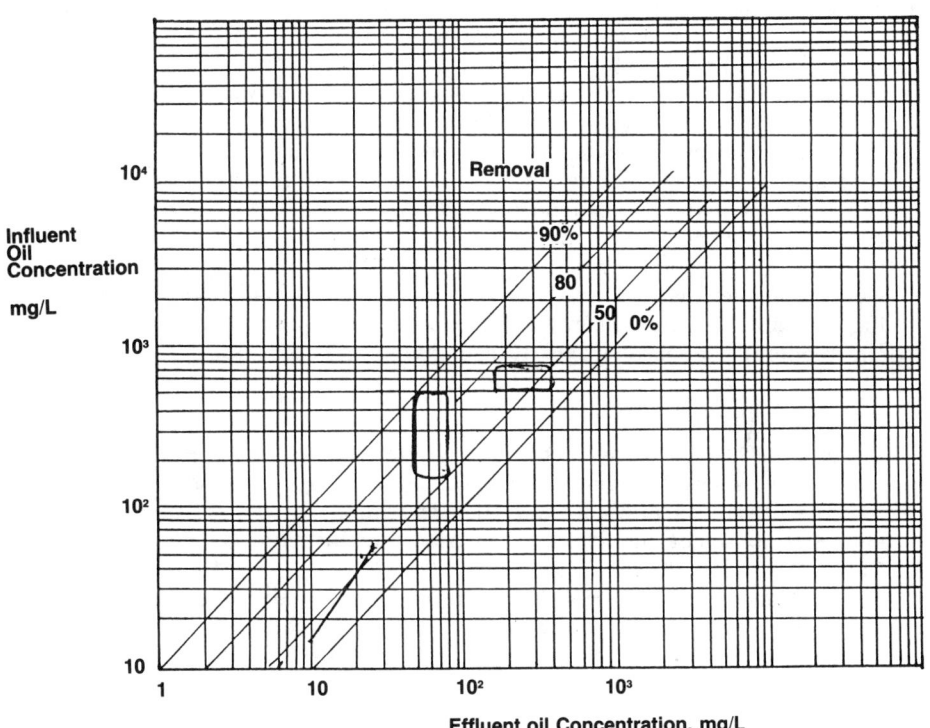

FIG. 42 General performance of gravity settlers.

V. CAUSING INSTABILITY: CHEMICAL DISPLACEMENT

When the emulsion is stabilized by surfactants and/or a film, a possible method for causing instability is to add a destabilizing surfactant that displaces the initial, stabilizing surfactant (mechanism 10). Alternatively, a solvent that might dissolve the film or a species that might change the solubility of the stabilizing surfactant might be used (mechanism 11). Consider each of these in turn.

A. Type of Surfactant

Usually, if a nonionic surfactant stabilizes the dispersion, then:

1. For an oil-in-water dispersion the surfactant has an HLB value greater than 9 and usually around 10 (we might try destabilizing it by adding a surfactant with an HLB less than 4).
2. For a water-in-oil dispersion, the surfactant has an HLB value of less than 6 and usually around 5 (we might try destabilizing it by adding a surfactant with an HLB greater than 9) (Lawrence and Killner, 1948; Lissant, 1983, p. 88; Osipow, 1962, p. 336).

Many naturally occurring systems are not as simple as this and consist of mixtures of surfactants and stabilizing films. Nevertheless, this general principle can be used as a guide in testing possible options. Lissant's (1983) survey of the application of these ideas to crude oil-water systems outlines the great variability, complexity, and lack of simple answers even though we have a reasonable understanding of the basic fundamentals. Some generalizations that illustrate the principles of mechanixm 10 are given in Table 19.

1. Determining the Chemistry

Lissant (1983) describes a bottle test procedure for field testing. In laboratory work, in general, a standardized procedure is chosen to mix the chemicals and to allow the separation to occur. Each researcher tends to select his or her own set of conditions, as illustrated in Table 19.

Different characteristics of the oils and waters should be noted to ensure that the test results can be transferred. For the water phase, for example, the ionic strength, the pH, and the calcium and/or magnesium concentration should be noted. For the oils, the asphaltene and free carbon concentrations are important because both seem to affect the stability.

To guide in the choice of test surfactants, Table 5 lists some surfactants and their HLB values. Other information about surfactants can be obtained from such publications as *McCutcheon's Deter-*

TABLE 19 Example Performance of Chemical Destabilizers

| | | Effluent removal (%) | Destabilizer[a] | | Conditions | | | | | |
| | | | | | Rapid mix | | | | | |
Scale	Feed		Name	HLB	rpm	Time	Settle	Other	Ineffective	Reference
Produced cold lake w/o										
Batch for w/o	Water Clay (25%)	63	Brij nonionic (N)+	>15	150	60 min	3 h	Added 2/1 w/w toluene	Brij HLB<15,	Cooper et al. (1980)
	Water Clay	40								
	Water Clay	100	BASF Pluronic (N)	4-19						
	Water Clay	55								
		95	Span	4-6					Span HLB<4,	
		55							Span HLB<6	
		50	Span 80 + Tween 20	≅9						
Conc. Emulsion: free C, 8 to 16%; asphaltene, 26 to 36; viscosity, 62 to 683										
Batch for o/w	Water (7 to 78%)	55-84	Amine 220 acetate (C)		1400	30 min	21 h		Acetic acid, salt of dimethylamino ethyl oleate, polyoxyethylene thioether	Pincus et al. (1951)
		51-79	Amine 220 (C)							
		76-78	Amine 220 (C) but ineffective on low % C and low viscosity							
		61-68	Amine 220 (C)							
		40-81	Amine 320 (C)							
Saybolts Batch for w/o	Water (27 to 48%)	Negligible to 60	Polyoxyethylene thioether (N)						Amine 220, Amine 320	

[a]N, Nonionic; C, cationic.

gents and Emulsifiers (J. W. McCutcheon, Inc., annual). For water-in-oil emulsions, Dow (1926) lists the following as being potential destabilizers to add: iron oxide, finely divided silica, clay, sodium soaps, resin and resin soaps, gums, starches, and organic sulfonates.

2. Equipment Selection, Sizing, and Results

Once a reasonable set of chemical conditions have been established, the process can be designed. Here there is great variability because of differences in the conditions. For the challenge of breaking emulsions produced underground from oil wells, the chemicals can be pumped directly into the well. For breaking emulsions in processing plants or waste treatment, the chemicals are usually mixed well with the process stream—say, by adding the chemicals to the suction of a pump—and then the phases are allowed to settle in a basin.

Example 14 Fuel oil has become contaminated with water and a stable emulsion of water in the oil has resulted. What might be done? The volume was about 17 m^3 and the water concentration is 70%.

Solution: Based on the general principles, for a water-in-oil emulsion we might try using a nonionic surfactant with an HLB of 8 to 10. From Table 19, Lawrence and Killner used an anionic surfactant, Teepol (A HLB 16), but here an optimum amount of reagent must be added based on the water content. Adding too much would stabilize the dispersion by means of Teepol; adding insufficient would mean that the asphalts adsorbed at the oil/water surface would not be displaced and the stabilizing membrane would remain. They used 0.13% of the volume of water in the emulsion. Thus the addition of about 15 L should be sufficient. The mixing would be done by circulating the contents through a pump and adding the Teepol to the suction. However, the temperature should be increased to about 43°C.

Comment: Lawrence and Killner (1948) were able to remove about 100% of the water when about 27 L had been added. The temperature was 58°C and the liquids were allowed to settle "overnight." Thus although the large-scale tests were relatively successful, only order-of-magnitude values could be predicted for the concentrations and temperature.

Other examples applications are given by Little and Patterson (1978) and Berkman and Egloff (1941).

B. Type of Solvent

Sometimes a solvent can be added to dissolve/displace the stabilizing agent (as suggested by mechanism 11). Gopal (1968) cites carbon disulfide and carbon tetrachloride as example solvents. Dow (1926) cites these and adds acetone, alcohol, and ether. Either small amounts of the solvent can be added, or almost equal volumes of solvent can be used so that the process essentially is solvent extraction. Examples are given in Table 20. If relatively large amounts of solvent are added to dissolve a surface membrane, the process can be sized like a traditional solvent extraction process.

C. Other Types of Displacing Additives

Microorganisms might play a role in destabilizing emulsions. Typically, microorganisms tend to stabilize emulsions. However, if the microorganisms removed the stabilizing membranes and did not themselves cause stability, they could be used to break emulsions. Angle and Hamza (1985) studied the stability of kerosene and bituminous emulsions in water. The microorganisms that they studied were *Nocardia amarae* and *Rhodococcus aurantiacus*.

VI. CAUSING INSTABILITY: CHEMICAL DESTABILIZATION-COAGULATION AND FLOCCULATION

The fundamental mechanisms that can be used to cause instability were given in Table 6. The most popular approaches are to change the pH (mechanism 2); to add an adsorbing ion (mechanism 4); and/or to add another component, such as bubbles, fibers, plates, or mesh or ground-up solids (mechanism 9). Although we can always consider pH adjustment, for most systems the oil drops are negatively charged when the water pH is around 7. Hence we trade off the expense of a pH excursion with the cost of the other options suggested by mechanisms 4 and 9.

The term "coagulation" usually applies to processes where chemical species are added. Alternatively, the use of "chemical destabilizers" usually results in coagulation or flocculation. [No agreement has been reached on a definition of the terms coagulation or flocculation. O'Melia (1972) suggests that coagulation relates to the "change in the charge" on the drops, while the term "flocculation" refers to the "transport" of the drops toward each other. LaMer and Healy (1963) use the term "flocculation" when the chemicals added are polymers, whereas, they refer to the inorganic chemicals as coagulants. In general, the terms "coagulant" and "flocculant" are used loosely and sometimes even refer to chemical additions that compress the double layer without adsorbing (mechanism 3).]

Processes that depend on mechanism 9 include dissolved air flotation (DAF), where bubbles are introduced; deep-bed filtration, where the dispersion flows through a bed of sand or similar material; and

TABLE 20 Example Performance of Solvents

Feed	Scale	Feed Name	Feed Conc	Feed Type	Effluent	Solvent	Conditions	References
Fuel oil in water		Water	70%	w/o	Effectively dewater	Aniline (2%)	Not given	Gopal (1968)
						Cyclohexanol (3%)		
						Benzyl alcohol (3%)		
Steel mill rolling mill waste	Pilot	Oil		o/w		Toluene		Zirimenya and Woods (1968)

coalescence promotors, where the dispersion flows through a mesh or network of plates. These processes are discussed later. Here the focus is on coagulation.

A. Types of Coagulants or Chemical Destabilizers

In this context, *coagulant* refers to any chemical added other than to alter the pH. The coagulants are usually inorganic or organic. Inorganic coagulants are usually salts of such metals as aluminum or iron. Because these tend to hydrolize in water, they are called hydrolizing coagulants. Inorganic, nonhydrolyzing coagulants are the salts or bases of such metals as calcium and magnesium.

Polymeric flocculants (often called emulsion breakers) are classified according to their charge rather than their structural unit: anionic, cationic, or nonionic if they carry negative, positive, or no charge, respectively. Some examples are listed in Table 15. Polymeric flocculants are commercially available in a variety of molar masses and charge densities. Often they are used in conjunction with inorganic coagulants to produce flocs that dewater and separate easily. When used in this context, polymeric flocculants are often referred to as coagulant aids, although often they are called destabilizing agents or emulsion breakers.

B. Determining the Chemistry

Deciding what to add and how much to add is often evaluated by the so-called "jar test." In this bench-scale method, equal volumes of the emulsion to be treated are placed in transparent jars. By trial and error, chemicals are added to the various jars and the resulting clarification of the emulsion is observed qualitatively. From this should come an acceptable set of chemical conditions. These results depend on the shape of the jars, the type of mixing/flocculation conditions used, and the criteria used to select the conditions.

In addition, zeta-potential measurements could be done to test the effect of chemical additions on the charge on the drops directly. Care should be taken in interpreting the batch results because the eventual process will include charge control (so that under equilibrium conditions the drops are unstable), rate phenomena that will cause the drops to come together to form a floc or to coalesce, and rate phenomena to allow the drops to separate from the continuous phase. In terms of hardware we think of a chemical addition or rapid-mix tank, a flocculation basin, and a settling basin, respectively. What happens in each of these units will affect the overall separation. For this reason, carefully designed, pilot-scale, continuous-operation equipment could also be used to select the chemicals and the conditions. Regardless, we can use the fundamentals to estimate the types of conditions that are most likely.

The pH could be adjusted to yield the ZPC (mechanism 2). This could be quite complicated, depending on the system. If there are no solids present and only relatively clean oil, then the ZPC is, from Fig. 17, at a pH of about 4 to 5. If solids are adsorbed at the interface, the ZPC of the solids affects the system as well. Figure 19 shows that for produced waters, which have a mixture of clay and oil, the two substances combine to yield a charged drop that has a relatively pH independent, negative charge. Here pH adjustment would not help much.

Chemicals can be added to compress the double layer (mechanism 3). From the pH and an estimate of the zeta potential we could estimate the critical coagulation concentration of counterions needed to create instability. Thus Eq. (22) or (23) would be appropriate.

Example 15 No. 3 crude oil is dispersed in refinery wastewater. The pH varies between 8.5 and 10.5. The electrolyte concentration is about 0.1 M NaCl. Estimate how much chemical would have to be added to create instability.

Solution: From Fig. 19, the zeta potentials are about -50 to -60 mV. This is a relatively large potential. For oil, from Table 1 the Hamaker constant is about 10 βJ. Since this is a negatively charged system, a cation is the effective counterion and we should choose the highest possible valence. If we chose alum, then from Fig. 32, the alum would not be in the Al^{3+} form. Hence we choose a nonhydrolyzing species, such as calcium. Thus, from Table 3, for ϕ 80 mV and $z = 2$ for calcium, the value of Y is 0.9149. Substitution of these values yields

$$CCC = \frac{8 \times 10^{-36} Y^4}{A^2 z^6}$$

$$= 890 \text{ mmol/L}$$

Hence we would need to add 890 mmol/L.

Although this example illustrated how a knowledge of the surface charge (and the mechanism we would like to invoke) can guide us in deciding on the amount, such estimates are only guidelines. The type and amount of coagulant needed depends on the type and amount of species present. The pH, in particular, has rather dramatic effects on the charge on both the dispersed phase (say, oil and suspended solids) and the charge on the coagulant/emulsion-breaking polymers.

For hydrolyzing, inorganic coagulants, for example, the pH controls the amount of each hydrolysis species formed during the hydrolysis of the metal coagulant. Optimum pH values for hydrolyzing coagulants are usually slightly acidic (pH between 4 and 6). For higher values of pH, increased dosages are required for effective coagulation. Inorganic coagulants such as alum, are more sensitive to pH changes than are cationic polymers, because the isoelectric point for inorganics is, in general, at a lower pH than that of polymers. Aluminum ionic species undergo charge reversal and carry a negative charge at alkaline pH and therefore are not very effective in coagulation (according to the adsorption-neutralization mechanism). This charge reversal of the coagulant at higher pH can cause operating problems. [For example, if the coagulated material is trapped in a deep-bed filter, the filter might unload at high pH (Grutsch and Mallatt, 1976).] The sensitivity of the removal of oil from water by alum to both ionic strength and the pH is illustrated in Fig. 43. In this figure, the three chemical concentration variables are the alum dosage, the concentration of electrolyte, and the pH. The dependent variable is the mass percent of oil removed; this is shown as a parameter at lines of constant removal. Thus for the system studied, better removals were obtained at the higher electrolyte concentration. Figure 32 helps to explain the pH-alum interaction. The effect of electrolyte is expected because of the suppression of the double layer (mechanism 3).

Enough about the warning of the complexity of the systems with which we are dealing. What about some general guidelines? The removal is a function of the chemicals added, the mixing conditions used, the equipment to waft the particles together, and the eventual phase-separating procedure. These options are listed in Table 21. The general removals, for two general types of feeds and for different components in feeds, are summarized in Table 22 for the general performance of coagulation for the three phase-separation methods: settling, dissolved air flotation, and filtering.

Polymer dosage requirements are, in general, lower than alum dosages for the usual refinery effluents. Sometimes a combination of alum and polyelectrolyte is used. The costs and the quality of the oil layer or sludge removed determines the choice.

Concerning the actual testing procedures, the hydrodynamic conditions in the jar should bear some relation to the large-scale conditions planned, but this may not be easy because of nonidealities on the larger scale and because of the ongoing research into what are the best conditions for mixing and coagulating. Table 23 summarizes some of the suggestions about the conditions for the jar test. For the design of a rapid mix/coagulation unit, the chemicals should be mixed in a standard jar at the appropriate conditions.

Other devices that better represent the scaled-down version of the proposed unit can be used. Ives (1978) describes the use of couette devices, baffled tanks, and tubes. Eisenlauer and Horn

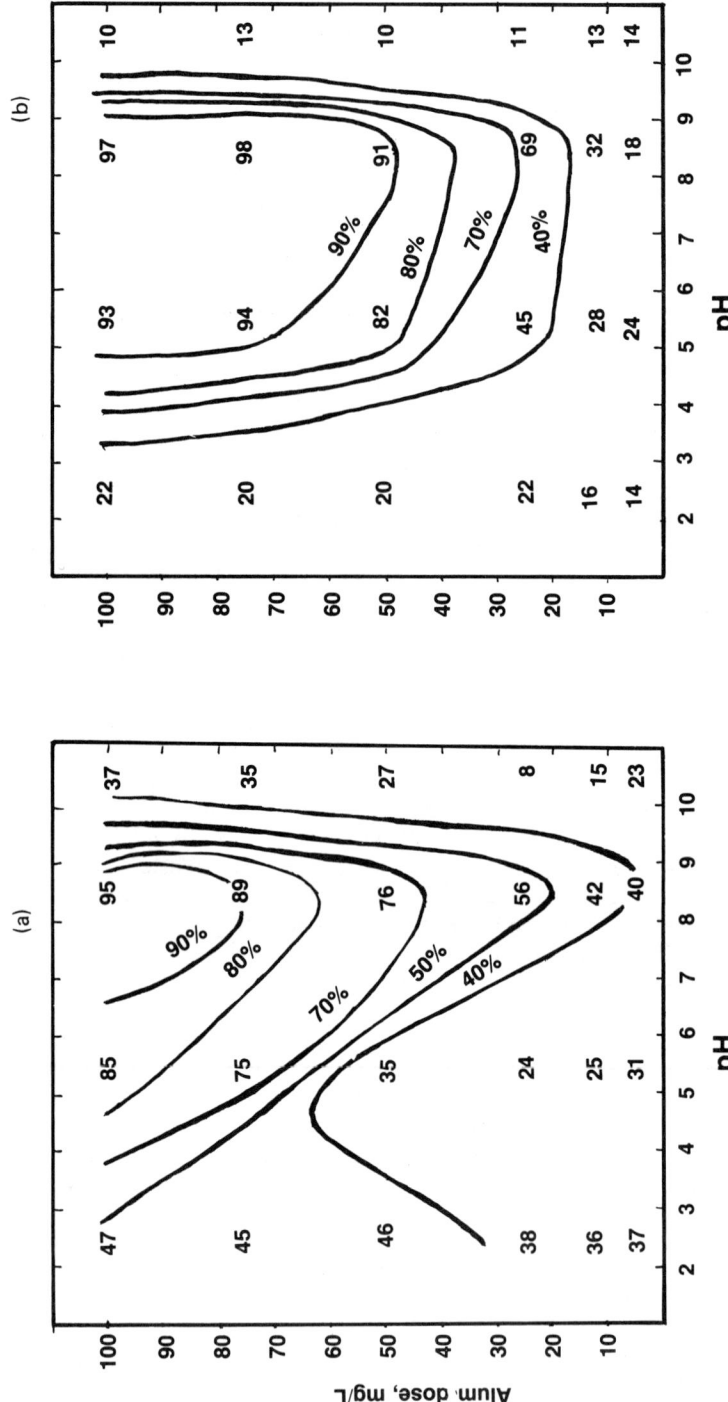

FIG. 43 Removal of oil from water as a function of aluminum dosage and pH.

TABLE 21 Options for Coagulation

	Mechanisms	Chemical addition in...	Preform?	Separate
"Coagulate"	4	Rapid mix ⟶	coagulation basin	⟶ gravity settler
		Rapid mix ⟶	coagulation basin	⟶ DAF § 7
				⟶ DAF § 7
		Mix with input		⟶ DAF § 7
				⟶ Filtration § 8
Polyelectrolytes to "break"	7, 8	Direct to ⟶	coagulation basin	⟶ gravity settler
		Direct to		⟶ DAF
		Direct to Feed to		⟶ Filter
Coagulate/ polymer flocculate combination	4, 8	Rapid mix ⟶ for coagulants polymer	coagulation basin	⟶ gravity settler
		Coagulants		⟶ DAF
		Polymer direct to		⟶ Filter

(1985) describe a continuous coagulation flow system where the coagulation basin is simulated by the flow through a tube. This allows continuous analysis of the stream and the effectiveness of the chemical additions. Szpak (1986) compares different jar and paddle configurations.

C. Equipment Selection, Sizing, and Results

As illustrated in Table 21, coagulation is followed by some form of phase separation. In this section the focus is on a separate coagulation basin followed by settling. The other options are discussed in Sections VII and VIII. The processing equipment needed usually includes a rapid mix (where the chemicals are intimately mixed with

TABLE 22 Example Chemical Conditions for Oil Removal

Type of Waste	Settling (mg/L)		DAF (mg/L)			Filtering (mg/L)		
	Alum	Cationic polymer	Alum	Polyelectrolyte Cationic	Polyelectrolyte Anionic	Alum	Polyelectrolyte Cationic	Polyelectrolyte Anionic
API effluent	40–80	0	25–75	—	—	10–30	—	0.1–0.4
	20	1 to 2	0	2–15	—	5–20	3–8	0.1–0.4
	0	2 to 4				0	4–10	0.1–0.4
High-strength wastewater	1000	—	NA	—	—			
	0	>1000	—	75–150	0–20			
Aerated lagoon effluent								
fresh water						5–20	3–6	0.1–0.4
Brackish water						15–20	10–20	0.2–0.4
Typical removals:								
Oil	88–100			68–99			64–95 <78>	
SS	65–99			70–98			50–95 <68>	
BOD	75–90			33–71			28–63 <45>	
COD	57–90			42–96			42–63 <55>	
TOC	69–91			39–72			17–62 <41>	

TABLE 23 Recommendations for Coagulation/Settling Systems

	Rapid mix		Precoagulation total			Coagulation total			Settling overflow time (L/m²·s)
	G (s^{-1})	t (s)	G (s^{-1})	t (min)	No. of basins	G (s^{-1})	t (min)	No. of basins	
Bratby (1981) for alum and turbidity removal	650	0.5–8	100	31.5	3	100 50 30	10 10 10		1
Alum in destabilization	16,000	1					30	3	
Alum in sweep floc	300	60				25	20		
King et al. (1983) (optimal)	110	120				25	20		
Churchill and Kaufman (1973) (optimal)						48	10		
Bloodgood (1952)		120				52	20		
Slezak and Woods (1973)		45				70 48 20	10 17 17 17	3	

the stream), a coagulation basin (where the principles from section C.2 are applied to gently bring the drops together), and a settling basin (where the flocs created can rise or settle out). This is illustrated in Fig. 44.

1. Rapid Mix

Sometimes the chemicals are added to the feedstreams into the coagulation basin; sometimes, a separate mixing tank is chosen. For a separate mixing tank upstream, the vessel is usually sized to provide a plug-flow, liquid residence time of 45 s. The mixer provides an impeller Reynolds number of about 10^4; this corresponds with a power input of 1.5 kW/m^3.

Bratby (1981) studied the conditions for the optimum removal of turbidity using alum. For conditions where the alum is approximately in the destabilization region (on Fig. 32) he recommended that the mixing conditions be $G = 650$ s^{-1} and $t = 0.5$ to 8 s. On the other hand, Amirtharajah and Mills (1982) recommended that for alum in the destabilization region, the values should be $G = 16,000$ s^{-1} and $t = 1$ s; for alum in the sweep floc region, the conditions in the rapid mix can be any combination to give a Gt of about 18,000 to 20,000 (for example, $G = 300$ s^{-1} and $t = 60$ s or $G = 1000$ s^{-1} and $t = 20$ s).

Sontheimer (1978) suggests slightly different values: liquid residence time of 120 to 300 s, mixer tip speed of about 5 m/s, and a power input of 0.5 to 2 kW/m^3. Other configurations, besides a mechanically mixed tank, can be used. Some use a baffled chamber or feed line; others just blend the chemicals together in the pipe.

A word of caution: Sometimes the chemicals, expecially polymers, are mixed in a pipe or channel too far upstream of the coagulation basin. As a result, the flocs begin to form, but because of the high shear in the pipeline, are torn apart. Thus the whole process fails to operate as hoped (Palmer and Averill, 1983).

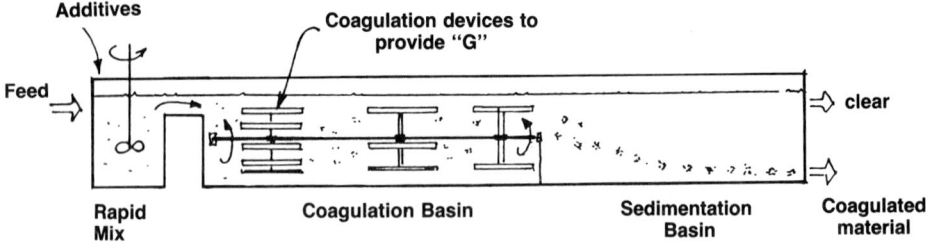

FIG. 44 Sketch of a coagulation separation system.

2. Coagulation Basin

The principles outlined in Section I.B.2.c are used to select the type of agitator, to estimate the velocity gradient G desired, and to provide the residence time to yield an acceptable Gt product. Amirtharajah and Mills (1982) used $G = 25$ s^{-1} and $t = 20$ min. Bratby (1981) recommended a three compartment precoagulation basin, each compartment with $G = 100$ s^{-1} and a total time for the three chambers of 20 min, followed by a tapered-gradient coagulation basin with three compartments in series with the $G = 100, 50,$ and 30 s^{-1} and a total time for these three chambers of 31.5 min. Neither of these studies dealt with the coagulation of oily wastes. Luthy et al. (1978) found that the G value in the coagulation basin for oil removal should be about 100 s^{-1}, whereas for turbidity removal it is usually 30 to 50 s^{-1}.

3. Settling Basin

For the settling basin, Bratby (1981) in designing alum/polyelectrolyte coagulation basins for turbidity removal recommended an overflow rate of 1 L/m^2·s to account for turbulence and short-circuiting. (He had noted that an uncorrected, ideal overflow rate would be 1.5 L/m^2·s.) For alum alone, the overflow rate was much smaller. Example conditions and results are summarized in Table 24. General relationships between the inlet oil concentration and the effluent concentration are given in Fig. 45.

VII. REMOVAL BY AIR ATTACHMENT WITHOUT AND WITH DESTABILIZATION

In the previous sections, especially Sections V and VI, the approach was to cause instability (primarily by mechanisms 3 and 4) and then use one of the three options to separate the coagulated flocs: settling, dissolved air flotation, or filtration. Here we consider dissolved air flotation both as a separating device on its own and in conjunction with coagulation.

The general principle is that an oil drop or a floc attached to an air bubble will separate more readily than it will without the presence of the bubble. The question then is: How best can we attach a bubble to the oil floc? Of the two options—have a bubble collide and attach or nucleate a bubble on the floc—nucleation of a bubble on the floc is the most attractive option. This is called dissolved air flotation (DAF). The approach is to take aside some of the water, supersaturate that water with gas (by putting a gas-liquid mixture under high pressure), and then pump that water containing the dissolved air out into a basin through which the contaminated water

TABLE 24 Example Data for Coagulation Followed by Settling or Coagulation Jar Tests

	Scale	Name	Feed Conc. (mg/L)	Effluent Conc. (mg/L)	Removal (%)	Chemicals Name
API effluent	Pilot	Oil	60		91	$FESO_4$
		SS	25		65	Activated Silica
		COD	140		57	Dolomite lime
		Turbid.	140			
	Pilot, 0.63 L/s	Oil	174	18	90	Alum alone
		SS	191	38	80	Alum plus
		BOD	164		70	Silica
		COD	744			Alum plus
		Turb.	505			Clay
		Oil	250–670			C WT 2635
		SS	49–64	6–15		C WT 2640
		Turb.	36–66			
	Pilot	Oil	729	10.5		C WT 2640
		Oil	450–1200	20		C WT 2640
	Bench					C WT 2860
						C Catfloc
						C Magnifloc 521C
						C WC 21 Tretolite
		Oil	271–444	15		Alum
						A WT 3000
						A WT 2900
						A WT 2700
						A WC-18
						N WT 2690
Synthetic Mixture	Batch	Oil	108		75	Lime
		SS	0			

Dose (mg/L)	Initial pH	Rapid mix G (rpm)	Rapid mix t (s)	Coagulation basin G (rpm)	Coagulation basin t (min)	Comment	Overflow settler (L/m²·s)	Reference
25	9						0.4	Austin and Vause (1951)
10								
150								
115	6.5–7.5	60			45		0.12	Weston (1950)
80								
20							0.24	
80								
200							0.24	
17								Luthy et al. (1978)
14	8			80–100	15			
15								
15	8			60	0.25			
11						Obtained zpc		
10								
17								
12								
20–30								
						Ineffective		
230		40	120	20	10		2 h	Bloodgood

TABLE 24 (Continued)

		Feed		Effluent		Chemicals
Scale	Name	Conc. (mg/L)	Conc. (mg/L)	Removal (%)		Name
	Oil	80		100		$FeCl_3$
						Lime
	Oil	250		100		$FeCl_3$
						Lime
	Oil	1,500		100		$FeCl_3$
						Lime
	Oil	134		100		Ferric hydroxide
Batch	Oil	1,000		Negl		
Batch	Oil	300		40		A Dow Purifloc A-22
	Clay SS	300		92		C Purifloc C-31
				96		Alum
				90		Dowmagnifloc 900N
Pilot	Oil	200		87		Alum
Batch	Oil	200		75		Alum
						A Petrolite WC-18
				16		A Petrolite WC-18
				80		Alum
						N. Petrolite WC-16
				77		Alum
						C Petrolite WC-14
Tar sand Wastewater	SS	3,000		99.5		Ferric chloride
	sol TOC	75–150		68–78		Hercofloc 855
	SS			99		Alum
	soluble TOC			69		
	SS			99		Lime
	soluble TOC			45		

Dose (mg/L)	Initial pH	Rapid mix G (rpm)	Rapid mix t (s)	Coagulation basin G (rpm)	Coagulation basin t (min)	Comment	Overflow settler (L/m^2·s)	Reference
5								(1952)
20								
10								
20								
20								
68								
38								
			45		50			Slezak and Woods (1973)
2–10	8.5–9.2							Humenick and Davis (1978)
2–10								
>30								
3								
20	8.8			52	20		0.15	Churchill and Kaufman (1973)
10	8.8							
0.1								
0.1								
10								
0.25								
10								
1								
300–400	4–5	110	120	48	10			King et al. (1983)
1.5								
1000	6							
1000	12.3							

TABLE 24 (Continued)

		Feed		Effluent		Chemicals
	Scale	Name	Conc. (mg/L)	Conc. (mg/L)	Removal (%)	Name
Steel mill rolling mill waste	Pilot	Oil	87 82 70 74 65		71 64 57 62 64	Ferric chloride
	Plant	Oil SS BOD COD TOC	7,200 590 3,250 18,000 5,200	80 70 320 1,800 520	99 88 90 90 90	"Chemicals"
Bio/ aerosol plant effluent	Bench	TOC COD TOC	533 800 540		37 37	Alum Alum
Oil Tanker Wastes		Oil SS BOD COD	55 8–53 11–30 400–710		83 ? Negl Negl Negl	Preadd anionic Alum Preadd anionic ferric chloride Lime Cationic polymers Nonionic polymers

flows. The air nucleates on the droplets. A sketch of the arrangements is given in Fig. 46. Whereas in processing solids we have the options of saturating all of the feed, a fraction of the feed, or some of the recycled effluent from the DAF basin, for oily water processing the preferred option is to saturate the recycled effluent. This is because the saturater pump operating on the feedstream will break down the drop size further. For clarification two terms need to be defined: the amount of liquid saturated and the amount of air added. The amount of liquid saturated is usually expressed as a percent of the effluent leaving the system and not as a percent of the liquid leaving the DAF unit. This is illustrated in Fig. 46. The amount

Dose (mg/L)	Initial pH	Rapid mix G (rpm)	Rapid mix t (s)	Coagulation basin G (rpm)	Coagulation basin t (min)	Comment	Overflow settler (L/m^2·s)	
50	9	45		70, 48, 20	50	Three-compartment tapered G	0.25	Slezak and Woods (1973)
150	9							
150	9							
79	7							
221	11							
1600	6.5–8.0					Suspect chemicals were demulsifiers (mechanism 9)		Katnik and Pavilcius (1979)
60	7.7–6.2							Gadjiev and Chian (1976)
1	6.8–7.5							Sadler et al. (1978)
20–50								
1						Need 10 to 30 min preaeration		
20–50								

of air added to the liquid passing through the saturation chamber depends on the temperature and pressure of the saturation chamber. The general range of values is 0.6 to 5 dm^3 per 100 L of liquid processed. However, the important information is the amount of air available for the total feed liquid to the DAF. This will include the unsaturated liquid plus the saturated liquid. Thus, depending on the recycle ratio, this will vary between 0.4 and 2 dm^3 per 100 L of total liquid feed to the DAF. For oily wastes we can use DAF directly without the addition of destabilizing chemicals upstream, or we can add chemical destabilizers—as described in Section VI—and use DAF downstream as the separating device.

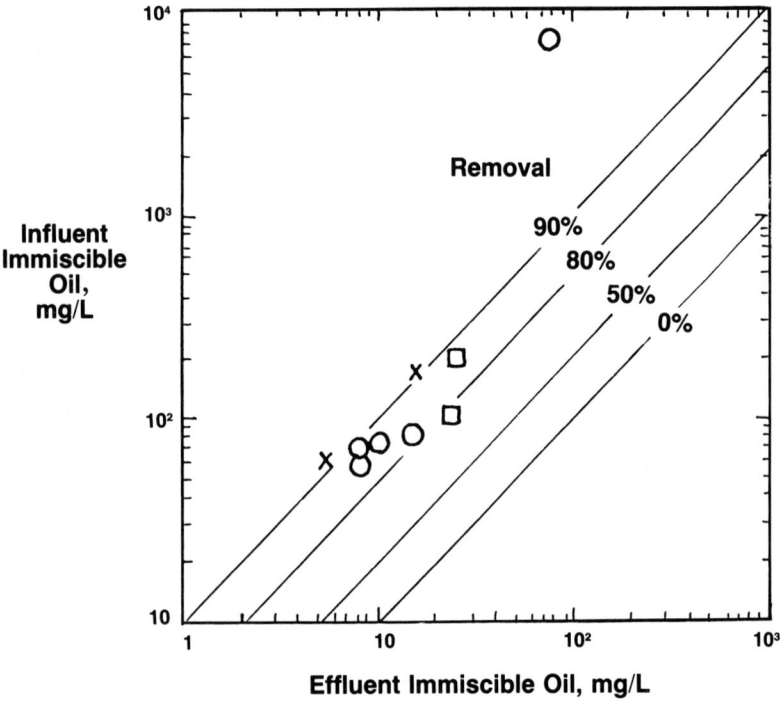

FIG. 45 Performance of coagulation settling.

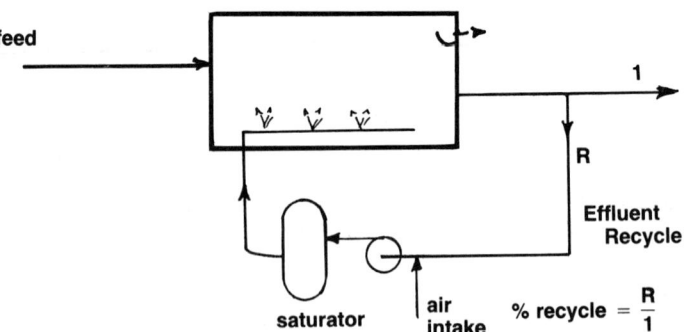

FIG. 46 Dissolved air flotation unit.

Dilute Oil-Water Emulsions and Dispersions

A. DAF Without Upstream Addition of Destabilizers

Mechanism 9, from Table 6, can be used where we put inserts—in this case bubbles—to remove the drops. The effectiveness of this approach is summarized in Table 25. The relationship between the influent oil and the effluent is illustrated in Fig. 47. Luthy et al. measured the drop size distribution for the stream leaving a DAF unit. The feed was API effluent. The drop size distribution was log-normally distributed with a geometric mass average diameter of 11 μm and a geometric standard deviation of 1.64.

Some general comments about the data are that most are for on-site plant operations in a refinery. Under these conditions the feed compositions and concentrations and the pH vary greatly (e.g., the oil concentration for Prather's work varied from 134 to 12,000 mg/L with an "average" value of 1500 mg/L). The conditions used for the removals are not very complete. There also is uncertainty as to what each author means by "oil removal." Only Luthy et al. identify and clearly distinguish between "immiscible oil"—which DAF should remove—and soluble oil. The others report simply "oil removal." Since the soluble oil concentration is about 18 mg/L, the impact on the results can be considerable. Only Katnik and Pavilcius (1979) and Pushkarev et al. (1983) report on the applications of DAF to feeds other than the effluent from refinery API separators. Pushkarev et al. describe the application of DAF for a cold-rolling mill lubricant-water system. No data are given. They present data for DAF-hydrocyclone treatment of rolling mill wastes with oil inlet concentrations of about 150 to 200 mg/L; the oil removal is about 70%.

A feature of DAF is that because we normally do not exceed 500 kPa pressure in the air saturation stage, there is a limit on the amount of air that can be dissolved. This in turn affects the amount of bubbles formed and hence the amount of particulates that can be removed. Steiner et al. (1978) for example, found that the optimum ratio of air to floatable material was about 0.06 to 0.08 kg of air per kilogram of immiscible materials. If the maximum air supplied via pressurization is 3 dm^3 per 100 L of total liquid fed to the DAF unit, the maximum immiscible component corresponds to about 400 to 600 mg/L. If half the loading is suspended solids, this limits the incoming oil concentration to 200 to 300 mg/L. (These calculations are sensitive to the recycle ratio and are meant to be illustrative of the limitations rather than to be definitive.)

Because of the turbulence created in pumping the liquid through the pressurization cycle, for immiscible liquid systems, effluent recycle saturation should be used.

Now consider briefly some of the key features of the data reported in Table 25 and Fig. 47. Rohlich studied the application of DAF to in-field API effluent. The stream showed great variability in pH and amount of oil. Hence, his results are averages. The variables he studied were

TABLE 25 DAF Separation Without Destabilizer Addition (API Effluent)

		Feed	Effluent		
Scale	Name	Conc. (mg/L)	Conc. (mg/L)	Removal (%)	pH
Pilot 4.7 L/s	Oil Oil	<157> <162> <131> <166> <260> <214>	<94> <50> <56> <98> <185> <131>	45 (40) 78 (69) 69 (62) 66 (57) 46 (41) 31 (29) 42 (39) 47 (39)	
Bench	Oil			48 (38) 53 (44) 47 (36)	6.4 6.5 7.6
Pilot	Oil SS TOC	<1500> <4500>	60–110	83 70	Variable
Pilot, 3.8 L/s	Oil SS	45		38 30	
Full	Oil Oil Oil SS	61 444 33	15 16 7	100 (75) 100 (96) 79 46	
Batch	Oil	356	108	73 (70)	
Pilot	Oil	<570>	77	64	7–8.5
Full scale	Oil SS			70–80 70–80	
Full scale, 170 L/s	Oil SS	18	6.3	65 55	
Full scale, 1100 L/s	Oil	100–125	30–40		6.5–10

Dilute Oil-Water Emulsions and Dispersions

Pressure (kPa)	Air dm³/100L Treated liquid	Air dm³/100L Total feed	Recycle %	Recycle Type	DAF overflow (L/m²·s)		Reference
344	1.23	0.41	50	ER	1.7	vert	Rohlich
344	4.87	1.61	50	ER	1.7	vert	(1953)
344	4.87	1.61	50	ER	1.5	API	(1954)
344	9.8	3.2	50	ER	1.7	vert	
344	1.23	0.41	50	F	1.15	vert	
344	0.61	0.61	0	F	1.15	vert	
344	1.85	1.85	0	F	1.15	vert	
344	4.9	4.9	0	F	1.0	API	
344			33				
482	5	1.8	59		1.35		Prather (1961)
			>15	ER	<4 3		Steiner et al. (1978)
378					0.47–0.56		Simonsen (1962)
			33	ER	1.74		
345			50	ER	1.3		Luthy et al. (1978)
375–445			50		4.05		Quigley and Hoffman (1966)
377			35				Hart (1970)
377							Stormont (1956)

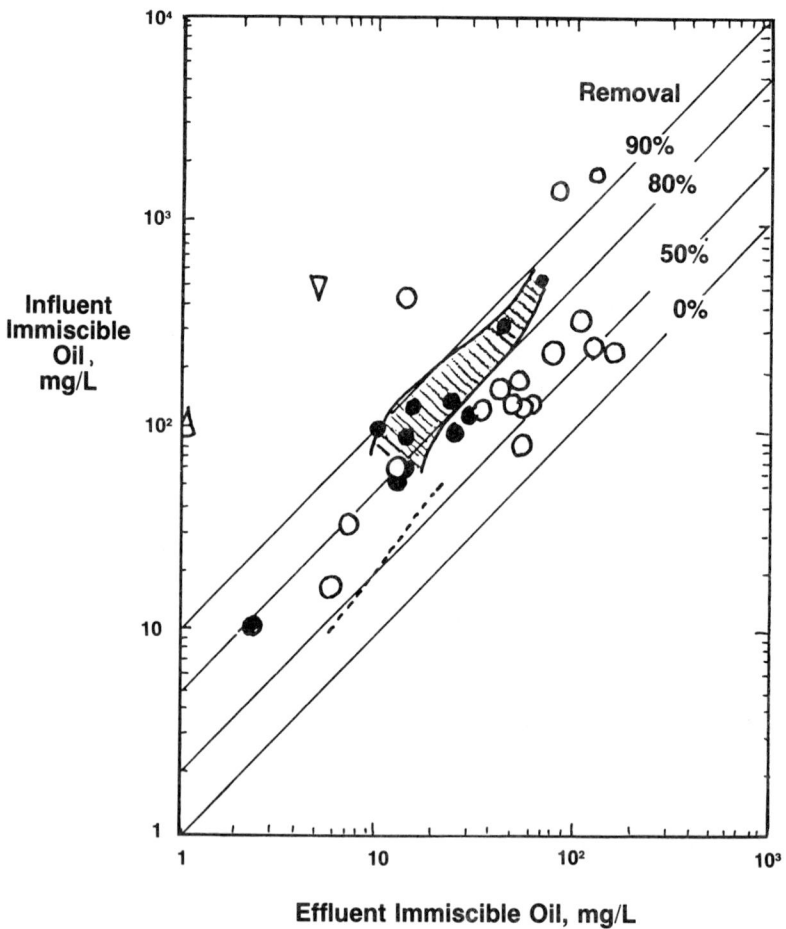

FIG. 47 Performance of DAF without destabilizers O, with destabilizers ●.

1. Configuration: A "vertical flow basin" where the DAF feed goes to the centerline along the full basin and then rises vertically and overflows along the sides, and an API where the feed comes in one end and goes out the other.
2. Amount of recycle: none (all is presaturated) and 1/3 the feed is presaturated.
3. Amount of air put in to presaturate the presaturated stream, 0.613 to 9.8 dm^3 per 100 L.
4. Total flowrate to the basin (2.5 or 1.68 L/s).

5. Type of recycle: effluent (his "recirculated") or feed (his "split flow").
6. Continuous pilot plant or batch kit.

He measured the oil by solvent extraction and so would measure both the soluble and immiscible oil together. Hence, his results were recalculated based on the soluble oil being about 18 mg/L. This is the measured value of Luthy et al. In Rohlich's work, he did get down as low as a 12 mg/L total so perhaps the 18 mg/L is too high for his work. Both results are given in Table 25.

Concerning the design configuration of the DAF, the vertical design gives about 78% removal as long as the air rate is about 1.6 dm^3 per 100 L of total liquid fed to the unit (including the recycled liquid). The API design gives about 10% lower removals (about 69%) even though the API provided a longer residence time of 12 min versus 9.5 min and a lower overflow velocity of 1.5 versus 1.7 L/m^2·s. The oil removals were relatively independent of pH for the range 6.4 to 7.6 based on the batch test work.

Prather (1961) also worked on-site with API effluent—again, he had to cope with great variability in pH and oil concentration. The exit concentrations were in the range 60 to 110 ppm. He, too, found that the removal was not affected by pH in the range 7 to 10.

Steiner et al. (1978), also worked on-site with API effluent. Their removals of both oil and suspended solids were low, 38% and 30%, respectively.

Simonsen (1962) reports on three full-size plant operations: the No. 2 Cleveland plant (with a self-contained DAF unit); a second plant where an API separator was converted to a DAF unit; and the Lima plant, which seems to be similar to the second plant configuration. The operation details are not given; the conditions in Table 25 are the best we could infer from the article. Again, the plant conditions varied extensively and average values are quoted. Simonsen's data for the full-scale plant with a feed concentration of 444 and an effluent concentration of 16 mg/L are not consistent with the other data.

Luthy et al. (1978) used a pilot-scale, on-site unit to treat API effluent. They controlled the pH to the DAF unit and clearly reported the oil concentrations and removals in terms of "immiscible" oil.

Campbell and Scoullar (1964) report on an industrial DAF that treats the effluent from an API. The feed comes from an equilizing basin and the pH has been adjusted to 8 to 8.5. A fraction of the feed is pressurized; no effluent is recycled. The inlet concentrations and removals are not reported.

Quigley and Hoffman (1966) describe a full-scale DAF unit following an API. They point out that initially, the DAF unit was getting <30% oil removals. This was attributed to insufficient recycle liquid being saturated. When the recycle ratio was increased, the removals increased to 65%.

In general:

1. The operating and design conditions are summarized in Table 26.
2. For no chemical addition, both Prather and Rohlich found that the removal was independent of the pH for the range 7 to 10.
3. The removal of suspended solids is about the same as for immiscible oil.
4. The removal tends to be in the range 50 to 80% when effluent saturation is used.
5. The vertical flow configuration is preferred to a long horizontal flow as provided by an API configuration.

B. DAF With Upstream Addition of Chemical Destabilizers

The removal of oil can be increased from the range 50 to 80% to 80 to 90% if chemicals are added to the feed upstream of the DAF unit. The same researchers as described above usually added destabilizing chemicals to try to improve the removals. The same general comments apply. Most worked on API effluent for refineries; they worked on-site and had to cope with great variability in the feed composition and concentration. The added complication was the care taken in designing an upstream coagulation basin. Some added the chemicals upstream of the DAF unit or to the DAF unit directly. On the other hand, some had carefully sized coagulation basins. Concerning the chemicals, some are characterized as a "polymer addition" or "chemical addition," so that interpretation is challenging.

The most comprehensive study was done by Luthy et al., who studied 13 chemicals. First, they measured electrophoretic mobilities and selected, on a bench scale, those chemicals that would reduce the charge on the oil droplets to zero (for fixed pH). They tested those that gave the best performance on a batch jar test. The jar test results confirmed the mobility measurements. The strongly cationic polydiallydimethyl ammonium compound (MW 2640) worked best. At pH 8, about 15 mg/L gave zero charge, for the jar test gave an optimum reduction in JTU turbidity. Overdosing causes restabilization (as we would expect from the fundamentals discussed earlier). When the same conditions were applied to the continuous, on-site plant, the best removals were also obtained for this set of chemical dosages. However, the optimum removal occurred when the velocity gradient, G, in the upstream coagulation basin was 80 to 100 s^{-1} (rather than the usual 30 to 50 s^{-1} applied for turbidity removal). The results were removals of about 96 to 99.6% with residual immiscible oil concentrations in the range 6 to 15 mg/L. When about the

TABLE 26 Suggested Design Criteria for DAF

	Value	
Parameter	Beychok (1967)	Ford and Elton (1977)
Recycle ratio	0.3–0.5	0.3–0.4
Air drum pressure (kPa)	172–380	
Air drum retention time (min)	2	
Air rate (dm^3/100 L of total liquid feed)		1.9–3.8
Flotation time (min for total liquid flow in)	15–20	20–40
Overflow rate ($L/m^2 \cdot s$, based on total liquid flow)	2.05	0.68–2.7
Tank depth (m)	1.8–2.5	

same doses of other chemicals were added, the effluent oil concentration was about double, 12 to 30 mg/L.

They found that when the influent oil concentration increased, with chemical addition, the removal rate was not constant; rather, the effluent concentration remained constant except when the solids concentration was greater than 1000 mg/L. For the latter conditions, the effluent oil concentration did vary with feed concentration.

Luthy et al. also considered a simpler design for an upstream coagulation basin: a long mixing pipe that supplied a Gt of 900, whereas the basin provided a Gt of 67,500. The pipe had G = 60 s^{-1} and t = 15 s. The result was an effluent concentration of 20 mg/L (instead of 10 mg/L). They found that for the full coagulation basin plus DAF, for an upstream pH of 7 to 8.5, an addition of 25 mg/L of alum gave an exit concentration of 15 mg/L (for a removal rate of 94 to 97%). The results from others are summarized in Fig. 47 and Table 27, with Luthy et al.'s results given in Table 24.

In general, alum or strongly cationic polymers applied upstream in a coagulation basin work well; pH should be controlled, preferably in the coagulation basin; overdosing causes restabilization; and the velocity gradient should be about 100 s^{-1} in the coagulation basin. Based on Luthy et al., we should be able to choose chemicals based

TABLE 27 DAF Separation with Coagulant Addition

		Feed		Coagulant			Effluent			Air			Recycle		Up-stream coag. basin?	DAF overflow $(L/m^2 \cdot s)$	Reference
	Scale	Name	Conc. (mg/L)	Name	Dose (mg/L)	pH	Name	Conc. (mg/L)	Removal (%)	Press. (kPa)	Treat liquid	Total feed	%	Type			
API effluent	Pilot, 4.7 L/s	Oil	106–134	Alum	4	9	Oil		79								Rohlich (1953)
	Bench	Oil	203	Alum	25	9	Oil		94								(1954)
		Oil	80–270	Alum	75	7–8	Oil		81–90								
									68–71								
									71–77								
	Full, 63 L/s	Oil	—	Alum	30–70	8.5–9.5	Oil		80–85								Quigley and Hoffman (1966)
		SS	—	Lime	75–150	10	Oil		68–96								
							SS										
	Full, 126 L/s	Oil	11	"Polymer"	2	—	Oil		79								Hart (1970)
		SS	37				SS		75.7								
		BOD	98				BOD		39.8								
		COD	224				COD		41.5								
	Pilot	Oil	45	Nalco 603 cationic polymer	7	—	Oil		82								Steiner et al. (1978)
		SS	33				SS		70								
		TOC	95				TOC		39								
Transport vehicle maintenance plant waste	Full, 13 L/s	Oil	490	Cationic polymer + Anionic polymer	150	9.5	Oil		99								Katnik and Pavil-cius (1979)
		SS	430				SS		98								
		BOD	520				BOD		33								
		COD	2400		0.25		COD		80								
		TOC	400				TOC		65								
Glass manufacturing waste	Full, 4 L/s	Oil	100	Cationic polymer + Anionic polymer	75	7.8	Oil		99								
		SS	150				SS		97								
		BOD	24				BOD		71								
		COD	470		2		COD		96								
		TOC	67				TOC		72								

C. Equipment Selection and Sizing

First, we need to decide whether the amount of separation required can be accomplished with DAF alone or whether destabilizing chemicals are needed and if, indeed, DAF is applicable. If we need destabilizing chemical addition, an upstream coagulation basin can be sized according to the principles in Section VI.

The DAF unit is designed to provide sufficient residence time in the basin for the nucleated bubbles to rise to the surface. General design values are summarized in Table 26. These can be used to size the unit. In addition to the DAF separation basin, the pressure saturation tank, the air compressor, and the pumps are needed. In sizing the pressure saturation tank, the pressure should be such that enough gas will be dissolved to provide an acceptable gas-to-solids ratio. In other words, we must have enough bubbles to lift up all the immiscible oil plus solids. Typical values for the air-to-solids ratio needed is about 0.005 to 0.1 kg/kg (Metcalf and Eddy, 1972; WPCF, 1977), although for oil separations Steiner et al. found the optimum to be 0.06 to 0.08 kg/kg. (This loading, divided by 1.2, gives units of liters of air needed per gram of solids removed.) A reasonable value is about 0.07 kg of air per kilogram of solids. From this, the operating pressure for the pressure tank and the required amount of effluent recycle can be estimated. Examples are given by Metcalf and Eddy (1972, p. 299). Usually, the pressure tank operates at 400 kPa pressure; the air dissolves quickly in water, so a residence time of about 30 to 60 s is all that is needed in the pressure saturation tank.

The sizing and selection procedure is as follows:

1. Check that DAF application seems feasible (see Fig. 41); if the maximum operating pressure is 400 kPa pressure and the recycle ratio is about 0.03 to 0.5/1, the maximum immiscible oil concentration in the feed is about 800 mg/L. Check that the oil concentration (and solids concentration) is small enough.
2. From Table 26, select an appropriate liquid overflow rate, about 1 L/m^2·s, and estimate the effluent recycle to be about 0.3 to 0.5/1. Then add the recycle flow to the feed flow rate and use the liquid overflow rate to estimate the horizontal cross-sectional area needed.
3. Estimate the depth needed to supply the necessary residence time; Rich suggests that a minumum depth is 1.8 m to minimize short circuiting and turbulence.
4. Estimate the exit concentration of solids not removed.

VIII. REMOVAL BY FILTRATION

In the previous sections, especially Sections V and VI, the approach was to cause instability (by mechanisms 3, 4, 8, and 10) or to add a material—gas bubbles—to capture the drops (by mechanism 9). In this section we extend consideration of adding a material to capture the drops. Here the material added includes filter cloths or membranes, beds of fibers, or beds of sand or crushed materials.

Four different types of filters can be used to separate oil-water dispersions: (1) a traditional cloth or screen filter to which a layer of fine precoat material has been added, (2) a membrane "ultrafilter," (3) a deep-bed (or gravity or sand) filter which can entrap the drops, and (4) a fibrous-bed filter that acts as a coalescer to increase the size of the drops for later processing. Table 28 summarizes the characteristics of the various alternatives. The general applicability in terms of drop size and liquid loading is illustrated in Fig. 48. The emphasis in this section is on the fibrous-bed and deep-bed filters.

A. Fundamental Principles and Characteristics

Whenever an object is placed in a flowing stream containing particulates, the particulates can hit the object and be "captured" if (1) the holes between the objects are so small that the particulates cannot squeeze through (this is unlikely for fluid drops because they just deform and pass through the opening), or (2) the drops move to the object, because of Brownian motion, density differences, inertial effects, or because they are carried by fluid movement. Figure 49 illustrates the order of magnitude of the "collection" and the probable dominant transport mechanism for different size of solid particles. Thus for sizes of less than 1 μm, Brownian motion is the major transport mechanism, with the amount of transport decreasing

TABLE 28 Filter Media

Type of filter	Usual medium
Precoat	Kieselguhr on cloth
"Deep" bed	Sand or anthracite
Coalescer	Glass or synthetic fibers
Ultrafilters	Membrane

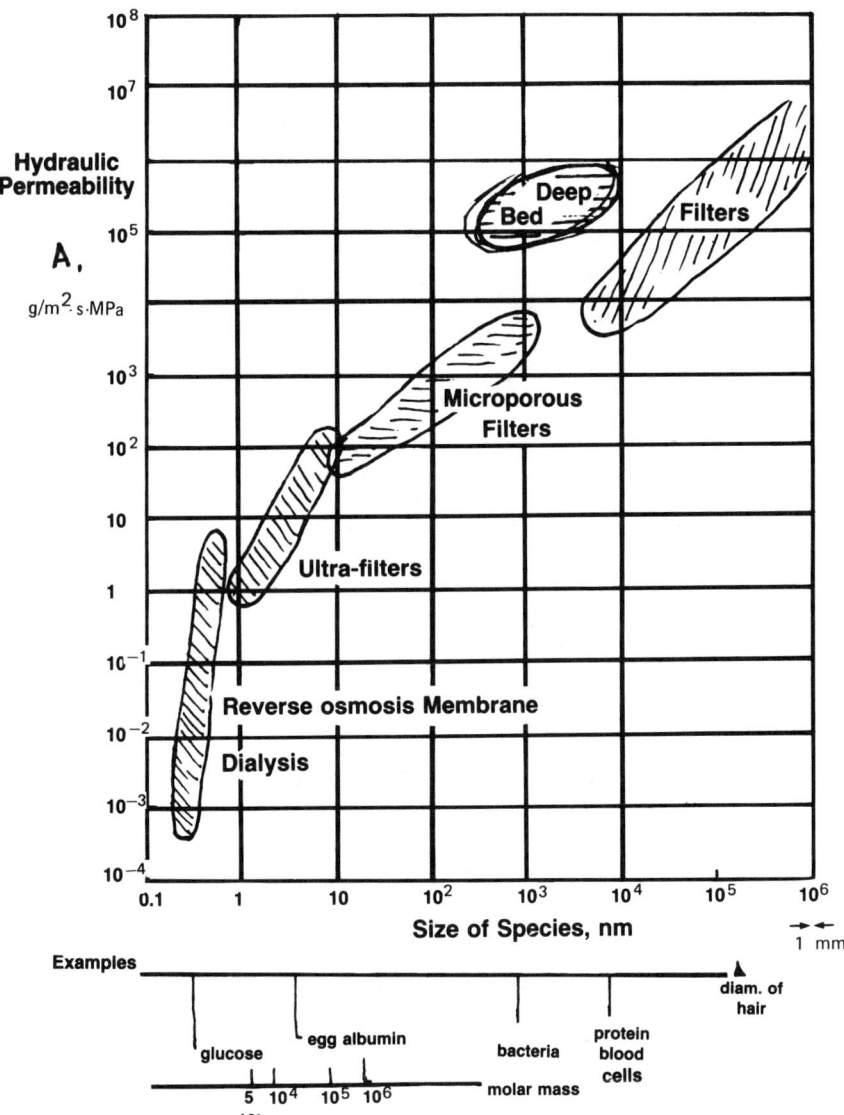

FIG. 48 Range of applicability of size separation devices [based on de Filippi (1977) and used with the permission of Marcel Dekker].

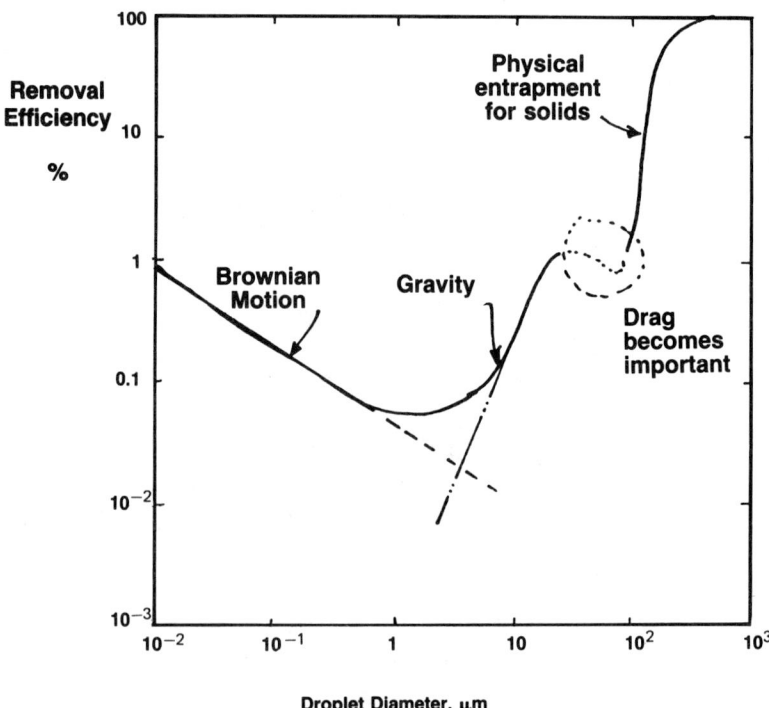

FIG. 49 Dominant removal mechanism as a function of droplet size [adapted from Payatakes (1977) with permission].

with increasing size. Above 1 μm, sedimentation tends to be the transport mechanism, with the rate of movement increasing with the size of the particle. For smaller-diameter drops, similar arguments probably apply. However, for larger-diameter fluid drops, the arguments are likely to be different because the drops can deform and break up. Once the drops are transported to the "collecting objects," hopefully, attachment occurs. Attachment might be considered, from an equilibrium point of view, to be described by the DLVO theory and energy-interaction diagrams. A simplified analysis for flow through a deep bed of media is illustrated in Fig. 50, which identifies the conditions under which capture would occur.

Figure 50 was based on the flow of a dispersion through a beds of spheres. Speilman and Fitzpatrick correlated their results in terms of three dimensionless groups: the van der Waals group, the zeta potential group, and the "size" group. Mathematically, the groups are

$$N_{Ads} = \text{van der Waals group} = \frac{Aa_c^2}{9\pi\mu U_\infty a_p^4} \frac{1}{A_s} \quad \text{dimensionless}$$

$$N_\zeta = \text{zeta potential group} = \frac{3\varepsilon_r \varepsilon_o \zeta_c \zeta_p a_p}{2A} \quad \text{dimensionless}$$

and

$$a^+ = Ka_p$$

where

a_c = radius of the media used as a collector
a_p = radius of the particle to be removed
A = Hamaker's constant between the particle and the collector
μ = viscosity of the fluid
U_∞ = superficial liquid velocity through the bed
A_s = dimensionless factor accounting for neighboring collectors and their effect on the target collector
ζ_c = zeta potential of the collector
ζ_p = zeta potential of the particle
K = inverse length of double layer

From their analysis they could identify regions where collection would or would not occur.

Example 16 Water from a nearby lake is to be sent through gravity sand filters to remove the fine clay particles present in the water. Assume that the zeta potential at pH 7 on the silica is -80 mV and on the clay is -100 mV. The diameter of the media is 0.5 mm and that of the clay particles is 2 μm. Assume that the water has a hardness of 500 mg/L as $CaCO_3$. The usual hydraulic loading is 2 U.S. gpm per ft^2 of cross-sectional area. Will the clay be removed if the major removal mechanisms are van der Waals force and electrochemical double-layer effects?

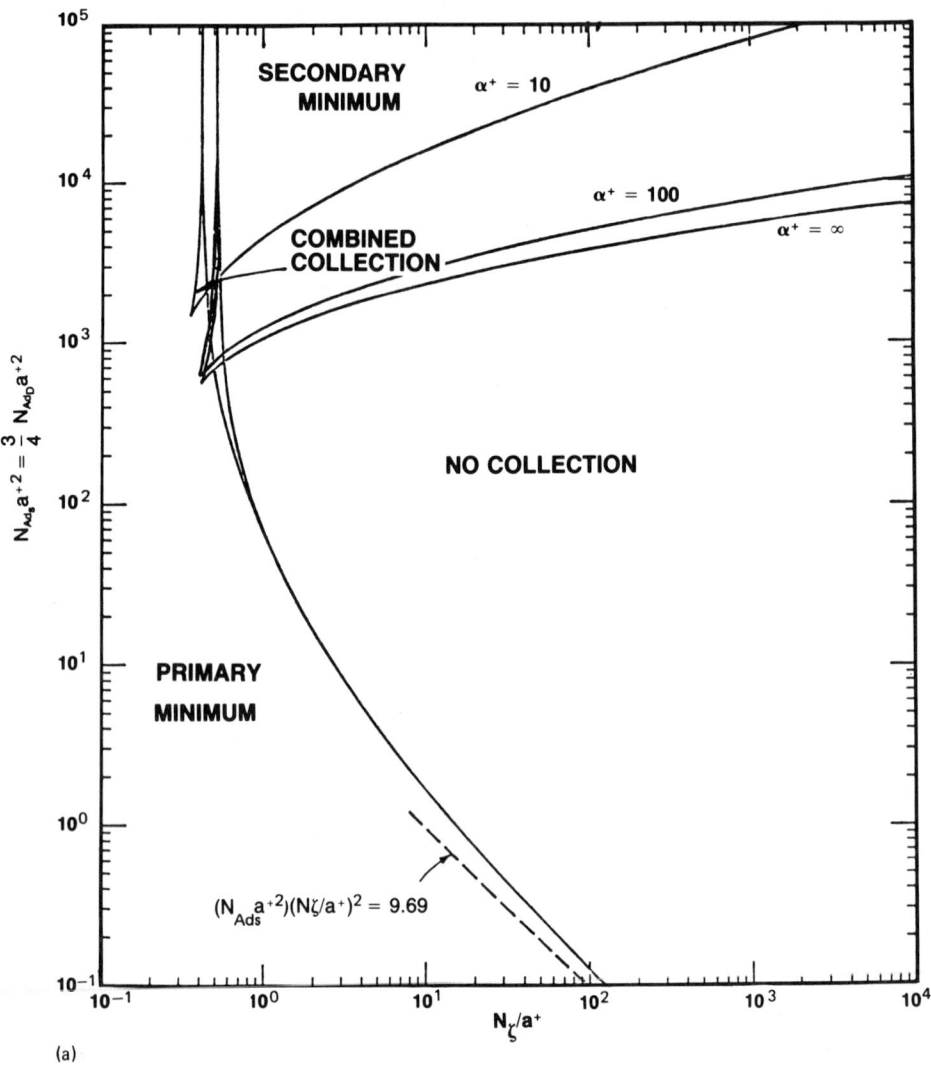

FIG. 50a Approach at constant potential. [From Speilman and Cukor (1973) and reprinted courtesy of J. Colloid and Interface Science.]

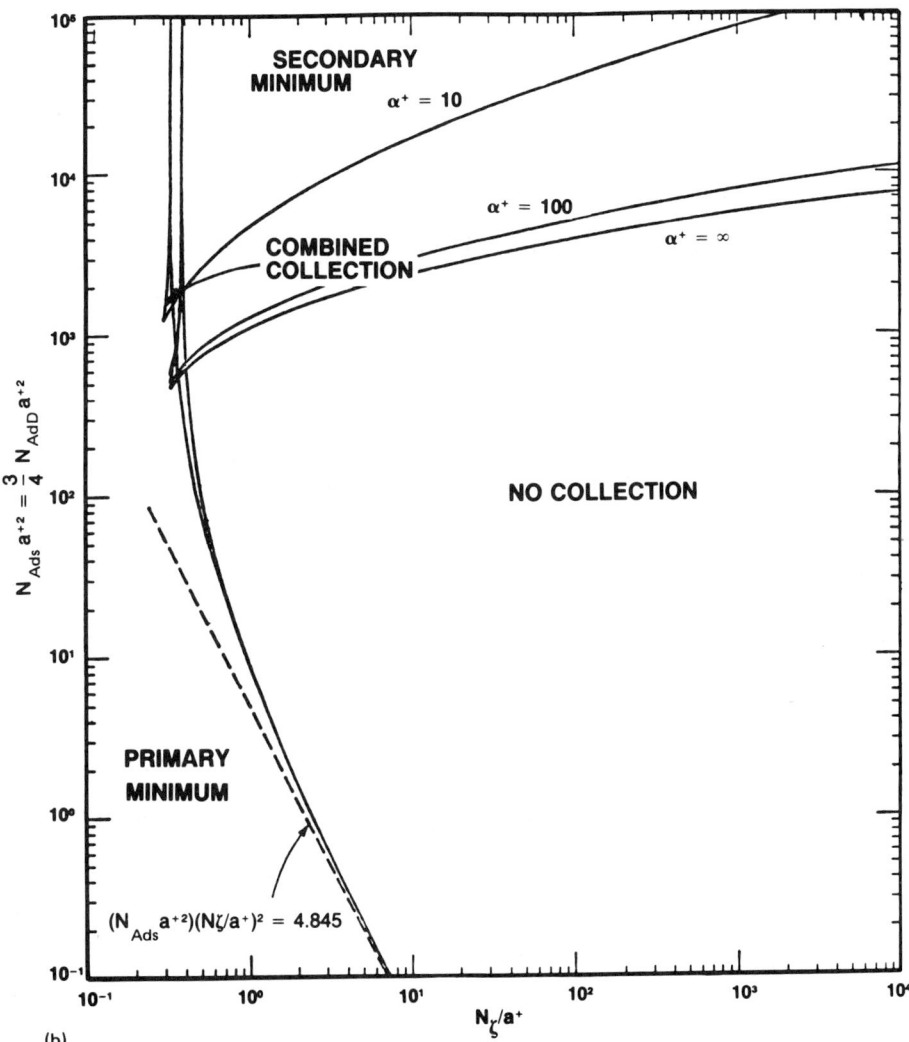

FIG. 50b Approach at constant charge. [From Speilman and Cukor (1973) and reprinted courtesy of J. Colloid and Interface Science.]

504 D. R. Woods and E. Diamadopoulos

Solution:

Unknown: removal or no removal

Ideas: We need to calculate the three dimensional groups. We are given quite a few of the data; however, we need Hamaker's constant, A_S, ε_r, liquid viscosity, and U_∞.

From Happel (1958a) the correction factor, A_s, is

$$A_s = 1 - \frac{(\gamma)^5}{1 - 3\gamma/2 + 3\gamma^5/2 - \gamma^6}$$

where $\gamma = (1 - \varepsilon)^{1/3}$ is the volume fraction packing. For this bed assume that ε = void volume = 0.35. Thus $\gamma = 0.86$.

$$A_s = \frac{1 - 0.49}{1 - 1.29 + 0.73 - 0.40}$$

$$= 12.75$$

Hamaker's constant: assume to be 10^{-20} J (10^{-13} erg); ε_r for water: assume to be 80; viscosity of water: assume to be 1 CP or 1 m Pa·s. The superficial velocity is the overflow velocity:

$$\frac{2 \text{ U.S gpm}}{\text{ft}^2} \left| \frac{3.785 \times 10^{-3} \text{ m}^3}{\text{min}} \right| \frac{\text{min}}{60 \text{ s}} \left| \frac{\text{ft}^2}{0.0920 \text{ m}^2} \right.$$

$$= 1.36 \times 10^{-3} \text{ m/s}$$

We need to know the Debye-Hückel reciprocal length. The definition is

$$K^2 = \frac{e^2}{kT} \varepsilon_r \varepsilon_o \sum z_i^2 n_i \bigg|_{\phi = 0}$$

or in consistent units for $T = 298$ K, $\varepsilon_r = 80$:

$$K^2 = \frac{(1.602 \times 10^{-19} \text{ C})^2}{(1.3805 \times 10^{-23} \text{ J/K}) 298 \text{ K}} \frac{1}{(80) 8.85 \times 10^{-12}} \frac{\text{J} \cdot \text{m}}{\text{C}^2} \sum z_i^2 n_i \bigg|_{\phi=0}$$

$$= 8.7 \times 10^{-9} \text{ m} \left(\sum z_i^2 n_i \right)$$

If we assume that all the ions are $CaCO_3$ (MW 100) and that it dissociates completely, then for 500 mg/L the number of molecules per cubic meter is

$$\frac{500 \text{ mg}}{\text{L}} \bigg| \frac{10^{-3} \text{ g}}{\text{mg}} \bigg| \frac{\text{mol}}{100 \text{ g}} \bigg| \frac{6.023 \times 10^{23} \text{ molecules}}{\text{mol}} \bigg| \frac{10^3 \text{ L}}{\text{m}^3}$$

$$= 30 \times 10^{23} \frac{\text{molecules}}{\text{m}^3}$$

Hence

$$K^2 = 8.7 \times 10^{-9} (4 \times 30 \times 10^{23} + 4 \times 30 \times 10^{23})$$

$$= \frac{2.08 \times 10^{17}}{\text{m}^2}$$

$$\frac{1}{K} = 2.2 \times 10^{-9} \text{ m} = 2.2 \text{ nm}$$

For convenience, for $T = 298$ K and $\varepsilon_r = 80$, the working equation is, for equivalent dissociation,

$$K = 2.28 z_i \sqrt{c_i} \quad [=] \frac{1}{\text{nm}}$$

for 1-1, 2-2, and 3-3 ionic dissociation, where c_i = mol/L with water as solvent. Hence for this problem, $1/K = 2.2$ nm. Calculating the van der Waals group yields

$$\frac{Aa_c^2}{9\pi\mu U_\infty a_p^4} \frac{1}{A_s} = \frac{10^{-20} \text{ J } (0.25 \times 10^{-3} \text{ m})}{9\pi(1 \times 10^{-3} \text{ Pa·s})} \frac{1}{1.36 \times 10^{-3} \text{ m/s}}$$

$$\frac{1}{(1 \times 10^{-6} \text{ m})^4 \, 12.75}$$

$$N_{Ads} = 1.27$$

$$a^+ = \frac{Ka_p}{}$$

$$= \frac{1 \times 10^{-6} \text{ m}}{2.2 \times 10^{-9} \text{ m}}$$

$$= 454$$

$$(a^+)^2 = 2 \times 10^5$$

Calculating the zeta potential group, $3\varepsilon_r \varepsilon_o \zeta_c \zeta_p a_p / 2A$, substitution yields

$$N_\zeta = \frac{3 \times 80 \times 8.85 \times 10^{12}}{2 \times 10^{-20} \text{ J}} \frac{C^2}{\text{J·m}} (-80 \times 10^{-3} \text{ V}) (-100 \times 10^{-3} \text{ V})$$

$$1 \times 10^{-6} \text{ m}$$

$$= 8.5 \times 10 \left(\frac{10^{-24}}{10^{-20}}\right)$$

$$= 850$$

Hence $N_A(a^+)^2 = 1.27 \times 2 \times 10^5 = 2.5 \times 10^5$,

$$\frac{N_\zeta}{a^+} = \frac{850}{454} = 1.87$$

From Fig. 50 this suggests that the clay particles will be captured in a secondary minimum mechanism regardless of whether the surface charges arise from constant-charge or constant-potential mechanisms.

In summary, surface principles and ideas are important for understanding deep-bed filtration.

Clayfield (1982) illustrates the importance of the fundamentals through his study of fibrous bed coalescence. He illustrated that the collection efficiency of clean glass fibers was about 50%. A check on the surface charges shows that just after cleaning, the glass fibers are about −100 mV. For neutral water, the oil droplets are negatively charged as well. Clayfield treated the glass fibers with alkoxysilanes to change the surface charge on the fibers. The results are summarized in Table 29.

Besides altering the charge on the fibers, improved collection efficiency was obtained by increasing the molar concentration of sodium chloride. For the glass fiber/oil system the collection efficiency started to increase above 50% at about 0.8 mM and was about 90% by 30 mN. Experiments were also done by adding surfactants. Adding an anionic surfactant such as sodium dodecyl sulfate decreased the size of the drops from about 8 μm to 3.8 μm and decreased the collection efficiency.

To sum up, both Fig. 50 and Clayfield's work illustrate the importance of the DLVO and energy interaction type of analysis when assessing the potential effectiveness of adding media. Thus the fun-

TABLE 29 Effect of Surface Charge on the Fibrous Media on Collecting Oil Drops Dispersed in Water[a]

Zeta potential on fibers (mV)	Collection efficiency (%)
+105 to +102	97.8
+70 to +63	97.0
+85 to −2	89.8
−11 to −22	59.2
−95 to −86	55.9
−110 to −80	51.2
−104 to −94	51

[a]Conditions: concentration, 400 mg oil/L; pH, approx 7; drop size, approximately 1 to 7 μm.

damental surface phenomenon of importance is the interaction between the droplets and the medium that is inserted into the system.

B. Characteristics of Deep-Bed Filtration

Now consider the characteristics of deep-bed filtration. A deep-bed filter is a deep bed of sand (or similar material) through which the oil/water mixture flows. The dispersed phase drops should be trapped on the sand particles or on the medium. Models for the movement and capture processes have been developed by Ives (1970), O'Melia (1967), and Payatakes (1977).

When the bed becomes loaded, it is brought out of filtering service and regenerated by removing the trapped dispersed phase. Thus, in sizing and selecting deep-bed filters, both the loading and the regeneration processes should be considered. Four different configureation are possible: downflow by gravity or pressure and upflow by either gravity or pressure. There is confusion about the terminology for this kind of filter. Sometimes they have been called rapid sand, sometimes deep bed, and sometimes gravity sand. None of these is best because sand is not always used as the medium; they are not necessarily operated by gravity flow and the depth of the bed varies. The so-called "rapid sand filters" have a depth of media that is less than 1 m. Usually, either one to two different types of media are used in the bed: for example, sand or sand and anthracite. Rapid sand filters have been used traditionally for water treatment to remove cloudiness from water. More recently, dual-media beds about 2 m deep have been used. The differences are illustrated in Table 30.

The advantages of the deeper bed are that they can be operated at higher liquid loading rates. For the deeper beds, to minimize the pressure drop, the top layer (for downflow systems) has larger-diameter grains with the bottom layer having smaller diameter. Thus we attempt to lower pressure drop and utilize the full depth of the bed for collection. Often, anthracite is used as the top medium.

The electrophoretic properties for sand media in tap water at neutral pH has been measured by Smith (1967) to be +27.5 mV. Other work measuring the charges on the media has been done by Wnek (1974).

C. Performance Without the Addition of Destabilizing Chemicals

Usually, filters are operated without the addition of destabilizing chemicals. Some general performance results are shown in Tables 22 and 31 and Fig. 51. About 75% of the immiscible phase is removed. Again, the data are not always easy to interpret because authors do not always clearly distinguish between "total oil," "im-

TABLE 30 Characteristics of Deep-Bed Filters

	Liquid load (L/m²·s)	max Δp (kPa)	Medium						Total depth (m)
			Layer	Material	Effective diam. (mm)	Uniformity coeff.	Density (Mg/m³)	Depth (m)	
Conventional "rapid"	1.5–3.5	15–24	Single	Sand	0.4–0.7	1.8	2.6	0.9	0.9
			1	Anthracite	0.5–2.0	1.8	1.6	0.3–0.60	
			2	Sand	0.4–2.0	1.4	2.6	0.3–0.60	
Ultra-fast "Deep"	3.5–17	30–70	1	Sand	2.0	N.R.	2.6	1.50	1.5–2.40
			2	Sand	0.5–1.6	1.3	2.6	0.6–1.80	
			1	Anthracite	1.2–6.3	N.R.	1.6	0.75–1.20	
			2	Sand	0.5–1.6	1.3	2.6	0.6–1.80	

Configurations:

	Downflow	Upflow
Gravity	√√ Usual	—
Pressure	Sometimes	Sometimes

TABLE 31 Bed Filters with Dilute Oil-Water Separations

	Press.		General oil			Additive		Hyd. load $(L/m^2 \cdot s)$	Depth (m)
			Feed						
Gra. down	Down	Up	Oil (mg/L)	SS (mg/L)	Oil/ SS	Name	Conc. (mg/L)		
X			60	0		—		13.6	0.9
						—		8.1	0.9
X			107	75	1.43	—		5.2	2.13
	X		44	56	0.78	Cat. pol	2	0.5	1.8
	X		100	100	1.0	Cat. pol	3	6.8	1.8
	X		100	100	1.0	Cat. pol	3	3.4	1.8
	X		100	100	1.0	Cat. pol	3	10.2	1.8
	X		100	300	0.33	Cat. pol	3	6.8	1.8
	X		100	500	0.2	Cat. pol	3	6.8	1.8
	X		300	300	1.0	Cat. pol	3	3.4	1.8
	X		300	300	1.0	Cat. pol	3	6.8	1.8
	X		300	300	1.0	Cat. pol	3	10.2	1.8
	X		300	100	3	—		6.8	1.8
	X		300	500	0.6	—		6.8	1.8
	X		500	500	1.0	—		10.2	1.8
	X		500	500	1.0	—		3.4	1.8
	X		500	500	1.0	—		6.8	1.8
	X		500	300	1.67	—		6.8	1.8
	X		500	100	5.0	—		6.8	1.8
X			125	0				20	0.6

Δp (kPa)	Scale Pilot	Scale Plant	Time (h)	Removal (%) Oil	TSS	TOC	Stop* run B,E, Δp	Bed load (kg/m³)	Reference
32	√		15	100					Gloyna et al. (1971)
32	√		66	100					
21		√	13.5	67	67			14	API 5,6,7 (1975)
?	√		2.5	70	67			0.18	API 8 (1975)
124	√		3.5	100	100		Δp	9.5	Humenick and Davis (1978)
124	√		8	100	100		Δp	10.8	
124	√		2.16	100	100		Δp	8.8	
124	√		1.41	100	100		B	7.6	
124	√		0.83	100	100		B	6.8	
124	√		3.5	100	100		B	14.2	
124	√		1.25	100	100		B	10.2	
124	√		0.25	100	100		B	3.06	
124	√		2.9	100	100		Δp	15.7	
124	√		0.75	100	100		B	8.2	
124	√		0.16	100	100		B	3.3	
124	√		1.3	100	100		B	8.8	
124	√		0.66	100	100		B	8.9	
124	√		1.08	100	100		B	10.9	
124	√		2.16	100	100		Δp	17.6	
32	√							75	Gloyna et al. (1971)

*Operation of the bed was stopped for one of three reasons: Δp, the pressure drop became excessive; B, a breakthrough occurred; or E, elapsed time is close to when either of these might occur.

TABLE 31 (Continued)

Gra. down	Press. Down	Press. Up	Feed Oil (mg/L)	Feed SS (mg/L)	Oil/SS	Additive Name	Additive Conc. (mg/L)	Hyd. load (L/m²·s)	Depth (m)
			General oil						
			125	0				20	0.6
			32	0				8.1	0.6
			32	0				8.1	0.6
			API effluent						
X			33	25	1.3	—		2.7–4.8	0.9
X			56	43	1.3	—		2.7–4.8	0.9
X			123	?		—		0.5–2.6	0.9
X			56	43	1.3	—		2.2–3.2	0.9
X			60	60	1.0	Cat. pol	2	4	2.13
						PEH	2		
	X		16	47	0.34	—		8.2	1.52
	X					—			1.52
	X					—			1.52
	X		16	47	0.34			12	2.38
			API → DAF →						
X			70	52	1.35	—		2.4	0.75
X			70	52	1.35				0.9
			266	196	1.35	—		1.36–4.0	0.8
			266	196	1.35	Polym A22	0.5	1.75	0.8
			14	15		—		1.75	

Dilute Oil-Water Emulsions and Dispersions

Δp (kPa)	Scale Pilot	Scale Plant	Time (h)	Removal (%) Oil	TSS	TOC	Stop run B,E, Δp	Bed load (kg/m³)	Reference
52	✓							89	Gloyna graphilter
32	✓							146	Gloyna sand
32	✓							209	Gloyna graphilter
17	✓		18–24	75	55		Δp	13	Kempling and Eng (1977)
?		✓	?	75	55		E		Kempling and Eng (1977)
14–24		✓	8	82	—		E	6.4	API 1 (1975)
50		✓	9	79	53		E	7.2	API 4 (1975)
21	✓		24	88	90			17	API 5,6,7 (1975)
?		✓	12	79	77		E	11.3	Brody et al. (1977)
?		✓	12	65	59		E		
?		✓	12	65	64		E		
?	✓		>5	79	77		B	>4.3	
7		✓	8–9	77	79			9.9	API 2 to 3 (1975)
	✓		?	84	90				
12	✓		1.25	95	99			5–10	API p.65 (1975)
	✓		1.25	93	99			4.4	API low conc. of oil (1975)
	✓		5	?	73			0.4	API low conc. of oil (1975)

TABLE 31 (Continued)

	API →DAF →							
	Press.		Feed			Additive		
Gra. down	Down	Up	Oil (mg/L)	SS (mg/L)	Oil/ SS	Name	Conc. (mg/L)	Hyd. load (L/m²·s)	Depth (m)
			14			Polym A-23	0.8– 0.3	1.75	
			14	20		—		3.4	
			14	20		polym A23	0.1– 0.2	3.4	
			20	?				2	
				10		Alum	25	2	

Produced waters							
X		29.8	19.4		—		12–20	2.85
X		9.8	10.65		—		10	2.55
X		25–50	25–50		—		7.4–10	2.4
X			?				6.8	2.55
Xª		30	20		—		11–20	2.1
Xᵇ		44	23				6–11	2.1

Aerated lagoon							
X		?	10		—		2.7–4.8	0.8
	X	50	117	0.42	—		4.5	2.28
X		77	121	0.63	Alum	20–30	3.4	1.00
X		6	66	0.1	Alum Polym	30 6	3.4	1.00
X		12	12	1.0	—		1.7	0.9
	X	27	80				3–4.3	1.5

[a]After upstream DAF treatment.
[b]After upstream flocculation treatment.

Δp (kPa)	Scale Pilot	Scale Plant	Time (h)	Removal (%) Oil	Removal (%) TSS	Removal (%) TOC	Stop run B,E, Δp	Bed load (kg/m³)	Reference
	√		5		?	87		0.4	API low conc. of oil
	√		7.8		?	74		1.8	API low conc. of oil
	√				?	70			API low conc. of oil
	√		5		?	?			API low conc. of oil
	√		5		?	96			API low conc. of oil (1975)
45	√		2–26	76	50			16	Wallace and Brown (1972)
45?		√	24	92	82			6	
47			27	77	62			16	
					56				
				76	50				Nebolsine et al. (1967)
				75	56				Nebolsine et al. (1967)
10	√		3–10 5	—	50		E	0.14	Kempling and Eng (1977)
?		√	12	50	50		?	7.2	API 5 (1975)
16		√	6	66	56		?	8.6	API 4
15		√	4.5	33	98		?	4.0	API 2
7		√	24	56	52		?	2.1	API 1
125	√		(32 calc)		40			25	Mohler and Clere (1977)

TABLE 31 (Continued)

	Press.		Feed			Additive		Hyd. load $(L/m^2 \cdot s)$	Depth (m)
Gra. down	Down	Up	Oil (mg/L)	SS (mg/L)	Oil/SS	Name	Conc. (mg/L)		
	Aerated lagoon								
		X		20				3–4.3	1.5
		X		37				3–4.3	1.5
		X				Alum			
		X	27	88	0.3	—		3–4.3	1.5
X				20				2.7	0.8
X								4.8	0.8
	Produce water before flotation								
	X		?	?				6.8	2.8
	Steel mill wastes								
X			229	86				4.9	2.1
X			215	120				10	2.1
X			229	86				5	2.1
X			230	106				10.7	2.1
X			229	86				5	2.1
X			230	106				10	2.1
X			229	86				5	2.1
X			230	106				10	2.1
X			229	86				5	2.1
X			230	106				10	2.1
X			229	86				5	2.1
X			230	106				10	2.1
X			229	86				5	2.1
X			230	106				10	2.1
X			20	50				10–4.7	$\cong 1$
			18	34				10	$\cong 2$

Δp (kPa)	Scale Pilot	Scale Plant	Time (h)	Oil	TSS	TOC	Stop run B,E, Δp	Bed load (kg/m³)	Reference
125	✓		?		90			25	Mohler and Clere (1977)
125	✓		?		90			25	
	✓			93–97				13–18	
125		✓	12	76	85			10.9	
	✓		10				Δp		Kempling and Eng (1977)
	✓		3				Δp		
<30			12	64	56				Wallace and Brown (1972)
		✓	12	87	72			27	Nebolsine (1970)
		✓	12	90	88			62	
		✓	12	94	82			29	
		✓	12	86	82			62	
		✓	12	91	64				
		✓	12	89	74				
		✓	12	86	71				
		✓	12	88	70				
		✓	12	86	72				
		✓	12	87	69			56	
		✓	12	86	72				
		✓	12	88	69				
		✓	12	88	71				
		✓	12	94	77				
				58	80				

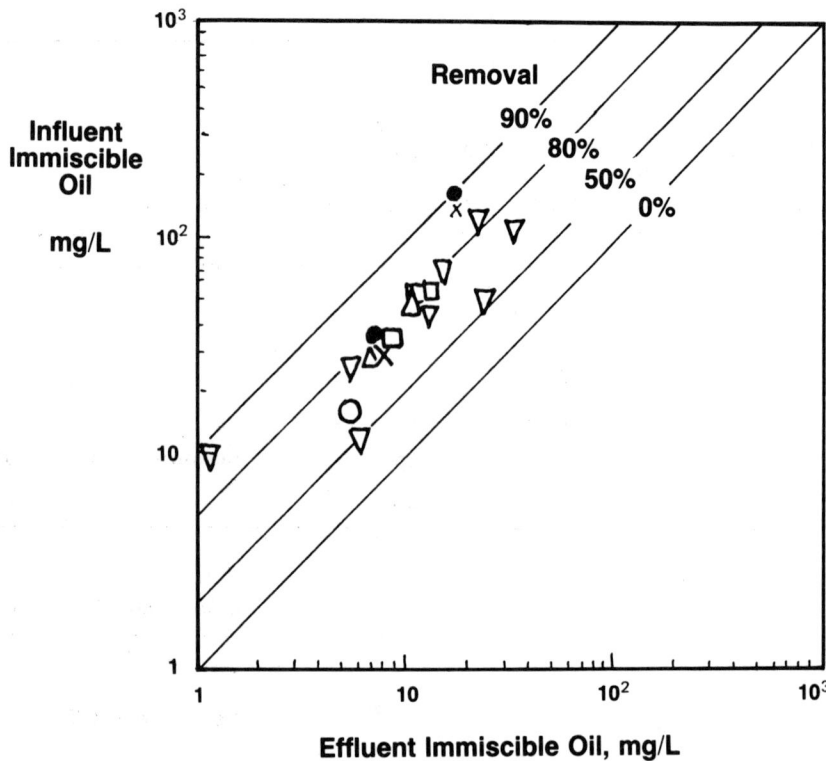

FIG. 51 Performance of filters.

miscible oil," and "soluble oil." They usually just talk about oil. A variety of solids may be present. In API refinery streams the solids tend to be corrosion products and catalyst; for steel mill wastes the solids are iron and steel particles; for produced waters, the solids are clay, and for biological liquids the solids are biological microorganisms. The reductions in BOD and COD in Table 31 are associated with the removal of the particulate matter.

Consider now the more detailed behavior of different types of feedstreams: synthetic oil/water systems; API effluents, API/DAF effluents; aerated lagoon effluents, produced waters, and steel mill wastes. The removals of oil and suspended solids are related, for different feeds, as illustrated in Fig. 52. The general conclusions are that an air scour must be included as part of the backwash; in scaling up bench-scale operations to full scale, multiply the removals by 0.94 and the loading time by 0.74; and the performance is sensitive to the solids concentrations.

Dilute Oil-Water Emulsions and Dispersions

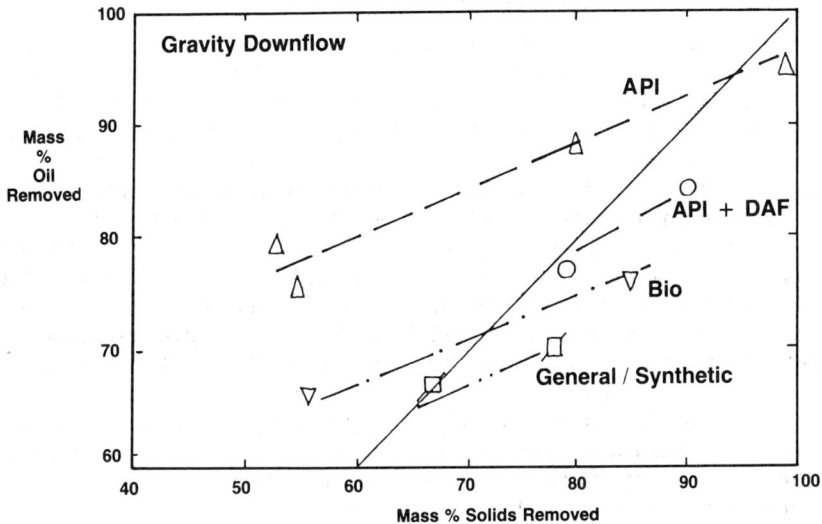

(a)

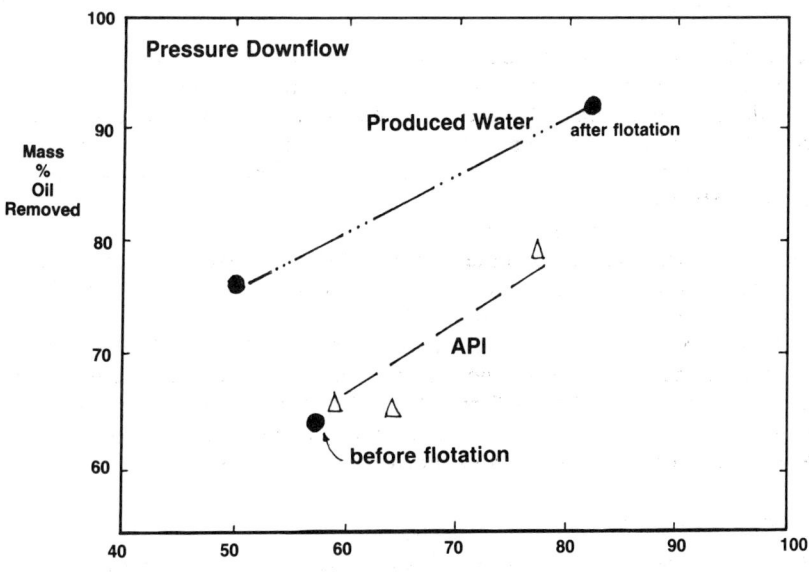

(b)

FIG. 52 General removals for deep bed filters. (a) Gravity downflow. (b) Pressure downflow.

1. Synthetic Oil-Water Systems

Humenick and Davis (1978) used different types of oil (slop oil, dry coker slop oil, and thermal cracker gas oil feedstock) with clay as the suspended solids. Tap water was the continuous phase. The concentrations ranged from 100 to 500 mg/L for both the solids and the oil. The removals were very sensitive to the oil/solid ratio. For higher concentrations of solids, the removals were unacceptable (based on the time until breakthrough). Chemical destabilizers were needed.

On the other hand, Gloyna et al. (1971), who worked with slop oil emulsions (about 60 mg/L) that contained no solids, found that the time until breakthrough for relative shallow beds of 0.6 to 0.9 m was acceptable at 15 to 66 h provided that the bed was preconditioned or aged. Initially, the oil channeled through the bed until the medium became wetted by the oil.

From the API survey (1975), for generally oily wastes, for full-scale plant operation, they could get 67% removals in 13.5-h runs provided that the top layer in the bed was a 0.76-m depth of coal. When no coal was present, the loading time was reduced to 5 h. Thus, in this example, coal was necessary. (In three other examples in the same report, the presence of a coal layer seemed to promote mudball formation.)

2. API Effluent

The concentrations of oil are in the range 20 to 125 mg/L. As long as an air scour is used as part of the backwash, the results given in Table 31 are essentially self-explanatory. Reasonable removals with acceptable loading times are achieved. Kempling and Eng (1977) note that the load time is inversely related to the incoming suspended solids concentration.

For effluents from refineries with high asphaltic crudes, the use of coal and sand media produced mudballs; the media clumped together and flushed out during backwash. However, acceptable operation occurred with 0.9 m of sand only (API, 1975).

Ford and Elton (1977) report that a pilot-scale, ultrafast bed of anthrafilt-sand-gravel reduced the influent oil (with an oil concentration of 35 to 178 mg/L) to 7 to 17 mg/L effluent concentration. For higher oil concentrations, they cite that the use of crushed graphite ore—instead of the sand—gave better oil removals.

3. Produced Waters

The feed concentration of oil and suspended solids is relatively small; the removal and loading times are acceptable. Wallace and Brown (1972) reported some plugging tendencies that they attributed to the asphaltenes. Their remedy was to backwash the bed with solvent every 4 months.

4. Biotreater Effluents

Primarily "shallow beds" have been reported for treating this effluent. The oil concentration is relatively small compared with the solids concentration; the removals are relatively small (in the range 30 to 50%) and very dependent on inlet concentrations.

6. Other Streams

Pushkarev et al. (1983, p. 160) cite filters as being applied for treating wastes from an electric pipe welding shop. The oil concentrations are 1200 mg/L with 155 mg/L of suspended solids. However, few technical and performance details are given.

5. Steel Mill Wastewaters

Very good removals have been reported for ultrafast filters by Nebolsine (1970), with example experimental data reported by Nebolsine and Sanday (1967). McGibbon (1968) described the use of pressure, downflow conventional rapid and ultrafast filters. The conventional, depth, anthracite-sand combination failed because the oil and the iron oxide formed mudballs. However, for a deep-bed, anthracite-sand unit, reasonably good removal occurred. When the solids concentration varied from 34 to 162 mg/L, the exit solid concentration remained 9 mg/L; changes in the inlet oil concentration, however, produced a corresponding change in the effluent concentration. McGibbon felt that for successful operation the inlet oil concentration should be kept below 20 mg/L.

D. Performance with the Addition of Destabilizing Chemicals

Destabilizing chemicals can be handled three different ways. An upstream coagulation basin can be designed—using the principles outlined in Section VI—and the filter handles the effluent from the coagulation basin. Second, the chemicals can be mixed with the feed just before it enters the filter so that the filter functions as a coagulation device and a filter. Ives (1970) relates the velocity gradient, G, one might expect in a filter to the head loss across the filter:

$$G = (\rho g v_0 H / \epsilon \mu L)^{0.5}$$

where ϵ is the void volume, H the head loss across the filter of depth L; and v_0 the superficial liquid loading. The third approach is to treat the medium in the bed with destabilizing chemicals during the backwash stage.

1. Synthetic Emulsions

Humenick and Davis (1978) studied the "high-rate," downflow-pressure, deep-bed filtration of the relatively concentrated (100 to 500 mg/L) synthetic mixtures. The addition of no destabilizing chemicals led to unsatisfactory performance; bed breakthroughs occurred within 5 to 10 min. To select feasible destabilizing agents for the dispersions, they performed coagulation jar tests for slop oil-clay emulsions with initial pH of 8.5 to 9.2. They found that:

1. *Anionic Dow Puriloc A-22*: oil removal about 40% for 0- to 8- mg/L dose
2. *Cationic Dow Puriloc C-31*: oil removal 90 to 92% for 2- to 10- mg/L dose
3. *Alum*: oil removal >96% for dose >40 mg/L
4. *Nonionic Dow Magnifloc 900N*: oil removal 80 to 90% for 2- to 8- mg/L dose

After a series of tests on the batch filter, they selected 3 mg/L of the cationic polymer to add to the feed. The best liquid loading was the lowest in the range they tried (3.4 to 10.2 L/m$^2 \cdot$s) and the run time could be extended to 80 min. For the cationic polymer the head loss was uniform throughout the bed; for the nonionic, the head loss was all within the top 18 cm of the 1.8-m bed. The results were very sensitive to the solids concentration. A fivefold increase in solids concentration inversely affected the time until breakthrough. In terms of removals, the solids and oil seemed to stay together, and the three different types of oil seemed to behave the same. Air scour was necessary for the unit to operate well.

Some data are reported by API (1975) on pilot-scale additions of destabilizers. On one unit, 2 mg/L of the cationic polymer was used but the loading time was short at 2.5 h. For this operation backwashing was difficult and was accompanied by a loss of the sand media.

2. API Effluent

One company reported, via API (1975), that 3 mg/L of pentaethylene hexamine plus 2 mg/L of cationic high-molar-mass polymer were added to yield 24-h load times for 60-mg/L feeds. These results were obtained on a pilot-scale unit. Ford and Elton (1977) report that the addition of 1 to 5 mg/L of "coagulant" did not improve the performance of a downflow, gravity ultrafast filter. Grutsch and Mallat (1976) suggest that because many of the API effluents are alkaline, alum is not a good choice because at this pH alum is amphoteric. They also suggest that combinations of pH and alum concentrations that are in the sweep flow region are not good because this clogs the filter very quickly. Furthermore, if alum is used and the pH increases to 9, the bed will unload and release the accumulated solids.

From Fig. 32 this is consistent with what we would expect because the range of alum dose rates they are discussing is 5 to 30 mg/L. Instead of alum alone, they suggest the use of three combinations of alum, cationic, and anionic polyelectrolytes as given in Table 22. The pH should be adjusted to 7 to 7.5. They relate the recommended liquid loadings to temperature of operation because at the lower temperature the viscosity increases should be counterbalanced by a decrease in liquid flow rate. The results are shown in Fig. 53.

3. Biotreater Effluents

The API (1975) reports some examples where alum is mixed with the feed. The concentration is 20 to 30 mg/L but no pH is reported for either of the cases given. Mohler and Clere (1977), working with upflow filters, tried the addition of alum with/without cationic polymer addition. On the pilot-plant scale, the addition of 20 to 30 mg/L of alum increased the oil removal to 93% to 97%. However, the amount of solids that could be retained in the bed decreased from 25 kg/m^3 to 13 to 18 kg/m^3. They did additional tests on polymers with a jar test screening program of 16 polymers. Details are not given.

Grutsch and Mallatt (1976) suggest that a combination of 5 to 20 mg/L of alum, 3 to 6 mg/L of cationic polyelectrolyte, and 0.1 to 0.4 mg/L of anionic polyelectrolyte be used for treating biocolloid waste-

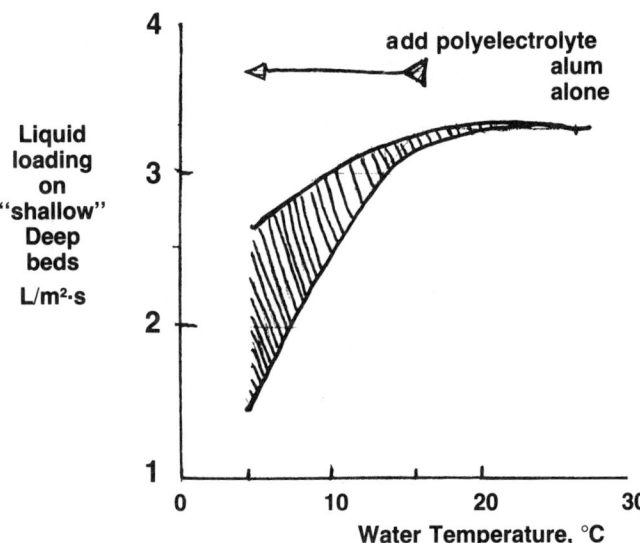

FIG. 53 Effect of temperature on liquid loading for filters [from Grutsch and Mallatt (1976)].

waters. If the salt concentration of these waters increases, then the dose rates—especially of the cationic polyelectrolyte—needs to be increased.

4. Steel Mill Wastes

Frank and Gravenstreter (1969) suggest that the operation is sometimes improved when a small amount of polyelectrolyte is added. Details are not given.

E. Regeneration of the Filter

Once the filter has been loaded, it is brought out of service and regenerated. The signs that the bed needs to be regenerated are that the pressure drop across the bed exceeds some norm, or that the oil or suspended solids break through the bed so that the effluent concentration of either increases dramatically, or that the elapsed time is close to when either of these two events might occur. Ideally, we would like to have both the pressure drop constraint and the breakthrough occur at the same time, but rarely does this happen. In Table 31 the code Δp means that the run was terminated because of the high pressure drop, B signifies a concentration breakthrough, and E means that either occurred. Usually, the bed is regenerated at specified time interfals. This is preferred especially when several filters are operated in parallel, and the filtration cycle of each bed can be controlled more easily with a timer so that none of the filters is overloaded. A shorter cycle time also seems appropriate when the filter grains tend to aggregate because of asphaltenes in the oil; more frequent backwashing results in cleaner filters (API, 1975). Most industrial installations have a dual backwash control activated by either a timed cycle or an excessive pressure drop across the bed. For oil filters, an air scour to turbulently scrub the media seems to be essential. The steps are:

1. Allow the bed to drain.
2. Apply an air scour (with/without water).
3. Apply a water-only backwash.
4. Restore the bed in preparation for in-service.

Usual values for these different steps are given in Table 32. For the air scour and the water-only backwash, the bed expands and becomes fluidized. The total regeneration cycle usually lasts 30 to 40 min. The total volume of water used in the backwash is about 2 to 7% of the water filtered.

The backwash water leaving the bed contains most of the recovered oil and solids. Typical concentrations are 200 to 2200 mg/L of oil and 1000 to 5000 mg/L of suspended solids. This water is usu-

TABLE 32   Typical Flow Rates and Duration of the Backwash Steps

	Drain	Air + H_2O	H_2O only	Total
Rate ($dm^3/m^2 \cdot s$ or $L/m^2 \cdot s$)	—	15–35 0–5	5–20	—
Time (min)	1.5–20	2–12	3–12	15–60

ally treated by sedimentation. This regeneration procedure usually restores the bed to its original condition; usually, the media is not washed away nor does it deteriorate.

F. General Selection/Sizing Procedure

The general removals are illustrated in Fig. 54, showing the oil removal as a function of both the liquid loading on the filter (and hence whether a conventional rapid-sand filter was used or if an ultrafast filter was used) and the total immiscible oil plus solids in the feed. The industrial applications are shown in the boxed region. Humenick and Davis's data are for the higher concentrations. In general, the oil removals achieved by both types are in the range 65 to 80% and are relatively independent of the type of filter. Hence it is advantageous to use the ultrafast filter with the greater liquid loading and capacity. From Fig. 54 the feed concentration does not seem to affect the oil removals [although individual researchers have varying comments. Some say the exit concentration of solids remains the same as the solids inlet concentration increases—and hence the removal of solids would have to increase with increasing feed concentration. The oil removal seems to be related to the solids removal (as illustrated in Fig. 52) and hence the oil removal would be expected to increase. It does not seem to.] This is consistent with the theoretical prediction of Yao et al. (1971) that for solids removal, the efficiency of the filter is independent of the feed concentration of solids.

Although the influent concentration does not seem to affect the removal, the pressure drop developed during filtration is very dependent on the feed concentration. Although more particles are deposited on the media, the pressure drop increases until the filter is inoperable. This is consistent with Humenick and Davis, who found that a fivefold increase in the concentration of suspended solids decreased the total volume of water that could be successfully treated by one-fifth. These trends can be converted into a loading term (total mass of solids or oil that can be held per unit volume of bed), an incoming suspended solids concentration, and the product of the liquid loading

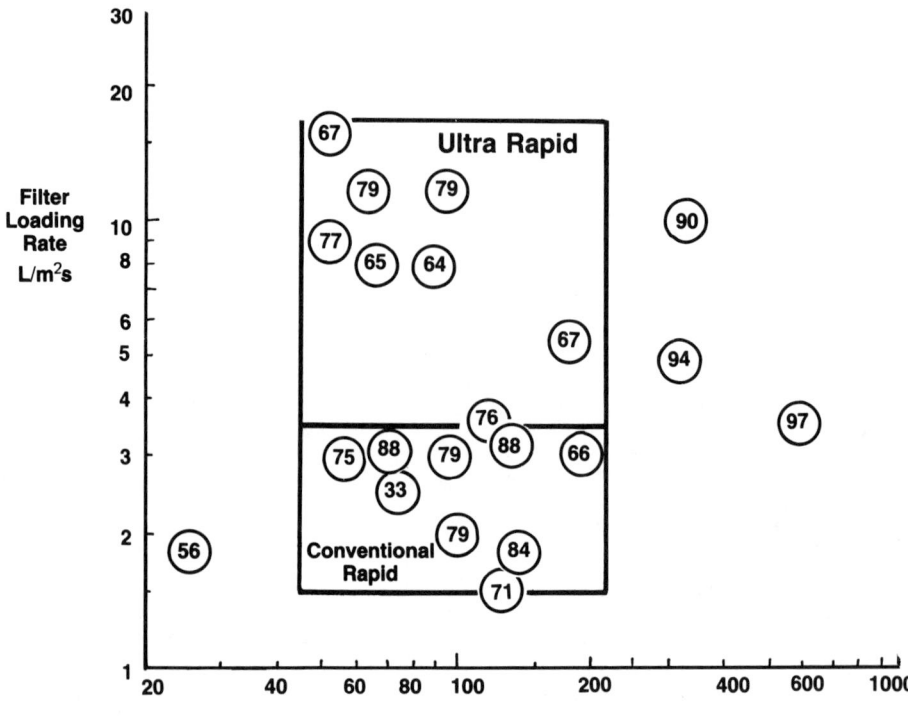

FIG. 54 Percentage of oil removal by filtration as a function of filter loading rate and total feed concentration.

rate and the time until the bed is "loaded." Figure 55 illustrates such interrelationships. To put the parameter (the solids loading) into perspective, if all the interstices of the bed were filled with oil, the loading would be 480 kg/m^3. If all the interstices were filled with solid particles, the maximum loading would be about 288 kg/m^3. Naturally, to minimize pressure drop and to allow the continuous phase to flow through the interstices, the loading is a fraction of the "full pore" volume that leads to the "upper limit" values above. Indeed, the usual loadings are in the range 5 to 30 kg/m^3.

Another set of useful information relates the total cycle time with the product of the liquid loading, the feed concentration of suspended solids, and the removal of the solids. The data shown in Fig. 56 show two distinctly different curves that depend primarily on the depth of the bed. Most of the data that cluster around the lines are for the treatment of general oily wastes, API effluent, and steel mill

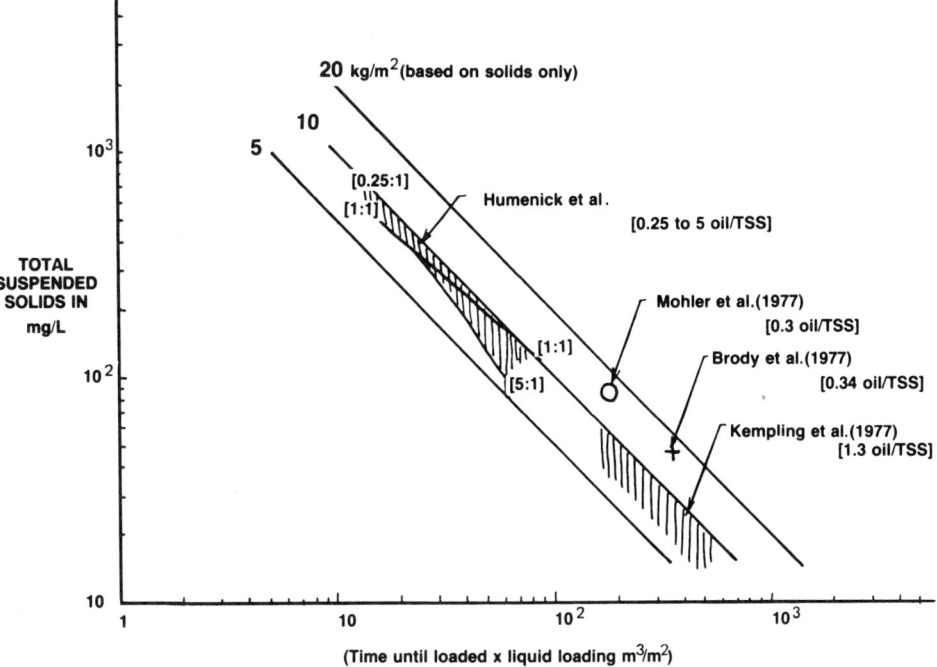

FIG. 55 Performance of deep bed filters.

wastes. The data for biotreater effluents are to the right of the lines for both depths of bed.

What is interesting about the data in Figs. 55 and 56, and in the general results, is that (although we are focusing on the removal of oil) the correlations seem to work best in terms of the removal of suspended solids.

A general selection/sizing procedure is as follows:

1. From Fig. 41, evaluate whether filtration might be a possibility. In general, the oil concentrations are usually very small, say in the range below 50 mg/L.
2. Unless dictated by other arguments, select a high-rate, bed thickness of >1.5-m filter. From Table 29 select a reasonable liquid loading.
3. From the desired oil removal, check that this seems feasible (from Fig. 41 or 51) and then, from Fig. 52, read off an anticipated solids removal (since all the sizing procedures are correlated in terms of the solids performance).

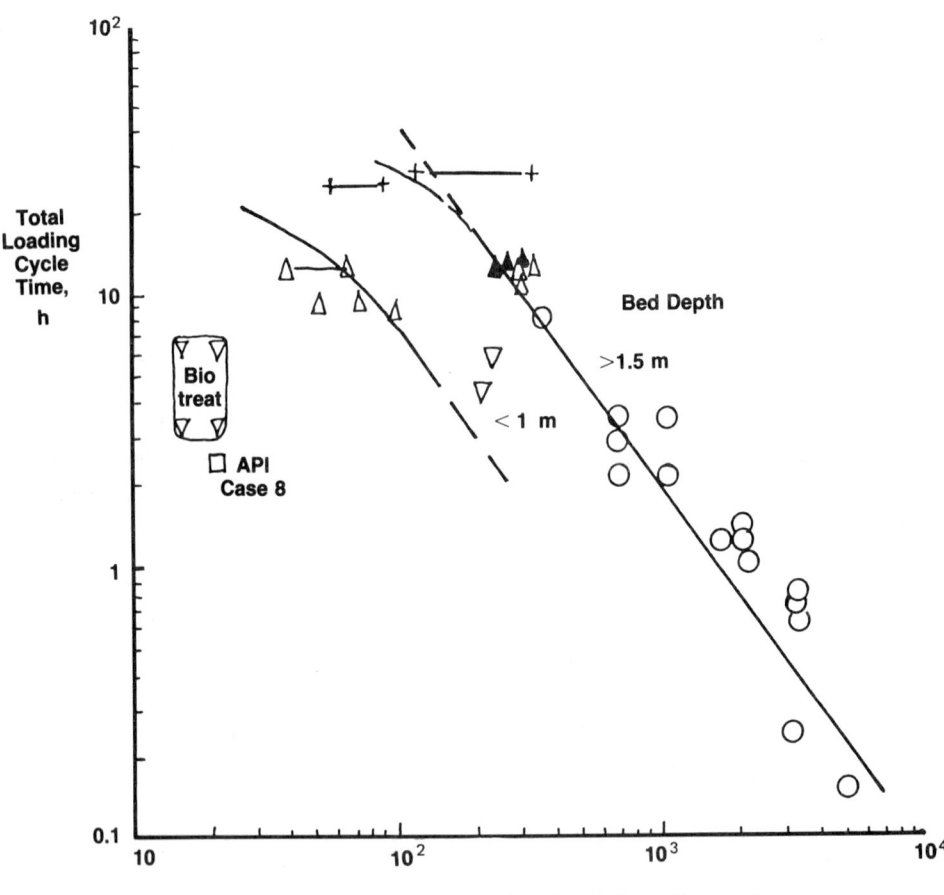

FIG. 56 Loading time for deep bed filters. Gravity: △ API [from Kempling et al. (1977)]; ○ [from Humenick et al. (1978)]; ▽ bioeffluent. Pressure: ▲ [from Brody (1977)].

4. From the liquid loading, the feed solids concentration and the anticipated solids removal, estimate, from Fig. 55, the total load time.
5. Confirm these values by cross-checking the values on Fig. 56.
6. Estimate the total cycle time by adding to the load time (obtained in steps 4 and 5) the regeneration time. Select appropriate values for this from Table 32.

Concerning the choice of media, the literature poses many contradictions, especially about the role of coal and anthracite media. Some report the formation of aggregates or "mudballs" because of an interaction between the media and the oil. Grutsch and Mattatt (1976) suggest that unexpectedly large slugs of oil coming to the filter can be particularly troublesome. Because of mudball formation, the horizontal cross-sectional area for filtration decreases, the pressure drop increases substantially, and early breakthrough occurs. The options seem to be, for these situations, to use sand only, to backwash frequently, and to try to eliminate slugs of oil by installing an equilization basin upstream. Other components in the feed that might cause problems include highly alkyllated, substituted aromatics, carboxyllic acids, and polyethylene-polypropylene glycols (Grutsch and Mallatt, 1976).

The issue of the role of pH and the charge interaction between the media and the emulsion has not received enough attention from researchers despite the fact that the surface phenomena fundamentals seem to point to this as being of key importance.

IX. SUMMARY

Some background details were given as to why oil-water dispersions would be stable and the techniques that can be used to destabilize such dispersions. Eleven factors affecting stability were outlined. Example calculations were given. An overview of the processing alternatives concentrated on gravity settling, the addition of destabilizing chemicals, coagulation, dissolved air flotation, and filtration. Details were given of studies done using these different approaches.

ACKNOWLEDGMENTS

We are pleased to acknowledge the financial support received from the Canadian Natural Sciences and Engineering Research Council, and the help of Helen Hadcock for some of the analysis.

REFERENCES

Adamson, A. W. (1967) *Physical Chemistry of Surfaces*, Interscience, New York.

Akitt, J. W., et al. (1972) J. Chem. Soc. London, 614.

Ali, L. H. (1978) Fuel, 57:357.

Amirtharajah, A., and Mills, K. M. (1982) J. Am. Water Works Assoc., 74:210.

Angle, C. W., and Hamza, H. A. (1985) paper presented at the 59th Colloid and Surface Science Symposium, Clarkson University, Potsdam, N.Y., June 24–28.

API (1963) *Manual on Disposal of Refinery Wastes*, API Series 6, No. 1, American Petroleum Institute, New York.

API (1975) *Granular Media Filtration of Petroleum Refinery Effluent Waters*, API Publication 947, American Petroleum Institute, New York.

Austin, R. J., and Vause, E. H. (1951) in Proc. 6th Ind. Waste Conf., Purdue University, West Lafayette, Ind., pp. 272–277.

Aveston, J. (1965) J. Chem. Soc.:4438.

Baes, C. F., and Mesmer, R. E. (1976) *The Hydrolysis of Cations*, Wiley, New York.

Baird, M. H. I., et al. (1981) *Surface and Interfacial Tension Measurements by Drainage Methods*, McMaster University, Hamilton, Ontario.

Balden, A. R. (1959) Sewage Ind. Wastes, 31:934.

Bargeman, D. and van Voorst Vader, F. (1972) J. Electroanal. Chem., 37:45.

Barnea, E. and Mizrahi, J. (1975) Trans. Inst. Chem. Eng., 53:61.

Barnea, E. and Mizrahi, J. (1977) CIM, Spec. vol. 21:374.

Bauer, D. (1976) Proc. Ind. Waste Cong., Purdue University, West Lafayette, Ind., pp. 816–822.

Behie, L. A., van Ham, C. J. M., and Berk, D. (1982) Water Pollut. Res. Can., 17:7.

Benger, M. (1964) Fluid Handl., June; cited by Beychok (1967).

Berg, J. C. (1981) The Fundamentals of Dispersion Stability, short course notes, personal communication, University of Washington, Seattle.

Berkman, S., and Egloff, G. (1941) *Emulsions and Foams*, Reinhold, New York.

Bersillon, J. L. (1983) Ph.D. thesis, McMaster University, Canada.

Beychok, M. R. (1967) *Aqueous Wastes from Petroleum and Petrochemical Plants*, Wiley, New York.

Bloodgood, D. E. (1952) Proc. 7th Ind. Waste Conf., Purdue University, West Lafayette, Ind., pp. 361–374.

Bottero, J. Y., et al. (1980) paper presented at the 10th International Conference on Water Pollution Research, Toronto.

Bratby, J. R. (1981) J. Am. Water Works Assoc. Res. Technol., 73: 318.

Brody, M. A., and Lumpkins, R. J. (1977) Chem. Eng. Prog., 73: (4):83.

Brosset, C., et al. (1954) Acta Chem. Scahd., 8:1917.

Brunsmann, J. J., Cornelissen, J., and Eilers, H. (1962) J. Water Pollut. Control Fed., 34:44.

Burrill, K. A. (1967) M. Eng. thesis, McMaster University, Canada.

Camp, T. R. (1946) Trans. Am. Soc., Civ. Eng., 111:895.

Camp, T. R. (1955) Trans. Am. Soc. Civ. Eng., 120:1.

Campbell, G. C., and Scoullar, G. R. (1964) Hydrocarbon Process. Pet. Refiner, 43(5):137.

Casassa, E., et al. (1985) personal communication, Carnegie-Mellon University, Pittsburgh, Pa.

Cayias, J. L., et al. (1975) in *Adsorption at Interfaces* (K. L. Mittal, ed.), ACS Symposium Series, 8, American Chemical Society, Washington, D.C.

Chieu, J. N., Gloyna, E. F., and Schechter, R. S. (1977) J. Environ. Eng. Div. Am. Soc. Civ. Eng., EE2:163.

Churchill, R. J., and Kaufman, W. J. (1973) *Water Processing Related Surface Chemistry of Oil Refinery Wastewater*, SERL Report 73-3, University of California, Berkeley, Calif.

Clayfield, E. J. (1982) paper presented at the AIChE Annual Meeting, Los Angeles, Nov. 20.

Cleasby, J. L. (1972) in *Physicochemical Processes for Water Quality Control* (W. J. Weber, Jr., ed.), Wiley, New York.

Cooper, D. G., et al. (1980) Can. J. Chem. Eng., 58:576.

Coulaloglou, C. A., and Tavalarides, L. L. (1976) AIChE J., 22:289.

Davies, J. T., and Rideal, E. K. (1963) *Interfacial Phenomena*, 2nd ed., Academic Press, New York.

Davies, G. A., Jeffreys, G. V., and Smith, D. V. (1971) AIChE J., 16:827.

de Chazal, L. E. M., and Ryan, J. T. (1971) AIChE J., 17:1226.

de Filippi, R. P. (1977) in *Filtration: Principles and Practice* (C. Orr, ed.), Marcel Dekker, New York.

DeHek, H., et al. (1978) J. Colloid Interface Sci., 64:71.

Dempsey, B. A., et al. (1984) J. Am. Water Works Assoc. Technol., 76:141.

Devereux, O. F., and de Bruyn, P. (1963) *Interaction of Plane Parallel Double Layers*, MIT Press, Cambridge, Mass.

Dow, D. B. (1926) *Oil Field Emulsions*, Bulletin 250, U.S. Dept. of Commerce, Washington, D.C.

Eckenfelder, W. W., Jr., (1966) *Industrial Water Pollution Control*, McGraw-Hill, New York.

Eisenlauer, J., and Horn, D. (1985) paper presented at the 59th Colloid and Surface Science Symposium, Clarkston University, Potsdam, N.Y., June 24–28.

Ford, D. L., and Elton, R. L. (1977) Chem. Eng., Oct. 17, pp. 49–56.

Fowkes, F. W. (1965) in *Chemistry and Physics of Interfaces* (S. Ross, ed.), American Chemical Society, Washington, D.C., 1–12.

Frank, V. P., and Gravenstreter, J. P. (1969) J. Water Pollut. Control Fed., 41:292.

Fraser, M. E., et al. (1971) Metall. Trans., 2:817.

Gadjiev, V. G., and Chian, E. S. K. (1976), Proc. 31st Ind. Waste Conf., Purdue University, West Lafayette, Ind., pp. 965–971.

Gewers, C. W. W. (1968) J. Can. Pet. Technol., 7(2):85.

Girifalco, L. A., and Good, R. J. (1957) J. Phys. Chem., 61:904.

Gloyna, E. F., Brady, S. O., and Marinez-Pereda, P. (1971) Proc. 26th Ind. Waste Conf., Purdue University, West Lafayette, Ind., pp. 308–317.

Golob, J., and Modic, R. (1977) Trans. Inst. Chem. Eng., 55:207.

Good, R. J., and Elbing, E. (1971) in *Chemistry and Physics of Interfaces* (S. Ross, ed.), American Chemical Society, Washington, D.C., pp. 71–96.

Good, R. J., et al. (1958) J. Phys. Chem., 62:1418.

Gopal, E. S. R. (1968) in *Emulsion Science* (P. Sherman, ed.), Academic Press, New York, pp. 1–73.

Gregory, J. (1969) Adv. Colloid Interface Sci., 2:396.

Grilc, V., Golob, J., and Modic, R. (1984) Chem. Eng. Res. Des., 62:48.

Grimes, C. B. (1977) Oil Gas J., 75(17):106.

Gruette, J. L. (1979) Pollut. Eng., 11(10):44.

Grutsch, J. F., and Mallatt, R. C. (1976) Hydrocarbon Process., 55: (7):113.

Hahn, H. H., and Stumm, W. (1968) J. Colloid Interface Sci., 28:133.

Hall, E. S., and Tollefson, E. L. (1983) Can. J. Chem. Eng., 60: 812.

Hamaker, H. C. (1936, 1937) Recl. Trav. Chim. Pays-Bas Belg., 55: 1015; 56. 3, 727.

Happel, John (1958a) Chemical Process Economics, Wiley, New York.

Happel, John (1958b) AIChE J., 4:197.

Harlow, B. D., and Doran, T. M. (1981) Ind. Wastes Chicago, 27:24.

Hart, W. B. (1947) Pet. Process., 2:282, 471.

Hart, J. A. (1970) in Proc. 25th Ind. Waste Conf., Purdue University, West Lafayette, Ind., pp. 406–413.

Hasiba, H. H., and Jessen, F. W. (1968) J. Can. Pet. Technol., 7:1.

Hayden, P. L., and Rubin, A. J. (1976) in Aqueous Environmental Chemistry of Metals (A. J. Rubin, ed.), Ann Arbor Science, Ann Arbor, Mich.

Hinze, J. O. (1955) AIChE J., 1:289.

Holmes, L. P., et al. (1968) J. Phys. Chem., 72:401.

Hooper, W. B., and Jacobs, L. J., Jr. (1979) in Handbook of Separation Techniques for Chemical Engineers (P. A. Schweitzer, ed.), McGraw-Hill, New York.

Hough, D. B., et al. (1980) Adv. Colloid Interface Sci., 14:3.

Hsu, P. H. (1977) in Minerals in Soil Environments (J. B. Dixon and S. B. Weed, eds.), Soil Science Society of America, Madison, Wis.

Hsu, P. H., and Bates, T. F. (1964) Mineral Mag., 33:749.

Humenick, M. J., and Davis, B. J. (1978) J. Water Pollut. Control Fed., 50:1953.

International Critical Tables of Numerical Data, Physics, Chemistry and Technology (1926), McGraw-Hill, New York.

ISEC 70, International Solvent Extraction Conference, paper 51.

Ives, K. J. (1970) Filtr. Sep., 7:700.

Ives, K. J. (1978) in The Scientific Basis of Flocculation (K. J. Ives, ed.), Sijthoff en Noordhoff, Alphen aan den Rijn, The Netherlands.

Jaisinghani, R. A., and Verdegan, B. M. (1982) *Separation of Oil/Water Emulsions by Deep Bed Filtration*, Nelson Industries, Stoughton, Wis.

Jaisinghani, R. A., et al. (1981) in *Theory and Practice and Process Principles for Physical Separations* (M. P. Freeman and J. A. FitzPatrick, eds.), Engineering Foundation, New York, pp. 579–594.

Jasper, J. J. (1972) J. Phys. Chem. Ref. Data, $1(4)$:841.

Jeffreys, G. V., and Hawksley, J. L. (1965) AIChE J., 11:413.

Jeffreys, G. V., Davies, G. A., and Pitt, K. (1970) AIChE J., 16:823.

Johanson, G. (1960) Acta Chem. Scand., 14:771.

Karabelas, A. J. (1978) AIChE J., 24:170.

Katnik, K. E., and Pavilcius, A. M. (1979) Proc. 33rd Ind. Waste Conf., Purdue University, West Lafayette, Ind., pp. 236–243.

Kawamura, S. (1981) in *Theory and Practice and Process Principles for Physical Separations* (M. P. Freeman and J. A. FitzPatrick, eds.), Engineering Foundation, New York, pp. 19–36.

Kaye, G. W. C., and Laby, T. H. (1966) *Tables of Physical and Chemical Constants and Some Mathematical Functions*, 13th ed., Wiley, New York.

Keith, F. W., and Hixon, A. N. (1955) Ind. Eng. Chem., 47:258.

Kempling, J. C., and Eng, J. (1977) Chem. Eng. Prog., $73(4)$:87.

King, P. H., et al. (1983) Proc. 37th Ind. Waste Conf., Purdue University, West Lafayette, Ind., pp. 35–40.

Kirby, A. W. W. (1964) Chem. Eng., (177), April; CE 76 cited by Beychok (1967).

Kolmogoroff, A. N. (1949) Dokl. Akad. Nauk SSSR, $66(5)$:825.

Kulowiec, J. J. (1979) Pollut. Eng., $11(2)$:49.

LaMer, V. K., and Healy, T. W. (1963) Rev. Pure Appl. Chem., 13:112.

Lawrence, A. S. C. (1948) Chem. Ind., London, Sept. 25, pp. 615–617.

Lawrence, A. S. C., and Killner, W. (1948) J. Inst. Pet. London, 34:821.

Lawson, G. B. (1967) Chem. Process Eng. London, 48:45.

Letterman, R. D., et al. (1973) J. Am. Water Works Assoc., 65:716.

Lifshitz, E. M. (1955) Zh. Eksp. Teor. Fiz., 29:94.

Lindsay, J. T., and Prather, B. V. (1977) J. Water Pollut. Control Fed., 49:1779.

Lissant, K. J. (1983) *Demulsification: Industrial Applications*, Surfactant Science Series, Vol. 13, Marcel Dekker, New York.

Little, R. C., and Patterson, R. L. (1978) Environ. Sci. Technol., 12:584.

Loeb, A., Wiersema, P. H., and Overbeek, J. Th. G. (1961) *The Electrical Double Layer Around a Spherical Colloid Particle*, MIT Press, Cambridge, Mass.

Luthy, R. G., Selleck, R. E., and Galloway, T. R. (1977) Environ. Sci. Technol., 11(13):1211.

Luthy, R. G., et al. (1978) J. Water Pollut. Control Fed., 50:331.

Mahanty, J., and Ninham, B. W. (1976) *Dispersion Forces*, Academic Press, New York.

Majid, A., and Farnand, J. R. (1980) *Investigation of Norcen Emulsions*, Chemistry Division Report C1081-80S, National Research Council of Canada, Ottawa.

Manchanda, K. D. (1966) M. Eng. thesis, McMaster University, Canada.

Manchanda, K. D., and Woods, D. R. (1968) Ind. Eng. Chem. Process Des. Dev., 7:182.

Marion, G. M., et al. (1976) Soil Sci., 121:76.

Matijevic, E., et al. (1961) J. Phys. Chem., 65:826.

Matijevic, E., et al. (1966) J. Colloid Interface Sci., 22:68.

May, H. M., et al. (1979) Geochim. Cos. Acta, 43:861.

McCutcheon's Detergents and Emulsifiers, Annual, J. W. McCutcheon, Inc., published by Allured, Ridgewood, N.J.

McGibbon, V. R. (1968) Iron Steel Eng., 45(4):93.

Meissner, H. P., and Chertow, B. (1946) Ind. Eng. Chem., 38(8):856.

Metcalf and Eddy, Inc. (1972) *Wastewater Engineering: Collection, Treatment and Disposal*, McGraw-Hill, New York.

Miranda, J. G. (1977) Chem. Eng., 84(3):105–107.

Mohler, E. F., Jr., and Clere, L. T. (1977) Chem. Eng. Prog., 73(4):74.

Napper, D. H. (1970) Ind. Eng. Chem. Process Des. Dev., 9:469.

Nebolsine, R. (1970) Proc. 25th Ind. Waste Conf., Purdue University, West Lafayette, Ind., pp. 885–891.

Nebolsine, R., and Sanday, R. (1967) Iron Steel Eng., 44(12):105.

Neumann et al. (1958) paper presented at API 23rd Midyear Meeting, May; cited by Beychok (1967).

Nilsson, H., Silvegren, C., and Tornell, B. (1985) J. Vinyl Technol., 7:112.

Ninham, B. W., and Parsegian, V. A. (1970) J. Chem. Phys., 53(9):3398.

Null, H. R., and Johnson, H. F. (1958) AIChE J., 4:273.

O'Melia, C. R. (1972) in *Physicochemical Processes for Water Quality Control* (W. J. Weber, ed.), Wiley, New York.

O'Melia, C. R. (1978) in *The Scientific Basis of Flocculation* (K. J. Ives, ed.), Sijthoff en Noordhoff, Alphen aan den Rijn, The Netherlands, p. 219.

O'Melia, C. R., and Stumm, W. (1967) J. Am. Water Works Assoc., 59(11):1393.

Osipow, L. I. (1962) *Surface Chemistry: Theory and Industrial Applications*, Reinhold, New York.

Osmond, D. W. J., Vincent, B., and Waite, F. A. (1973) J. Colloid Interface Sci., 42:262.

Overbeek, J. T. G. (1968) in *Colloid Science* (H. R. Kruyt, ed.), Elsevier, Amsterdam.

Padday, J. F. (1973) in *Surface and Colloid Science* (E. Matijevic, ed.), Wiley, New York.

Palmer, A. J., and Averill, D. W. (1983) paper presented at Purdue Industrial Waste Conference, May.

Parks, G. A. (1972) Am. Mineral., 57:1163.

Parsegian, V. A. (1975) in *Physical Chemistry: Enriching Topics from Colloids and Surface Science* (H. van Olphen and K. J. Mysels, eds.), IUPAC Commissions I.6, Therox, La Jolla, Calif.

Payatakes, A. C. (1977) *Deep Bed Filtration: Theory and Practice*, course notes for the ASEE Summer School for Chemical Engineering Faculty, Snowmass, Colo.

Pincus, I., et al. (1951) Ind. Eng. Chem., 43:521.

Prather, B. V. (1961) Pet. Refiner, 40(5):177.

Probstein, R. F., et al. (1981) in *Theory and Practice and Process Principles for Physical Separations* (M. P. Freeman and J. A.

FitzPatrick, eds.), Engineering Foundation, New York, pp. 19–36.

Pulfer, K., and Kramer, J. R. (1982) *Aluminum Hydrolysis Review*, Environmental Geochemistry Report 1, McMaster University, Hamilton, Ontario.

Pursell and Miller (1961) Proc. 16th Ind. Waste Conf., Purdue University, West Lafayette, Ind.; cited by Beychok (1967).

Pushkarev, V. V., Yuzhaninov, A. G., and Men, S. K. (1983) *Treatment of Oil-Containing Wastewater*, Allerton Press, New York.

Quigley, R. E., and Hoffman, E. L. (1966) Proc. 21st Ind. Waste Conf., Purdue University, West Lafayette, Ind., pp. 527–533.

Rich, L. G. (1961) *Unit Operations of Sanitary Engineering*, Wiley, New York.

Rohlich, G. A. (1953) Proc. 8th Ind. Waste Conf., Purdue University, West Lafayette, Ind., pp. 368–381.

Rohlich, G. A. (1954) Ind. Eng. Chem., 46:304.

Sadler, A. B., Jr., et al. (1978) Proc. 32nd Ind. Waste Conf., Purdue University, West Lafayette, Ind., pp. 709–715.

Selker, A. H., and Sleicher, C. A. (1965) Can. J. Chem. Eng., 43:298.

Shackleton, L. R. B., Douglas, E., and Walsh, T. (1960) *Pollution of the Sea by Oil*, The Institute of Marine Engineers, London, pp. 1–20.

Sharma, P. K., and Chand, T. (1979) J. Indian Chem. Soc., 56:1206.

Sherony, D. F. (1969) Ph.D. thesis, Illinois Institute of Technology.

Sherony, D. F., Kintner, R. C., and Wasan, D. T. (1978) in *Surface and Colloid Science* (E. Matijevic, ed.), Plenum Press, New York.

Shinnar, R. (1961) J. Fluid Mech., 10:259.

Shinoda, K. (1978) *Principles of Solution and Solubility*, Marcel Dekker, New York.

Siffert, B., et al. (1984) Fuel, 63:834.

Simonsen, R. N. (1962) Hydrocarbon Process. Pet. Refiner, 41(5):145.

Sleicher, C. A. (1962) AIChE J. 8:471.

Slezak, M. A., and Woods, D. R. (1973) Journal Water Pollut. Control Fed., 45:2138.

Smith, C. V., Jr. (1967) J. Sanit. Eng. Div. Am. Soc. Civ. Eng., SA 5:91.

Smith, D. V., and Davies, G. A. (1970), Can. J. Chem. Eng., 48:628.

Sontheimer, H. (1978) in *The Scientific Basis of Flocculation* (K. J. Ives, ed.), Sijthoff en Noordhoof, Alphen aan den Rijn, The Netherlands, pp. 193–217.

Speilman, L. A., and Cukor, P. M. (1973) J. Colloid Interface Sci., 43:51; Correction, 44:392 (1973).

Steiner, J. L., et al. (1978) Chem. Eng. Prog., 74(12):39.

Stormont, D. H. (1956) Oil Gas J., 10:26.

Strassner, J. E. (1968) J. Pet. Technol., 20:303.

Sullivan, J. H., and Singley, J. E. (1968) J. Am. Water Works Assoc., 60:1280.

Szpak, J. (1986) M. Eng. thesis, McMaster University, Canada.

Thompson, C. S., Stock, J., and Mehta, P. L. (1972) Oil Gas J., 70:53.

van Ham, N. J. M., Behie, L. A., and Sorcek, W. Y. (1983) Can. J. Chem. Eng., 61:541.

Verwey, E. J. W., and Niessen, K. F. (1939) Philos. Mag. Ser., 7(27):435.

Verwey, E. J. W., and Overbeek, J. Th. G. (1948) *Theory of the Stability of Lyophobic Colloids*, Elsevier, Amsterdam.

Vijayan, S., (1978) personal communication, McMaster University, Hamilton, Ontario.

Vijayendran, B. R. (1980) in *Polymer Colloids*, Part II (R. M. Fitch, ed.), Plenum Press, New York.

Vincent, B. (1973) J. Colloid Interface Sci., 42:270.

Vold, M. J. (1961) J. Colloid Sci., 16:1.

Wahl, J. R., Jr., (1980) Proc. 34th Ind. Waste Conf., Purdue University, West Lafayette, Ind., p. 719 (Extension Series 1979).

Wallace, J. T., and Brown, J. S. (1972) Proc. 27th Ind. Waste Conf., Purdue University, West Lafayette, Ind., p. 151.

Weber, W. J. (1972) *Physicochemical Processes for Water Quality Control*, Wiley, New York.

Weston, R. F. (1950) Ind. Eng. Chem., 42:609; cited in Beychok (1967).

Weston, R. F., et al. (1949) Proc. 4th Ind. Waste Conf., Purdue University, West Lafayette, Ind., pp. 290–313.

Wnek, W. (1974) Filtr. Sep., May/June, pp. 237–242.

Woods, D. R. (1984) *Colloids, Surfaces and Unit Operations*, McMaster University, Hamilton, Ontario.

WPCF (1977) *Wastewater Treatment Plant Design*, MOP 8, Water Pollution Control Federation and the American Society of Civil Engineers, New York.

Yao, K., Habibian, M. T., and O'Melia, D. R. (1971) Environ. Sci. Technol., 5:1105.

Zirimenya, J. B. K., and Woods, D. R. (1968) *Purification of Oily Waste Water Using a Solvent Dilution Method*, Parts I and II, Report to AISI Project 182, McMaster University, Hamilton, Ontario.

Index

A

Acetylene tetrabromide, surface tension for, 441
Acidic extractants, 144–147
Acrolein, water of solubility of, 283
Acrylic acid, heat of polymerization of, 275
Acrylonitrile
  heat of polymerization of, 275
  water of solubility of, 283
Activator herbicide adjuvants, 239
Admicellar chromatography, 115–120
  adsolubilization, 115, 116
  fixed-bed operation, 117–120
Adsolubilization, 115, 116
Adsorbed polymers (steric stabilization), 404–408
Adsorbing colloid flotation, 196
Adsorption of the flotation surfactants, 201–225
  chain-chain interactions, 209, 210
  chemical forces, 210–212
  electrostatic factors, 201–208

[Adsorption of the flotation surfactants]
  surface chelation, 219–222
  surface precipitation, 212–215
  surfactant solution chemistry, 222–225
  xanthate interactions, 215–218
Agricultural herbicides, 241
Alcohol dehydrogenase, 111
Alkylaryl ethoxylates/sulfates, chemical structure of, 307
Alkyl benzene sulfonates, chemical structure of, 307
Alkyl ethoxylates/sulfates, chemical structure of, 307
Alkylphosphates, 204
  chemical structure of, 307
Alkylsulfates, 204
  chemical structure of, 307
Alkylsulfonate, 204
n-Alkyl xanthates, 205
Aluminum chemistry, kinetics of, 432, 433
Aluminum coagulants, destabilizing oil-water dispersions with, 433, 434
Aluminum monomers, aqueous, 427–431

Aluminum polymers, aqueous, 431, 432
Amine salts, 204, 205
Amyl acetate, surface tension for, 441
Amyl alcohol, surface tension for, 441
α-Amylase, 111
i-Amyl butyrate, surface tension for, 441
Amyl chloride, surface tension for, 441
Anatase, PZC of, 207
Aniline, surface tension for, 441
Anisole, surface tension for, 441
API effluents
　filtration with destabilizing chemicals, 522, 523
　filtration-separation of, 520
Applications of surface dilatational properties, 14–24
　emulsion stability, 23–25
　enhanced oil recovery, 17–20
　foam rheology, 14–17
　foam stability, 20–23
Aqueous aluminum monomers, 427–431
Aqueous aluminum polymers, 431, 432
Aqueous process streams, treatment of, 77–125
　admicellar chromatography, 115–120
　　adsolubilization, 115, 116
　　fixed-bed operation, 117–120
　extraction into reverse micelles, 111–115
　foam separations, 98–106
　　colligend flotation, 104, 105
　　continuous-flow foam separation process, 105, 106
　　particulate flotation, 99–104
　　precipitate flotation, 105

[Aqueous process streams, treatment of]
　micellar-enhanced ultrafiltration, 78–98
　　heavy metal removal, 94–98
　　organic compound removal, 78–94
　surfactant-enhanced carbon regeneration, 106–111

B

Barite, PZC of, 207
Basic extractants, 144–149
Benzaldehyde, surface tension for, 441
Benzene, 119
　rejection in MEUF of, 87
　surface tension for, 441, 453
Benzyl alcohol, surface tension for, 442
Biomedical applications of microemulsions, 353
Biotreater effluents
　filtration with destabilizing chemicals, 523, 524
　filtration-separation of, 521
Bromobenzene, surface tension for, 442
Bromoform, surface tension for, 442
α-Bromonaphthalene, surface tension for, 442
o-Bromotoluene, surface tension for, 442
Butadiene, water of solubility of, 283
1,3-Butadiene, heat of polymerization of, 275
n-Butanol, surface tension for, 442
Butyl acetate, surface tension for, 442
i-Butyl alcohol, surface tension for, 442

Index

sec-Butyl alcohol, surface tension for, 442
Butyl acrylate
 heat of polymerization of, 275
 surface tension for, 442
n-Butyl acrylate, water of solubility of, 283
Butyl benzoate, surface tension for, 442
i-Butyl chloride, surface tension for, 442
tert-Butyl chloride, surface tension for, 443
Butyl methacrylate
 heat of polymerization of, 275
 surface tension for, 443
4-tert-Butylphenol, rejection in MEUF of, 87
Butyronitrile, surface tension for, 443

C

Calcite, PZC of, 207
Caprylic acid, surface tension for, 443
Carbon disulfide, surface tension for, 443
Carbon regeneration, surfactant-enhanced, 106–111
Carbon tetrabromide, surface tension for, 443
Carbon tetrachloride, surface tension for, 443
Cassiterite, PZC of, 207
Cationic surfactants in micellar-enhanced ultrafiltration, 82
Cetylpyridinium chloride monohydrate (CPC), 82, 84–88, 93, 94
Chelating agents as collectors, 219–222
Chemical destabilization-coagulation and flocculation in causing oil-water dispersion instability, 471–481

Chemical displacement in causing oil-water dispersion instability, 468–471
Chloroacetone, surface tension for, 443
Chlorobenzene, surface tension for, 443
Chloroform, surface tension for, 443
α-Chloronaphthalene, surface tension for, 443
Chloroprene
 heat of polymerization of, 275
 water of solubility of, 283
Chromite, PZC of, 207
α-Chymotrypsin, 111
Cocoanut fatty acid, surface tension for, 443
Collectors
 characteristics of, 204, 205
 chelating agents as, 219–222
Colligend flotation, 104, 105
Colloidal stability of latex particles, 294–306
 basis of stability, 294–297
 effect of added surfactants, 207–302
 effect of soap on dispersion stability, 302–306
Colloid flotation, 196
Commercial surfactants, chemical structure of, 307
Constant-flow-rate experiments, saturation data for, 18
Constant-pressure experiments, saturation data for, 18
Continuous-contact distillation devices, 31, 33
Continuous-flow foam separation process, 105, 106
Copper recovery from ore leachers, estimated cost of, 143
Corn oil, surface tension for, 443
Corundum, PZC of, 207
Cosurfactants role in microemulsion formation, 326
Cotton seed oil, surface tension for, 443

m-Cresol
  rejection in MEUF of, 87
  surface tension for, 443
Critical micelle concentration (CMC), 78, 80
  of sodium, potassium, and lithium alkyl sulfates in water, 176, 177
Cuprite, PZC of, 207
Cyclohexane, 119
  rejection in MEUF of, 87
  surface tension for, 443
Cyclohexanol, surface tension for, 444
Cyclohexanone, surface tension for, 444
p-Cymene, surface tension for, 444

D

Deactivators in flotation systems, 225, 227–230
n-Decane, 119
  surface tension for, 444
Deep-bed filtration for separation of oil-water dispersion, 508, 509
Demulsification, 132, 133
Deosycorticosterone, 117
Depressants in flotation systems, 225–227
Design characteristics of microemulsions, 332
Dialysis, 128
Di-bromoethane, surface tension for, 444
Di-n-butyl amine, surface tension for, 444
1,1-Dichloroacetone, surface tension for, 444
β,β-Dichloroethyl sulfide, surface tension for, 444
Diethyl carbonate, surface tension for, 444

Diethyl ether, surface tension for, 444
Diethyl phthalate, surface tension for, 445
Diisoamyl, surface tension for, 445
Diisoamylamine, surface tension for, 445
Diisobutylamine, surface tension for, 445
Diisobutylcarbinol, surface tension for, 445
Diisobutylene, 119
Diisopropyl ketone, surface tension for, 445
Dilatational modulus
  foam half-life and equilibrium foam height versus, 24
  thin-film drainage time versus, 22
Dilatational properties of surfactant interfaces, 1–28
  applications, 14–24
    emulsion stability, 23–25
    enhanced oil recovery, 17–20
    foam rheology, 14–17
    foam stability, 20–23
  dynamic surface properties (surface rheology), 2–10
  surface dilatational properties, 11–13
Dilute oil-water emulsions and dispersions, 369–539
  characterizing a dispersion, 440–462
    predicting the continuous phase, 461, 462
    predicting the drop size, 440–461
    surface-interfacial tension, 440
  chemical destabilization-coagulation and flocculation, 471–481
    determining the chemistry, 473–477

Index 545

[Dilute oil-water emulsions and dispersions]
  equipment selection, sizing, and results, 477–481
  types of coagulants or chemical destabilizers, 473
  chemical displacement, 468–471
    type of solvent, 470, 471
    type of surfactant, 468–470
  factors that keep dispersion stable, 383–419
    equilibrium considerations, 383–408
    fluid dynamics considerations, 419
    rate considerations, 411–419
  overview for separating, 463
  physical separation, 464–467
  reasons why dispersion should be unstable, 370–383
  removal by air attachment without and with destabilization, 481–497
    DAF with upstream addition of destabilizers, 494–497
    DAF without upstream addition of destabilizers, 489–494
    equipment selection and sizing, 497
  removal by filtration, 498–529
    characteristics of deep-bed filtration, 508, 509
    fundamental principles and characteristics, 498–508
    general selection/sizing procedure, 525–529
    performance with addition of destabilizing chemicals, 521–524

[Dilute oil-water emulsions and dispersions]
  performance without addition of destabilizing chemicals, 508–521
  regeneration of the filter, 524, 525
  role of inorganics, 427–434
  role of polymers, 434, 435
  role of surfactants, 421–427
  solubility, 436–440
2,3-Dimethylpentane, 119
3,3-Dimethylpentane, 119
Dimethylstyrene, water of solubility of, 283
Dipropyl amine, surface tension for, 445
Discrete-stage distillation devices, 31, 32
Dispersants in flotation systems, 225, 230
Dissolved air flotation (DAF), 481–497
  equipment selection and sizing, 497
  with destabilization, 494–497
  without destabilization, 489–494
Distillation, surface-active agents in, 29–76
  distillation process, 30–33
  effect of surface tension on flow structure, 33–38
  surface-tension gradients, 38–59
    stability analysis of Wang et al., 53–59
    supported-area equipment, 47–53
    unsupported-area equipment, 41–47
  surfactant effects, 59–73
    supported-area equipment, 65–73
    unsupported-area equipment, 60–65
Distillation columns, 31–33

n-Dodecane, 119
Dolomite, PZC of, 207
Dynamic surface properties (surface rheology), 2–10

E

Electrochemical double layer, dilute oil-water dispersions and, 383–404
Electrodialysis, 128
Emulsification, spontaneous, 319–326
Emulsion polymerization, 263–314, 346, 347
  colloidal stability of latex particles, 294–306
    basis of stability, 294–297
    effect of added surfactants, 297–302
    effect of soap on dispersion stability, 302–306
  emulsion polymerization plant, 311, 312
  examples of, 308–310
  growth of particles, 274–294
    heterogeneous nucleation, 274–284
    homogeneous nucleation, 285–294
  historical perspective, 266, 267
  polymerization process, 267–274
    chemical reactions, 269–274
  practical considerations for surfactant selection, 306–308
Emulsion separation technology
  practical aspects of, 173, 174
  theoretical aspects of, 175–186
Emulsion stability, 23–25
Engineering applications of microemulsions, 353

Enzymic reactions in microemulsions, 349–350
Estradiol, 117
Estrone, 117
Ether, surface tension for, 445
Ethyl acetate, surface tension for, 445
Ethyl acetoacetate, surface tension for, 445
Ethyl acrylate
  heat of polymerization of, 275
  surface tension for, 446
  water of solubility of, 283
Ethyl benzene, 119
  surface tension for, 446
Ethyl bromide, surface tension for, 446
Ethyl caproate, surface tension for, 446
Ethyl chloroacetate, surface tension for, 446
Ethyl cinnamate, surface tension for, 446
Ethylene
  heat of polymerization of, 275
  water of solubility of, 283
Ethylene bromide, surface tension for, 447
Ethylene dibromide, surface tension for, 447
Ethyl ether, surface tension for, 446
Ethyl hydrocinnamate, surface tension for, 446
Ethyl iodide, surface tension for, 446
Ethyl isovalerate, surface tension for, 446
Ethyl mercaptan, surface tension for, 447
Ethyl methacrylate, heat of polymerization of, 275
Ethyl nonylate, surface tension for, 447
Ethyl oleate, surface tension for, 447

Index

Ethyl propyl ketone, surface tension for, 447
Extractants (for metal extraction by means of liquid surfactant membranes), 144–151
  acidic extractants, 144–147
  basic extractants, 144–149
  neutral extractants, 150
  synergistic effect between extractants and surfactants, 150, 151
Extraction into reverse micelles, 111–115

F

Facilitated transport mechanisms, 130–132
  enhancement of, 151–160
Fatty acid soap, 204
Filtration, separation of oil-water dispersion by, 498–529
  characteristics of deep-bed filtration, 508, 509
  fundamental principles and characteristics, 498–508
  general selection/sizing procedure, 525–529
  performance with addition of destabilizing chemicals, 521–524
  performance without addition of destabilizing chemicals, 508–521
  regeneration of the filter, 524, 525
Fixed-bed operation for admicellar chromatography, 117–120
Flotation, 195–235
  flotation techniques, 197–201
    classified on basis of separation mechanism and size of material separated, 196

[Flotation]
  modifying agents, 225–230
  surfactant adsorption, 201–225
    chain-chain interactions, 209, 210
    chemical forces, 210–212
    electrostatic factors, 201–208
    surface chelation, 219–222
    surface precipitation, 212–215
    surfactant solution chemistry, 222–225
    xanthate interactions, 215–218
Flow structure, effect of surface tension on, 33–38
Fluoroapatite, PZC of, 207
Foam flotation, 196
Foam fractionation, 196
Foam rheology, 14–17
Foam separations, 98–106
  colligend flotation, 104, 105
  continuous-flow foam separation process, 105, 106
  particulate flotation, 99–104
  precipitate flotation, 105
Foam stability, 20–23
Frother usage, 202, 203
Froth flotation, 197
  of nonpolar minerals, 196
Furfural, surface tension for, 447

G

Geometric aspects and structure of microemulsions, 326–333
Glycol diacetate, surface tension for, 447
Goethite, PZC of, 207

H

Heat of polymerization of common monomers, 275

Heavy metal removal, micellar-enhanced ultrafiltration for, 94–98
Hematite, PZC of, 207
n-Heptane, 119
　surface tension for, 447
n-Heptanol, rejection in MEUF of, 87
Heptylic acid, surface tension for, 447
Herbicide dispersions, 237–261
　historical development, 241, 242
　new horizons, 252–258
　surface properties of importance, 238, 239
　surfactant choice, 242–252
　　relationship to vehicle, 250, 251
　　surface activity, 242–250
Heterogeneous nucleation in emulsion polymerization, 274–284
Hexane, surface tension for, 453
n-Hexane, 119
　rejection in MEUF of, 87
　surface tension for, 448
n-Hexanol, rejection in MEUF of, 87
n-Hexyl acrylate, water of solubility of, 283
Homogeneous nucleation in emulsion polymerization, 285–294
Hydrophilic-lipophilic balance (HLB)
　for microemulsions, 325, 326
　of some surfactants, 404, 406
Hydroxyoximes, structure and water solubility of, 220, 221
Hydroxysteroid dehydrogenase, 111

I

Ideal dispersions, mechanisms for, 409, 410
Inorganics, role in oil-water dispersion instability of, 427–434
Interfacial tensions, ultralow microemulsions and, 320, 321
　middle-phase microemulsions and, 343–345
Interfacial viscosity, effect on thin-film drainage time of, 21
Iodobenzene, surface tension for, 448
Ion flotation, 196

K

Kaolinite, PZC of, 207
Kerosene, surface tension for, 448

L

Latex particles, colloidal stability of, 294–306
　basis of stability, 294–297
　effect of added surfactants, 297–302
　effect of soap on dispersion stability, 302–306
Liquid-liquid systems, surface tensions for, 441–454
Liquid surfactant membranes for metal extraction, 127–168
　enhancement of facilitated transport, 151–160
　extractants, 144–151
　　acidic extractants, 144–147
　　basic extractants, 144–149

Index

[Liquid surfactant membranes for metal extraction]
  neutral extractants, 150
  synergistic effect between extractants and surfactants, 150, 151
  metal extraction processes, 133–143
  process economics of liquid membrane extraction, 141–143
  published systems, 134, 136, 137
  typical results of laboratory and pilot tests, 134–141
  principles of, 129–133
  demulsification, 132, 133
  facilitated transport, 130–132
  general description of liquid surfactant membrane, 129, 130
Lithium alkyl sulfate, CMC in water of, 177
Lysozyme, 111

M

Macroemulsions, 319
Mathematical modeling of liquid surfactant membrane systems, 155–160
Mercury, surface tension for, 448
Mesitylene, surface tension for, 448
Mesityl oxide, surface tension for, 448
Metal extraction, liquid surfactant membranes for, 127–168
  enhancement of facilitated transport, 151–160
  extractants, 144–151

[Metal extraction, liquid surfactant membranes for]
  acidic extractants, 144–147
  basic extractants, 144–149
  neutral extractants, 150
  synergistic effect between extractants and surfactants, 150, 151
  metal extraction processes, 133–143
  process economics of liquid membrane extraction, 141–143
  published systems, 134, 136, 137
  typical results of laboratory and pilot tests, 134–141
  principles of, 129–133
  demulsification, 132, 133
  facilitated transport, 130–132
  general description of liquid surfactant membrane, 129, 130
Metal removal, heavy, micellar-enhanced ultrafiltration for, 94–98
Metal xanthates, solubility products for (at 20°C), 217
Methacrylic acid, heat of polymerization of, 275
Methyl acrylate
  heat of polymerization of, 275
  surface tension for, 448
  water of solubility of, 283
Methyl butyl ketone, surface tension for, 448
Methyl *tert*-butyl ketone, surface tension for, 449
Methylcyclopentane, 119
Methylene chloride, surface tension for, 449
Methylene iodide, surface tension for, 449
Methyl ethyl ketone, surface tension for, 449
Methyl hexyl carbinol, surface tension for, 449

Methyl hexyl ketone, surface tension for, 449
Methyl isobutyl ketone, 119
  surface tension for, 449, 453
Methyl methacrylate
  heat of polymerization of, 275
  water of solubility of, 283
Methyl propyl ketone, surface tension for, 449
Micellar-enhanced ultrafiltration (MEUF), 78–98
  heavy metal removal by, 94–98
  organic compound removal by, 78–94
  qualitative effect of operating variables on performance of, 92
Microemulsions, 315–367
  advantages of microemulsion technology, 353
  as media for chemical reactions, 345–353
    enzymic reactions, 349, 350
    photochemical reactions, 347–349
    polymerization, 346, 347
    precipitation, 351–353
  definition of, 319
  formation, structure, and properties of, 319–345
    experimental studies and properties, 342–345
    geometric aspects and structure, 326–333
    solubilization and phase equilibria, 333–342
    spontaneous emulsification and thermodynamics stability, 319–326
  novel applications of, 353
Microfiltration, 128, 196
Middle-phase microemulsions, 343–345
Modifying surfactant agents, 225–230

Molecular flotation, 196
Monosodium methanearsonate (MSMA), 241

N

Neutral extractants, 150
Nitrobenzene, surface tension for, 449, 453
Nitromethane, surface tension for, 449
m-Nitrotoluene, surface tension for, 449
o-Nitrotoluene, surface tension for, 449
Nonionic surfactants, evaluation of, 240
Nonyl alcohol, surface tension for, 449
Nujol, surface tension for, 450

O

i-Octane, surface tension for, 450
n-Octane, 119
  surface tension for, 450
Octanoic acid, surface tension for, 450
n-Octanol, 119
  rejection in MEUF of, 87
n-Octyl acrylate, water solubility of, 283
n-Octyl alcohol, surface tension for, 450
Octylamine, 119
Oil production, use of liquid surfactant membranes for, 128
Oil-water emulsions and dispersions, dilute, 369–539
  characterizing a dispersion, 440–462
    predicting the continuous phase, 461, 462

[Oil-water emulsions and dispersions, dilute]
  predicting the drop size, 440–461
    surface-interfacial tension, 440
  chemical destabilization-coagulation and flocculation, 471–481
    determining the chemistry, 473–477
    equipment selection, sizing, and results, 477–481
    types of coagulants or chemical destabilizers, 473
  chemical displacement, 468–471
    type of solvent, 470, 471
    type of surfactant, 468–470
  factors that keep dispersion stable, 383–419
    equilibrium considerations, 383–408
    fluid dynamics considerations, 419
    rate considerations, 411–419
  overview for separating, 463
  physical separation, 464–467
  reasons why dispersion should be unstable, 370–383
  removal by air attachment without and with destabilization, 481–497
    DAF with upstream addition of destabilizers, 494–497
    DAF without upstream addition of destabilizers, 489–494
    equipment selection and sizing, 497
  removal by filtration, 498–529
    characteristics of deep-bed

[Oil-water emulsions and dispersions, dilute]
    filtration, 508, 509
    fundamental principles and characteristics, 498–508
    general selection/sizing procedure, 525–529
    performance with addition of destabilizing chemicals, 521–524
    performance without addition of destabilizing chemicals, 508–521
    regeneration of the filter, 524, 525
  role of inorganics, 427–434
  role of polymers, 434, 435
  role of surfactants, 421–427
  solubility, 436–440
Oil-water-oil (o-w-o) system in liquid surfactant membranes, 129
Oil well-control fluid, use of liquid surfactant membranes in, 128
Oleic acid, surface tension for, 450
Oleic and stearic acids, separation of, 169–193
  emulsion separation technology
    practical aspects, 173, 174
    theoretical aspects, 175–186
  optimization of theoretical and practical aspects of separation by aqueous solutions of surfactants, 186–188
One-stage MEUF process compared to two-stage MEUF process, 89, 91
Organic compound removal, micellar-enhanced ultrafiltration for, 78–94
Organic pollutants, comparison of rejection of, in MEUF, 87

P

Packed column (supported-area) distillation equipment, 31
Paraffin oil, surface tension for, 450
Particle growth in emulsion polymerization, 274–294
  heterogeneous nucleation, 274–284
  homogeneous nucleation, 285–294
Particulate flotation, 99–104
i-Pentane, surface tension for, 451
Pentetole, surface tension for, 451
Pepsin, 111
Perchlorethylene, surface tension for, 451
Peroxidase, 111
Phase equilibria of microemulsions, 337, 338
Phenetole, surface tension for, 451
Phenol, rejection in MEUF of, 87
Phenyl ether, surface tension for, 451
Phenyl isothiocyanate, surface tension for, 451
Photochemical reactions in microemulsions, 347–349
Physical separation of dilute oil-water emulsions and dispersions, 464–467
Plate column (unsupported-area) distillation equipment, 31
Point of zero charges (PZCs) of oxides, 206, 207
Pollutants, organic, comparison of, rejection of, in MEUF, 87
Polydimethylsiloxane, surface tension for, 451
Polymerization in microemulsions, 346, 347
Polymers, role in oil-water dispersion instability of, 434, 435
Potassium alkyl sulfate, CMC in water of, 177
Precipitate flotation, 105, 196
Precipitation in microemulsions, 351–353
Primary amine salts, 204
Process economics of liquid membrane metal extraction, 141–143
Process streams, aqueous, treatment of, 77–125
  admicellar chromatography, 115–120
    adsolubilization, 115, 116
    fixed-bed operation, 117–120
  extraction into reverse micelles, 111–115
  foam separations, 98–106
    colligend flotation, 104, 105
    continuous-flow foam separation process, 105, 106
    particulate flotation, 99–104
    precipitate flotation, 105
  micellar-enhanced ultrafiltration, 78–98
    heavy metal removal, 94–98
    organic compound removal, 78–94
  surfactant-enhanced carbon regeneration, 106–111
Progesterone, 117
Propyl acrylate, surface tension for, 451
i-Propyl ether, surface tension for, 451
Protein extraction by reverse micelles (PERM), 112–115

Index 553

Published systems of metal extraction by liquid surfactant membranes, 134, 136, 137

Q

Quartz, PZC of, 207
Quaternary amine salt, 205

R

Reverse micelles, extraction into, 111–115
Reverse osmosis, 128
Ribonuclease, 111
Ricinoleic acid, surface tension for, 451
Rutile, PZC of, 207

S

Saturation data for constant-pressure and constant-flow-rate experiments, 18
Secondary amine salts, 205
Separation of oleic and stearic acids, 169–193
  emulsion separation technology
    practical aspects, 173, 174
    theoretical aspects, 175–186
  optimization of theoretical and practical aspects of separation by aqueous solutions of surfactants, 186–188
Shape fluctuations of microemulsions, 332, 333
Shell Sol K, surface tension for, 451
Sodium alkyl sulfate, CMC in water of, 177

Sodium dodecyl sulfate (SDS), 95
Solubilization and structure of microemulsions, 333–335
Soybean oil (SBO), 241
Spontaneous emulsification, 319–326
Spray modifier herbicide adjuvants, 239
Spreaders, 239
Stability analysis by Wang et al. of surface-tension gradients in distillation, 53–59
Stearic and oleic acids, separation of, 169–193
  emulsion separation technology
    practical aspects, 173, 174
    theoretical aspects, 175–186
  optimization of theoretical and practical aspects of separation by aqueous solutions of surfactants, 186–188
Steel mill wastewaters
  filtration with destabilizing chemicals, 524
  filtration-separation of, 521
Steric stabilization (adsorbed polymers), 404–408
Sticker/spreaders, 239
Structural dynamics of microemulsions, 332, 333
Styrene
  heat of polymerization of, 275
  surface tension for, 451
  water of solubility of, 283
Sulfosuccinates, chemical structure of, 307
Supported-area distillation equipment
  schematic of, 31
  surface-tension gradient for, 47–53
  surfactant effects, 65–73
Surface-active agents in distillation, 29–76
  distillation process, 30–33
  effect of surface tension on flow structure, 33–38

[Surface-active agents in distillation]
  surface-tension gradients, 38–59
    stability analysis of Wang et al., 53–59
    supported-area equipment, 47–53
    unsupported-area equipment, 41–47
  surfactant effects, 59–73
    supported-area equipment, 65–73
    unsupported-area equipment, 60–65
Surface chelation, 219–222
Surface dilatational properties, 11–13
Surface rheology, 2–10
Surface tensions for liquid-liquid systems, 441–454
Surfactant-enhanced carbon regeneration (SECR), 106–111
Synergistic effect between extractants and surfactants, 150, 151
Synthetic oil-water systems, filtration-separation of, 520

T

Talc, PZC of, 207
Tenorite, PZC of, 207
Tertiary amine salts, 205
Testosterone, 117
Tetrabromoethane, surface tension for, 452
Tetrachloroethane, surface tension for, 452
Tetrachloroethylene, surface tension for, 452
Thermodynamic aspects of flotation, 198–200
Thermodynamic stability of microemulsions, 319–326
Thin-film drainage time
  effect of interfacial viscosity and interfacial tension gradient on, 21
  versus dilatational modulus, 22
Toluene, 119
  rejection in MEUF of, 87
  surface tension for, 452
1,2,3-Tribromo propane, surface tension for, 452
Tributyl phosphate, surface tension for, 452
Trichloroethylene, surface tension for, 452
1,2,4-Trimethylcyclohexane, 119
Trimethyl ethylene, surface tension for, 452
Trimethyl pentane, surface tension for, 452
Trypsin, 111
Two-stage MEUF process compared to one-stage MEUF process, 89, 91
Type 1 facilitated transport mechanism, 130
Type 2 facilitated transport mechanism, 130

U

Ultrafiltration, 128
  micellar-enhanced, 78–98
    for heavy metal removal, 94–98
    for organic compound removal, 78–94
Ultraflotation, 196
Ultralow interfacial tension
  microemulsions and, 320, 321
  middle-phase microemulsions and, 343–345
Undecylinic acid, surface tension for, 452

Unsupported-area distillation equipment
  schematic diagram of, 31
  surface-tension gradient for, 41–47
  surfactant effects, 60–65
Uranium recovery
  capital cost estimates for, 142
  operating cost estimates for, 143

V

i-Valeric acid, surface tension for, 452
i-Valeronitrile, surface tension for, 453
Vinyl acetate
  heat of polymerization of, 275
  surface tension for, 453
  water of solubility of, 283
Vinyl chloride
  heat of polymerization of, 275
  surface tension for, 453
  water of solubility of, 283
Vinylidene chloride
  heat of polymerization of, 275
  water of solubility of, 283
Vinyl monomers, water solubility of, 283

Vinyltoluene, water solubility of, 283

W

Water-in-oil emulsions, coalescence rate data for, 26
Water-oil-water (w-o-w) system in liquid surfactant membranes, 129
Water solubility of vinyl monomers, 283
Well-control fluid, use of liquid surfactant membranes in, 128
White oil, surface tension for, 453
Winsor-type microemulsion systems, phase behavior of, 338–340

X

Xanthate interactions, 215–218
o-Xylene, surface tension for, 453
p-Xylene, 119
  surface tension for, 453

Z

Zinc (Zn^{2+}), removal from wastewater of (with MEUF), 96, 97
Zircon, PZC of, 207